SILICON PROCESSING

A symposium
sponsored by ASTM
Committee F-1 on Electronics,
National Bureau of Standards,
and Stanford University
San Jose, Calif., 19–22 Jan. 1982

ASTM SPECIAL TECHNICAL PUBLICATION 804
D. C. Gupta, Siliconix Incorporated, editor

ASTM Publication Code Number (PCN)
04-804000-40

ASTM 1916 Race Street, Philadelphia, Pa. 19103

Library of Congress Catalog Card Number: 82-83529

NOTE

The Society is not responsible, as a body,
for the statements and opinions
advanced in this publication.

Printed in Baltimore, Md. (b)
June 1983

Foreword

The Symposium on Silicon Processing was held in San Jose, California, on 19–22 January 1982. ASTM Committee F-1 on Electronics, the National Bureau of Standards, and Stanford University sponsored the symposium, and Dinesh C. Gupta, Siliconix Incorporated, presided as chairman. The Advisory Board consisted of Kenneth E. Benson, Bell Laboratories; W. Murray Bullis, Fairchild Advanced R&D Laboratory; Robert E. Lorenzini, Siltec Corporation; Samuel L. Marshall, Solid State Technology; and James D. Meindl, Stanford University. The Technical Committee consisted of Theodore I. Kamins, Hewlett-Packard Laboratories; Eric Mendel, International Business Machines; Krishna C. Saraswat, Stanford University; Joseph Stach, Penn State University; and Fritz G. Vieweg-Gutberlet, Wacker Chemitronic; and was headed by Paul H. Langer, Bell Laboratories. The Arrangements and Publicity Committee consisted of Austin R. Blew, Lehighton Electronics Incorporated; Edward E. Gardner, IBM-GTD; Gilbert A. Gruber, Siliconix Incorporated; Charles Koehler, Fairchild Advanced R&D Laboratory; Donald J. Levinthal, Semiconductor International; Philip L. Lively, ASTM; David G. Mead, Nicolet Instrument Corporation; J. Timothy Raab, Rockwell International Recticon; Donald G. Schimmel, Bell Laboratories; and William R. Wheeler, Tencor Instruments; and was headed by Carl A. Germano, Motorola Incorporated. The Spouse Committee, consisting of Lou Ann Gruber, Vijay Gupta, and Carol Koehler, was headed by Kathleen Bullis; and the Registration Committee, consisting of Elaine Cohen, National Bureau of Standards, and Jan Meighan, San Jose Convention Bureau, was headed by Mr. Lively. We appreciate their efforts.

Appreciation is also extended to the following session and workshop chairmen: James T. Clemons, Bell Laboratories; James R. Ehrstein, National Bureau of Standards; Carl A. Germano, Motorola Incorporated; Gilbert A. Gruber, Siliconix Incorporated; Theodore I. Kamins, Hewlett-Packard Laboratories; William H. Kroeck, Western Electric Company; John W. Lampe, Martin Marietta Aerospace; Paul H. Langer, Bell Laboratories; Eric Mendel, International Business Machines; C. W. Pearce, Western Electric Company; R. Fabian W. Pease, Stanford University; Mike Powell, Motorola Limited; Sy Prussin, TRW Systems; Krishna C. Saraswat, Stanford University; Joseph Stach, Penn State University; and Fritz G. Vieweg-Gutberlet, Wacker Chemitronic.

Related ASTM Publications

Laser-Induced Damage in Optical Materials: 1979, STP 726 (1980), 04-726000-46

Lifetime Factors in Silicon, STP 712 (1980), 04-712000-46

Laser-Induced Damage in Optical Materials: 1978, STP 689 (1979), 04-689000-46

Optical Interferograms—Reduction and Interpretation, STP 666 (1979), 04-666000-46

Annual Book of ASTM Standards, Volume 10.04: Electronics (I), 01-100483-46

Annual Book of ASTM Standards, Volume 10.05: Electronics (II), 01-100583-46

A Note of Appreciation to Reviewers

The quality of the papers that appear in this publication reflects not only the obvious efforts of the authors but also the unheralded, though essential, work of the reviewers. On behalf of ASTM we acknowledge with appreciation their dedication to high professional standards and their sacrifice of time and effort.

ASTM Committee on Publications

Preface

The papers in this volume were presented at the Symposium on Silicon Processing held in San Jose, California, on 19–22 January 1982. The symposium was sponsored by ASTM Committee F-1 on Electronics, the National Bureau of Standards, and Stanford University. In addition to the technical presentations, the symposium included well-attended workshops on each technical topic; impressions of these workshops are provided in Appendix II.

The trend towards very-large-scale integration (VLSI) is usually described as increasing device count on an integrated circuit chip. The associated trends towards smaller feature size, larger chip size, and larger wafer diameter present new problems in processing technology and control to the semiconductor device manufacturer. The symposium was designed to identify the future directions in materials and processing technology required for VLSI and areas where additional measurement development and standardization are required to accelerate the efficient implementation and control of these technologies.

The symposium opened with a series of brief talks on the role of standards and on the work performed by the Stanford Center for Integrated Systems, the National Bureau of Standards, and the National Science Foundation in relation to the semiconductor industry. These talks were followed by an overview projection of future VLSI technology which emphasized the need for rigorous and disciplined control of processes to achieve the potential of VLSI. These talks are presented in Appendix I.

The opening general session included a discussion on semiconductor business and education requirements in the Eighties by a panel of prominent individuals in industry, academia, and government. Brief summaries of the panel responses to key questions are provided in Appendix II.

The response to the symposium was extremely favorable, suggesting a continuing need for a regular forum to discuss technology topics in the context of measurement and control. Accordingly, the symposium sponsors have initiated plans both to repeat the symposium at two-year intervals and to broaden industry involvement in the organization and conduct of future symposia. The present symposium has established a consistent theme for these symposia as they evolve: understanding and control of the complex process technologies required for VLSI and other advanced device concepts. The problem areas and standardization needs identified in this and subsequent symposia will provide the feedback to the research community and voluntary standards system essential for the future growth of the industry.

We appreciate the cooperation and support of the ASTM staff in the formulation of these proceedings. The assistance of W. Murray Bullis is acknowledged. Finally, we are indebted to our industrial, government and university colleagues who contributed to the content of the symposium and the proceedings.

Contents

DIELECTRICS AND JUNCTION FORMATION

DEFECTS, OXYGEN PRECIPITATION, AND CARRIER LIFETIME

MEASUREMENTS FOR PROCESS CONTROL

APPENDIXES

Appendix I

Introduction

This volume is organized into six sections: (*1*) Crystal Growth and Material Shaping; (*2*) Microelectronics Imaging and Patterning; (*3*) Silicon Chemical Vapor Deposition (CVD) Techniques; (*4*) Dielectrics and Junction Formation; (*5*) Defects, Oxygen Precipitation, and Carrier Lifetime; and (*6*) Measurements for Process Control. The papers emphasize day-to-day problems in silicon processing, from its simplest use through advanced and emerging applications to very-large-scale-integrated (VLSI) circuits.

Crystal Growth and Material Shaping

Crystal growth and shaping technologies are sharply influenced by a combination of new discoveries in material properties and final device requirements. The five papers in this section deal with the problems associated with tighter controls, large-charge growing, and automated shaping techniques.

Microelectronics Imaging and Patterning

Lithography is a dominant technology for fabrication of device structures. It requires new means to generate submicrometer structures. The three papers in this section describe the advantages and limitations of various methods and materials used in pattern generation, emphasizing their suitability and impact on the geometries required for VLSI.

Silicon CVD Techniques

Since its inception two decades ago, silicon chemical vapor deposition has made great strides. Not only does it form an integral part of bipolar processing, but new applications are emerging in MOS processing as well. For example, the use of epitaxial silicon is now employed in the fabrication of random-access 64K and 256K memory circuits for the reduction of external noise by separating the unwanted charge generated by alpha particles and permitting the use of a heavily doped substrate that sweeps away excess carriers such as electrons in P/P+ structures. Intricacies associated with silicon epitaxy and other chemical vapor depositions such as polysilicon and silicon oxides are discussed in this section.

Dielectrics and Junction Formation

The manufacturing yields, throughputs, and device performance, for the most part, are still limited by material, equipment, and process technologies rather than by the limitations on the device concepts. Papers in this section describe the junction and dielectric process technology constraints and the requirements for the fabrication of VLSI circuits.

Defects, Oxygen Precipitation, and Carrier Lifetime

The need for the generation of sophisticated circuitry and the larger area of silicon required imposes a new demand on the quality of silicon used for its fabrication. The role of intrinsic and process-induced defects must be understood. The subject matter is complicated when gettering mechanisms and the interactive effects of multiple processing on defects and defect growths are included. An area of tremendous research in material and process technology is discussed in this section.

Measurements for Process Control

This section discusses a most important requirement in semiconductor fabrication: in-process measurements and process modeling for the diagnosis and prediction of the process. Theoretical concepts applied to most measurement technologies are so complex that it becomes imperative to calibrate the systems with known standards in order to achieve better accuracies. ASTM Committee F-1 on Electronics has therefore stressed the need to expand the Standard Reference Material Program of the National Bureau of Standards to provide reference standards for the semiconductor industry. Modeling concepts developed by the Stanford Center through SUPREM programs are revolutionizing the processing technology, and yet we see the need for better measurement controls for the present generation of electronic circuits and for the high degree of integration yet to be introduced.

Dinesh C. Gupta
Siliconix Incorporated, Santa Clara, California; symposium chairman and editor

Crystal Growth and Material Shaping

Anthony C. Bonora[1]

Silicon Crystal Growth and Processing Technology: A Review

REFERENCE: Bonora, A. C., "**Silicon Crystal Growth and Processing Technology: A Review,**" *Silicon Processing, ASTM STP 804,* D. C. Gupta, Ed., American Society for Testing and Materials, 1983, pp. 5–23.

ABSTRACT: Crystal growing and shaping technologies are being influenced by a combination of new discoveries in materials properties and new requirements from the marketplace. Key elements of change in crystal growth include control of carbon and oxygen, coupled with a transition to computerized large-charge growers, and investigation of magnetic field influence.

Materials shaping is being driven by stringent quality requirements coupled with newly available processing technologies including laser marking, gettering, and edge profiling. New instruments for inspection of particles, material composition, and wafer flatness are contributing to an overall trend towards automation of critical processes. Planarity of the polished slice continues to be a significant concern, since some complex lithography processes show performance gains with ultraflat wafers.

KEY WORDS: silicon, crystal growth, materials processing, shaping, laser marking, inspection instruments, wafer quality

Crystal Growing

Virtually all semiconductor-device technology employs a crystalline material as its starting point. The term *crystal* refers to a material that is highly ordered on a long-range atomic basis. The process of converting a random-ordered or polycrystalline material into one of long-range order is known as crystal growing.

The physics of semiconductor-crystal growth generally is based on establishing thermal and geometrical conditions such that a liquid is solidified in a controlled interface. Most electronic crystal production is currently centered on silicon and utilizes the technique invented by Czochralski (CZ)[2] and developed by Teal and Little[3].

In the CZ process, crystal growth begins with the melting of a charge of

[1] Vice President and General Manager, Siltec Instrument Division, Menlo Park, Calif. 94025.

[2] Gatos, H. C. in *Crystal Growth: A Tutorial Approach,* W. Bardsley et al, Eds., Proceedings of Third International Summer School on Crystal Growth 1977, North-Holland, The Netherlands, 1979.

[3] Teal, G. K., *IEEE Transactions,* Vol. ED-23, No. 7, July 1976, pp. 621 39.

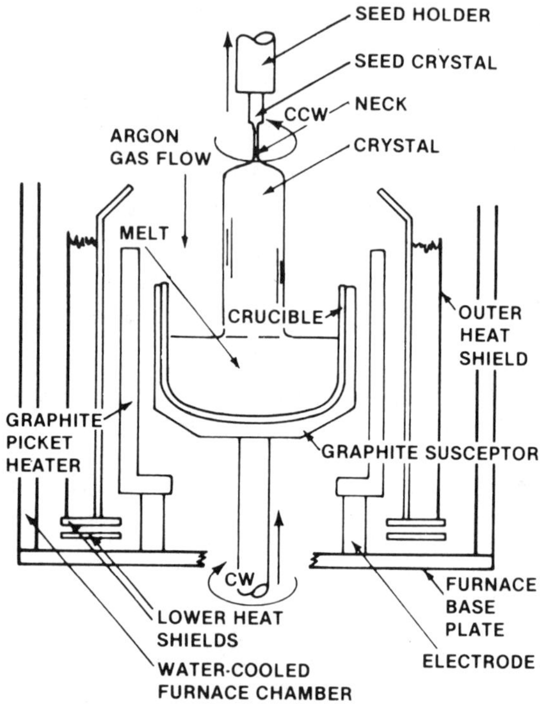

FIG. 1—*Schematic cross section of hot zone in a typical furnace for growing single-crystal silicon by the Czochralski process, showing relationships among melt, crystal, heater, and heat shields.*

polycrystalline material in a suitable crucible, using surrounding heater(s) and heat shields to establish carefully controlled axial and radial temperature gradients in the melt. A typical furnace for implementing this process is shown schematically in Fig. 1. A seed crystal with the desired crystallographic orientation is dipped into the melt at a temperature near the solidification point. If conditions are correct, it will be possible to begin raising the seed crystal in such a manner that the material makes the transition to a solid-phase crystal at the solid-liquid interface. The newly solidified material replicates the crystalline orientation of the seed. By regulating the temperature and seed-pulling rate, the diameter of the growing crystal can be controlled.

Growth Operations

A prepared polycrystalline charge is loaded into a clean furnace, after which the assembly is checked for leaks and filled with an inert gas, usually argon. With applied power ranging from 50 to 100 kVA, meltdown of the charge proceeds. When the melt reaches "dip-in" temperature (approximately 1420°C), the seed is dipped, and the necking and shouldering procedures follow. Operator skill remains a crucial factor throughout these operations. Once the crystal reaches the diameter desired, the automatic-diameter-control (ADC) system is turned on and closed-loop controls regulate the diameter.

The process continues through body growth. As the crystal is pulled, the melt level in the crucible drops relative to other hot-zone components. Because this changing melt level creates undesirable changes in thermal profiles at the growth interface, an electromechanical crucible-lift system is normally provided to elevate the crucible continuously during the growth process. The rate of lift, in turn, affects other growth parameters, so the furnace operator must monitor visual indications of structure on the growing crystal to assure that crystal pull speed and crucible lift rate are appropriate.

Body growth proceeds at pull speeds ranging from 50 to 100 mm/h, until only a small volume of melt remains in the crucible; at this point the round-off procedure begins. Structure must be maintained meticulously during round-off, because crystal defects such as "slip" can propagate up through the still-plastic crystal body. After round-off, power to the furnace is turned off and the crystal is allowed to cool. The crystal may later be heat treated to stabilize electrically active oxygen.

Crystal Evaluation

The completed crystal is immersed in an etchant designed to highlight defects such as slip. When the shoulder and round-off portions have been cropped with a diamond saw, resistivity and type measurements are made and recorded for the crystal ends. The cropped ingot is ground to the desired diameter, and may be rechecked for axial profile by taking resistivity readings along the length of its cylindrical surface.

At a location determined by atomic orientation, a flat is ground along the length of the ingot; this will serve as a reference plane for later operations. The crystal may now be sliced into wafers or cropped into segments for specific end users.

CZ Material Characteristics

Much of the current research on silicon is focused on the exceedingly complex relationships among certain impurities and defects in silicon crystals, because point defects in wafer substrates exert an increasingly strong influence on final die yields. This surge of interest in crystal defect formation arises from the shrinking die and circuit-element size used in very-large-scale-integration (VLSI) technology. The complexities lie partially in the fact that impurity concentration alone does not constitute defect formation; other determining factors include thermal gradients at the time of solidification, localized melt convection, instantaneous growth rates, melt–remelt phenomena, and thermal conditions after solidification.

Current technical literature on the subject of crystal defects is plentiful, but its utilization in production practices presents many challenges. To determine accurately the influence of process changes in crystal growth upon device yields in a particular fabrication line may involve an elapsed time of three to six months, and requires detailed communication between wafer supplier and user.

Oxygen and Carbon

Control of oxygen and carbon in the growing crystal is an active area of research. Oxygen is present in CZ silicon as a result of dissolution of the silica crucible wall, and generally is higher in concentration at the top end of an ingot. Reported concentrations range from <10 to >40 ppm. Oxygen influences wafer strength, resistance to thermal warping, minority-carrier lifetime, and intrinsic impurity gettering. There is evidence that the level of oxygen content influences die yields, and that appropriate levels for maximum yield depend upon the particular device production technology.

Carbon is present in CZ silicon as a result of contributions from the original polycrystalline material and from graphite components of the hot zone. Reported concentrations range from 0.1 to 7 ppm, with levels usually rising throughout the length of the ingot. Carbon is strongly implicated in the generation of point defects; its role appears to be interrelated to the presence of oxygen.

Carbon and Oxygen Measurement/Control

Control of carbon and oxygen must be achieved primarily by the materials producer, and substantial effort is being expended in related applied research. Oxygen can be partially controlled by means of growth techniques that minimize the rate of crucible-wall dissolution and reduce the incorporation of dissolved oxygen into the growing crystal. Control of carbon is related to gas flow patterns, hot zone materials, furnace leak integrity, and starting material.

The primary technique for measuring oxygen and carbon concentrations employs a transmission/absorption-type infrared spectrophotometer. Lengthy and careful sample processing is necessary to obtain usable results, although the recent introduction of Fourier transform infrared equipment facilitates rapid measurement of oxygen and carbon on polished slices.

A subjective indication of the influence of carbon and oxygen on CZ material is provided by a variety of "swirl" tests. This technique employs high-temperature oxidation of a polished sample, followed by decoration etching and visual examination. Swirl is evidenced as a cloudy appearance, often having a swirled form, that is caused by the high-density formation of microscopic etch pits at sites where concentrated microdefects occur. Because swirl has been correlated to areas of high device-yield loss, CZ growth technologies that provide low or zero swirl may improve die yields.

New Developments in Crystal Growing

The field of CZ growth is experiencing the emergence of several new technologies in such diverse areas as control of oxygen by magnetic fields, liquid-silicon recharging, high resistivity, and furnace automation. An overview of these areas will provide a general idea of work currently under way.

Magnetic CZ

In ordinary CZ growth, thermal gradients set up convection currents that, when combined with crucible and crystal rotation, give rise to a general flow pattern such as that in Fig. 2. This convection-induced flow transports oxygen through the melt into the liquid-crystal interface via dissolution of the quartz crucible.

The convection currents also give rise to regularly occurring temperature transients at the melt-crystal interface, creating what is termed the melt–remelt phenomenon. This repetition of incremental solidification and melting is implicated in the development of microdefects.

Standard CZ pullers afford limited capability for oxygen control through such processes as changing rotation and pull speeds and hot-zone geometry. Recently it has been found that oxygen content and crystal striations may be reduced significantly by applying a transverse magnetic field to the silicon melt area in order to damp the liquid convection currents.[4] Several silicon

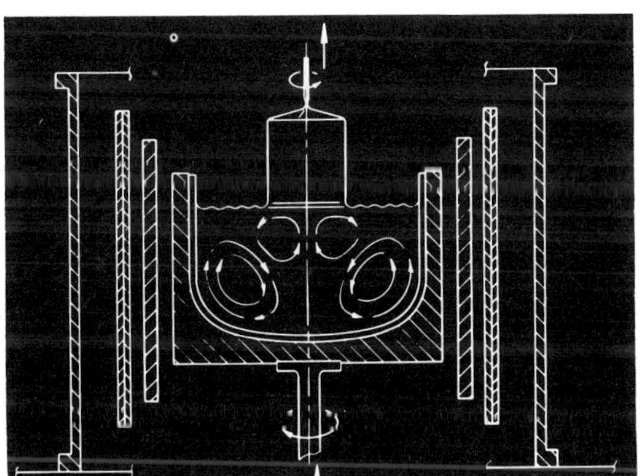

FIG. 2—*Schematic representation of melt flow patterns induced by thermal convection and crucible rotation.*

[4] Suzuki, T., Isawa, N., Okubo, Y., and Hashi, K. in *Semiconductor Silicon 1981,* H. R. Huff et al, Eds., Proceedings of Fourth International Symposium on Silicon Materials Science and Technology, The Electrochemical Society, Pennington, N. J., 1981.

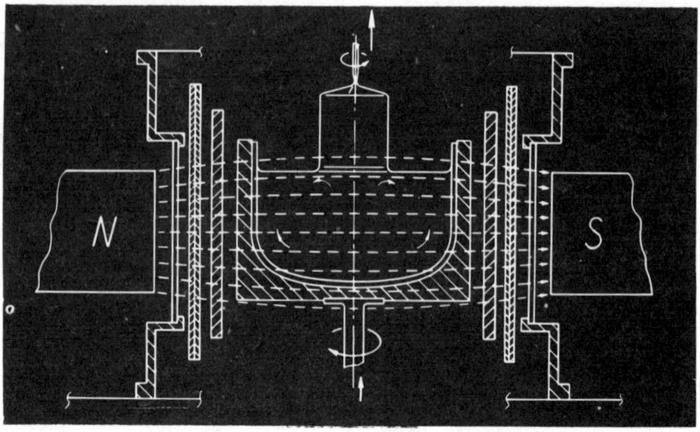

FIG. 3—*Diagram of magnetic flux lines and inhibited thermal convections.*

producers are investigating the effects of this technique, known as magnetic CZ (MCZ).

The MCZ effect offers a direct means of regulating oxygen transport and temperature transients. In principle, any motion of the highly conductive molten silicon is impeded when it crosses the magnetic flux field (Fig. 3). This provides a thick, stable boundary layer between the crucible wall and the bulk silicon melt, so that oxygen transport is limited to diffusion through a stable liquid medium. At the same time, the temperature transients at the growth interface are suppressed; the result is a relatively steady instantaneous growth rate.

Continuous Liquid-Silicon Feed

The concept of continuous liquid-silicon feed (CLF) has been under development for several years. CLF employs a heated quartz tube to transfer liquid silicon from a meltdown chamber to a growth chamber, allowing for control of dopant concentration in the feed material, and for semicontinuous replenishment of polycrystalline material and crystal solidification. Advantages include axially uniform crystal properties, improved productivity, and decreased crucible and power costs. Individual crystal sizes up to 65 kg have been achieved, and the process has shown capability for growing zero-defect-quality material. Operation of the CLF system is technically complex, however, and additional furnace automation is required for production feasibility. A schematic representation of a CLF system is shown in Fig. 4.

High-Resistivity CZ

Historically, most requirements for resistivities greater than 25 Ω·cm have been filled with material grown by the float-zone process, which uses no crucible. Increasing demands for resistivities as high as 70 Ω·cm have re-

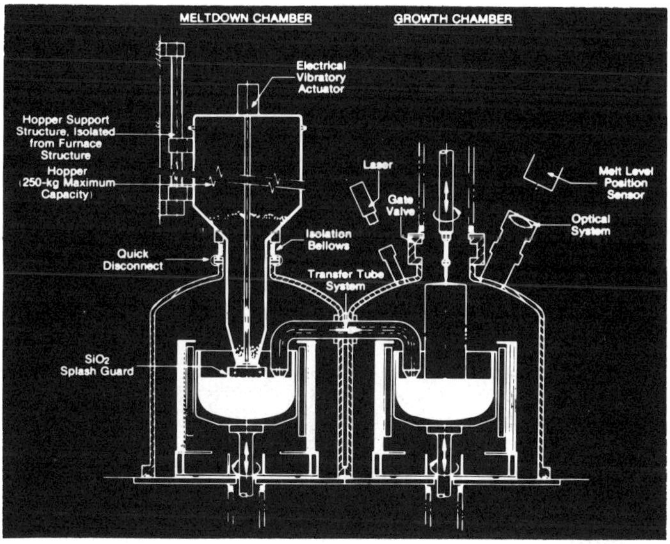

FIG. 4—*Cross section of furnace incorporating continuous liquid feed of silicon during growth.*

cently been satisfied with CZ material. Production experience to date indicates that extra care in doping and charge preparation can keep yields close to levels expected for standard product, provided that high-purity polycrystalline material is used. General availability of the product may afford additional flexibility in the design of high-speed memory devices.

Furnace Automation

To free the crystal growth process from operator judgment, it is desirable to have a computer monitor the process status and effect necessary changes. Improved furnace mechanical design, coupled with the uniform consistency of digital control and new process transducers, will lead to substantial automation, and preprogrammed process changes will give better run-to-run uniformity.

Present activities emphasize effective use of linked automation and development of "recipe" loading systems that will facilitate end-user process development without the need for software and computer system skills. Figure 5 shows a state-of-the-art computer-controlled furnace.

Crystal Processing

Slicing

Conversion of ground ingots into slices (wafers) is accomplished primarily by inside-diameter (ID) diamond sawing. Machine technology has changed to accommodate larger ingots as well as new accuracy requirements. Typical

FIG. 5—*General view of a typical computer-controlled furnace with 60-kg charge size capability.*

new machines incorporate hydraulically or mechanically tensioned blade heads in the 685.6- to 762-mm-diameter range, using blades with 12- and 15-mil kerf widths and 203.2- to 254-mm bores. Machine types currently available include horizontal and vertical spindles with rolling-element and air-bearing construction. All types are successfully in use by leading silicon producers, and automatic control-feedback features are becoming available. All saws allow adjustment of the ingot axis to permit cutting on the desired crystal orientation. Typical productivity figures for 100-mm wafers are 150 to 220 wafers per 8-h shift, and blade life is in the range of 3500 to 7000 cuts.

Control of planarity during the slicing process has become critical in the last three years, since repeated thermal processing of wafers has been found to exaggerate the distortions of wafer bow. Flatness of the polished wafer product also is influenced by bow, because the subsequent grinding/lapping

FIG. 6—*Close-up view of ID saw, showing automatic blade dressing feature.*

processes remove material from bowed wafers in an unpredictable manner. Most saws incorporate a means to facilitate bow control by measuring the deflection of the blade during each wafer cut.

Generally, a blade that cuts without lateral deflection will deliver slices with minimal "cut-in" bow, but bow induced by nonsymmetrical surface stresses on opposing faces may still exist. The saw operator normally monitors both blade deflection and general wafer bow trends, and will correct the cutting action by a combination of blade-dressing techniques and blade tensioning.

Figure 6 shows a prototype saw that allows the cut ingot face to be automatically raised during slicing. This semiautomatic device also allows the measurement for cut-in bow, and for initiation of a programmed blade-dressing sequence indicated by the direction and magnitude of the bow.

Another feature that is becoming prevalent is "single-slice recovery", wherein a retrieval mechanism recovers each wafer and loads it into a conventional cassette. This feature eliminates drag of the blade on the return stroke, with improved safety and ease of operation.

Wafer slicing and cleaning remain critical first fabrication steps, and a wafer with excessive bow, lack of planarity, or work damage creates problems in further processing.

Lapping

Typically, the as-cut slice from an ID saw is not flat enough to meet polished slice requirements, and a free-abrasive double-sided machining process (lapping) is used to planarize both slice surfaces and provide uniform finish.

Planetary-motion carriers move the wafers between opposed, rotating, metal lapping plates. Silicon carbide and alumina abrasive powders, typically of 10- to 15-μm particle size, remove 15 to 40-μm of stock per side, depending on the wafer-starting conditions and particular process.

Commercially available lapping machines are loaded manually and process 20 to 40 100-mm wafers in a 5- to 15-min cycle.

Edge Profiling

After lapping (or sometimes before), the slice is processed on a machine that removes the square-cornered slice periphery, which tends to be laden with microcracks from slicing and is very fracture prone. Edge-contoured wafers have the following advantages:

1. Wafer is more fracture resistant.
2. Line contamination from flakes/particles is reduced.
3. Quartz boat loading is easier.
4. Mask life in proximity/contact printing is improved.
5. Photoresist edge buildup is minimized.
6. Epicrown effect is reduced.

Some difficulties associated with edge contouring have included:

1. Loss of diametral accuracy.
2. Rounding of the flat cusps.
3. Angular shifting of the wafer flat.
4. Random scalloping of the wafer edge.
5. Insufficient control of top and bottom contours.

Various equipment types and processes have been used to produce contoured edges, including flexible abrasive disks, abrasive blasting, manual grinding, and high-speed grinding with a preformed wheel. A common method uses a grooved wheel rotating at high speed, which is held against the periphery of a slowly rotating wafer by a constant-force mechanism. An alternative method (Fig. 7) uses a multistep sequence to grind the outside diameter, top edge, and bottom edge on a cam-controlled profile grinder. Advantages include (*1*) accurate final diameter, independent of wafer-starting dimensions; (*2*) accurate primary-flat and secondary-flat generation; and (*3*) independent selection of top and bottom edge contours.

Etching

The application of etching media to as-cut or as-lapped wafers removes surface layers laden with microcracks (work damage). If not removed, these

FIG. 7—*Wafer-edge profiler that dimensions wafer diameter, shapes primary and secondary flats, and cuts top and bottom chamfers.*

cracks tend to propagate farther into the stressed wafer. Two processes currently in use (acid and caustic etching) remove 10 to 30 μm per side.

Acid etching uses mixed hydrofluoric, acetic, and nitric acids. Uniformity of removal, surface finish, and removal rate are greatly affected by acid composition and temperature. In addition, reaction by-products affect the process, which makes it difficult to achieve steady-state conditions in a production-line setting. Under elevated temperatures, removal rates can go as high as 10 μm/s, but thickness control is difficult, and more moderate cycles of 30 s to several minutes are commonly used. Difficulties with acid etching include control of noxious oxides of nitrogen evolved, operator hazards, and a tendency towards nonuniform etch removal which leads to a nonplanar wafer surface.

For these and other reasons, caustic etching in high-temperature (70 to

130°C) potassium or sodium hydroxide has become commonplace in the last two years. Although time cycles are considerably longer than with acid, material removal tends to be more uniform, and the etched wafer flatter. Process characteristics are more uniform throughout bath life, and both etchant costs and air pollution are reduced.

Rinsing is particularly important after caustic etching, and several stages of deionized (DI) water rinse may be necessary. The wafer surface tends to be grainier or rougher after caustic etch, which leads to occasional problems with backside contamination. Residual sodium-potassium contamination has not presented problems.

Gettering

Controlled residual work damage on the backside or at strain sites in the silicon lattice acts to "getter" various impurities during thermal processing, preventing their migration upward into active device regions. Common methods of introducing controlled backside residual damage include:

1. *Abrasive blasting*—Controlled-diameter particles of alumina or other hard material are directed at high velocity against the wafer backside in a process similar to sandblasting.
2. *Laser beam*—A small-diameter focused laser pulse of appropriate wavelength and energy strikes the surface and creates localized melting or material removal. A matrix of damage sites in the desired pattern is achieved by repeated pulsing.
3. *Direct surface abrasion*—The wafer is rotated on a vacuum chuck and an abrasive-coated cloth or paper is moved radially in point or line contact with the wafer, leaving a series of concentric damage sites.
4. *Liquid honing*—A high-velocity stream of liquid containing abrasive particles is traversed across the wafer backside.

In all systems of this type, major considerations are uniformity of damage depth and freedom from residual contamination. In addition, certain damage processes leave deep fissures which tend to pick up contaminants that cannot be easily detected or cleaned, and which diffuse through the wafer at subsequent steps. These processes all have different characteristics in terms of damage intensity and ease with which damage sites anneal out and lose effectiveness.

A variety of thermal and chemical means are also available to improve gettering. One technique involves processing a wafer in a manner so as to achieve a sharp gradient of oxygen concentration at the polished surface, thus creating a "denuded zone" which contrasts with relatively high oxygen content in the bulk of the wafer. Other techniques to achieve oxygen gradients are being explored experimentally. The concept of the silicon wafer substrate as a homogeneous entity, polished on one side, is rapidly giving way to the recognition that many desirable properties can be deliberately and methodically created.

Introduction of backsurface gettering is normally done following the etching operation, and rigorous cleaning is necessary prior to polishing.

Intermediate Wafer Inspection

After the slice/lap/etch operations, integrated automatic equipment (Fig. 8) is used to inspect and sort the wafers for resistivity, thickness, bow, flatness, and surface appearance. Visual inspection of as-etched wafer surfaces is performed to check for etch stains, residual saw marks, flakes and chips, edge contour, and uniformity of appearance. Although this operation relies heavily on human visual skill and judgment, automatic wafer-handling equipment can assure consistency of wafer inspection time and viewing conditions. Automatic handling also eliminates wafer breakage and chipping which occur with coin stacking and with tweezer or vacuum pencil pickup.

Laser Marking

The concept of using a focused laser beam to create alphanumeric or bar code identification marks on wafers has been in the embryonic stage since 1975. Commercial equipment with good production capability is available from several manufacturers, and some silicon producers are shipping pilot quantities of marked wafers. Despite adoption of the Semiconductor

FIG. 8 *Partial view of resistivity/thickness sorter, incorporating 24 receiver stations.*

FIG. 9—*Self-contained cassette-type wafer-handling module with remote computer interface capability.*

Equipment and Materials Institute (SEMI) standard and universal recognition that an identification code is desirable, the transition to widespread use is slow because of the many challenges involved in effectively utilizing code information.

The marking performed by the silicon manufacturer is normally done as part of, or after, slice sorting, because the correct resistivity will be known at that point. Laser marking may raise a cratered edge and leave slag on the slice surface, and the producer may prefer to mark prior to etching, particularly if the mark is to be on the wafer backside. In this event, the sorting steps are repeated prior to polishing.

During the next several years, it can be expected that equipment manufacturers and device line planners will integrate the use of marked information into an overall concept to aid process control and material flow. This ambitious undertaking will involve the availability of highly reliable code readers and wafer-buffering systems. Truly hands-off material flow in both silicon and device plants will require production control systems with computer-controlled feedback to regulate processes and to allow for equipment downtime and nonuniform production rates in some operations. Multipurpose cassette-type wafer handlers (Fig. 9) probably will be utilized for transient inventory (buffering) as well as for machine loading.

Polishing

For almost two decades, one-sided wafer polishing has been accomplished by caustic colloidal-silica slurry and a rotating pad for stock removal (10 to

30 μm) followed by finishing—a two-step chemical-mechanical system designed to produce a highly reflective planar surface, free from subsurface work damage, haze, and microscratches. Traditional wax-mounting or resin-mounting techniques are prone to leave dimples, waves, or surface contamination, and are being replaced by templates, surface tension, friction, and solid film bonding. These waxless mounting processes have improved yields and reduced processing costs, but industry-wide flatness goals for nonlinear thickness variation remain a challenge.

Flatness requirements are the major driving force and the figure of merit for developing polishing technology. Improvements in flatness continue, and wafers currently can be produced to less than 3-μm TIR (peak-to-peak total indicated reading). The complete spectrum of requirements for polished wafer quality dictates a highly engineered system to deliver satisfactory results. Important polished-wafer parameters such as flatness, taper, and quality of surface finish are primarily affected by:

1. Composition, flow rate, radial distribution, and temperature of the slurry.
2. Surface speed, temperature distribution, relative diameters, rigidity, and geometrical contours of the main platen and carriers.
3. Materials and procedures for wafer mounting.
4. Compressibility and slurry retention of polish pads.
5. Unit content pressure, cycle times, rinse sequence, and ambient conditions.

The majority of equipment sold today (whether wax or waxless) uses multiple mounting of wafers on a carrier, with variations between manufacturers based on the number, type, and size of carriers. A commonly used polishing machine (Fig. 10) uses four motor-driven carriers, each of which handles seven 100-mm wafers. The main platen for the polishing pad is driven by a 20-hp motor, and a typical stock removal cycle is 10 to 20 min. A similar machine, with different pad and run conditions, would be used for finish polishing in a 1- to 5-min cycle.

Automation of multiwafer polishing currently focuses on handling and loading of wafers onto the carrier, and on demounting. The machine shown in Fig. 10 uses a medium-pressure pulsed jet of DI water to demount wafers at the rate of 2 s per wafer.

Wafer Cleaning

The demounted wafers go through precleaning and final cleaning steps. The preclean process is matched to the characteristics of the polishing system (wax or waxless). When the wafer is removed from preclean, its surface is either entirely hydrophilic or hydrophobic, depending on the presence or absence of an oxide film on the polished silicon surface. Since the precleaning is

FIG. 10—*Wafer polisher, showing main polish platen and four heads mounted with 100-mm wafers.*

generally the last cleaning step for the wafer backside, it is important to use no chemicals having nonvolatile residues and to control water purity for rinsing.

Final cleaning generally is accomplished on automated wafer scrubbers in a clean-room atmosphere. It entails sequential chemical scrubs, rinsing, and spin drying.

Final Inspection

A rigorous final inspection, in many cases at a 100% level, is performed following final cleaning. Parameters to be checked visually include particulates, haze, stains, waves, and backside cleanliness. In addition, a sample-level check is run on measurable parameters, including bow, flatness, resistivity/thickness/type, flat length and position, oxygen and carbon content, swirl level, and gettering effect.

FIG. 11—*General-purpose inspection station for visual detection of wafer-surface quality.*

The visual inspection requires experienced production personnel and is generally subjective. To diminish subjectivity, efforts are being made to automate portions of the process by the use of an incident-angle laser beam scan for detecting particulates, haze, and stains.

Mechanized handling (Fig. 11) is being introduced in a variety of inspection operations to ensure consistency of inspection conditions and to reduce contamination and breakage.

Flatness measurement is of special interest because it has a major influence on the practical limits for shrinking line widths and increasing circuit complexity. The depth of field of optimum yield/pattern definition in projection alignment is small, and therefore the degree of precision in predicting the surface plane of the wafer directly affects the entire photolithographic process.

In the area of flatness inspection, several different types of equipment are in use, each with substantially different principles of measurement:

1. Grazing-angle laser interferometry with fixed wafer and chuck
2. Ultrasonic probe with spiral scan and mechanical wafer rotation.
3. Multipoint capacitance probe with fixed wafer.
4. Single-axis laser scan with mechanical wafer rotation.
5. Multipoint optical probes with mechanical wafer traverse.

An additional measurement system has been introduced recently that expands flatness measurement to include a new parameter—local slope. This

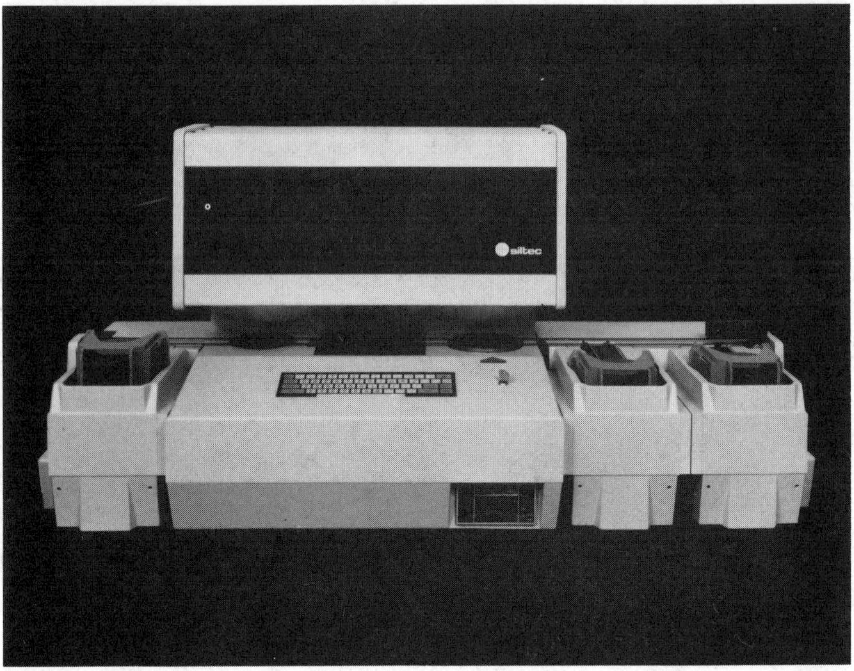

FIG. 12—*Surface measurement system that employs a laser-beam scan, displays a three-dimensional topographical readout, and calculates percentage of a wafer that meets a specified flatness standard.*

system (Fig. 12) uses a two-dimensional laser scan with a solid-state sensor to determine the angular deviation of the reflected beam. In effect, slope is measured directly, with conventional flatness data obtained by the digital integration of slope data over distance traversed. In this manner, complete surface topology is obtained. Tests indicate reliable operation with both polished and patterned wafers.

The issue of slope-flatness measurement on wafers after several thermal processing steps may be of interest, since thermally induced distortion can become severe enough to prevent total wafer pulldown on a vacuum chuck. Thus a wafer may be perfectly flat at incoming inspection and fail in subsequent lithography steps owing to lack of flatness.

A tendency towards thermal distortion is a material variable influenced by intrinsic oxygen, grown-in stresses, and backside treatment, as well as by oxidation/diffusion temperatures and push/pull rates.

The interactive effects between silicon wafer manufacturing processes and device line performance typify the complex issues facing wafer manufacturers and users. It is apparent that technology efforts are expanding in the development of silicon materials. New material parameters likely to be studied for possible effects on device yield and performance include:

1. Interstitial oxygen (including thickness gradient effects).

2. Nondopant trace-metal content.
3. Nonvisual surface-state conditions.
4. Residual lattice stress.
5. Carbon, precipitate formation, and point defect distribution.
6. State of oxygen in the lattice.

Control of these factors may in turn contribute to the development of new-generation semiconductor end-products.

Wen Lin[1] *and David W. Hill*[1]

Large-Diameter Czochralski Silicon Crystal Growth

REFERENCE: Lin, W., and Hill, D. W., **"Large-Diameter Czochralski Silicon Crystal Growth,"** *Silicon Processing, ASTM STP 804,* D. C. Gupta, Ed., American Society for Testing and Materials, 1983, pp. 24–38.

ABSTRACT: Some practical aspects of the engineering and science of large-diameter crystal growth are presented. Growth characteristics and problems associated with thermal stress and crystal shape are discussed. Oxygen distribution in the melt and its incorporation into the crytal are investigated. Finally, a double-crucible technique is presented and its application to N^+ silicon growth is discussed.

KEY WORDS: silicon, crystal growth, Czochralski growth, large-diameter silicon, oxygen in crystals, double-crucible technique

The demand for high-capacity Czochralski (CZ) crystal growers has increased significantly with the demand for 100-mm-diameter wafers in the semiconductor industry. Typically, the new growers are the scale-up versions of the last generation designs, and they are operated at reduced pressure. The increase in melt size and crystal diameter and the associated modifications in the grower's hardware have resulted in changes in many aspects of growth. The thermal characteristics resulting from the scale-up have vital impact on the growth and yield performance. The effects of temperature gradient and ambient pressure need to be understood. The requirements of very-large-scale integration (VLSI) put stringent demands on silicon material properties. For example, the oxygen concentration of CZ wafers is shown to be intimately related to the intrinsic gettering capability of the wafers during device processing. It is desirable to have wafers with a uniform oxygen distribution and a desired concentration level to facilitate the gettering treatments. An understanding of the oxygen incorporation mechanism is essential for developing processes for controlled oxygen incorporation. To meet the materials requirements, and to improve growth yield in some cases, special growth processes need to be explored, by either changing growth parameters or by using specially designed apparatus or both.

[1] Member of Technical Staff, Semiconductor Materials Group, Bell Laboratories, Allentown, Pa. 18103.

This paper discusses large-diameter silicon growth characteristics in terms of their relationship to ambient pressure, crystal surface cooling, and melt thermal gradient. Next, an investigation of the oxygen distribution in the silicon melt is presented and its incorporation into the crystal during growth is discussed. Finally, a technique for double-crucible growth of silicon is presented. The advantage of the technique in suppressing severe solute segregation and improving growth yield for N^+ silicon is demonstrated.

Growth Characteristics of Large-Diameter Silicon Crystals

The perfection, shape, and other properties of large-diameter crystals are determined by the nature of the heat flow about the growing interface. The interface shape and isotherms in the growing crystal are controlled by crystal cooling conditions and melt radial temperature gradient as well as by the solidification rate of the crystal. The surface cooling is by gas convection and radiation, while the melt radial temperature gradient is determined by the melt aspect ratio, heater geometry, and other hot-zone components.

One of the characteristics of growth under atmospheric pressure is the use of a collimated inert purging gas to prevent condensation of SiO vapor in the vicinity of the growing crystal. The collimation is carried out by a tube concentric with the crystal shaft, with a diameter somewhat larger than the growing crystal. The collimated gas provides convection at the crystal surface and at the solid-liquid meniscus region. When the crystal diameter is small, the crystal may be considered as a one-dimensional conductor. The surface cooling rate would, at most, affect the growth rate. During large-diameter silicon growth, however, the shape of the interface and isotherms in the crystal becomes highly dependent on the surface cooling and growth rate. When the surface cooling is high—for instance, under high purging gas velocity—and the growth rate is also high, the interface will be highly concave (Fig. 1*a*). Significant radial thermal stress can develop [*1,2*] and slip dislocations can be generated when the thermal stress exceeds the critical resolved shear stress.[2] This usually occurs in the region right about the growing interface where the temperature is in the plastic deformation region (>900°C). Once a dislocation is generated, it is propagated toward the interface, and continued growth will not be dislocation-free. Thermal-stress-induced perfection loss is usually accompanied by a highly concave interface shape. Our 100-mm-diameter growth experiments at atmospheric pressure have shown that a highly collimated gas stream (with a purging tube) has a significant cooling effect on the crystal surface as well as on the meniscus region. This cooling condition and a shallow melt radial gradient can readily result in significant thermal stress and thus in perfection loss. The stress is most significant near the peripheral region of the crystal. By reducing the purging gas velocity (either by reducing the gas flow rate or by relaxing the collimating

[2] The italic numbers in brackets refer to the list of references appended to this paper.

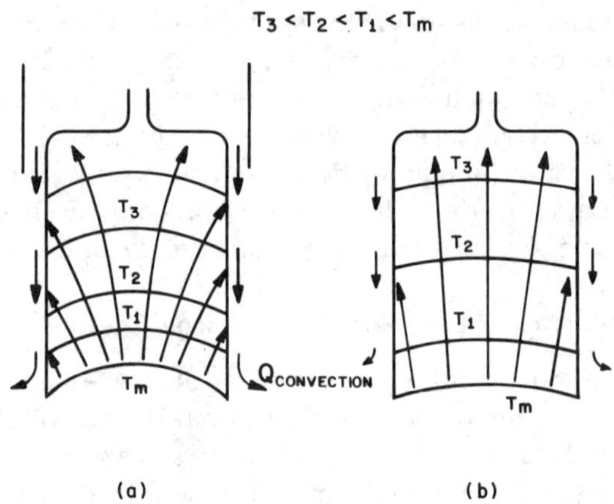

FIG. 1—*Patterns of heat flow and related interface and isotherm shapes during growth of large diameter CZ silicon.* (a) *High radial thermal stress configuration.* (b) *Low thermal stress configuration.*

effect of the purge tube) or by increasing the melt radial temperature gradient, the thermal stress may be minimized. The improvement may be observed by the reduction of the interface curvature and lateral preferred growth.

Reduced pressure growth eliminates the need for collimation of the inert gas; thus the excessive cooling due to collimated purging gas is avoided. The influence of the gas convection on crystal surface cooling and melt radial gradient is minimal. The crystal surface cooling is mainly due to radiation in this case. This cooling condition and a sharp melt radial gradient would give a desirable growing condition, corresponding to that shown in Fig. 1*b*. This hot zone interface and isotherm shape would produce minimum radial thermal stress. In large-diameter silicon growth, it is essential to have a hot zone that displays "tolerant" thermal characteristics. This hot zone includes a rather steep melt radial temperature gradient that extends over a significant vertical distance such that a slight change in the melt level position, with respect to the heater, would result in an insignificant change in radial gradient. A good test of the radial gradient of a growing system is to grow ⟨111⟩ crystals and to examine their growth morphology, such as "flat" formation and other preferred growth features.

The noncylindrical crystal shape frequently observed in ⟨111⟩ crystal growth is the result of retarded growth in ⟨211⟩ directions. This retarded growth perpendicular to the crystal periphery is caused by the formation of flats. The flats are the results of excessive supercooling that is caused by a shallow radial temperature gradient in the melt or by high velocity gas convection at the meniscus. The schematic in Fig. 2 shows the stacking of the close-packed layers associated with (111) and $(\overline{1}\overline{1}1)$ facet growth. The zigzag structure in the flat along the growth direction is caused by temperature

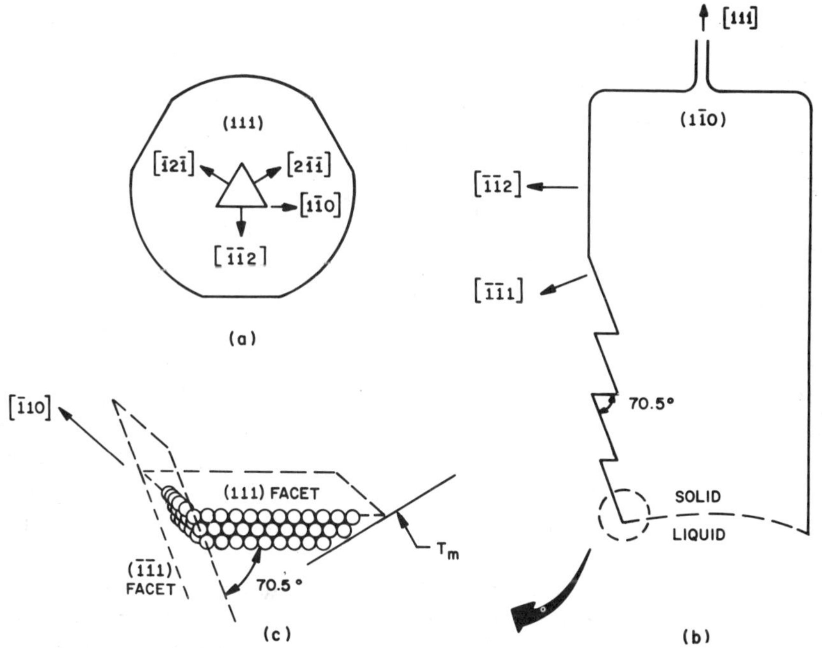

FIG. 2—*Flat formation in* ⟨111⟩ *silicon growth.* (a) *Transverse cross-sectional view.* (b) *Zig-zag nature of the flat formation along growth direction, viewing at* (110) *cross section.* (c) *Microscopic view of relationship between* (111) *and* $(\bar{1}\bar{1}1)$ *facet formation near the meniscus.*

fluctuations, which alternates domination between the horizontal (111) growth in the $[\bar{1}\bar{1}2]$ direction and the nearly vertical $(\bar{1}\bar{1}1)$ facet mode. When the growth thermal environment is steady, the flat width may be maintained constant by this zig-zag action. However, when supercooling is excessive or the crystal is in a more favorable orientation, the "side" facet growth mode will dominate and it continues to extend along a direction inclined 19.5 deg toward the crystal axis. At the same time, the flat width is enlarged, which results in a triangle-shaped crystal cross section.

In the past, seed misorientation has been observed to affect flat formation. Recent investigation indicates that the predominant controlling factor is the thermal characteristic of the growing system, primarily the melt radial temperature gradient. When the thermal characteristics of the growth system are poor the ⟨111⟩ seed misorientation becomes important. Slight misorientation can activate flat formation and amplify the problem. However, when the thermal characteristics of the growth system are more favorable, the flat formation is suppressed and the system is not sensitive to the seed misorientation. Crystal growth experiments with 100-mm-diameter crystals have been carried out with ⟨111⟩ seed misorientations ranging from 3 to 20 minutes of arc, in a reduced pressure system having good thermal characteristics. No significant correlation has been observed between the crystal shape and the extent of seed misorientations. In fact, the same growing system affords the growth of crystals with seeds oriented 3 deg off the ⟨111⟩ axis without flat formation.

The ⟨111⟩ facet formation and dislocations do not coexist during crystal growth. The introduction of dislocations in the facet region at the growth interface would interrupt the facet growth, which consists of fast sheet growth of perfect close-packed layers, and cause the facet to shrink in size. Eventually the facet growth is eliminated, even though the interface shape and temperature are favorable. Energetically, the growth on dislocated sites by the normal "stepwise" atom stacking mechanism is favored over the facet sheet growth mechanism. It is estimated that the former mechanism requires 1°C supercooling, while the latter can only be initiated at 9°C supercooling [*3*]. Therefore, once the dislocation is introduced at the interface, the facet mode growth will cease to continue. Empirically, the disappearance of the edge facet is used as an indication that the growing crystal has lost its dislocation-free feature for both ⟨111⟩ and ⟨100⟩ crystals. For the same reason, non-dislocation-free ⟨111⟩ crystals would not develop the noncylindrical crystal shape caused by facet formation.

It should be noted that the aforementioned shape problems due to flat formation are common to lightly doped ⟨111⟩ silicon growth. In the growth of N^+ ⟨111⟩ crystals, such as antimony-doped, with concentrations in the range of 10^{18} cm^{-3}, the situation is quite different. The preferred growth in this case is in ⟨211⟩ directions and {111} side facet growth is retarded. The result is a cam-shaped crystal cross section. The effect of dopant type and concentration on the silicon growth habit is not entirely clear.

Oxygen Incorporation

The oxygen incorporation during CZ growth is determined by the interplay between thermal convection, forced convection, free melt surface evaporation, and crucible dissolution. These functions are shown schematically in Fig. 3; they have been analyzed and discussed elsewhere [*4*]. Because of the interplay among these controlling factors, the axial oxygen distribution and concentration level of a grown silicon crystal depends on the melt geometry, crystal diameter, ambient pressure, etc. The incorporation behavior varies from one growing system to another. Figure 4 gives an example where the oxygen axial distributions of 100-mm-diameter crystals due to two growing systems are shown. It is noted that the slope of the [O_i] distribution is significantly larger for the atmospheric growth than that of reduced pressure. It has been previously proposed [*4*] that thermal convection is the main controlling mechanism for oxygen incorporation in large melt growth when the melt aspect ratio is high. The vigorous thermal convection flow causes a high crucible dissolution rate (thinner crucible-melt diffusion boundary layer) as well as fast oxygen transport to the melt center, following the path

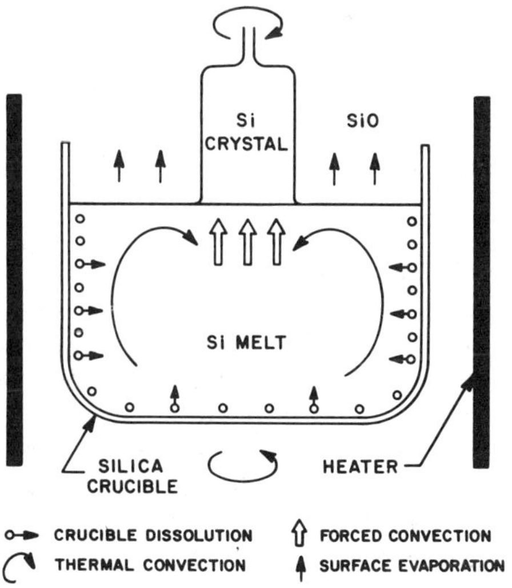

FIG. 3—*Schematic of a CZ growth system, showing relationship between oxygen controlling functions.*

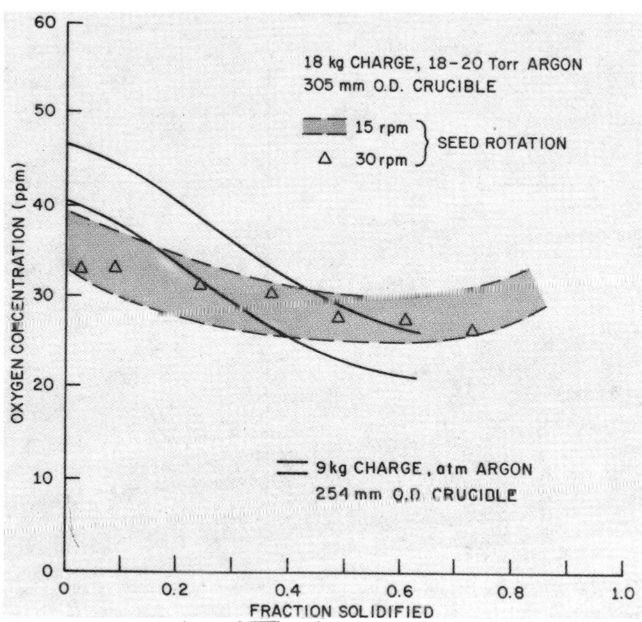

FIG. 4—*Comparison of axial* [O_i] *distributions for 100-mm-diameter crystals grown by using two different growing systems.*

shown in Fig. 3. Under atmospheric pressure growth the [O_i] in the flow reaching the melt center region is high, although some oxygen is evaporated when the flow is near the surface. The high [O_i] melt flow results in an inhomogeneous oxygen distribution near the growing crystal, which can be modulated by forced convection. Increased crystal rotation will draw oxygen-rich melt toward the growing interface and enhance oxygen incorporation [*4,5*].

When reduced pressure is applied in crystal growth, surface evaporation dominates. Oxygen evaporation is enhanced, while the flow is near the melt surface before reaching the crucible center region. This results in a substantially lower initial incorporation level observed for reduced pressure growth. Since the magnitude of thermal convection decreases when the melt level drops, the effect of reduced pressure also decreases. In the extreme case of very low aspect ratio melt configuration, the incorporation would reach a common level independent of the growing ambient, provided all other growth parameters are held the same [*4*]. This is not obvious in Fig. 4, where the two distributions are not due to the same growth system, but merely demonstrate the qualitative difference due to ambient effect. The radial oxy-

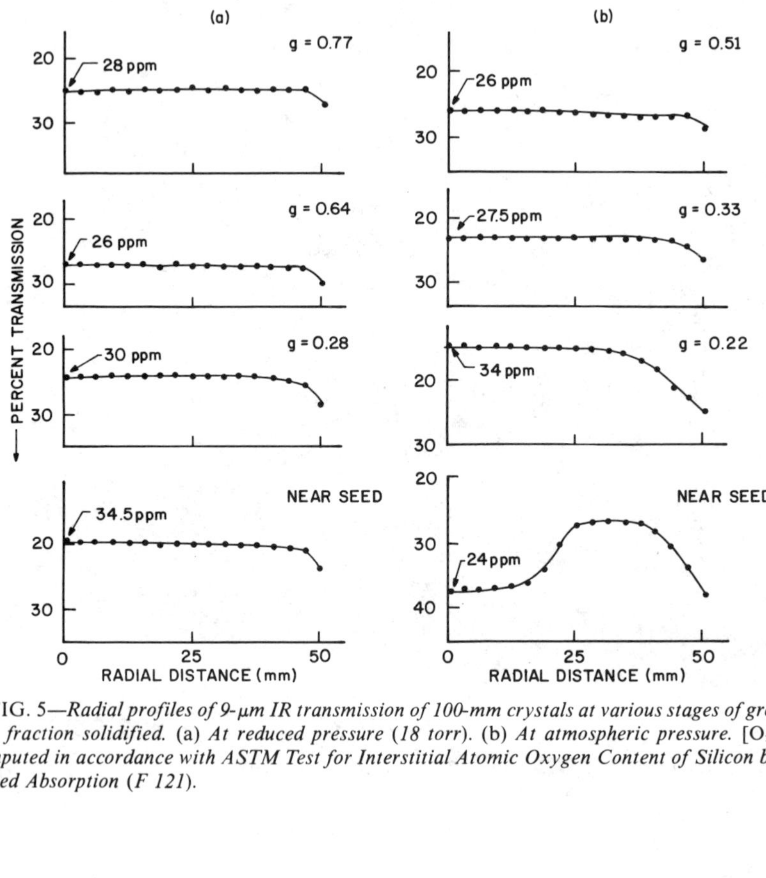

FIG. 5—*Radial profiles of 9-μm IR transmission of 100-mm crystals at various stages of growth;* g = *fraction solidified.* (a) *At reduced pressure (18 torr).* (b) *At atmospheric pressure.* [O_i] *are computed in accordance with ASTM Test for Interstitial Atomic Oxygen Content of Silicon by Infrared Absorption (F 121).*

gen uniformity of 100-mm-diameter crystals is shown in Fig. 5 for reduced and atmospheric pressure growth. A comparison of radial 9-μm infrared (IR) transmission profiles clearly indicates that crystals grown under reduced pressure exhibit better radial uniformity. Note that the oxygen concentration is inversely proportional to IR transmission at 9 μm. As discussed previously [*4*], the dopant elements exhibit complete mixing in large melt growth. Therefore the radial uniformity of dopant incorporation is largely controlled by interface properties, that is, the uniformity of diffusion boundary layer thickness across the interface. In the case of oxygen, the crystal radial uniformity will also depend on the oxygen uniformity in the melt near the growing interface. It is shown, from the analysis of crystal rotation experiments [*4,5*] and evidence to be presented in the following paragraphs, that the melt near the growing interface tends to be more homogeneous in [O_i] at reduced pressure than at atmospheric pressure. The observed superior radial oxygen uniformity may be attributed to the more homogeneous nature of oxygen distribution in the melt at reduced pressure.

When the melt is homogeneous in oxygen, an increase in forced convection is not expected to enhance the oxygen incorporation level. This is shown in Fig. 4, where the oxygen distribution of a 100-mm crystal with 30 rpm seed rotation is compared with those rotated at 15 rpm. All other growth parameters are held the same. There is no obvious evidence of enhancement effects in oxygen incorporation throughout all stages of growth by increasing seed rotation from 15 to 30 rpm. This result further confirms the observation that the melt tends to be more homogeneous under reduced pressure, and forced convection is not effective in oxygen enhancement. However, the fast seed rotation does seem to improve the radial oxygen gradient, by increasing the oxygen level at the crystal peripheral region. The low oxygen region normally observed at the crystal periphery has been thought by many to be due to oxygen out-diffusion at high temperature after solidification. The present result indicates that this may not be the only controlling factor, as the oxygen incorporation level at the crystal periphery is also sensitive to the melt hydrodynamics at the growing interface.

To further investigate the nature of the oxygen distribution near the surface of a large silicon melt and its dependence on ambient pressure, a series of oxygen measurements were carried out on the "square shoulder" of several crystals. Figure 6*a* shows schematically the convex cross-section of a square shoulder during initial growth. The positions of the 9-μm IR transmission measurements are indicated. During the shoulder growth, a very low pull rate is used (~0.2 mm/min); however, the lateral growth rate is substantial (~1.3 mm/min). The advancing solid-liquid interface at the crystal edge is nearly perpendicular to the melt surface. The freezing front continuously interfaces the top layer of silicon melt. Therefore the oxygen incorporated during the lateral shoulder growth represents, to a first approximation, the oxygen distribution at the melt surface. Although the true fluid flow of

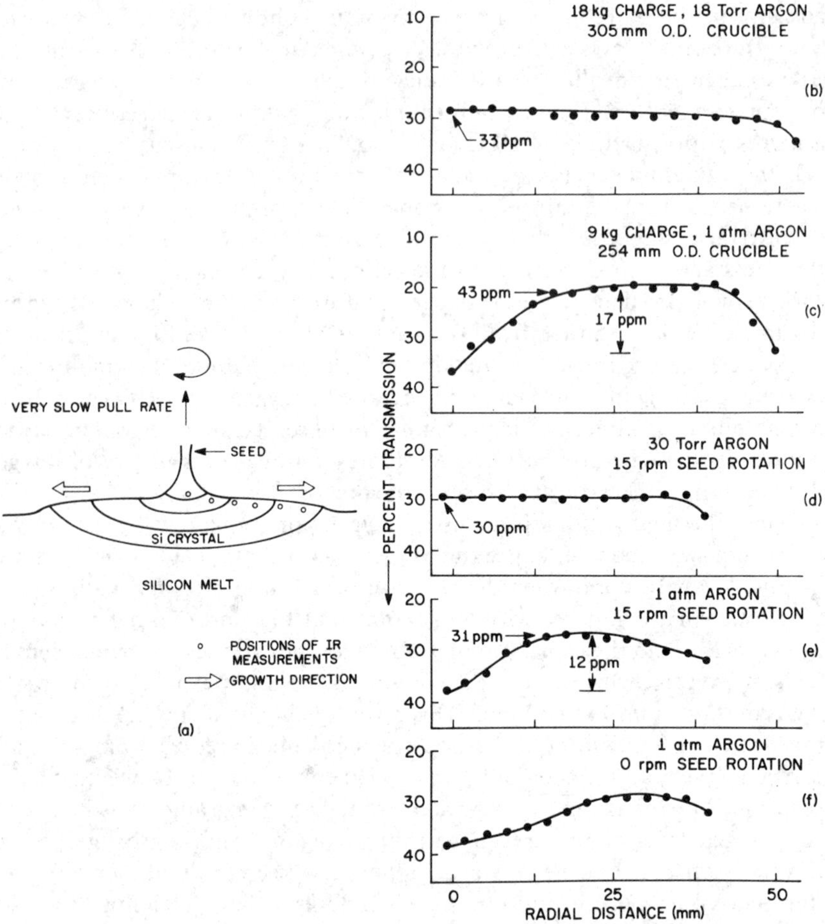

FIG. 6—(a) *Schematic of crystal's longitudinal view, showing experimental conditions.* (b) (c) *IR transmission profiles for 100-mm crystals.* (d) (e) (f) *IR transmission profiles for 75-mm crystals grown from 6-kg charges in 203-mm O. D. crucibles.*

oxygen-carrying flow may be modified in the presence of a growing shoulder, the effect is minimal near the melt surface, where the solidification of the advancing shoulder takes place.

Figure 6*b* shows a radial 9-μm IR transmission profile measured along positions indicated in Fig. 6*a* for a 100-mm diameter crystal grown at reduced pressure (2.4 kPa [18 torr]). The distribution is very uniform, with a slightly lower oxygen level near the crystal periphery. This distribution may be compared with that grown at atmospheric pressure, as is shown in Fig. 6*c*. The latter profile shows a distinct radial nonuniformity. There exists a low [O_i] region near the melt center with a diameter of approximately 40 mm. This observation seems to agree with the suggestion regarding thermal convection flow pattern [*6,7*] that there exists a stagnant central region where flow velocity is low. It is reasonable to assume that this low oxygen central region is the result of velocity stagnation. To ensure that the observed difference between Figs. 6*b* and 6*c* is not due to the effects of melt size or a particular melt aspect ratio, similar analyses were made on 75-mm crystals grown in smaller crucibles (203-mm outside diameter). Similar results are obtained, as shown in Figs. 6*d* and 6*e*. Furthermore, in order to obtain a better representation of the melt surface oxygen profile, and to minimize possible force convection effects, a 75-mm diameter crystal was grown with no seed rotation. The corresponding 9-μm transmission profile is shown in Fig. 6*f*. With no seed rotation, a slight shift in oxygen distribution is seen. However, the characteristic low oxygen central region remains the same. From these experimental results, it may be concluded that under atmospheric pressure the melt oxygen distribution near the surface is not uniform. There exists a stagnant region of low oxygen concentration that conforms well with the suggested thermal convection flow pattern. The region of low oxygen concentration is eliminated when the melt surface compared to atmospheric pressure is subjected to reduced pressure. It is believed that the evaporation of SiO is greatly enhanced due to a large oxygen concentration gradient across the gas-liquid boundary layer at the melt surface. A large portion of the oxygen carried along by thermal convection flow is evaporated when it is near the surface, resulting in a lower oxygen concentration and a better uniformity across the melt surface compared to atmospheric pressure. Hence, under reduced pressure, the role of thermal convection is reduced and that of diffusion is enhanced.

Based on the aforementioned experimental evidence, physical models of oxygen distributions in large melts at atmospheric pressure and reduced pressure are proposed. They are shown schematically in Fig. 7. It is shown that thermal convection is the main oxygen transport mechanism when the melt level is high. The oxygen-rich flow conforms with the thermal convection pattern shown in Fig. 3. Under atmospheric pressure, the oxygen distribution is not uniform near the melt center. The nature of this nonuniformity is displayed in the crystal's radial [O_i] distribution near the seed end, as shown in Fig. 5*b*. As the crystal growth progresses, the forced convection due to crystal rotation modulates the oxygen distribution and results in local mixing, and the gross nonuniformity near the stagnant region is largely diminished. However, some inhomogeneity exists until the melt level is low. On the other hand, the [O_i] distribution is more uniform near the melt surface under reduced pressure. The crystals possess better radial uniformity. This is shown in Fig. 5*a*.

When thermal convection is not significant, as in the case of a low aspect ratio melt configuration, the melt [O_i] is both diffusion rate- and crucible dissolution rate-dependent. The melt tends to be uniform in oxygen. Both enhanced forced convection and ambient pressure have little effect on the oxygen incorporation.

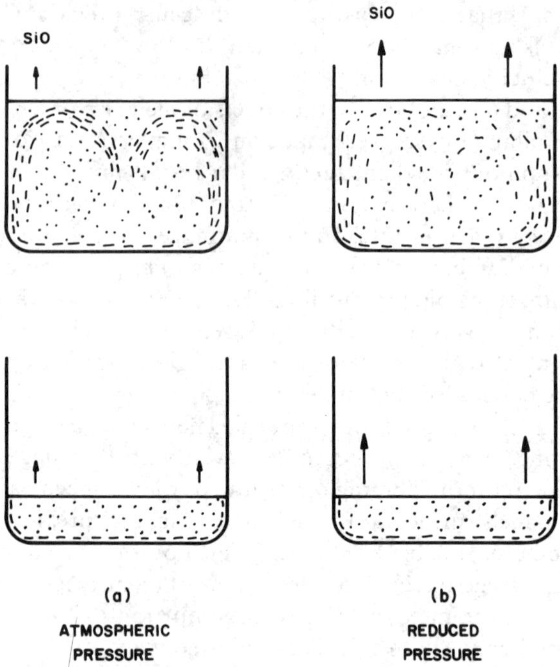

FIG. 7—*Schematic representations of oxygen distributions in silicon melt at high and low melt level configurations. The dots and lines represent oxygen concentration.* (a) *At reduced pressure.* (b) *At atmospheric pressure.*

By understanding the basic roles of the controlling functions of oxygen incorporation, one can manipulate one or more functions to control the incorporation level. Examples of controlled oxygen growth methods include enhanced forced convection for high [O_i] [*5*], suppression of thermal convection by internally [*8*] or externally [*9*] applied magnetic field for low [O_i] incorporation, and balanced incorporation functions by an inner crucible-like baffle for uniform oxygen incorporation at a medium level [*10*].

Double-Crucible Growth

Among commonly used dopant elements, Sb has a rather small solid solubility in silicon. The distribution coefficient k is also small ($k = 0.023$). These factors limit the growth yield in the growth of highly doped silicon crystals. Following normal freezing behavior, the severe segregation of Sb would quickly build up melt dopant concentration such that the constitutional supercooling and associated interface breakdown can occur before the solubility limit is reached [*11*]. If the starting Sb concentration of the crytal is in the range of 10^{18} cm^{-3}, the growth yield is limited to approximately 60 to 65%. To prevent dopant concentration build-up and associated constitutional supercooling, the double-crucible technique was applied in the growth of

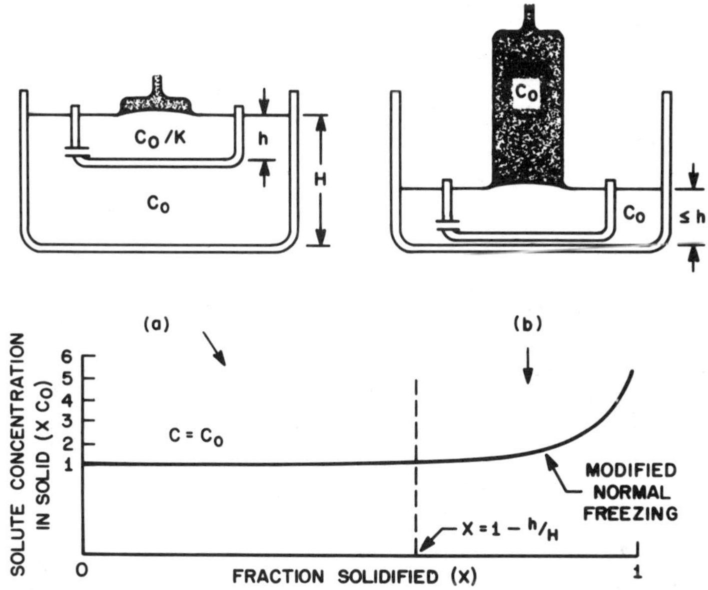

FIG. 8—*Schematic of the current double crucible growth arrangements and corresponding dopant concentration profiles.* (a) *Uniform incorporation.* (b) *After two crucibles make contact.*

highly doped silicon. The present process employs a fused silica crucible assembly and is designed for growth based on the "constant volume" principle. A smaller diameter crucible with an opening in the bottom is submerged in the melt held by a large outer crucible (Fig. 8*a*). If the desired crystal dopant concentration is C_0, then the dopant concentration in the inner crucible should be C_0/k to account for the segregation effect. As the melt in the inner crucible is converted into crystal, the melt is continuously replenished from the outer crucible through the opening with melt of concentration C_0 to maintain both volume and concentration constant. This may be achieved by raising the outer crucible such that the melt level remains unchanged with respect to a reference point. When the dopant concentration of the melt in both crucibles is so maintained, a crystal with uniform axial dopant concentration may be obtained. Thus, the dopant segregation effect is suppressed. The uniform dopant concentration growth ends when two crucibles make contact. The fraction (X) of the crystal with uniform dopant incorporation is therefore $X = 1 - h/H$, where h/H is the ratio of the height of the initial melt levels in the inner and outer crucible. When the contact is made, the inner crucible will rise with the outer crucible, and dopant concentration of the crystal increases in a manner similar to that of normal freezing. This behavior is schematically illustrated in Fig. 8*b*. To maximize the uniform dopant portion, a combination of shallow inner crucible and a large melt is preferred. Figure 9 shows the resistivity profile of a silicon crystal heavily doped

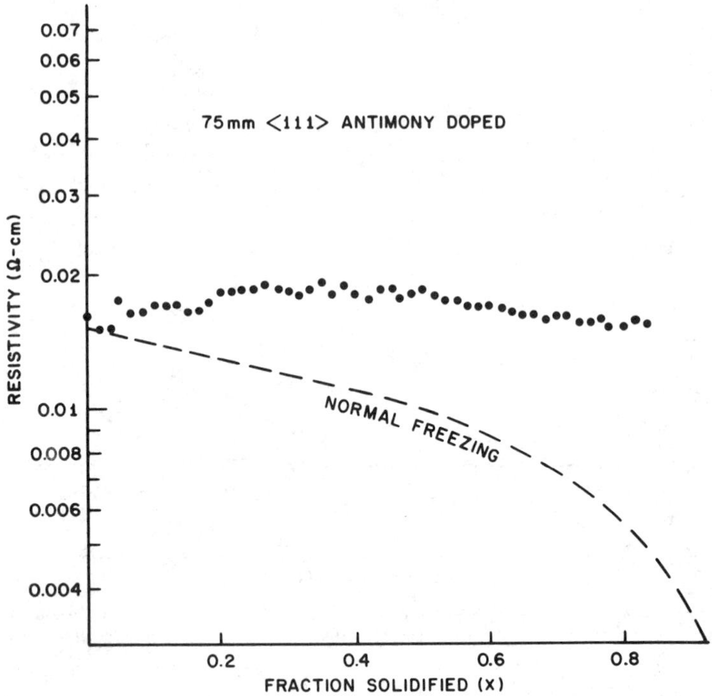

FIG. 9—*Axial resistivity profile of an antimony-doped crystal grown by the double-crucible method.*

with antimony, grown by the double-crucible arrangement. In this case, the h/H used is approximately ¼.

When applying double-crucible technique in growing N^+ silicon, the main consideration is the dopant back diffusion from the inner to the outer crucible. The double-crucible method may be viewed as a two-container system linked by an interconnecting tube (Fig. 10). The melt in the two containers is doped at two concentration levels, and a concentration gradient ($dc/d\ell$) exists across the length of the connecting tube. This gradient is inversely proportional to the tube length and directly proportional to the doping level when the concentration ratio of the two containers is fixed. At the static (no flow) condition, dopant diffusion may be approximated as

$$\text{dopant atom transfer rate } \alpha = \left(D \frac{dc}{d\ell}\right) A_t \tag{1}$$

where A_t is the cross section of the interconnecting tube, and D is the diffusion coefficient of dopant in the melt. The controlling factors, then, would be the tube dimensions and doping level. A low doping level and therefore a low concentration gradient results in less diffusion than in the case of high doping levels.

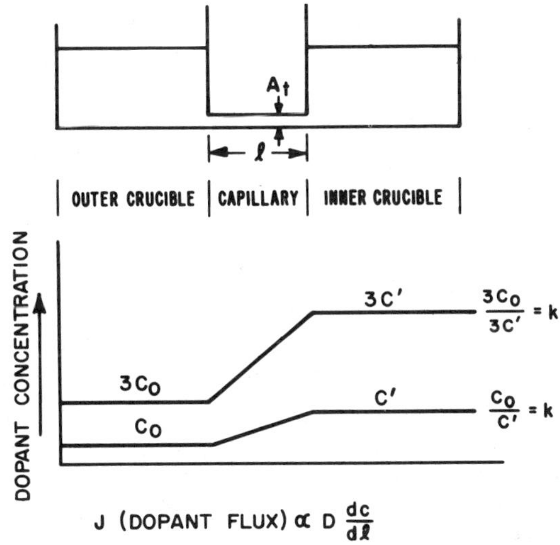

FIG. 10—*Schematic showing the dependence of dopant back diffusion on interconnecting tube length and doping level in a double-crucible arrangement.*

From a simple consideration of diffusion length, the average time for a dopant atom to jump across the interconnecting tube of length ℓ is related to the diffusion coefficient: $\ell = \sqrt{(Dt)}$ or $D = \ell^2/t$. When a crystal is pulled from the inner crucible—that is, the melt is withdrawn—the pressure difference in the two crucibles drives the melt from the outer to the inner crucible with a finite capillary flow rate V_t, which is opposite to the diffusion direction. The capillary flow, therefore, helps to suppress the dopant back diffusion. To sufficiently suppress the dopant transfer, the criterion would clearly be

$$\ell V_t \gg D \tag{2}$$

The capillary flow rate V_t is related to crystal pull rate V_1, crystal cross section A_s, and the tube cross-section A_t, by $V_t = V_1(A_s/A_t)$. Then the inequality in Eq 2 may be written as

$$\ell V_1 \left(\frac{A_s}{A_t}\right) \gg D \tag{3}$$

The D-value is on the order of 10^{-4} cm^2/s [*12*] for commonly used dopants in silicon. For the pull rate used in the 50- to 100-mm-diameter crystals, realistic and convenient values of ℓ and A_t may be found to sufficiently satisfy this criterion. In the current work, we have not detected evidence of significant back diffusion during crystal growth. However, the more important diffusion consideration is when there is no capillary flow. This corresponds to the necking and shouldering part of the growth process where no significant material is withdrawn from the inner crucible. Here diffusion control depends on a sufficiently long and narrow connecting tube to minimize dopant transfer for the duration of the "no-flow" part of the growth process. Avoidance of prolonged necking and shouldering is also essential. The relationship in Eq 1 and empirical findings are used as guidance for the interconnecting tube design.

Summary

In the growth of crystals from large melts, the crystal surface cooling and melt radial temperature gradient are shown to have profound effect on thermal stress and on crystallographic preferred growth. Excessive surface cooling and poor radial thermal gradient can result in structure loss and enhance flat formation during ⟨111⟩ crystal growth. The flat formation and its relations to facet growth mechanism, temperature, and seed crystal misorientation are analyzed. The melt radial temperature gradient is observed to be the primary controlling factor.

The oxygen distribution near the melt surface in a large melt is found to be inhomogeneous. The oxygen-rich flow seems to conform with the thermal convection pattern. When under reduced pressure, however, enhanced oxygen evaporation causes diffusion to play a more important role in oxygen transport near the melt surface, resulting in a more uniform oxygen distribution in that region. Finally, a double crucible technique for silicon growth is discussed. The application to N^+ silicon growth is demonstrated, where severe dopant segregation and associated constitutional supercooling are suppressed.

References

[*1*] Billig, E., *Proceedings of the Royal Society,* Vol. A235, No. 1200, April 1956, p. 37.
[*2*] Jordan, A. S., Caruso, R., and Von Neida, A. R., *Bell System Technical Journals,* Vol. 59, No. 4, April 1980, p. 593.
[*3*] Abe, T., *Journal of Crystal Growth,* Vol. 24/25, Oct. 1974, p. 463.
[*4*] Benson, K. E., Lin, W., and Martin, E. P. in *Semiconductor Silicon,* The Electrochemical Society, Pennington, N.J., 1981, p. 33.
[*5*] Katz, L. E. and Hill, D. W., *Journal of the Electrochemical Society,* Vol. 125, No. 7, July 1978, p. 1151.
[*6*] Carruthers, J. R., *Journal of the Electrochemical Society,* Vol. 114, No. 9, Sept. 1967, p. 959.
[*7*] Kobayashi, N., *Journal of Crystal Growth,* Vol. 43, No. 3, April 1978, p. 357.
[*8*] Hoshikawa, K., Hirata, H., Nakanishi, H. and Ikuta, K. in *Semiconductor Silicon,* The Electrochemical Society, Pennington, N.J., 1981, p. 101.
[*9*] Suzuki, T., Isawa, N., Okubo, Y., and Hoshi, K. in *Semiconductor Silicon,* The Electrochemical Society, Pennington, N.J., 1981, p. 90.
[*10*] Lin, W., and Pearce, C. W., *Journal of Applied Physics,* Vol. 51, No. 10, Oct. 1980, p. 5540.
[*11*] Kodera, H., *Japanese Journal of Applied Physics,* Vol. 2, No. 9, Sept. 1963, p. 527.
[*12*] Kodera, H., *Japanese Journal of Applied Physics,* Vol. 2, No. 4, April 1963, p. 212.

Ashok Murgai[1]

Impurity Incorporation in Czochralski Silicon Grown for VLSI

REFERENCE: Murgai, A., **"Impurity Incorporation in Czochralski Silicon Grown for VLSI,"** *Silicon Processing, ASTM STP 804,* D. C. Gupta, Ed., American Society for Testing and Materials, 1983, pp. 39–49.

ABSTRACT: Tight control of oxygen concentration (within ±0.3 ppma) is attained with a new method in the Czochralski growth of large-diameter silicon (86 to 125 mm). Programmed variation in growth parameters is employed during growth to compensate for the normal decrease in oxygen concentration otherwise present. The presence of Al and B impurities in vendor-supplied silica crucibles (up to 100 and 1 ppm by weight, respectively) is shown to significantly alter the apparent dopant segregation coefficient. The nondetection limits of various other metallic impurities, such as Mo, V, Ti, etc., being large (~5 ppm by weight) in the silica crucibles, the concentrations of these impurities can easily exceed undesirable levels in the crystal (up to 10^{12} atoms/cm^3). Fluctuations in growth parameters can raise the concentrations of these impurities further (by approximately two times) over microscopic regions of growth.

KEY WORDS: silicon, Czochralski, crystal growth, oxygen

While the detailed material requirements for very large scale integration (VLSI) silicon are still being identified, it is generally agreed that tight control of all impurities—oxygen, carbon, dopant, and others—is of prime importance. Both macro and micro scale control is required. The underlying mechanisms and source of oxygen are different than those of carbon. In the Czochralski method of silicon growth, large amounts of oxygen [*1–7*][2] (concentration levels: 20 to 40 ppma, as per ASTM Test for Interstitial Atomic Oxygen Content of Silicon by Infrared Absorption [F 121]) are unavoidably incorporated through dissolution of the silica crucible [*8*] in which the melt is contained. Small amounts of various other metallic impurities may also be present in the silica crucible. These impurities are also incorporated in the crystal. On the other hand, carbon comes from various graphite parts in the puller, such as the heater, crucible parts/supports and insulation package, etc. The concentration level of carbon in silicon varies from nondetection limits to a few ppma, depending primarily on the nature of argon gas flows employed.

[1] Advisory Engineer, IBM General Technology Division, Hopewell Junction, N.Y. 12533.
[2] The italic numbers in brackets refer to the list of references appended to this paper.

In the following investigation, a method of macroscopic oxygen control is presented during large-diameter growth of Czochralski silicon. The limits placed by VLSI requirements of homogeneity on certain growth parameters are examined. From a comparison of the experimental and theoretical resistivity profiles, the simultaneous incorporation of other impurities from the silica crucible are identified. Also reviewed is the effect of microsegregation on the latter impurities.

Experimental Procedure

Crystal Growth

Both small- and large-diameter (~20 mm and 80 to 130 mm), dislocation-free silicon single crystals were pulled along the ⟨100⟩ direction by the Czochralski method. Experimental details related to the growth and characterization of the small diameter crystals may be found in Ref *9*. Interface demarcation was applied during growth of these crystals for determining the microscopic growth rates by transmitting current pulses (18-A, 50-ms duration at 0.5-s intervals) across the crystal-melt interface.

The large-diameter crystals were pulled from 7, 9, 12, or 20 kg melts doped for boron dopant concentrations of 10^{14} to 10^{15} atoms/cm^3, from 203, 254, or 305 mm vendor-supplied silica crucibles.

Crystal pull rates in the range 60 to 115 mm/h were employed with ±15% pull rate variations applied for automatic diameter control. The crystals were grown under partial vacuum in an argon purge (pressures: 1.99 to 6.66 kPa [15 to 50 torr]) with crystal-crucible counter rotation rates applied in the range 1 to 35 rpm.

Measurements: Macroscale

The axial measurements were made at the center of cross-sectional slices taken at periodic intervals along the crystal length. Roughly 1.5-mm-thick slices were chem-etched, heat treated at 650°C for 1 h to annihilate oxygen related thermal donors [*9*], and then silica-polished to a mirror finish on both the surfaces. Oxygen measurements were made with an FTIR spectrophotometer. The measurements were reduced to oxygen concentration in ppma by multiplying 9.63 times the 9-μm absorption coefficient corrected for internal reflections, in accordance with the procedure given in ASTM F 121. Macroscopic resistivity measurements were made with a 4-point probe in locations on the slice that were sandblasted to ensure good electrical contact to the probes.

Measurements: Microscale

The microscale measurements were made on ⟨100⟩ longitudinal slices (~1 to 2 mm thick) cut along the growth direction. These slices were also chem-etched and silica-polished as above. Oxygen striations were characterized from (MoKα_1, (400)) transmission X-ray topographs. The details on longitudinal spreading resistance measurements and characterization of swirl-microdefects may be found in Ref *9*.

Results and Discussion

The detailed mathematical model on non–steady-state, axial variations in oxygen concentration was developed elsewhere [*10*]. There it was shown that normally the effect of decreasing crucible-melt interface area as the melt height is lowered during growth is to result in a significant decrease in the flux rate of oxygen introduced in the melt. Correspondingly, the oxygen concentration in the crystal decreases with the fraction of melt solidified. However, since the dissolution rate of the silica crucible may be expected to increase with increased mixing of the melt, the above decrease may also be expected to be compensated by increasing the rates of crucible rotation during growth. Uniform oxygen concentration may thus be attained.

Figure 1 shows the results of such experiments. Three axial oxygen concentration profiles are shown: (*1*) for an 86-mm-diameter crystal pulled from 12 kg melt and constant crystal-crucible counter-rotation rates of 22 and 15 rpm, pulled at 76 mm/h; (*2*) also for an 86-mm-diameter crystal pulled under similar growth parameters as above except that a linear increase in crucible rotation rate was applied during growth (+1 rpm for every 38 mm of crystal

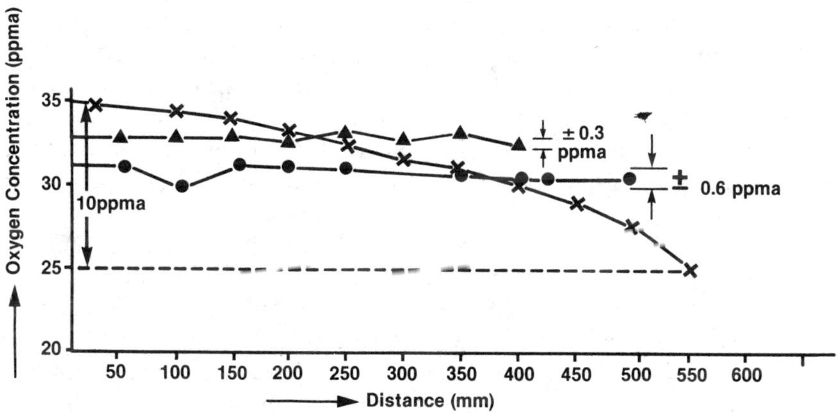

XXX *86-mm-diameter crystal; 12 kg melt; constant crystal-crucible counter-rotation rates. 22 and 15 rpm respectively.*

●●● *Growth parameters same as above except that crucible rotation rate was increased (1 rpm)/(38 mm of crystal grown) after crystal shoulder was attained.*

▲▲▲ *125-mm-diameter crystal; 18 kg melt; constant crystal rotation rate: 28 rpm; crucible counter-rotation rate: 10 rpm + average increase of (0.44 rpm)/(25.4 mm of crystal grown).*

FIG. 1—*Axial oxygen concentration variations under constant and variable crucible rotation rates applied during growth.*

length grown); (*3*) for a larger diameter (125 mm) crystal for which a nonlinear increase in crucible rotation rates was applied (average rate of increase of 0.44 rpm for every 25.4 mm of crystal grown). As can be seen, a significant decrease in oxygen concentration seed-tail (10 ppma) is present for the 86-mm-diameter crystal for which the crystal-crucible rotation rates were held constant during growth. However, with the linear increase in crucible rotation rates applied during growth for the other 86-mm-diameter crystal, uniform oxygen concentration can be seen to be attained, with seed-tail variations in concentration ±0.6 ppma [*11,12*]. Tighter control in oxygen concentration is attained for the 125-mm-diameter crystal (concentration variations ±0.3 ppma), for which the nonlinear increase in crucible rotation rates compensated further for the variation in the last profile. The foregoing method of attaining uniform oxygen concentration may be applied, in general, to a wide range of crystal-crucible diameters.

Furthermore, the aforementioned experiments were conducted with the commonly available hot zones in vendor-supplied Czochralski pullers. No attempt was made to increase the thermal gradients in the melt for growth of the larger diameter crystals. As can be seen, the latter necessitated a decrease in crystal pull-rates with increasing crystal diameter (57 mm/h for 125-mm-diameter crystal versus 76 mm/h for 86-mm-diameter crystal) for attaining uniform oxygen concentration. Thus, the requirement of oxygen homogeneity places a maximum limit on the crystal pull-rates that may be employed.

The effect of short period and long period pull-rate fluctuations (periods less than a few minutes or greater than several minutes, respectively) on oxygen concentration variation is shown in Fig. 2. Such variations in pull-rate are normally present with the commonly used automatic diameter control systems. In these systems, variations in pull-rate and heater power are employed in a closed loop to control the crystal diameter. Figure 2 shows four axial oxygen concentration profiles, all for 125-mm-diameter crystals for which the same nonlinear increase in crucible rotation rates was applied during growth.

The two lower concentration profiles correspond to growth in the presence of long period pull-rate fluctuations. On the other hand, the two upper profiles correspond to growth in the presence of only short period fluctuations. The numbers adjacent to each measurement point on a profile denote the corresponding oxygen concentration value. The horizontal straight lines in the two lower plots were drawn to emphasize the otherwise small variations in oxygen concentration for these plots.

As can be seen, the oxygen concentration variations (up to ±1.8 ppma) in the two lower plots due to the long period pull-rate fluctuations are minimized (up to ±0.5 ppma) in the two upper plots coresponding to only short period fluctuations. Evidently, the long period increase in pull-rate results in a large enough depletion of oxygen in the melt ($ke = 1.25$ [*13*] for oxygen) with a corresponding decrease in concentration in the crystal and vice versa.

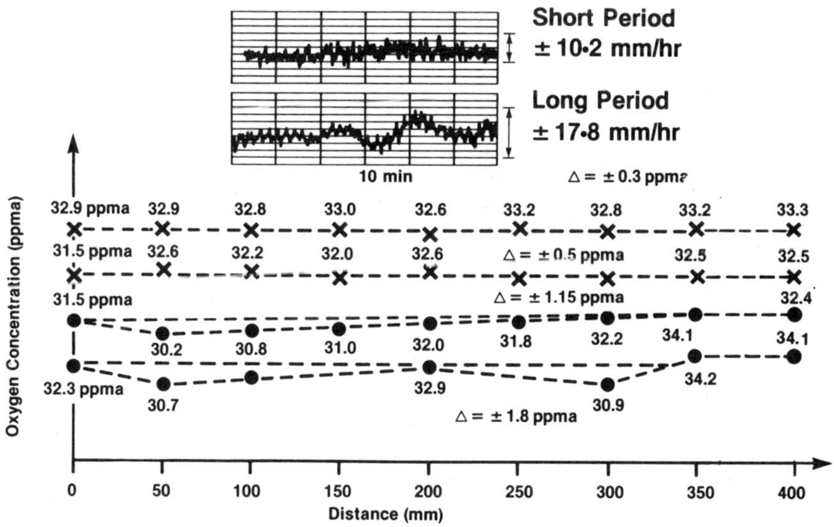

●●● *Long period pull-rate variations were present during growth.*

XXX *Only short period pull-rate variations present.*

FIG. 2—*Axial oxygen concentration variation in the presence of short period and long period pull-rate variations present under automatic diameter control. Growth parameters for all crystals same as for in Fig. 1.*

Thus, the maximum acceptable variations in oxygen concentration also place a maximum limit on the acceptable pull-rate variations.

The presence of other silica-crucible impurities in the crystal were identified from a direct comparison of the axial experimental and theoretical resistivity plots shown in Fig. 3. The experimental plot was measured for a typical 86-mm-diameter crystal pulled from 7-kg melt in a 203-mm-diameter (2 γ_c) vendor-supplied crucible. The melt was boron doped to attain 15 Ω-cm resistivity at the seed end of the crystal. Prior to the measurements the silicon slices were heat treated at 650°C for 1 h to annihilate the oxygen thermal donors. The upper, theoretical plot is a normal-freeze plot ($C_S = C_0 \cdot ke \cdot (1 - g)^{ke-1}$, in the usual notation) for boron ($ke = 0.838$ determined from the B.P.S. relation [*14*] by employing $ko = 0.8$, $\delta/D = 130$ [*15*] and $V = 20$ μm/s). This would be expected to prevail if only the boron dope powder added before growth were present in the melt. However, a significant discrepancy can be seen to be present between the experimental and theoretical resistivity profiles.

In order to identify the reasons for the discrepancy, some of the vendor-supplied Silica Crucible Materials 1, 2, and 3 were analyzed for impurities through emission spectroscopy (see Table in Fig. 3). Significant amounts of Al and B impurities (100 and 1 ppm by weight, respectively) were identified in a silica crucible, Material 1, similar to that employed during the growth of the experimental crystal. The theoretical resistivity profile, taking into ac-

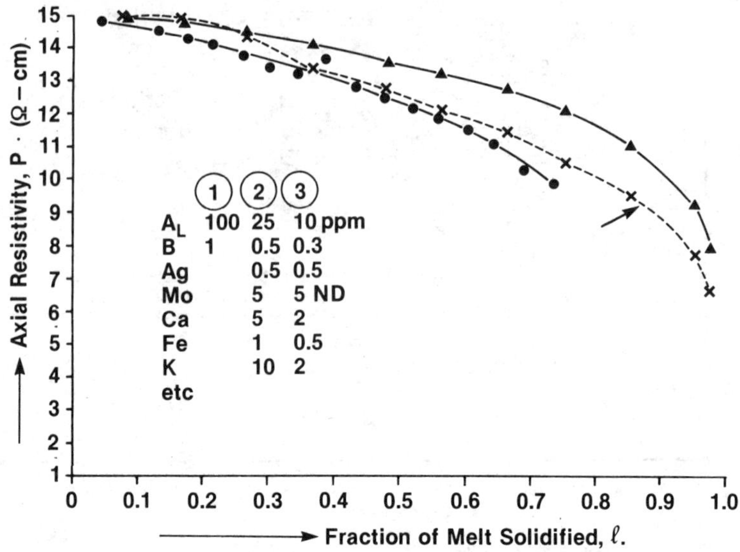

▲▲▲ *Theoretical, normal freeze plot corresponding to* $Cs = Co \cdot ke \cdot (1 - \ell)^{ke-1}$, *disregarding the presence of silica crucible impurities.* ke *for B = 0.838 (see text).*

XXX *Theoretical resistivity plot taking into account the presence of 1 ppm by weight of B in the silica crucible. Initial silica crucible dissolution time period was assumed to be 1 h.*

●●● *Experimental resistivity plot. 86-mm-diameter crystal; 203-mm-diameter crucible; 7 kg melt.*

FIG. 3—*Axial resistivity variation. Seed-end resistivities were targeted for 15 Ω-cm. Inset table shows impurities identified through emission spectroscopy for Silica Crucible Materials 1, 2, and 3. Material 1 was analyzed only for Aℓ and B.*

count the presence of 1 ppm by weight of B in the silica crucible, is also shown in Fig. 3, in accordance with an equation similar to that of Bradshaw [*8*], rederived in the present form by Patrick [*16*]:

$$\frac{Q}{M} = \frac{Qo}{Mo} \cdot ke \cdot (1 - \ell)^{ke-1} + \frac{ke}{2 - ke} \cdot k2 \cdot Mo \cdot (1 - \ell) \cdot [(1 - \ell)^{ke-2} - 1] + \frac{ke}{1 - ke} \cdot k3 \cdot [(1 - \ell)^{ke-1} - 1] \quad (1)$$

where Q/M is the fractional impurity content in the crystal with fraction ℓ of melt solidified, Qo/Mo is the fractional impurity content incorporated at the start of growth, ke is the effective segregation coefficient, $k2 = 2C_c\Upsilon/V' P_\ell \gamma_c$ and $k3 = (C_c/V') \cdot (\Gamma + \Upsilon \cdot \Pi \cdot \gamma_c^2)$, C_c being the fractional impurity content by weight in the silica crucible material, Υ being the crucible dissolution rate (g/cm^2-s), V' being the crystal growth rate (g/s), P_ℓ being the density of liquid silicon melt (g/cm^3), and Γ being the dissolution rate of the silica crucible near the melt surface (g/s). In Eq 1, the experimentally observed values of Υ

and Γ determined by Patrick [*16*] on Crucible 1 were employed. These were determined from weight decrements of silica plates cut from the crucible and submerged in silicon melts for varying intervals of time. Furthermore, the impurity concentrations as determined through Eq 1 were reduced to resistivity *p*-values by employing Irwin's curves [*17*].

Two important features may be noted from the plot in Fig. 3 incorporating the effect of B impurity present in Crucible 1: (*a*) The plot exhibits a bump in resistivity due to the combined effects of normal freeze of the initial boron dopant present in the melt as well as of the continual additions from the silica crucible during growth. (*b*) The plot indicates a faster decrease in resistivity than present due to normal freeze alone. The combined effects of the presence of varying amounts of other impurities such as Al and P are shown in Fig. 4. Figure 5 shows the plots of $\ln p$ versus $\ln (1 - \ell)$ corresponding to the plots in Figs. 3 and 4. The deviation in the slopes of the latter plots from that corresponding to the normal freeze ($ke = 0.838$ for B dopant) can be seen to increase with increasing impurity content present in the crucible. This may be regarded as an apparent decrease in the boron dopant segregation coefficient (to values as low as 0.7).

Apart from resistivity control, from the VLSI standpoint, the foregoing

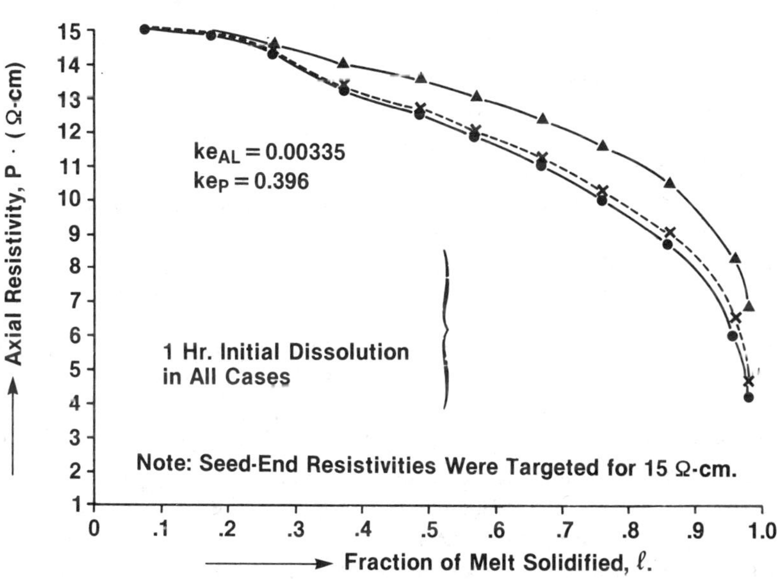

▲▲▲ *Plot for silica crucible material containing 0.22 ppm (by weight) B + 17 ppm Aℓ. Resistivities were targeted by adjusting weight of initial dope added to melt to attain the desired resistivity at seed-end.*

●●● *Plot for silica crucible material containing 1 ppm B + 100 ppm Aℓ.*

XXX *Plot for silica crucible material containing 1 ppm B + 100 ppm Aℓ. Also 2 ppba P present in polysilicon charge.*

FIG. 4—*Theoretical axial resistivity variation.*

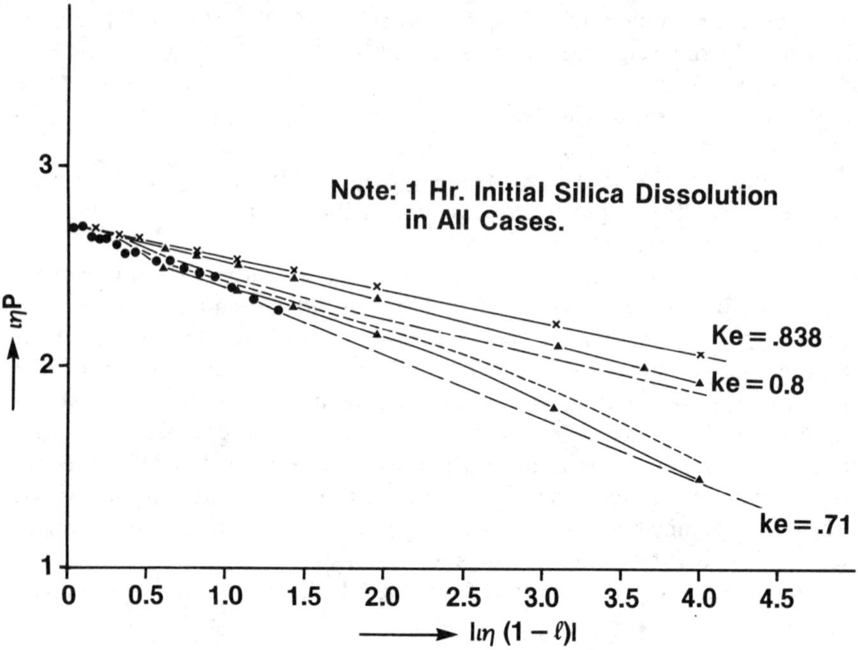

XXX *Normal-freeze plot.*
Upper ▲▲▲ *0.22 ppm B + 17 ppm A in silica crucible.*
— - — *1 ppm B in silica crucible.*
------ *1 ppm B + 100 ppm A in silica crucible; 2 ppba P in polysilicon.*
Lower ▲▲▲ *1 ppm B + 100 ppm A in silica crucible.*
●●● *Experimental plot. Apparent decrease in segregation coefficient to 0.71.*
— — *Straight line representing limits for* ke = *0.71.*

FIG. 5—*Apparent decrease in B dopant segregation coefficient (seen from increasing negative slopes of ln* p *versus ln [1 − ℓ]) due to crucible/poly-silicon impurities.*

analysis highlights a more serious problem—the possibility of contamination from other metallic impurities such as Mo, V, Ti, etc., which are known to be very effective killers of electrical carrier lifetime. These impurities are normally present below detection limits in the silica crucible material (the detection limits of the impurities being of the order of a few ppma). However, since their segregation coefficients are very low, their concentrations can easily exceed levels of 10^{12} atoms/cm^3 for the last portions of melt solidified. This is seen from comparison with Al segregation (also *P*), which also has a low segregation coefficient, in Table 1. The Al concentration increases by a factor of six or seven times the initial concentration with 70% melt solidified. A further increase in concentration can also occur over microscopic regions of growth (up to approximately two times) due to the normally present fluctuations in growth parameters related to thermal asymmetry/melt thermohydrodynamics. For the latter, the reader is directed to Refs *4* and *9*. In these

TABLE 1—*Increase in aluminum and phosphorus concentrations with fraction of melt solidified.*[a]

	Aluminum Concentration, atoms/cm^3		Phosphorus Concentration, atoms/cm^3	
ℓ	100 ppm, 1 h	22 ppm, 1 h	1 ppm, 1 h	1 ppm, 2 h
0.070	0.087×10^{14}	0.015×10^{14}	0.527×10^{14}	0.593×10^{14}
0.169	0.128	0.022	0.638	0.708
0.267	0.176	0.030	0.769	0.845
0.366	0.235	0.040	0.927	1.010
0.464	0.312	0.053	1.126	1.218
0.563	0.419	0.071	1.389	1.493
0.661	0.579	0.098	1.762	1.883
0.760	0.861	0.146	2.353	2.502
0.858	1.516	0.258	3.521	3.726
0.957	5.072	0.862	7.948	8.366
0.982	12.028	2.044	13.848	14.550

[a] Aluminum is introduced into melt from silica crucibles containing 100 ppm or 22 ppm by weight of aluminum. A 1-h initial dissolution of these silica crucibles is considered before crystal growth. The phosphorus is introduced from both the silica crucible and the polysilicon charge (with concentrations 1 ppm by weight and 2 ppba respectively). Initial silica crucible dissolutions of 1 and 2 h are considered for the latter.

references the impurity concentration variation was related to microscopic growth rate variation and to the presence of striated distributions of microdefects.

The oxygen striations in macroscopically uniform oxygen concentration crystals (oxygen and boron concentrations: 10^{18} versus 10^{14} atoms/cm^3) seen earlier were revealed through X-ray topography [(400) reflection] on longitudinal slices taken from the seed-end cone sections of the crystals. Figure 6 shows a topograph for a typical 125-mm-diameter crystal for which the crystal rotation rate and pull-rate were varied during the cone growth. As can be seen, the oxygen striations (instantaneous crystal-melt interface shapes) are convex towards the melt for crystal rotation rates below 15 rpm. As the rotation rate is increased, the striations become flatter till around 30 rpm

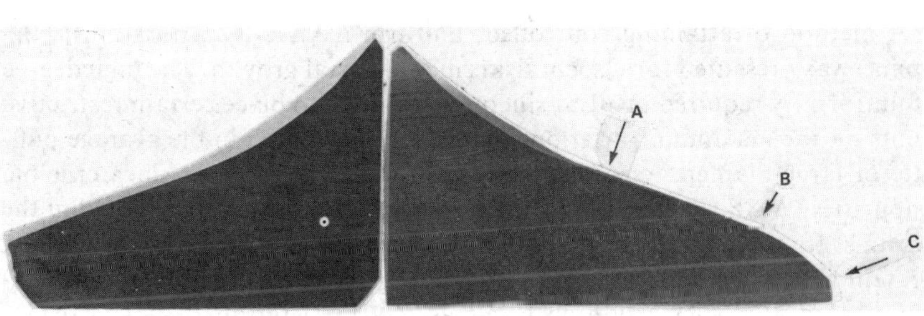

FIG. 6—*X-ray topographs of cone section of 125-mm-diameter crystal showing oxygen striations. Mo — $K\alpha_1$↑ (400) reflection. Crystal rotation rate: 15 rpm; pull-rate: 25 mm/h for growth above* A. *For growth below* A, *crystal rotation rate 30 rpm. Pull-rate was raised to 203.2 mm/h at* B *and at* C *lowered to 57.2 mm/h at* C.

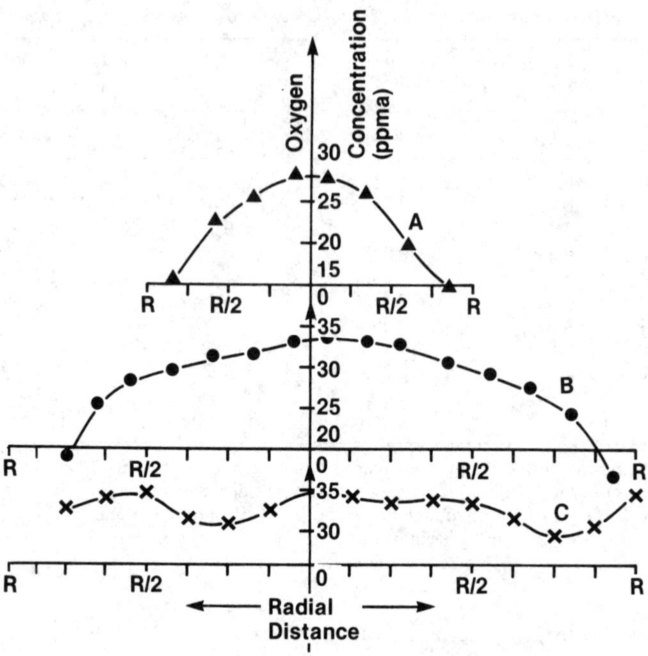

FIG. 7—*Radial variations in oxygen concentration in cone section of the 125-mm-diameter crystal in Fig. 6, at locations around* A, B, *and* C.

they are below the resolution limit, in the lower part of the topograph. From Fig. 7, a corresponding decrease can be seen in the radial variation in oxygen concentration with the increasing crystal rotation rates. Thus, the latter analysis indicates a direct interrelation prevailing between crystal-melt interface shapes, melt hydrodynamics, and radial oxygen concentration variation.

Summary

A method of attaining controlled, uniform oxygen incorporation (±0.3 ppma) was presented for Czochralski silicon crystal growth. The high degree of uniformity required in VLSI silicon was shown to place certain restrictive limits on the maximum variations that can be permitted in the average pull-rate of large-diameter crystals. Significant amounts of other silica crucible impurities (Al, B, P, etc.) were shown to be nonuniformly distributed in the crystals. It was pointed out that impurities such as Mo, W, Ti, etc., normally present below detection limits in vendor-supplied silica crucibles, may be incorporated at sufficiently high levels as to be deleterious to VLSI device yields. Higher purity crucibles and better methods of impurity characterization will be needed in the future.

References

[*1*] Kaiser, W. and Keck, P. H., *Journal of Applied Physics,* Vol. 28, No. 8, Aug. 1957, pp. 882–887.

[*2*] Goss, A. J. and Addlington, R. E., *Marconi Review,* Vol. 22, 1959, pp. 28–43.

[*3*] Yatsurugi, Akiyama, N., Endo, Y., and Nozaki, T., *Journal of the Electrochemical Society,* Vol. 120, No. 7, July 1973, pp. 975–979.

[*4*] Murgai, A., Chi, J. Y., and Gatos, H. C., *Journal of the Electrochemical Society,* Vol. 127, No. 5, May 1980, pp. 1182–1186.

[*5*] Abe, T., Kikuchi, K., and Shirai, S. in *Semiconductor Silicon 1977,* H. Huff and E. Sirtl, Eds., The Electrochemical Society, Pennington, N. J., 1977, pp. 95–104.

[*6*] Murgai, A., *The Electrochemical Society Extended Abstracts,* Vol. 80, No. 2, 1980, pp. 1333–1335.

[*7*] Turovskii, B. M., *Izvestiya Akademii Nauk SSR, Inorganic Materials,* Vol. 11, No. 4, April 1975, pp. 505-507.

[*8*] Bradshaw, S. E. in *Solid State Physics in Electronics and Telecommunication,* Desirant and Michels, Eds., Academic Press, New York, 1960, Vol. 1, Part I, pp. 44–60.

[*9*] Murgai, A., Gatos, H. C., and Westdorp, W. A., *Journal of the Electrochemical Society,* Vol. 126, No. 12, Dec. 1979, pp. 2240-2245.

[*10*] Murgai, A. in *Semiconductor Silicon 1981,* H. Huff, R. J. Kriegler, and Y. Takeishi, Eds., The Electrochemical Society, Pennington, N. J., 1981, pp. 113–125.

[*11*] French, U.S. Patent applied for.

[*12*] Murgai, A., Patrick, W. J., Combronde, J., and Felix, J. C., *IBM Journal of Research and Development,* Vol. 26, No. 5, Sept. 1982.

[*13*] Yatsurugi, Y., Akiyama, N., Endo, Y., and Nozaki, T., *Journal of the Electrochemical Society,* Vol. 120, No. 7, July 1973, pp. 975–979.

[*14*] Burton, J. A., Prim, R. C., and Slichter, W. P., *Journal of Chemical Physics,* Vol. 21, 1953, p. 1987.

[*15*] Runyan, W. R. in *Silicon Semiconductor Technology,* McGraw-Hill, New York, 1965, pp. 57 and 108.

[*16*] Patrick, W. J., private communication.

[*17*] Irvin, J. C., *Bell System Technical Journal,* Vol. 41, No. 2, March 1962, p. 387.

Kedar P. Gupta[1] *and Robert O. Gregory*[2]

Dependence of Silicon Float-Zone Refining Parameters on Frequency

REFERENCE: Gupta, K. P. and Gregory, R. O., "**Dependence of Silicon Float-Zone Refining Parameters on Frequency,**" *Silicon Processing, ASTM STP 804,* D. C. Gupta, Ed., American Society for Testing and Materials, 1983, pp. 50–61.

ABSTRACT: Float-zone refined silicon is produced by establishing a molten zone in a silicon rod, vertically situated and passing the zone through the rod. Induction heating is the typical conventional means for producing the molten zone. Since power levels are very high with large-diameter rods, the efficiency of the heating process is a concern. One of the key parameters is the frequency of the radio frequency current in the induction coil. Silicon rods of 65-mm diameter were zone-refined, at a series of frequencies in the 200 kHz to 3.5 MHz range. It was found that frequencies lower than 500 kHz are not suitable because they produce undesirable surface melting characteristics. On the other hand, higher frequencies tend to increase the susceptibility for arcing. Indeed, the suitable range appears to be between 2 and 3 MHz. The presentation also highlights the major differences between the float-zone and Czochralski processes, and explains the reasons why large-diameter float-zone technology has lagged behind Czochralski technology.

KEY WORDS: silicon, zone refining, float-zone, crystal growth, induction heating

There are two major techniques employed for commercial production of large-diameter silicon crystals. One method, known as Czochralski or crystal pulling, uses a large-diameter quartz crucible in which a silicon charge is melted by employing a radio frequency (rf) or a resistance heater. Crystals are pulled by dipping a seed from the top and pulling it upward in a controlled manner. In the other method, termed float-zone refining, a polycrystalline silicon rod is zone-melted by passing it through an induction coil. The energy source is an rf generator. Float-zone technology has lagged behind Czochralski technology where large-diameter crystal production is concerned (Fig. 1). For some obvious economical reasons, large-diameter Czochralski crystals are progressively in demand and hence produced in large quantities, filling more than 80% of total silicon needs. Float-zone crystals still remain the required material for high-power rectifiers and

[1] Engineering Manager, Semiconductor Division, Cincinnati Milacron, Maineville, Ohio 45039.

[2] Professor, Electrical Engineering Department, Washington University, St. Louis, Mo. 63130.

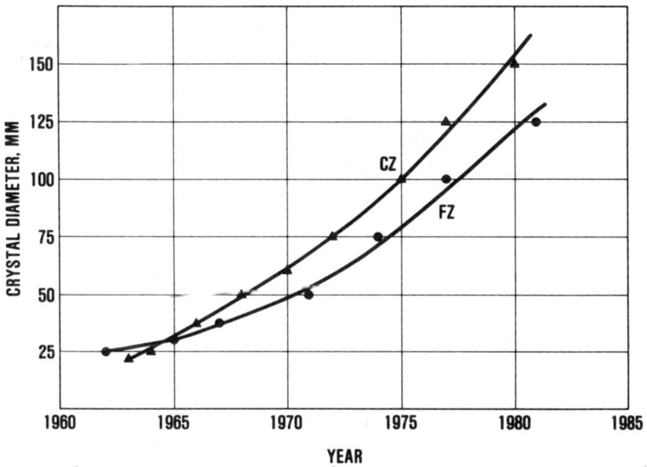

FIG. 1—*The historical trend in Czochralski and float-zone crystal diameters.*

thyristors where high-resistivity material is necessary. Although some float-zone material has been used to prepare integrated circuits, there is no clear indication that it has any significant advantage over the Czochralski material.

The ease in producing larger diameter Czochralski material is only one of several reasons for its large usage relative to float-zone material. Table 1 describes some of the major differences between these two techniques.[3] The growth of large-diameter crystals is easier in the Czochralski method because the mechanical growth conditions are more favorable as a result of self-stabilization of the hanging crystals. In the float-zone technique, the seed, being on the bottom, supports the weight of the crystal, which increases as the crystal is grown. The system is mechanically unstable, and is very sensitive to any mechanical or electrical vibrations.

Recently there have been attempts to mechanically hold the float-zone crystal during the growth process, which appears to be a critical requirement for growing long, 75-mm diameter or larger crystals. Other critical factors that make float-zone a difficult process are the requirements for good quality, smooth surface poly feed rods of large diameters and for an ability to melt these feed rods efficiently without any mechanical or electrical interruptions. The straightforward method of resistance heating, which is used for the Czochralski process, has not been used for float-zone applications. The energy source is primarily an rf generator using an induction coil to create the molten zone. Large-diameter crystals with large feed charges require that the power output capability of the rf generators be increased substantially without any danger of corona formation or arcing between the

[3] Keller, W. and Muhlbauer, A., *Floating-Zone Silicon*, Marcel Dekker, New York, 1981, p. 211.

TABLE 1—*Comparison of Czochralski and float-zone growth methods.*

	Method	
Characteristic	Czochralski	Float-Zone
Charge Method	Chunks of polysilicon	Poly rod with good surface quality
Container	Quartz crucible	No container used
Seeding Method	From top	From bottom (top seeding method limited to about 35 mm diameter)
Heating Technique	Mostly resistance	rf induction heating
No. of Passes Needed	One	Generally two
Mechanical Growth Conditions	Gravity stablizes melt and crystal	Gravity destablizes melt and crystal. Rotation rates, vibrations critical
Crystal Size	Large diameters easy Large weights 125-mm production	Large diameters possible Reasonable weights 100-mm production
Thermal Growth Conditions	Small gradients Minor transients Large melt volume	Large gradients Large transients Small zone volume
Factors Influencing Economics	Large charge size (30 kg) Low pull rate (2 mm/min) Good yield Lower poly cost Crucible and graphite parts Reasonable labor and technical skill required Commercial equipment and processes available	Reasonable weight Fast pull rate (>3 mm/min), but two passes are needed Reasonable yield High poly cost/No remelt No crucible High labor and technical skills needed Limited availability of equipment and processes

coil legs or between the coil and the feed rod. Figure 2 describes the power requirements for various crystal diameters with a normal zone for a needle-eye coil where the molten zone is traveling at the rate of 4 mm/min.[4] The selection of rf frequency is a key parameter in achieving efficient, productive zone refining. This paper examines some of the theoretical and experimental aspects of the dependence of float zone refining on the rf frequency.

Theoretical Relationship Among Critical Generator Parameters

An equation can be developed relating the melt current in the rod to the frequency and the power. A simplified arrangement is shown in Fig. 3*a*, where the molten silicon zone is treated as a cylinder of diameter *Ds*, skin

[4] Kramer, H., "Monsanto Report: The Historical Trend in Float-Zone Crystal Diameters and Power Requirements," Monsanto Corporation, St. Peters, Mo., Sept. 1981, p. 7.

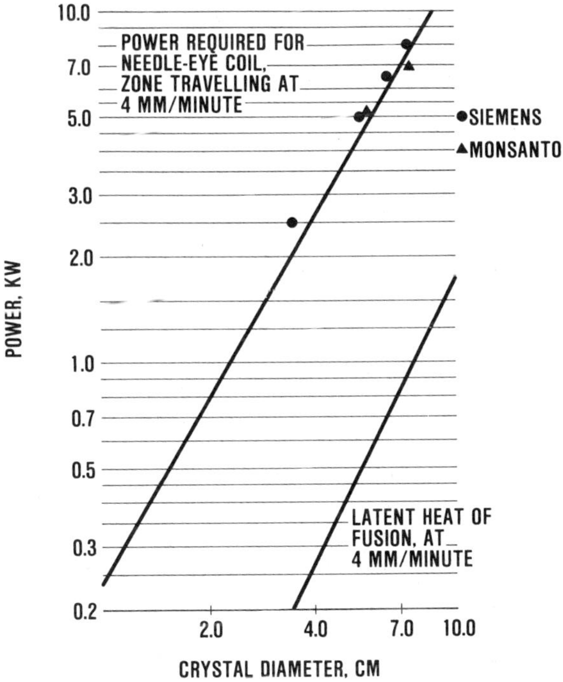

FIG. 2—*Power requirements for various crystal diameters during zone-refining.*

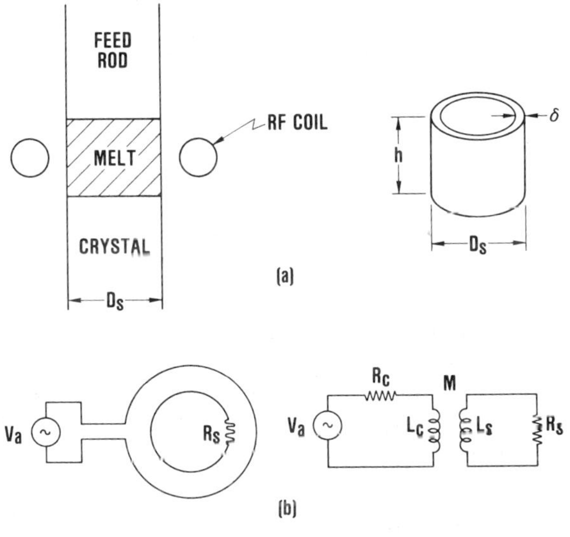

FIG. 3—(a) *Schematic view of rf heating of silicon rod, and a simplified shape of the molten zone.* (b) *Equivalent circuit diagrams.*

depth δ, and height h. The resistance of the melt Rs can be expressed as

$$Rs = \frac{\Pi Ds}{\sigma E h \delta} \tag{1}$$

where σE is electrical conductivity of molten silicon. An equivalent circuit of the above arrangement is shown in Fig. 3*b*. The Kirchhoff equations for Fig. 3*b* can be written as

$$Va = RcIc + j\omega LcIc + j\omega MIs \tag{2}$$

$$0 = j\omega MIc + j\omega LsIs + RsIs \tag{3}$$

where mutual inductance $M = k(LsLc)^{1/2}$, δ = skin depth of the melt $= \sqrt{2}(\omega\mu 0\sigma E)^{-1/2}$, Va = work coil applied voltage, Ic = work coil current, Is = melt current, Rs = melt resistance, Rc = work coil resistance, Lc = work coil inductance, Ls = melt inductance, k = coefficient of coupling, $\omega = 2\Pi f$ (frequency), $\mu 0$ = magnetic permeability, and $j = \sqrt{-1}$.

From the foregoing equations, one obtains

$$Is = \frac{\omega MIc}{\sqrt{Rs^2 + \omega^2 Ls^2}} \simeq \frac{M}{Ls} Ic \tag{4}$$

if $Rs \ll \omega Ls$, and for power to the melt:

$$P_{\text{melt}} = Is^2 Rs = \frac{\omega^2 M^2 Ic^2}{Rs^2 + \omega^2 Ls^2} Rs \tag{5}$$

$$\simeq \frac{M^2}{Ls^2} Ic^2 Rs$$

and

$$Va \simeq K(P_{\text{melt}})^{1/2} f^{3/4} \tag{6}$$

where K is a constant. Equation 6 also depends on the assumptions that $Rs \ll \omega Ls$ and $Rc \ll \omega Lc$.

From Eq 6 it is clear that the voltage applied between the coil terminals, for the same power level, depends on the oscillating frequency. Since corona formation is a result of a large coil voltage impressed across a relatively small gap between the coil and the rod, and between coil terminals, a substantial improvement should result from lowering the frequency. The higher power output required for larger diameter crystals should be obtainable without any arcing at a lower frequency. But work by Gupta et al[5] indicates that zone refining at low frequencies, in the 200 to 400 kHz range, has some serious problems. The increased electrical skin depths at the lower frequencies make it very difficult for the outer surface to melt properly. Commercial practice uses frequencies for float-zone application in the range of 2 to 3.5 MHz.

For high rf frequencies, we can assume $Rs \ll \omega Ls$; Eqs 1 and 5 will yield, for constant P_{melt}, a relationship between coil current and frequency f:

$$Ic \propto f^{-1/4} \tag{7}$$

Hence if the power level, or what amounts to the same thing, the zone shape and size, are maintained for a particular size of rod being zone-refined, the current in the coil should vary as a function of $f^{-1/4}$. This paper reports on a number of experiments that were conducted to verify this relationship.

Equations 5, 6, and 7 all depend, in their approximate form, on the assumption that $Rs \ll \omega Ls$. To verify this for the frequency range of interest, we may consider the ring of Fig. 3*a*. The resistance is

$$Rs = \frac{\Pi Ds}{\sigma E \delta h}$$

With $Ds = 6$ cm, $h = 3$ cm, $\sigma E = 10^4\ \Omega/\text{cm}$ (typical for molten silicon), and $\delta = 0.2$ cm gives $Rs \simeq 0.003\ \Omega$. The inductance Ls of such a structure is $Ls = (0.4) FDs\ \mu\text{H}$,[6] where $F =$ form factor $= 0.025$ and $Ds = 6$ cm.

This gives $Ls \simeq 0.06\ \mu\text{H}$; at a frequency of 2 MHz, $\omega Ls \simeq 0.75\ \Omega$. So, the assumption that $Rs \ll \omega Ls$ is well justified for the frequency range 2 to 3.5 MHz, if Rs is assumed to be the resistance of a molten silicon ring.

Experimental Results

Polysilicon rods selected for this study had low base levels of less than 0.3 ppb *N*-type impurities. These rods were 6.5 cm in diameter and 50 cm long, and were selected from the same reactor run. Two passes were made on each rod. All measurements were made during the second pass in a steady-state condition, during the middle of the run. The second pass was needed to eliminate any variation resulting from the feed rod. No attempt was made to grow single crystal in the bottom. The coil used was a one-turn copper coil with an opening smaller than the feed rod, and it was concentric to the top and bottom rods.

The experiments were run between 2 and 3 MHz. Every attempt was made

[5] Gupta, K. P., Gregory, R. O., and Rossnick, M., *Journal of Crystal Growth*, Vol. 44, No. 5, Dec. 1978, pp. 526–532.

[6] *Reference Data for Radio Engineers*, Howard W. Sams, Indianapolis, 1969, p. 6–1.

to maintain the size and shape of the molten zone. Current in the work coil was monitored by measuring the rf voltage in the capacitive portion of the tank circuit. Knowing the tank circuit capacitive reactance, one can calculate the tank circuit current, which is equal to the work coil current.

Figure 4 describes the relationship between rf frequency and the coil current measured experimentally. The curve does not show the theoretical relationship predicted in Eq 7. Experiments were run to determine the exact shape and size of the molten zone. A molten zone of silicon was solidified suddenly by interrupting the rf power. A longitudinal cross-section of the zone is shown in Fig. 5. It is clear that on the feed side, only a thin layer of silicon was in the molten form. The whole top portion can be visualized as having a thin layer of molten silicon running downward to form a molten pool in the bottom section of the rod.[5] The thickness of the melt layer was determined by polishing the longitudinal section of the zone and measuring the melt thickness at various places at high magnification. It was found that the melt layer thickness was less than the skin depth for the melt at those frequencies. The measured relationship between the melt-thickness *d* and the skin depth is given in Fig. 6.

Discussion and Conclusions

As Fig. 4 shows, the measured work coil current variation with frequency does not follow the anticipated pattern. Equation 7 predicts that the work coil current, under the stated assumptions, should vary as $f^{-1/4}$. Experimentally, it was observed that the current, in the frequency range of 2 to 2.5 MHz, decreases more than Eq 7 predicts, while in the frequency range 2.5 to 2.88 MHz the coil current appears to increase rather than decrease with frequency.

One of the assumptions made in deriving Eq 7 was that *Rs*, the effective re-

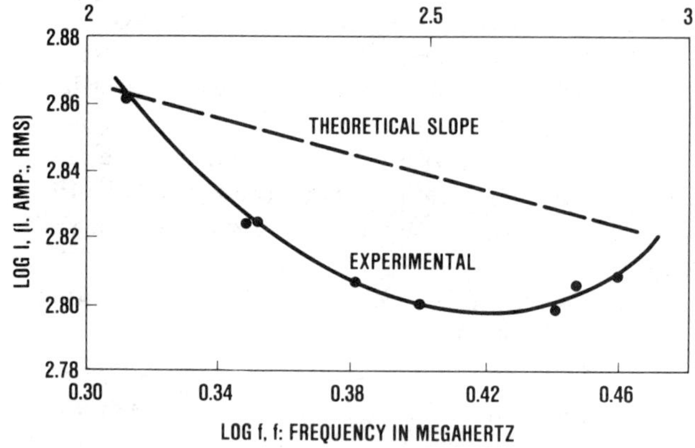

FIG. 4—*Variation of rf current in the coil as a function of frequency for same zone shape.*

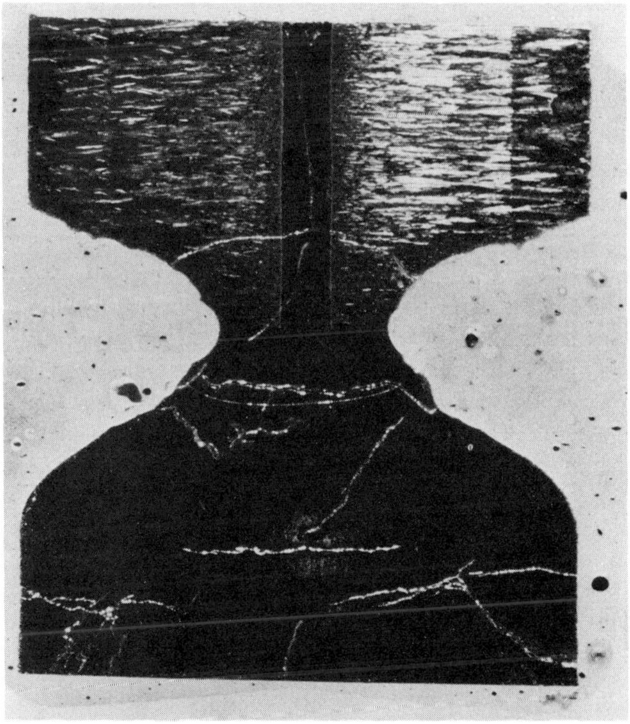

FIG. 5—*Longitudinal cross-section of the molten zone.*

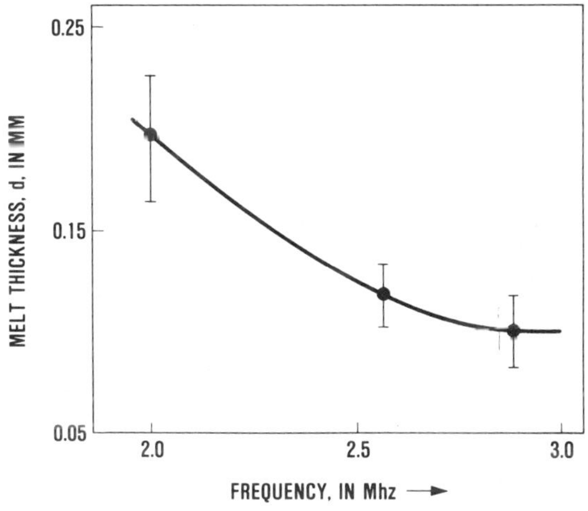

FIG. 6—*Variation of melt thickness as a function of frequency.*

TABLE 2—*Effect of frequency on melt thickness and skin depth.*

Serial Number	Frequency, MHz	Melt Thickness d, mm	Melt Skin Depth δ, mm	d/δ	Ic, A, rms, Experimental
1	2.03	0.185 ± 0.042	0.36	0.51	762
2	2.53	0.117 ± 0.015	0.32	0.36	631
3	2.88	0.090 ± 0.017	0.300	0.30	645

sistance of the melt, would be computed on the basis of a ring of melt, substantially thicker than the electrical "skin-depth." In light of the results shown in Fig. 6, which gives the measured melt film thickness on the top rod as a function of frequency, it appears that the top rod melt thickness is actually smaller than the skin depth. This is shown in Table 2. Melt film thickness varies from about 0.185 mm to 0.09 mm as the frequency varies from 2.03 to 2.88 MHz. In this same frequency range, the ratio of melt thickness to skin depth varies from 0.51 to 0.3, so a more accurate representation of the melting top rod is required. To obtain a more accurate representation, the resistance/unit area of a semi-infinite conductor layered with a film of different conductivity was computed. The geometry is shown in Fig. 7. The expression for resistance per unit area is derived in the Appendix, and is given by

$$\frac{R/\square}{R_{\text{melt}}/\square} = \frac{1 + 1.382\, e^{-2d/\delta} \sin \dfrac{2d}{\delta} - (0.691)^2 e^{-4d/\delta}}{1 - 1.382\, e^{-2d/\delta} \cos \dfrac{2d}{\delta} - (0.691)^2 e^{-4d/\delta}}$$

Figure 8 shows the result of this calculation. As the frequency varies from 2 to

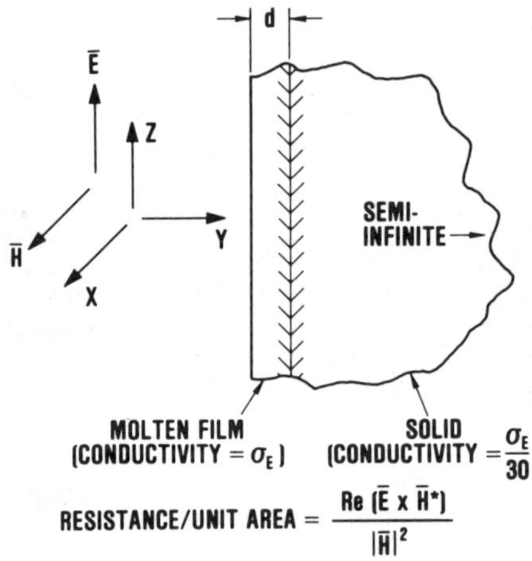

FIG. 7—*A schematic view of the melting interface.*

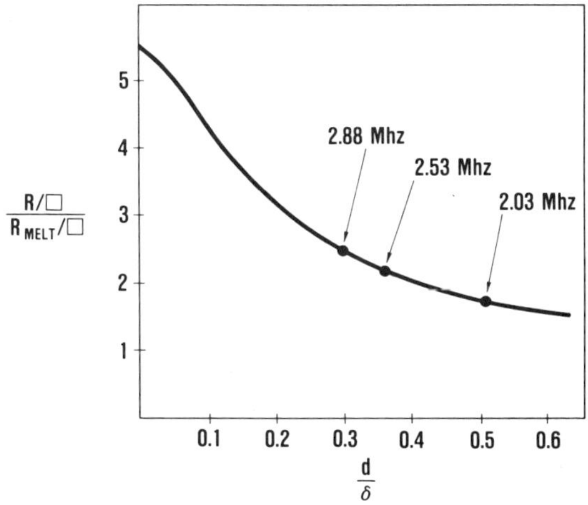

FIG. 8—*Resistance per unit area as a function of* d/δ.

3 MHz, using the measured ratios of film thickness to skin depth, it appears that the resistance per unit area varies by a factor of about 1.7 rather than the factor of 1.22 that would be predicted as a result of skin depth variation only.

The top rod model of a solid covered with a thin film of melt seems to explain reasonably well why the current should decrease more with frequency than is predicted by Eq 5. That the film thickness should decrease as frequency increases appears rational in view of the fact that heat is generated in the top rod, at the surface, in a thin region approximately equal to the skin depth in thickness. As was pointed out by Gupta et al,[5] there will be a temperature gradient to support heat radiation from the surface. This effectively implies that the film thickness should be thicker at lower frequencies, where its value ultimately would be determined by surface tension if the skin depth were quite large. But at frequencies below 500 kHz another phenomenon is observed where surface melting of the rod is quite irregular.[5]

There appears to be little theoretical justification, at this time, for the measured minimum in coil current at a frequency of about 2.6 MHz. Not enough data are available at higher frequencies to determine if a true minimum exists or not. Although further measurements at frequencies in excess of 3 MHz on rods of 65 mm diameter will be difficult to make without encountering problems from excessive coil voltage, additional measurements in this range, and also on rods of different diameters, will be required to clarify this anomalous result.

Acknowledgments

We are thankful to Monsanto Company, Electronics Division, St. Peters, Missouri, for allowing us to publish the experimental work contained in this paper. We express our gratitude to Messrs. James Lenk and Dan Ramey for running the zone refiners.

APPENDIX

Resistance of Solid Covered with Thin Film of Melt

If the melt thickness is d (Fig. 7), then the problem is equivalent to finding Re (the real value of) (Z), where Z is the input impedance into a lossy line of length d, terminated in Zs, where:

$$Zs = \sqrt{\frac{j\omega\mu 0}{\sigma s}}$$

where

σs = conductivity of solid, and
$\mu 0$ = free space permeability.

The lossy line has characteristic impedance Zm and propagation constant Γ, where

$$Zm = \sqrt{\frac{j\omega\mu 0}{\sigma E}}$$

where σE = conductivity of melt, and

$$\Gamma = \sqrt{j\omega\mu\sigma E}$$

If we define $\delta = \sqrt{(2/\omega\mu 0\sigma E)}$, and compute reflection coefficient $\gamma = (Zs - Zm)/(Zs + Zm)$, we have

$$\gamma = \frac{\sqrt{\frac{\sigma E}{\sigma s}} - 1}{\sqrt{\frac{\sigma E}{\sigma s}} + 1}$$

Using $(\sigma E/(\sigma s) = 30$ as representative for silicon, this gives $\gamma = 0.691$.

The input impedance Z into the lossy line is

$$Z = \sqrt{\frac{j\omega\mu 0}{\sigma E}} \left\{ \frac{1 + \gamma e^{-2d/\delta(1+j)}}{1 - \gamma e^{-2d/\delta(1+j)}} \right\}$$

which is:

$$Z = \sqrt{\frac{j\omega\mu 0}{\sigma E}} \left\{ \frac{1 - \gamma^2 e^{-4d/\delta} - j2\gamma e^{-2d/\delta} \sin 2d/\delta}{1 - 2\gamma e^{-2d/\delta} \cos 2d/\delta + \gamma^2 e^{-4d/\delta}} \right\}$$

and

$$Re(Z) = \sqrt{\frac{\omega\mu 0}{\sigma E}} \left\{ \frac{1 - 2\gamma e^{-2d/\delta} \sin 2d/\delta - \gamma^2 e^{-4d/\delta}}{1 - 2\gamma e^{-2d/\delta} \cos 2d/\delta + \gamma^2 e^{-4d/\delta}} \right\}$$

Thus the expression for resistance per unit area is given by:

$$\frac{R/\square}{R_{\mathrm{melt}}/\square} = \left\{ \frac{1 + 1.382\, e^{-2d/\delta} \sin 2d/\delta - (0.691)^2 e^{-4d/\delta}}{1 - 1.382\, e^{-2d/\delta} \cos 2d/\delta - (0.691)^2 e^{-4d/\delta}} \right\}$$

Michael H. Christ[1] *and Bryan S. Maurantonio*[2]

Influence of Laser Marking on Silicon Wafer Properties

REFERENCE: Christ, M. H. and Maurantonio, B. S., "**Influence of Laser Marking on Silicon Wafer Properties,**" *Silicon Processing, ASTM STP 804,* D. C. Gupta, Ed., American Society for Testing and Materials, 1983, pp. 62–82.

ABSTRACT: The effects of laser marking on silicon wafers were studied by oxidation and preferential etching for structure characteristics. Scanning electron microscopy (SEM) and optical microscopy were used to study cleanliness and epitaxial growth characteristics. Laser marking not followed by etching is seen to cause dislocations and slip patterns in the vicinity of the mark after an oxidation. If the wafer is etched after marking in either caustic (KOH) or acid (HF/nitric/acetic) no dislocations or slip patterns develop after an oxidation cycle. Marking done in dot matrix mode exhibits slag around the dots which is above the plane of the polished surface. It is shown that epitaxial growth on this slag is slower than on the polished surface. No epitaxial spikes are observed on marks engraved before polishing. No contamination is observed on finished polished slices. However, at high magnification (×4000 and ×20 000) particulate matter is seen after a photoresist has been applied, baked, exposed, developed, and stripped.

KEY WORDS: laser marking, contamination, structure damage, epitaxial spike, process automation

Identification codes marked on individual wafers permit lot traceability to avoid mixes, development of production history to improve yields, and more accurate inventory control. Until recently, however, the most widely used method of marking has been hand scribing with a diamond stylus. This method suffers from low productivity, increased wafer breakage, and contamination. The result is that hand marking is mostly limited to engineering samples, while large volume runs must rely on antiquated lot card systems.

Laser-marking equipment able to mark silicon wafers has been available since the late 1970s. Unfortunately it did not have automatic cassette feed. Nevertheless, some large users of silicon wafers recognized the benefits and incorporated manually-fed marking units in their processes [*1*].[3]

Within the last two years several companies have developed automated

[1] Product Development Director, Dynamit Nobel-SMIEL, Cupertino, Calif. 95014.
[2] SMIEL Spa Dynamit Nobel Group, Novara, Italy.
[3] The italic numbers in brackets refer to the list of references appended to this paper.

cassette-feed marking systems [*2,3*]. Parallel to the development of automated marking equipment the Semiconductor Equipment and Materials Institute (SEMI) Wafer Standard Committee has developed the Alpha Numeric Marking Specification M1.1.14, which was approved in September 1981.

With automated equipment and a standard marking specification now available it is expected that large volumes of marked wafers will soon be on the market. In fact, most major silicon vendors now have automated marking systems in-house for process development. However, the effects of marking on wafer properties have not been thoroughly studied and reported. Some early work indicated that at the very least there may be some problems with splashing of molten silicon, which results in high densities of particles in the vicinity of the mark [*3,4*]. Another problem surfaced at the May 1981 meeting of the SEMI Wafer Standards Committee where it was reported that epitaxial spikes had been observed on marked characters.

This paper reports results of our studies concerning contamination, crystal structure degradation, and epitaxial growth characteristics on marked wafers.

Experimental Procedure

P(111) and P(100) wafers of 100 mm diameter were prepared by a normal wafer manufacturing sequence, that is, slicing, lapping, etching, and polishing. At various stages of manufacturing, wafers were marked with a Nd-YAG, Q-switched system which operates at a wave length of 1.06 μm. Table 1 details the experimental matrix for preparing wafers used in this study.

Structure Study

Polished wafers representing each element in the matrix of Table 1 were decoration etched with and without oxidation. Oxidation was at 1100°C for 60 min in steam. These wafers were then examined by high-intensity light and optical microscopy for the presence or absence of dislocations or slip in the vicinity of the marked characters. Decoration etches were Sirtl [*5*] for (111) wafers and Wright [*6*] for (100) wafers.

Epitaxial Characteristics

Eight wafers were selected from each element in the matrix of Table 1. Epitaxial (epi) layers were grown on two wafers each to thicknesses of 5, 12, 25, and 50 μm. The 5 and 12 μm epi was grown in radiant-heated barrel reactors. The 25 and 50 μm epi was grown in RF-heated barrel reactors. Wafers were examined by scanning electron microscopy (SEM) at low angles (80 to 88 deg) for epitaxial spikes and at higher angles (up to 60 deg) for character definition.

TABLE 1—*Experimental matrix for preparing wafers used in this study.*

Wafers Marked in Lapped Form Followed By Etching Then Polishing (Engrave Mode)			
Orientation	P(111)	P(100)	P(100)
Etch	acid[a]	acid[a]	caustic[b]
Energy density	18 J/cm^2	18 J/cm^2	35 J/cm^2
Pulse width	120 ns	120 ns	120 ns
Spot size	60 μm	60 μm	60 μm
Spot overlap	87.5%	87.5%	87.5%
Wafers Marked in Etched Form Followed By Polishing (Engrave Mode)			
Orientation	P(111)	P(100)	P(100)
Etch	acid[a]	acid[a]	caustic[b]
Energy density	35 J/cm^2	35 J/cm^2	35 J/cm^2
Pulse width	120 ns	120 ns	120 ns
Spot size	60 μm	60 μm	60 μm
Spot overlap	87.5%	87.5%	87.5%
Wafers Marked in Polished Form (Dot Matrix)			
Orientation	P(100)		
Energy density	10 to 40 J/cm^2		
Pulse width	. . .		
Spot size	60 μm		

[a] HF/nitric/acetic.
[b] KOH.

Cleanliness

Wafers selected from each element in the matrix of Table 1 were examined by SEM after polishing and after oxidation (700 Å) and photoresist processing to see if any residues remained in the mark after stripping. The photoresist process included oxidation, spin on, bake, expose/develop, and strip. Table 2 gives the photoresist process conditions.

Results and Discussion

Structure

Figures 1*a* and 1*b* illustrate the smoothing action of acid etching on engraved marks in (111) and (100) wafers. Figure 1*c* illustrates the anisotropic action of caustic etch. In each case the grooves making up the characters are open and have internal surface characteristics that should permit effective cleaning.

TABLE 2—*Photoresist process conditions.*

	Positive Resist	Negative Resist
Resist used	HPR204, Hunt Chemical	Waycoat SC, Hunt Chemical
Thickness	1.2 μm	2.5 μm
Bake	105°C/15 min	105°C/30 min
Strip	isoclean	isoclean

Wafers marked in engrave mode before etching exhibited no dislocations or slip by decoration etch either before or after oxidation. Wafers marked in engrave mode after etching exhibited no dislocations in the vicinity of the mark by decoration etch. After an oxidation cycle, however, decoration etching revealed slip lines originating at the marked characters. The slip array occurs on each character. It is exactly the width of each character and extends about 2.5 cm from the edge to the center of the wafer.

Wafers marked in dot matrix mode after polishing exhibited no dislocations in the vicinity of the mark by decoration etch. After an oxidation cycle, however, decoration etching revealed random dislocations around each character in densities 100 to $4000/cm^2$. This cloud of dislocations did not extend more than 0.5 mm above or below the characters. This is within the 2-mm window specified in SEMI Alpha Numeric Marking Specification M1.1.14.

In order to realize the advantages of encoded wafers manufacturers of silicon wafers will probably want to do the marking as early in the process sequence as possible. As long as marking is done before the etching stage there should be no problem. In any event, marking on the etched wafer is excluded due to the structure damage incurred.

Users of the wafers must mark after polishing. For current automated equipment this will be done in the dot matrix mode. Local stresses in the lattice result in development of dislocations in the vicinity of the marked characters. However, since the dislocations seem to be limited to the immediate 0.5-mm area around the characters, there may be little or no effect on device yield. More thorough investigation is needed. In addition, it must be emphasized here that the manufacturers of marking equipment are attempting to reduce the local damage caused by dot matrix marking.

Cleanliness

Although examination by optical microscope did not detect contamination in the engraved marks on polished wafers, the SEM micrographs in Fig. 2 did show particulate matter. In Fig. 2*a* the large particle is probably dust acquired during preparation for the SEM; however, the submicrometre particles seen in Figs. 2*b* and 2*c* are of unknown origin.

Many of the SEM micrographs show particulate matter in and around the marks. Since no particulate was observed by optical microscopy we feel that the particles were generated during specimen preparation for the SEM. Further investigations will observe rigorous procedures to minimize contamination generated in specimen preparation.

A major concern of wafer users is whether engraved marks harbor contaminants from process chemicals and photoresist. Figure 3 comprises SEM

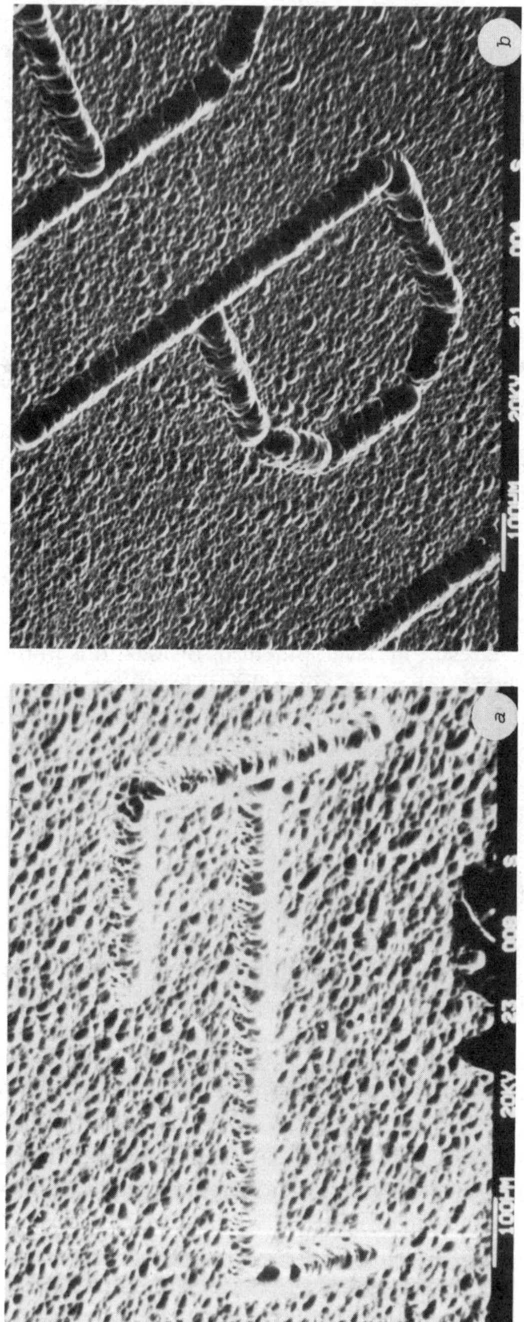
b
a

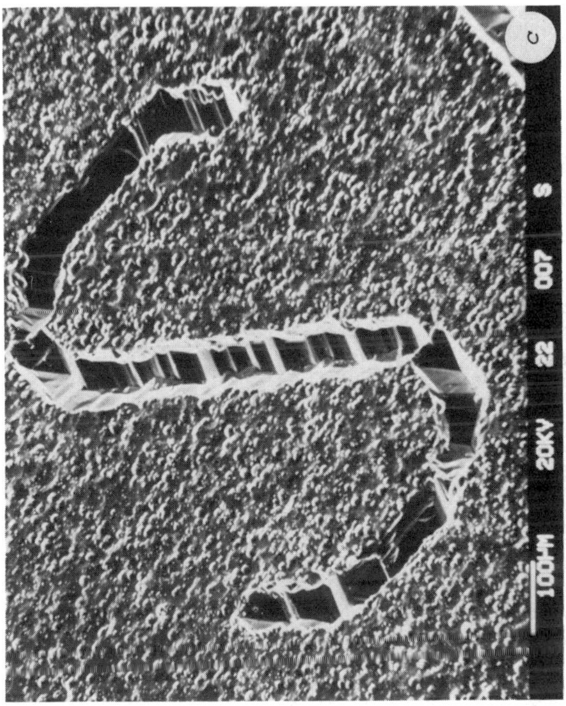

FIG. 1—*Wafers marked in lapped form followed by etching.* (a) *P(111) acid.* (b) *P(100) acid.* (c) *P(100) caustic.*

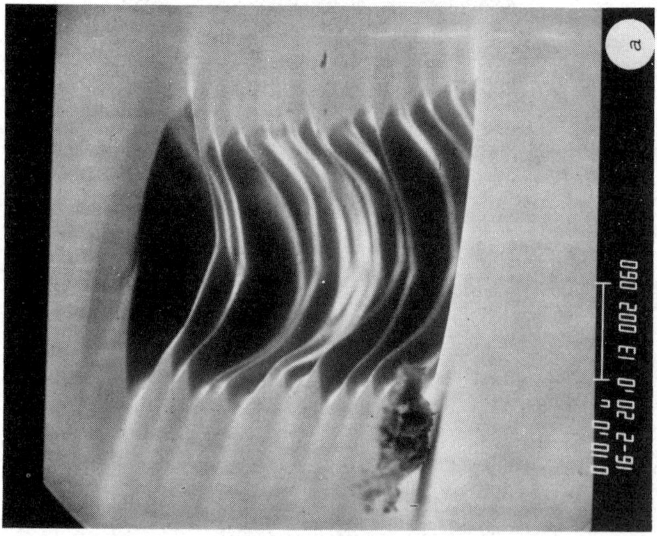

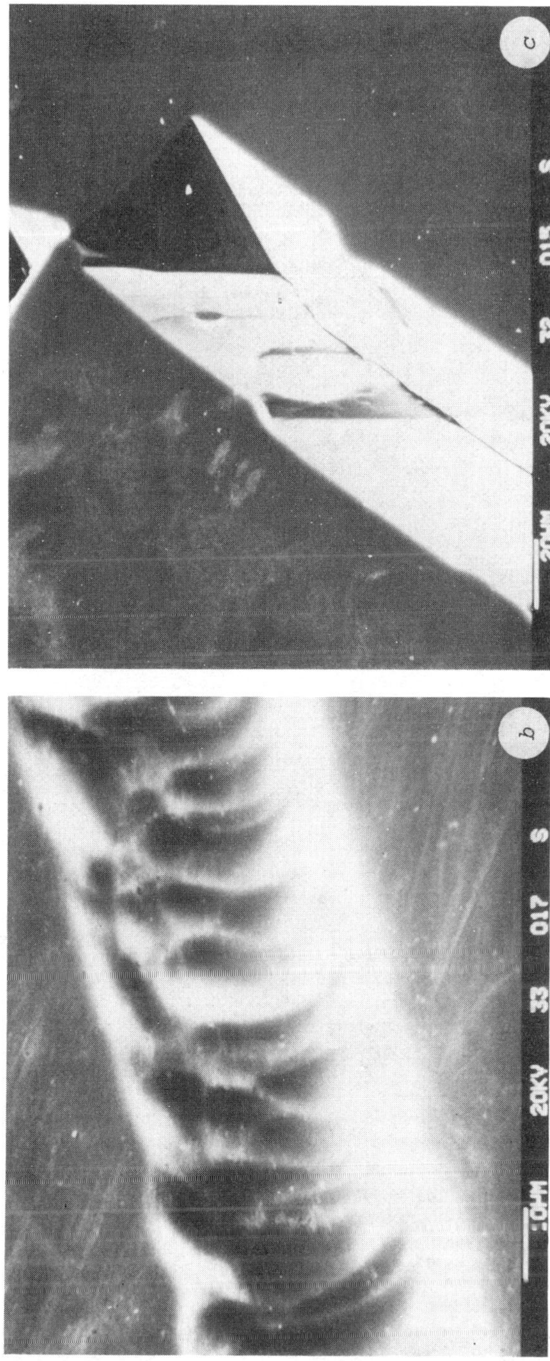

FIG. 2—*Wafers engraved in lapped form, etched then polished.* (a) *P(111) acid.* (b) *P(100) acid.* (c) *P(100) caustic.*

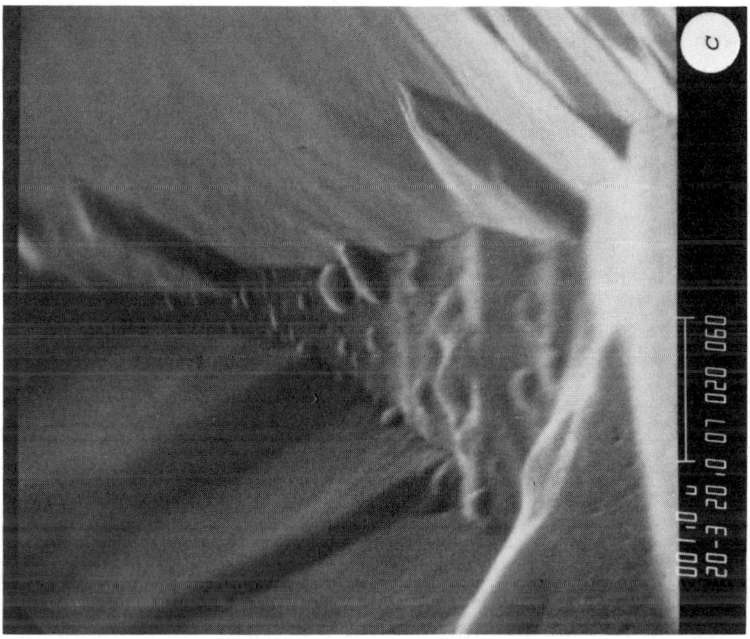

FIG. 3—*P(100) wafers after photoresist contamination test.* (a) *Negative photoresist after stripping.* (b) *Negative photoresist after stripping.* (c) *Positive photoresist after stripping.*

micrographs made on wafers that have been photoresist processed as given in Table 2. Figure 3*a* shows a section of engraved mark apparently free of residue at ×1000. Figure 3*b* shows another section of engraved mark on the same wafer. This view at ×4000 reveals a large number of submicrometre particles. Figure 3*c* shows another mark at ×23 000. At this extreme magnification submicrometre particles can be seen at the bottom of the groove. We do not know what these particles are. As mentioned earlier, further investigation will include rigorous procedures to minimize contamination in specimen preparation.

In the dot matrix mode molten silicon is blasted away from the polished surface. Most of this debris is cooled before falling back to the surface and is easily removed in standard wafer-scrubbing systems; however, some of the material is still molten when it falls back and becomes firmly attached to the surface. Figures 4*a* and 4*b* are worst-case illustrations. To be fair, manufacturers of marking systems have greatly reduced the marking debris in dot matrix mode. Yet some nonremovable slag still remains around the dots. This nonremovable slag is of concern in contact printing and epi applications.

We have already seen that wafers marked after etching suffer from structure degradation in oxidation. Figures 5*a* and 5*b* show a mark made in the etched surface followed by polishing. Many cracks and fissures can be seen along with fragile-appearing sections which may break off in later processing. Although particulate contamination is not visible in these micrographs, the very rough character of the mark is clearly not suitable for ultra-clean processing.

Epitaxial Characteristics

In Fig. 6 we can see that there is a smoothing of the rough features of engraving done after etching. Although the mark is too rough to be considered for a production process, it is interesting that no epi spikes are present.

Marks done in dot matrix mode are the only ones that developed epi spikes. Apparently these spikes are grown on the rigidly adhering slag at the rim of each dot. Figure 7 shows a full view of the character "L" with several spikes clearly visible. The closer views shown in Fig. 8 reveal the faceted nature of the spikes. The maximum height of spikes we observed was about 8 μm. Of course, this is a function of the original height of the slag. An interesting phenomenon concerning spike height is seen in a sequence of micrographs (Figs. 8*a* to 8*d*). The spike height is about the same in 5- 12-μm-thick epi layers. At 25- and 50-μm-thick epi layers, however, the spike height is less than half as high. Given that maximum slag height was about the same on all the characters studied it would appear that spikes are less of a problem on thick epi than on thin epi. For processes sensitive to protrusions above the surface plane spikes on dot matrix marked wafers may be a problem.

Marking-system manufacturers have made significant improvements in

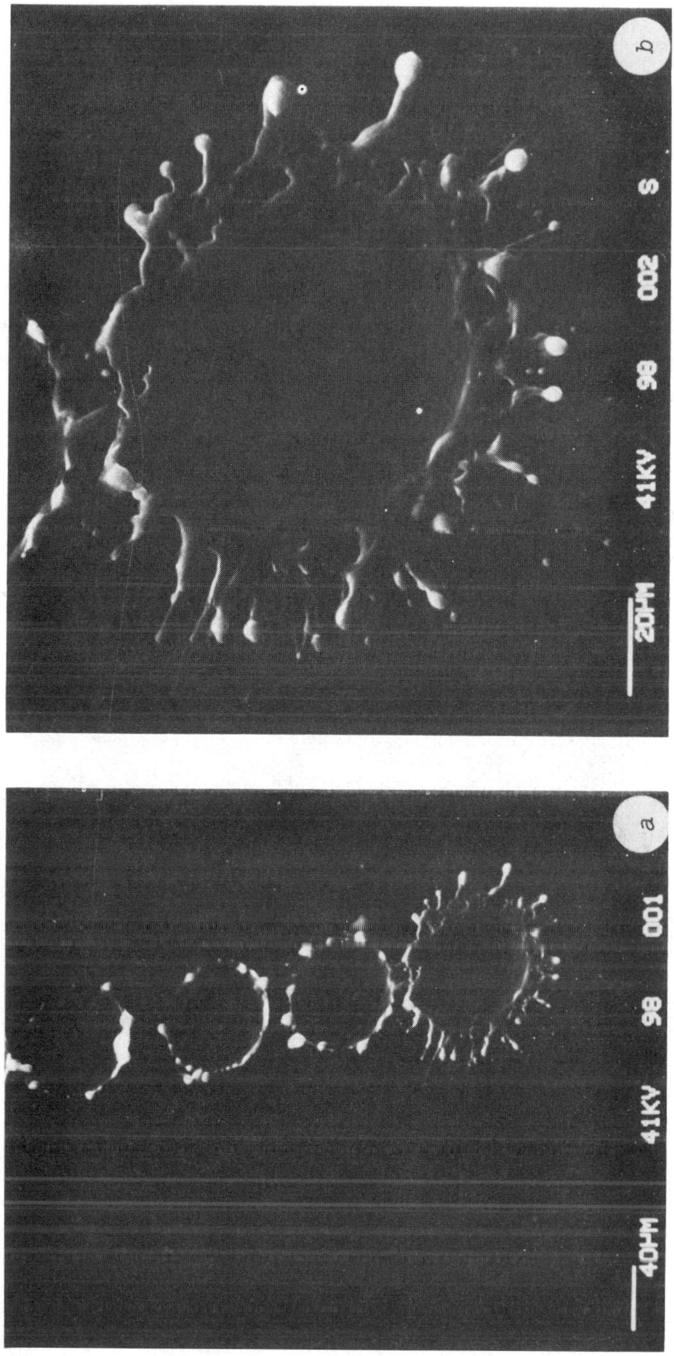

FIG. 2—*P(100) wafers marked in dot matrix mode on the polished surface.* (a) *After cleaning.* (b) *Enlarged view of* (a).

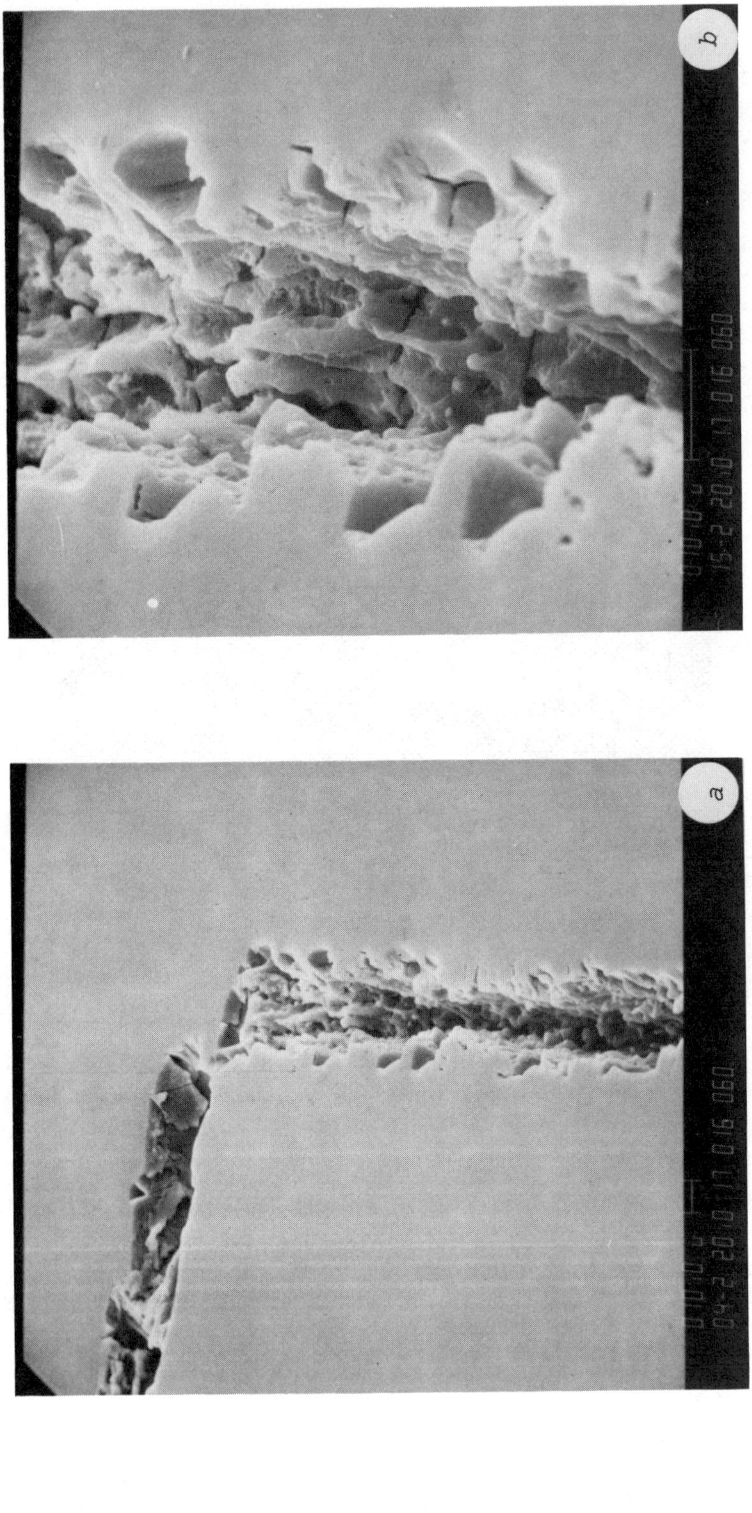

FIG. 5—*P(100) engraved in etched form.* (a) *After polishing.* (b) *Same as* (a) *but at a higher magnification.*

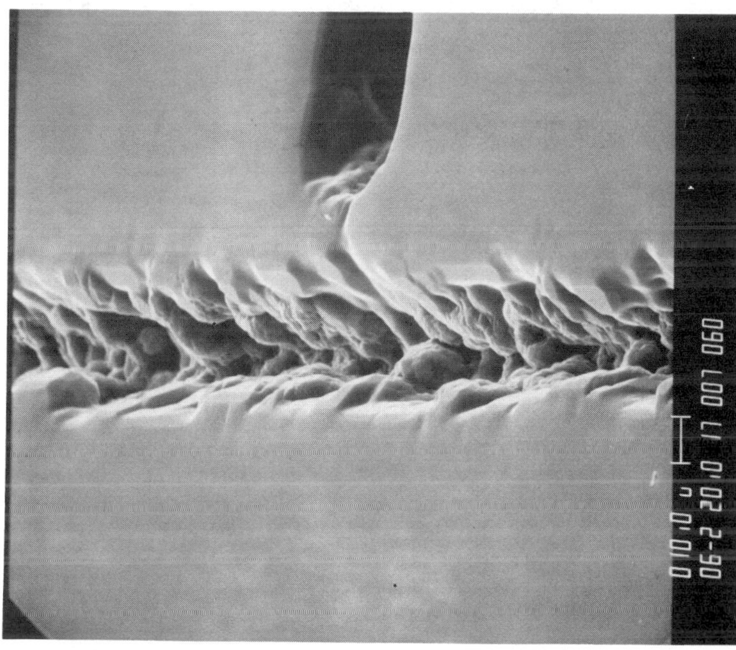

FIG. 6—*P*(100) *wafer with 5-μm epi grown on polished surface marked after etching.*

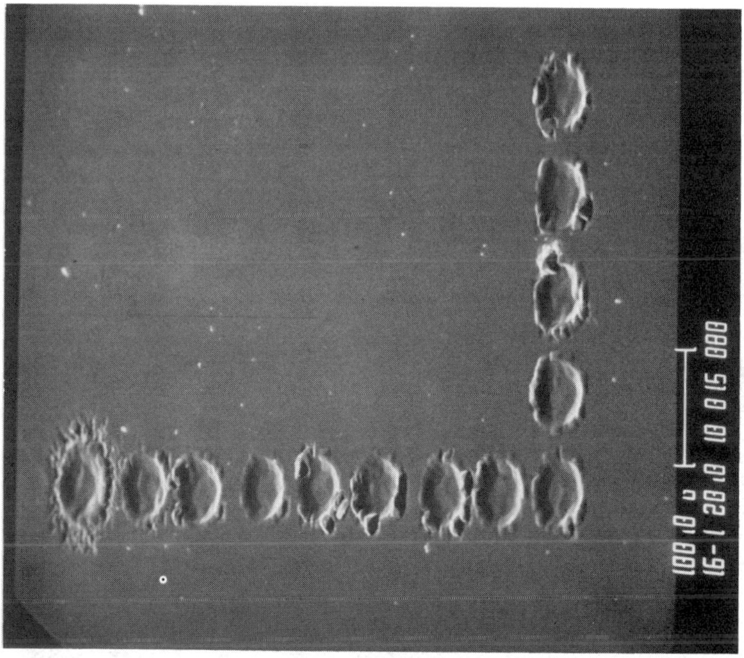

FIG. 7—*P*(100) *wafer with 12-μm epi grown on polished surface marked in dot matrix mode.*

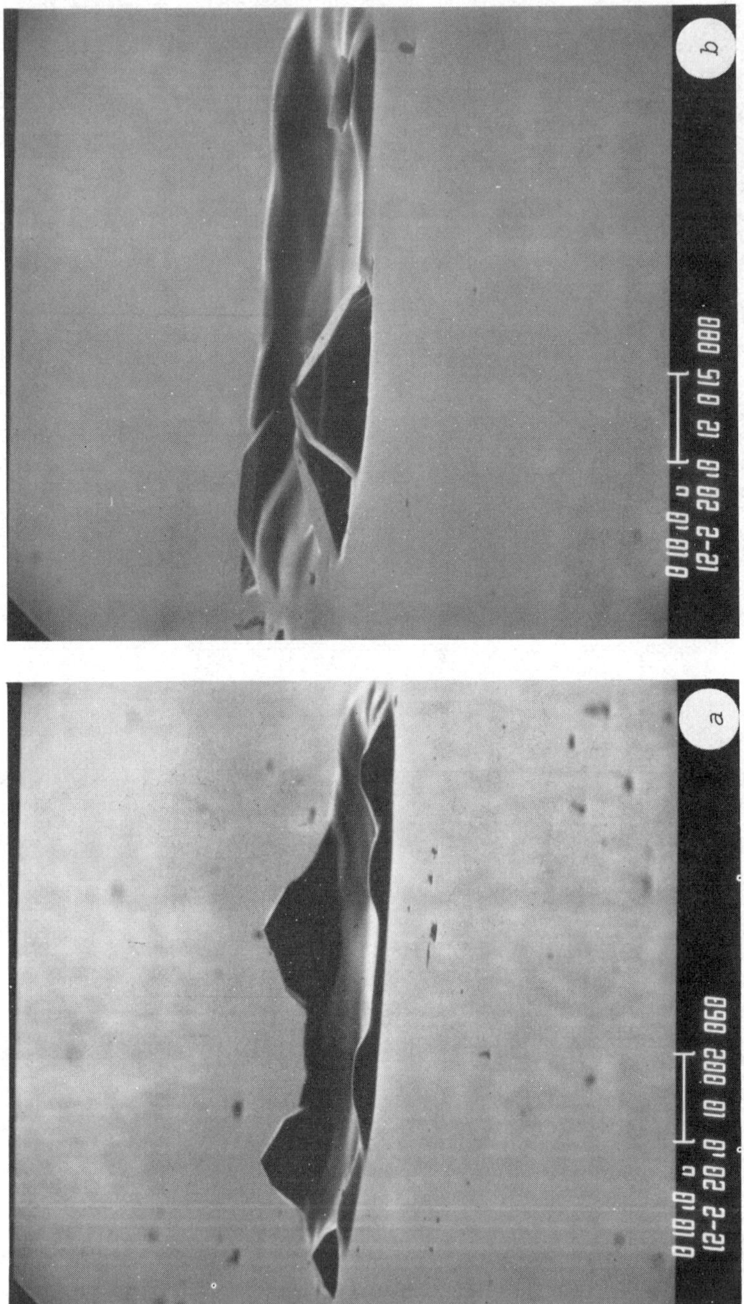
b
a

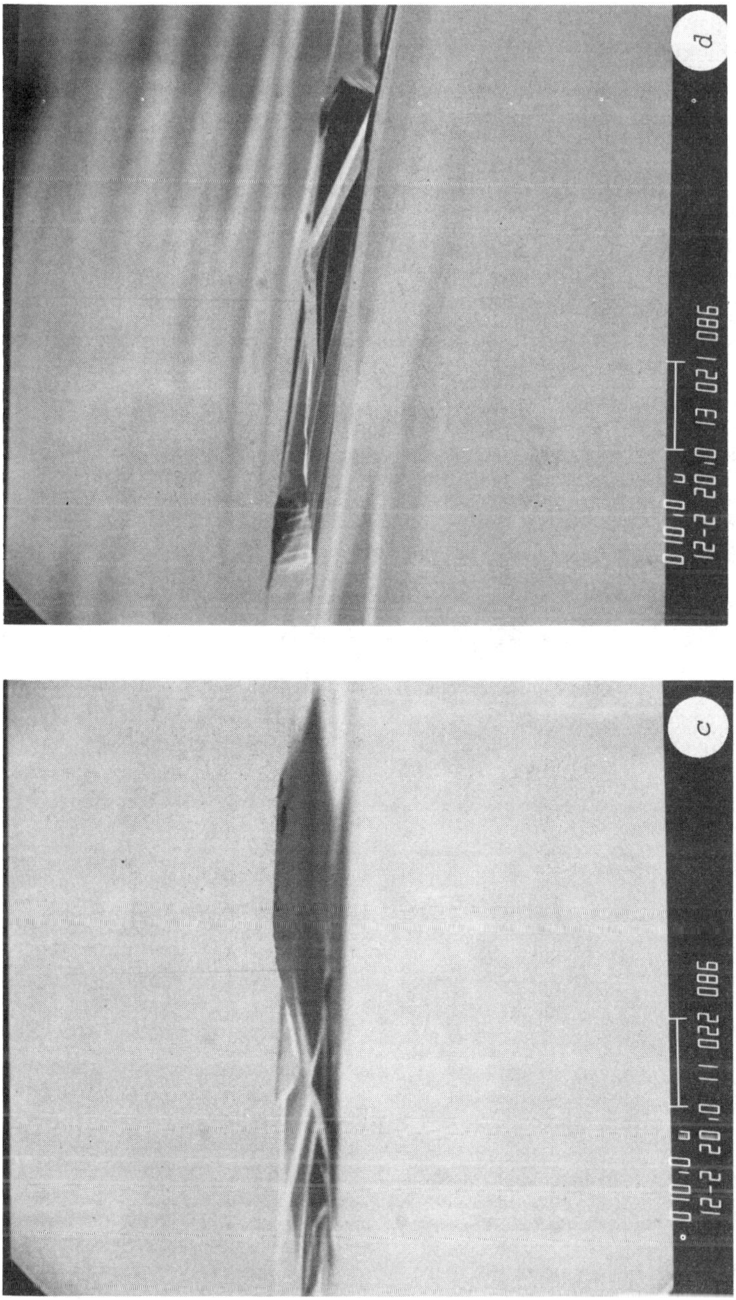

FIG. 8—*P(100) wafers marked in dot matrix mode with the epi layer grown on the polished surface.* (a) *5-μm epi.* (b) *12-μm epi.* (c) *25-μm epi.* (d) *50-μm epi.*

reducing slag. We will follow these developments and report results at an appropriate time.

Wafers engraved before etching developed no epi spikes at 5, 12, 25, or 50 μm epi thickness. This holds true for both P(111) and P(100) wafers etched after engraving in either acid or caustic. Figures 9 and 10 show the softening of contours and complete lack of epi spikes, or crown, at the edge of the marks.

Micrographs in Fig. 11 show a range of epi thickness grown on characters caustic etched after engraving. No epi spikes or crown is observed. On the 25 and 50 μm epi layers faceted growth at the edges of the grooves tends to widen the effective width of the mark. The width of these marks is well within the limits specified in SEMI Alpha Numeric Marking Specification M1.1.14. Therefore no problem is expected from this widening effect.

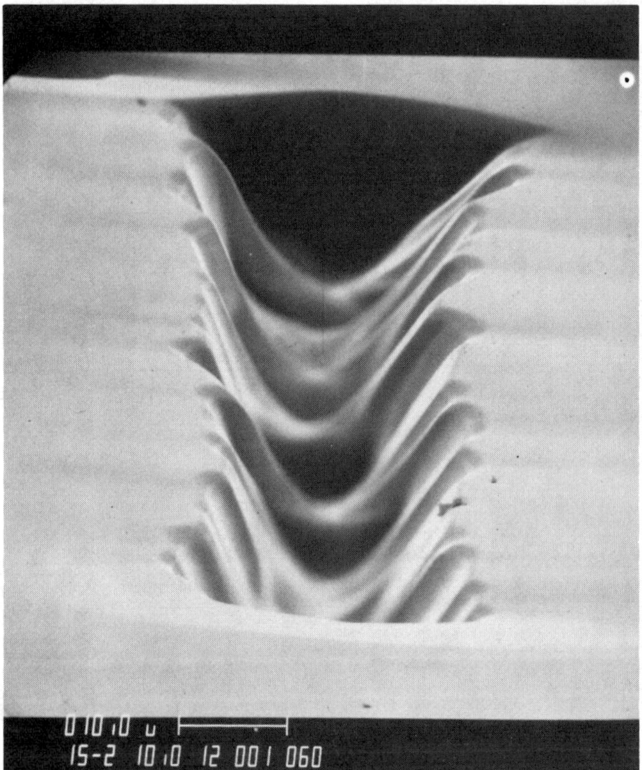

FIG. 9—*P(111) wafer engraved in lapped form, acid etched, polished, and then 12-μm epi layer grown.*

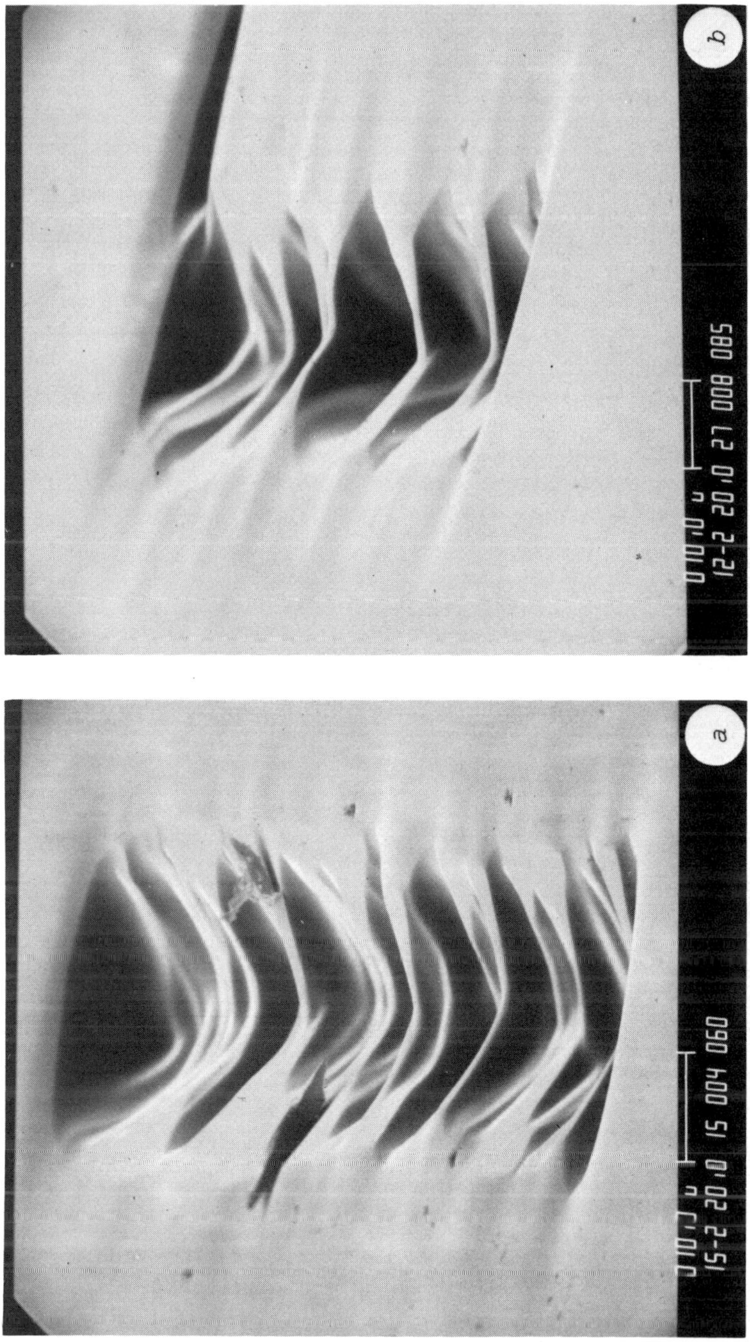

FIG. 10—*P(100) wafers engraved in lapped form, acid etched, polished, then epi layer grown.* (a) *5-μm epi.* (b) *12-μm epi.*

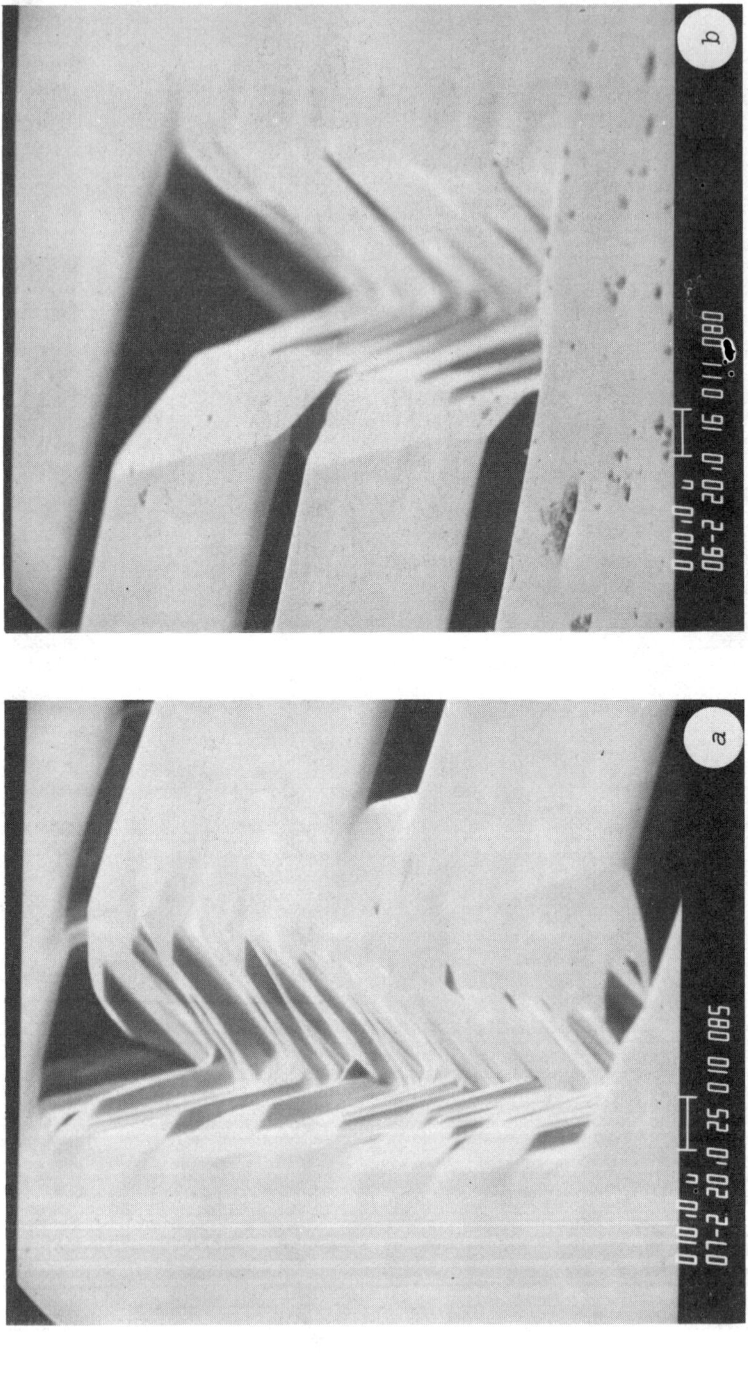
b
a

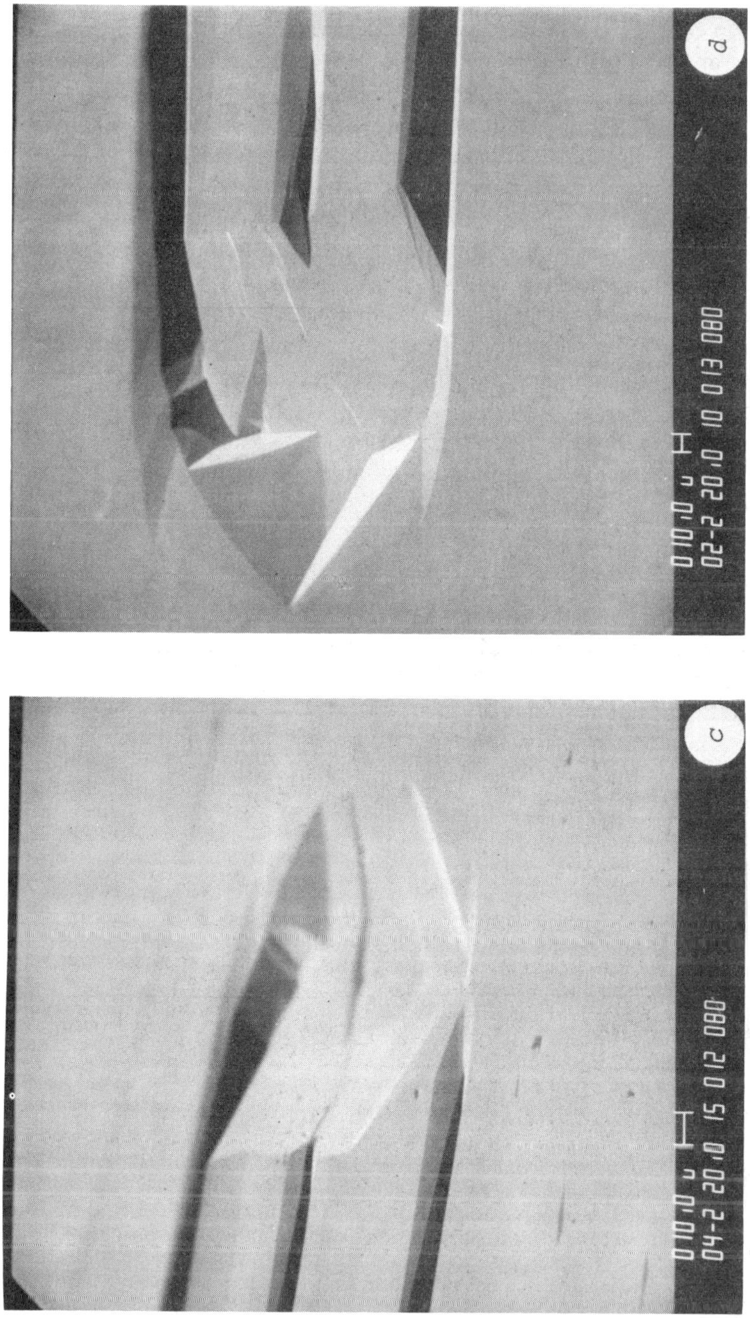

FIG. 11—P(100) *wafers engraved in lapped form, caustic etched polished, then epi layer grown.* (a) *5-μm epi.* (b) *12-μm epi.* (c) *25-μm epi.* (d) *50-μm epi.*

Conclusion

Marking in engrave mode does not degrade crystal structure as long as it is followed by an etch. Improvements being made by marking-system manufacturers may eliminate any problems from dot matrix structure degradation. Further investigation is required in this area.

It is our opinion that marks engraved before etching do not retain contamination from process chemicals or photoresist. Further investigations will observe rigorous procedures to avoid contamination in specimen preparation.

Debris generated in dot matrix marking is removable by standard wafer-scrubbing processing and does not present a problem.

Engrave mode marking after etching is not acceptable as a marking process. It creates very rough features with cracks and fissures which can be expected to harbor contaminants. No epi problems were observed on engrave mode marks.

Epi spikes observed on dot matrix marks may be a problem in sensitive processes. Improvements being made by marking-system manufacturers may eliminate this problem.

Acknowledgments

We wish to thank E. Dall'Oco for oxidation and decoration etch work and R. Schindler for his support. We also wish to thank for their assistance in this study: A. Armigliato, C.N.R. Institute–Lamel Bologna, Italy, and Wendy Douglas, National Semiconductor, Santa Clara, California, for SEM work; A. Casadei, SGS, Milan, Italy, for 5 and 12 μm epi; V. Aurino, Timesa, Biasca, Switzerland, for 25 and 50 μm epi; and A. Panchieri, SGS, Milan, Italy, for photoresist.

References

[*1*] Billeau, D. and Baker, P., *Semiconductor International,* April 1980, p. 77.
[*2*] O'Neill, T. G., *Semiconductor International,* July/Aug. 1980, p. 81.
[*3*] Bernard, B. and Hansen, R. L., "An Automated Wafer Measurement and Identification System," SEMICON East Technical Presentation, Sept. 1980.
[*4*] Duncan, G., "Wafer Marking Quality," Task Force Report, SEMI Wafer Standards Committee, Sept. 1980.
[*5*] Sirtl, E. and Adler, A., *Zeitschrift für Metallkunde,* Vol. 52, 1961, p. 529.
[*6*] Jenkins, M. W., *Journal of the Electrochemical Society,* May 1977, p. 757.

Microelectronics Imaging and Patterning

Daryl Ann Doane[1]

A Review of Optical Lithographic Techniques for VLSI

REFERENCE: Doane, D. A., "**A Review of Optical Lithographic Techniques for VLSI,**" *Silicon Processing, ASTM STP 804,* D. C. Gupta, Ed., American Society for Testing and Materials, 1983, pp. 85–107.

ABSTRACT: Optical lithography using UV radiation (200 to 450 nm) has long been the workhorse in pattern generation for fabricating integrated circuit devices on silicon. Various techniques employing projection and proximity printing using electron-beam-generated photomasks have been developed to reduce feature sizes, to improve device yields, and to lower device fabrication costs. In this review, some optical lithographic techniques available for fabricating semiconductor devices are described. The advantages and limitations of the various methods are discussed with an emphasis on their suitability for the geometries required for VLSI.

KEY WORDS: optical lithography, projection and proximity printing, linewidth control, ultraviolet lithography, electron beam/X-ray lithography

To review possible lithographic techniques for VLSI requires a discussion of available lithographic techniques as well as a definition of VLSI.

VLSI (very-large-scale integration) refers to the increase in circuit density for both MOS and bipolar integrated circuits. An accepted definition for VLSI has evolved as a level of integration of at least 100K gates, or at least 100K memory bits per circuit [*1–4*].[2] Such a level of integration requires the definition of 1-μm features in some levels of the device structure. This review will concentrate on the processing improvements required to achieve VLSI, in general, without specifying the exact level of integration.

Both horizontal and vertical scaling of devices are required for VLSI. Horizontal scaling includes, for example, the decrease in feature size, with the accompanying increase in die or chip size, that accounts primarily for the increase in the number of circuit components/chip. Horizontal-scaling techniques to decrease feature size involve lithographic and etching processes for pattern definition. Improvements in horizontal scaling will dominate in the goal to achieve more functions/chip.

Vertical scaling of devices is also required for the production of a higher

[1] Member Technical Staff, Bell Laboratories, Murray Hill, N.J. 07974.
[2] The italic numbers in brackets refer to the list of references appended to this paper.

level of integration. Vertical scaling refers to the production of thinner films of oxides, nitrides, or polysilicon, and to the use of materials other than aluminum for interconnections and gate materials. A reduction in vertical dimensions is critical in achieving the desired electrical properties of devices, as well as in attaining a high reliability in their performance. Although important in any consideration of the fabrication of very large-scale integrated circuits, vertical scaling techniques will not be discussed further in this review.

The requirements in the fabrication of the photomasks necessary to support any device production strongly influence any progress in horizontal device scaling. Some factors in the production of acceptable photomasks which become more critical include (*1*) a higher resolution capability with more stringent control of linewidths over larger areas of mask substrate, (*2*) a lower defect density with an accompanying reduction in the size of critical defects, (*3*) a higher degree of accuracy in the position placement of features, and (*4*) more accurate registration between features on different mask levels. The introduction of electron-beam-generated master masks provided a real breakthrough in improving resolution and linewidth control as well as in lowering defect densities [*5,6*]. In addition, there was a reduction in the turnaround time necessary for the production of a set of masks. The use of projection printing methods of exposure demanded these higher-quality photomasks with lower tolerances. For the more recently introduced step-and repeat exposure systems, which require the use of reticles, the emphasis in mask fabrication shifted to a reduction in defect level rather than to a higher resolution capability. A single defect on a reticle may produce zero yield for that device level, while effective linewidth control is improved (by a factor related to the exposure system magnification) relative to that produced on a reticle.

The trend towards the use of larger-diameter wafers to improve fabrication cost-effectiveness also affects progress in horizontal scaling. Printing smaller feature sizes requires flatter surfaces at every patterning step, for example, whether such a patterning step is defined over an extended surface (for full-wafer printing) or over a field defining a single chip (as occurs for reduction projection printing with step-and-repeat exposure systems). Because of the various thermal effects inherent in device-processing procedures, strain-producing warpage occurs in wafers. Such warpage also causes subsequent pattern distortion.

Lithographic imaging steps are often regarded as the "rate-limiting" steps in the goal of increasing the scale of integration in device fabrication. In pursuit of a submicron lithographic technology, four major areas of development have occurred:

1. Optical/UV lithography.
2. Electron-beam lithography.
3. X-ray lithography.
4. Ion-beam lithography.

Optical lithography, using UV radiation (exposing wavelengths in the range 200 to 450 nm), is the present dominant technology [*7*]. It is a well-established technology, relatively low in cost and with a high wafer throughput. Until recently, contact printing methods occupied the major position as the lithographic technique of choice. Then projection printers offered dramatic improvements in yield by virtue of the separation between mask and wafer during exposures [*8–11*]. A further improvement in the resolution capability of these printers was made possible through the use of special coatings on the optical components, which offered a far-UV (310 nm) capability [*12*].

Along with the improvement in the design of these projection printers, a resurgent interest in proximity exposure methods has also occurred [*13,14*]. Claims are made for a 4-μm resolution capability in the near-UV (330 to 450 nm) region of the spectrum, and for a 2-μm capability in the deep-UV (200 to 260 nm) region, for proximity gaps on the order of the 20-μm required to separate mask and wafer [*15*].

To overcome some of the problems associated with full-wafer imaging, step-and-repeat optical techniques have been introduced. These systems provide a high-resolution printing capability which takes into account wafer distortions but at the expense of a reduction in throughput [*16–18*].

All optical-exposure systems are diffraction limited. Their resolution capability is proportional to a quantity λ/NA, where λ is the wavelength of the source illumination and NA is the numerical aperture of the optics. Independent of such effects as the resist processing latitude or the topography of the substract surface, then, there will be an inherent limitation on feature size definition on the order of 0.3 μm. In a production environment, however, it is not expected that feature sizes below 0.5 μm will be realized.

Direct writing on wafers using electron beams (e-beams) or ion beams, as well as proximity printing with X-rays, have been successful in achieving feature sizes below 1 μm [*19,20*]. Commercial e-beam exposure systems are presently available, mainly for uses related to the fabrication of photomasks [*5,21,22*]. Several in-house systems also exist for exposure of patterns directly on wafers, but generally with throughputs that are not cost-effective [*23,24*]. While no commercial X-ray exposure systems are currently available, it is thought by some [*20*] that X-ray lithography will have an effective impact in the latter half of this decade. Finally, ion-beam lithographic technologies using a focused ion beam for direct writing on a substrate, or using an ion-beam mask in close proximity to the substrate, are still in too early a stage of development to predict their practicality. Excellent recent reviews of these newer lithographic techniques give the current status of developments in these areas [*20,25*].

This review will concentrate on recent developments in optical lithography and explore some near-UV (330 to 450 nm), UV (260 to 330 nm), and deep-UV (200 to 260 nm) options.

Optical Lithographic Systems

To evaluate or characterize lithographic performance or to make predictions concerning future advancements in lithographic technology requires a consideration [*11*] of

1. The quality of the optical image independent of the resist system.
2. The nature of the resist image in terms of its profile after development.
3. The reproducibility of the resist image over the entire wafer surface, which takes into account feature size variations with position as well as feature placement accuracy.

Such considerations can be made in terms of a generalized lithographic system consisting of a photomask, an exposure tool, and a resist-coated wafer (Fig. 1). The typical IC photomask consists of a thin, optically opaque film (usually chromium) on a transparent substrate (typically different types of glass substrates). The pattern information defined on the mask is to be transferred to the resist-coated wafer by means of a lithographic exposure system. The exposure system forms an image of the mask near the wafer surface, called the aerial image. This aerial image consists of a spatially dependent light intensity pattern near the wafer. It is this aerial image that really describes the performance of an exposure system.

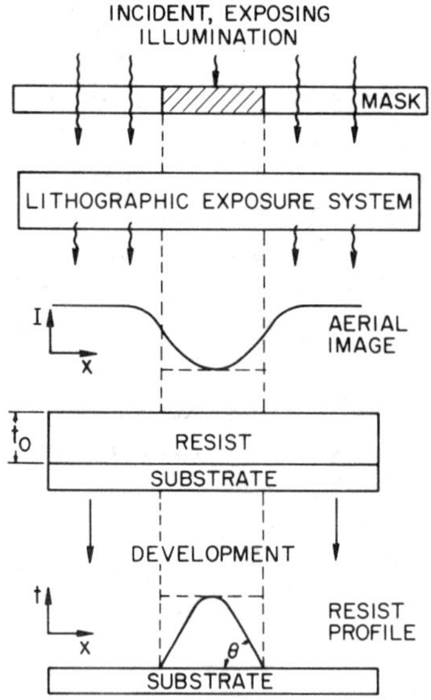

FIG. 1—*An ideal lithographic exposure sequence, including photomask and resist-coated substrate* [11].

Exposure Systems

Practical exposure systems for any VLSI lithographic technology on multilevel device structures must eliminate mask-to-wafer contact. Typical projection systems employing scanning reflective optics for full-field wafer exposures, and step-and-repeat systems having demagnifying refractive optics, are available to meet this requirement. In general, the refractive lens systems have a higher near-UV resolution capability, while the reflecting lens systems offer an economic advantage with respect to throughput in a manufacturing environment. In addition, lens systems using reflective optics have a further potential improvement in resolution capability through the exploitation of the far-UV (below 330 nm) region of the spectrum.

For optical projection systems, the resolution capability of the "lens" system can be described in terms of its numerical aperture (NA). For diffraction-limited imaging, the relations between minimum feature size, numerical aperture, wavelength of the exposing radiation, and the depth of focus are given in Fig. 2 [*7*]. The numerical aperture (NA) defined in Fig. 2 is the parameter used to describe both the resolution capability of the optics and the system depth of focus. From the expressions in Fig. 2, it can be seen that setting the resolution requirements (minimum feature size, MFS) determines the "system" depth of focus. This criterion for the limit of resolution of a feature, however, does not take into account such parameters as feature size control or dimensional uniformity across an extended area such as a photomask or wafer.

The trade-off between resolution capability and depth of focus can be optimized in device patterning by taking into account the processing latitude of

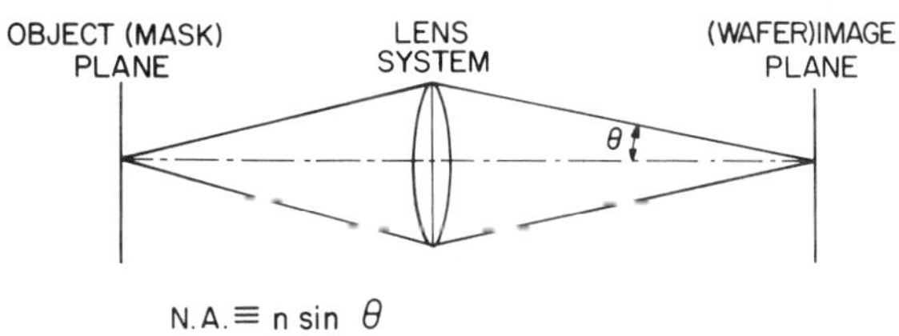

$$\text{N.A.} \equiv n \sin \theta$$

$$F \equiv f/\# = \frac{1}{2\text{N.A.}}$$

$$\text{M.F.S.} \sim \frac{\lambda}{\text{N.A.}}$$

$$\delta \sim \pm \frac{\lambda}{2(\text{N.A.})^2}$$

FIG. 2—*Characteristic parameters for a simplified lens system. Assuming diffraction-limited imaging, the system numerical aperture* (NA), *the system effective f-number* (F), *the minimum feature size resolvable* (MFS), *and the associated depth of focus* (δ) *are defined.* n *is the refractive index, which in air is one.*

the resist. The amount of tolerable defocus in a process depends also on the linewidth control required for a given device design. Attempts to characterize lithographic exposure system performance in terms of exposed test structures on wafers suffer from difficulties in isolating the effects of resist processing parameters from the effects of variables associated with the exposure system. Traditionally, then, exposure systems have been characterized in terms of their modulation transfer function (MTF).

The concept of MTF is useful because it is a measure of the quality of the aerial image of an exposure system rather than a measure of the image content recorded in a resist. The MTF is independent of the resist system. In principle, when the MTF for an exposure system is known, the spatial-frequency dependence of the intensity in the aerial image can be calculated.

The MTF is defined in Fig. 3 as the ratio of the modulation in the image plane (which locates the aerial image) to the modulation in the object plane (which locates the mask) for a given exposure system. The quantity I_{max}/I_{min} defines the system contrast.

The MTF for an exposure system depends on the NA of its optics, the spatial coherency [*26*] of the source illumination, the wavelength of the illumination, and the frequency of the object to be imaged [*27*]. The MTF curve giving the frequency dependence of the MTF (Fig. 4) characterizes the exposure system. Each such system is characterized by a limiting spatial frequency, which is defined as the frequency at which the MTF is zero, and which is determined primarily by the NA of the system. The system resolution limit is the inverse of this limiting frequency.

The wavelength dependence of the MTF enters through the ratio λ/NA (Fig. 2). Decreasing the wavelength of the source illumination increases the modulation in the image at a given frequency [*28*].

The effect of spatial coherency in the source illumination of an exposure system is also reflected in its MTF curve. Spatial coherency refers to the effective size of the source relative to the aperture of an optical system. A point source gives rise to spatially coherent illumination, whereas an infinite (or extended) source leads to incoherent illumination. Partially coherent illumination occurs when the effective size of the source is less than the size of the

$$M(\nu) \equiv \frac{I_{max} - I_{min}}{I_{max} + I_{min}}$$

$$\longrightarrow \mathrm{MTF}(\nu) \equiv \frac{M_{IMAGE}(\nu)}{M_{OBJECT}(\nu)}, \text{ FOR SINUSOIDALLY-VARYING OBJECT}$$

$$\sigma \equiv \frac{\text{N.A.}_{CONDENSER}}{\text{N.A.}_{OBJECTIVE}} \longrightarrow \sigma \begin{cases} = 0 & \text{COHERENT ILLUMINATION} \\ = \infty & \text{INCOHERENT ILLUMINATION} \end{cases}$$

FIG. 3—*Defining relations for the modulation in an object plane,* $M_0(\nu)$, *and the corresponding quantity in the image plane,* $M_i(\nu)$, *when a sinusoidal line grating of spatial frequency* ν *is defined in the object plane. The modulation transfer function for the exposure system, MTF is given by the ratio of this modulation in the image and object planes.*

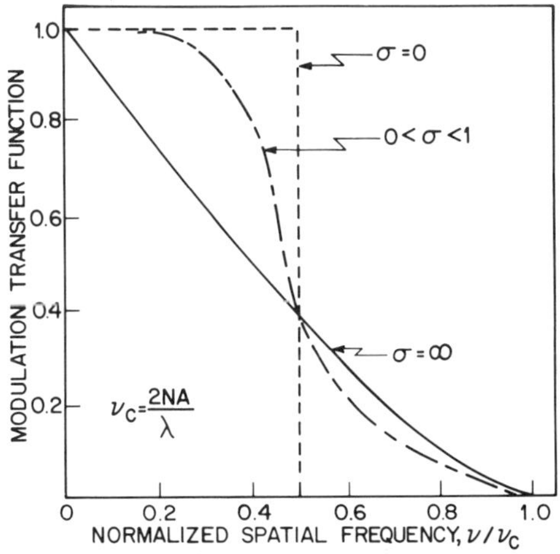

FIG. 4—*Variation of the modulation transfer function with spatial frequency for systems having coherent illumination, incoherent illumination, and partially coherent illumination.*

system aperture. The degree of coherency in the source illumination can be described in terms of the parameter σ given in Fig. 3. For a coherent system, $\sigma = 0$; for an incoherently illuminated system, $\sigma = \infty$. Decreasing the value of σ ($0 \leq \sigma \leq 1$) increases the partial coherency of the source illumination. The difference between the incoherent case when the exit pupil is overfilled ($\sigma = \infty$) and when the source illumination just fills the exit pupil ($\sigma = 1$) is small. Because higher orders of diffracted light can be collected by the system optics for partially coherent or incoherent illumination, the MTF curve for an incoherent, or partially coherent, system will decrease monotonically out to a cutoff frequency which turns out to be greater than that occurring for a coherent optical system (Fig. 4).

Since partially coherent illumination increases the MTF at a given frequency, the instrument resolution capability is improved relative to incoherent illumination [*28*]. Partially coherent illumination also increases the edge gradient of an image relative to incoherent illumination [*29*], which contributes to the overall improvement in effective instrument resolution in terms of resist edge profiles, and linewidth control across an extended area. This improvement in image quality can be related to the intensity in the aerial image as a function of position for partially coherent illumination relative to incoherent illumination. The dependence of the intensity in the aerial image on spatial frequency can be obtained from the MTF curve for an exposure system when some simplifying assumptions are made [*11*].

Resist Systems

In the discussion relating to the performance of an exposure system it was assumed that the variation with position in intensity of the aerial image was replicated in the resist system. This will occur for the delineation of features on "flat" substrate surfaces, where the resist coatings are uniform. When resists are applied on profiled surfaces, however, such as occurs in devices having steps in a previously defined level, there will be a variation in the final film thickness. There will be a loss in resist film thickness over the top edge of the step, with an accompanying increase in thickness at the bottom edge. Assuming a given external exposure level, thickness variations in photoresist lead to exposure intensity variations within the resist. There will be additional intensity variations, which depend on the reflectivity of the substrate, such as standing wave interference phenomena. These intensity variations lead to nonuniformities in linewidth that ultimately limit the feature size that can be defined over a step using a single-level resist process. Standing wave effects, which refer to the spatial variations in exposure intensity within a given resist coating, are common to all exposure systems in optical lithography and have been dealt with extensively elsewhere [*7,9*].

To overcome the difficulties associated with resist thickness variations, innovative multilayered resist structures have been developed that provide simultaneously high resolution, good step coverage, and linewidth control.

There are two broad categories of multilayered resist schemes [*30*]. Some schemes involve the use of resists in at least two layers; both layers are exposed and developed. In other schemes, only the top layer is used as the pattern-defining resist. The top layer is then used as a mask for the selective removal of the other layers. When the differential etch rate between the top and bottom polymer layers is large enough, no third, intermediate layer is required.

The double-layer resist system shown in Fig. 5 is typical of the schemes used. A thin, top layer resist (0.2 to 0.3 μm) is used as the pattern-defining resist. This top resist makes it possible to exploit fully the lithographic performance of the exposure system. The thick, bottom layer resist is used to planarize the substrate surface so that thickness variations caused by step coverage or other variations in topography are eliminated. This thick, bottom layer can be any resist or polymer that remains inert to the processing associated with the top-layer resist.

The inorganic resist systems based on the Ag-Se-Ge glass chalcogenides have been used successfully in a bilevel process by Tai, et al [*31–33*]. They

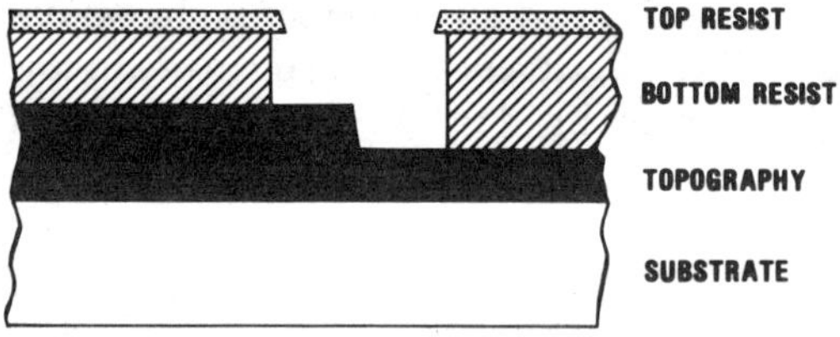

FIG. 5—*Double-layer resist scheme, where thick bottom resist layer is used to planarize surface topography.*

used the $Ag_2Se/GeSe_2$ resist system. Typically, submicron geometries were obtained in 0.3-μm-thick coatings. The Ag-Ge-Se resist systems, which exhibit extremely high contrast, were also highly resistant to the O_2 plasmas required for the reactive-ion etching of the bottom layer. Standing wave interference effects were not observed with these resist films due to their high absorbance at wavelengths where they are photosensitive ($\lambda < 400$ nm). This high absorbance also permits the use of thinner resist coatings than is possible with typical photoresists.

In forming the bilevel structure, thick coatings of the positive photoresist HPR 206 (2.5 μm) were spun on to form the bottom layer, providing the necessary flat surface. Then 0.2-μm-thick films of $GeSe_2$ were deposited. Developed patterns in these films serve as a mask in the etching of the bottom coating of resist. A schematic of this bilevel scheme is shown in Fig. 6.

Using this bilevel structure, the authors were able to define steep-walled patterns in the photoresist at submicron levels, examples of which are shown in Fig. 6. Figure 7*a* demonstrates the ability to resolve 1-μm features over 1-μm polysilicon steps in a typical 16K MOS RAM structure using standard contact exposures. Resist layers over highly reflecting aluminum substrates yielded comparable linewidths (Fig. 7*b*). In these cases, 16K MOS RAM de-

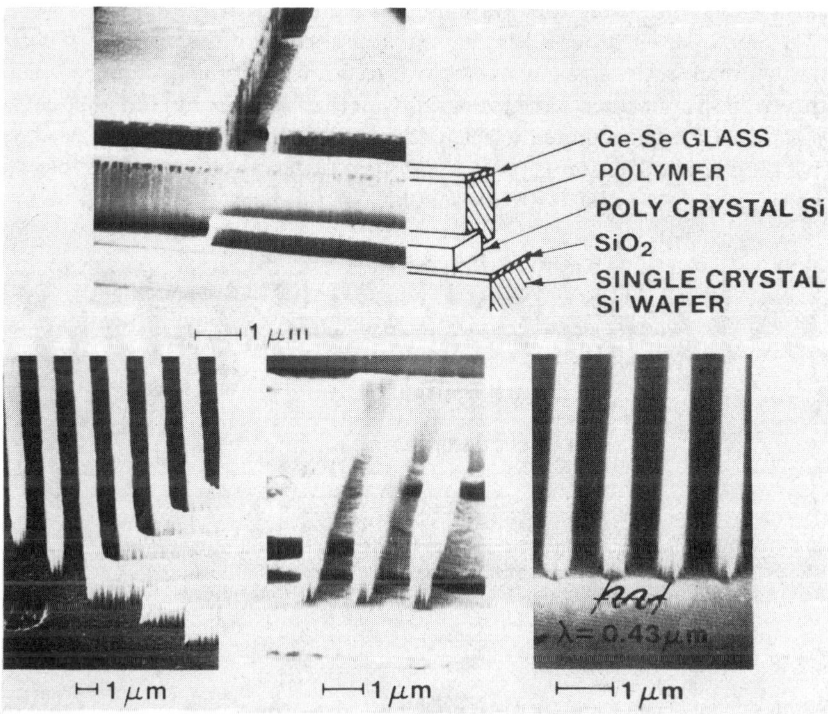

FIG. 6—*Use of inorganic resist/polymer bilevel scheme to define 1-μm and submicron features (courtesy of K. Tai, Bell Laboratories).*

FIG. 7—*SEM micrographs of typical 2.5-μm-thick GeSe/HPR 206 bilevel structures (less than 1 μm in width) over* (a) *1-μm steps in a polysilicon level and* (b) *on an aluminum substrate* [32].

vice wafers processed through the polysilicon level were used. The results demonstrate the high-resolution capability of a two-level resist structure for surfaces having severe topographies and high reflectivity.

A new tri-level process [*34*] can be described with the aid of Fig. 8. Initially a thick resist coating (typically 2 to 3 μm) is applied to the substrate; this coating conforms to the topography of the substrate but has a planar top surface. An intermediate dielectric layer (SiO_2, Si_3N_4 or B_2N_3) is used as a masking layer. The top layer is a *thin* film of resist. (Photoresists, electron resists, and X-ray resists are all suitable for top-layer patterning.)

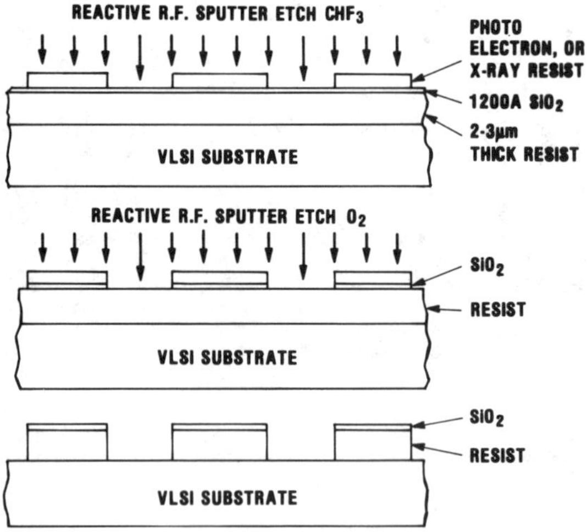

FIG. 8—*Schematic of steps needed to form the trilevel resist structure.*

The high resolution capability of the trilevel structure follows from the fact that fine features defined in the top layer can be replicated in the thick bottom coating with steep sidewalls. The top layer of resist serves as a mask for the plasma, or reactive-ion, etching of the dielectric layer. This layer, in turn, acts as a mask for the etching of the bottom resist coating. Steep wall profiles with very little undercut are produced in the bottom resist film using reactive-ion etching procedures. Feature sizes defined in the top layer are reproduced in the thick bottom layer. Typically, 1-μm linewidths and spacewidths can be defined over 1-μm high steps. No measurable variation in the 1-μm linewidths was observed across such steps [*34*].

The trilevel resist process has been used successfully in the e-beam fabrication of devices with submicron dimensions (about 0.5-μm gate lengths) and precise level-to-level registration tolerances of less than ± 0.25 μm [*35*].

Image Positioning

In addition to the resolution capability of an exposure system, lithographic techniques for VLSI must satisfy stringent requirements in the positioning of images at each level in the fabrication of devices. In the approach to patterning geometries 1 μm and below in size, positional errors must be reduced by the same factor that dimensions are reduced, in order to maintain functional devices. Accuracy in the placement of an image is determined by distortion in the image field as well as by the accuracy in level-to-level registration, the so-called overlay accuracy.

Distortions in the image field can result from distortions in the lens system, and thus depend on the particular optics of the exposure system. Optical distortions in the lenses result from manufacturing imperfections and can be important, even with lens systems that are nominally diffraction limited. These distortions can also have a directional dependence.

Overlay in imaging refers to the ability to register a given structure with previously defined patterns [*36*].

Submicron overlay accuracy is needed for devices having 1-μm design rules. Generally, it is required that a given device level overlay previous levels to a tolerance of better than 20% of the minimum feature size. Thus for 1-μm design rules on 5-in. silicon wafers, an overlay accuracy of better than ± 2 ppm would be required for feature registration within ± 0.2 μm over the entire wafer. Such overlay accuracy represents a formidable challenge as geometries decrease in size and wafer substrates increase in size.

Overlay error can be induced by either the process or the particular exposure tool. Other contributions to overlay error relate to properties of the pattern generated on a mask or reticle, properties of the pattern replicated on wafers, and mask-wafer interactions.

Process-induced contributions to the overlay error result from variations in the edge positions of the etched pattern in a given level relative to a previously defined level. Such variations result from thermal cycling during device fabrication, for example. These variations in processed edge position are a major, if not the largest, source of overlay error (as much as ±0.5 μm/edge), and represent a severe limitation in the fabrication of devices with submicron feature sizes [*37,38*].

Generally a set of fiducial marks on a given level in a device structure is used for referencing the pattern on subsequent levels. The error in such an alignment process contributes to overlay error. Very precise (to better than ±0.1 μm) automated optical and electron beam edge detection systems have been developed to reduce alignment errors [*39*]. Instrument overlay accuracy depends mainly on the alignment optics of the exposure system. Thermal gradients within an exposure system also contribute to the instrument overlay error. Final limitations in overlay accuracy involve the "run-out" error associated with the patterns generated on a mask, as well as the differential thermal expansion and bending (and other deformation) of mask and wafer. Quartz masks have come into more common use to minimize overlay errors due to differential thermal expansion between mask and substrate materials.

The entire tradeoff between full-wafer scanning exposure systems and step-and-repeat reduction systems for VLSI applications will be determined by the degree to which overlay errors can be minimized in the scanning systems.

Comparison of Exposure Systems

Reflective, Full-Wafer Scanning Projection Systems

The optical system in a 1:1 reflective, scanning projection aligner represented a major milestone in the development of a lithographic capability for VLSI.[3] This optical system uses spherical, reflecting mirror surfaces to project the image of the mask features onto a wafer [*8,11*]. A scanning, rather than a full-field exposure, mode was chosen to minimize distortions in the image. In the basic optical system (Fig. 9) light from the mask is imaged onto the wafer using the mirror surfaces. The mask is illuminated through a slit. The mask and wafer move together by means of a continuous scanning mechanism. By keeping the scan direction within the zone of good optical correction,[4] the aberrations of the mirror system are effectively zero. A folded projection system (composite block of the mirror surfaces) ensures that the image on the wafer is a mirror image of the mask and is comparable with the inversion that occurs with contact printing. The design of the condenser system was optimized using the theory of images with partially coherent illumination.

[3] The discussion on 1:1 reflective, scanning projection printing refers to the Perkin-Elmer MICRALIGN series mask aligners, although other comparable systems are available, at least in the near-UV spectral region.

[4] The zone of good correction is a narrow zone concentric with the optic axis within which the system aberrations are nearly zero.

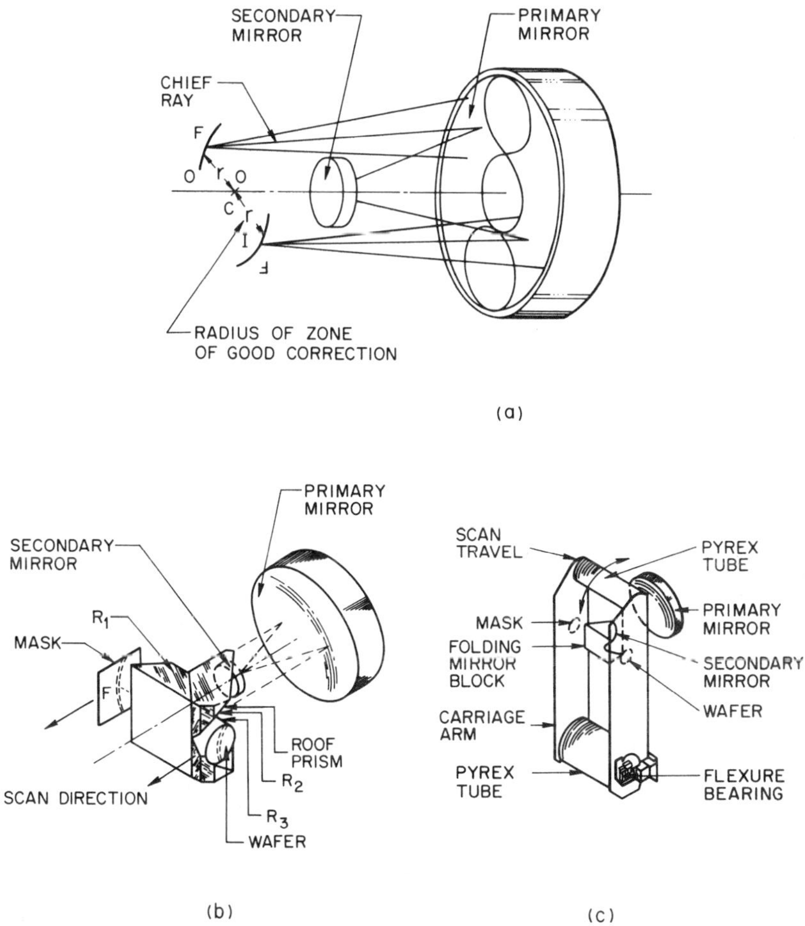

FIG. 9—*Basic projection optical system for Perkin-Elmer MICRALIGN mask aligner showing* (a) *ray path between the two concentric spherical mirrors, as well as the zone of good correction;* (b) *folded mirror configuration needed to form an inverted image on the wafer; and* (c) *scanning carriage mechanism.* r *is the radius to the zone of good correction for the object* O (*mask*) *and image* I (*wafer*).

Using illumination at shorter wavelengths in reflective projection printing systems to improve the resolution has also been successful. Decreasing the wavelength of the exposing radiation rather than increasing the numerical aperture of the optical system is advantageous.

Optical mirror systems, such as in a Perkin-Elmer aligner, have been coated and used with filters to eliminate wavelengths above 330 nm from the mercury source [*12*]. Such a system has been found to remain diffraction limited at these shorter wavelengths.

The latest version of these projection mask aligners has been announced recently for the full-field exposure of 125-mm wafers with a throughput on the order of 100 wafers/h, with limiting resolution capability in the range of 0.9 to 1.25 μm (for corresponding wavelength regions between 240 to 450 nm) and with an overlay accuracy of ±0.25 μm.[5,6] This version still uses a mirror lens system having NA = 0.16, so that the resolution capability is similar to the earlier models. In addition to the UV-A (3900 Å) and the UV-B (3200 Å) regions of the spectrum, previously available as options, a UV-C (2500 Å) region is planned, providing a deep-UV capability. The performance of this aligner as specified by Perkin-Elmer is given in Table 1.[5]

The major improvements in this system include a reduction in the image distortion to less than 0.25 μm over a 125-mm field, an increase in the overlay accuracy to ±0.25 μm, and the possibility of a variable projection magnification to compensate for variations arising from in-plane wafer distortions. The reduction in image distortion can be traced to the high quality of the optical surfaces and to the precise positioning of the optical components. The mirror surfaces are optimized to within 0.01λ, for example, even for λ = 2500 Å. The mechanical and environmental control has also been improved relative to earlier models by using new designs.

The basic projection optics is derived from the ring-field optical systems used in earlier models. Two such systems are placed back-to-back (Fig. 10) and are coupled optically using a pair of weak and a pair of strong refractive elements called shells. These weak and strong shells are made of fused quartz. The back-to-back design of the mirror system serves to eliminate aberrations present in earlier models, which results in diffraction-limited images over a 4-mm wide annulus. The shells also compensate for aberrations that arise when the mask and wafer are asymmetric with respect to the optic axis [*40*].[6] Primary and secondary mirrors having diameters larger than in earlier models are used in the optics. The larger mirrors produce an arc image field having a radius of curvature sufficient to cover a 125-mm-diameter wafer.

TABLE 1—*Perkin-Elmer Model 500 performance specifications.*[a]

Throughput	100 wafers/h
Resolution, 400 nm	1.25-μm-wide lines/spaces
Resolution, 300 nm	1.0-μm-wide lines/spaces
Resolution, 260 nm	0.9-μm-wide lines/spaces
Depth of focus	±6 μm for 1.5 μm geometry
Image field	125 mm diameter
Uniformity of illumination	3.0%
Distortion plus magnification tolerance	0.25 μm, 3 σ
Alignment accuracy	$(\lvert\bar{X}\rvert + 3\sigma) \leq 0.5$ μm
Spectral range	240 nm through visible

[a] Courtesy M. C. King, Perkin-Elmer Corporation, Norwalk, Conn.

[5] King, M. C., Perkin-Elmer Corporation, Norwalk, Conn., private communication.
[6] Cuthbert, J. D., Bell Laboratories, Allentown, Pa., private communication.

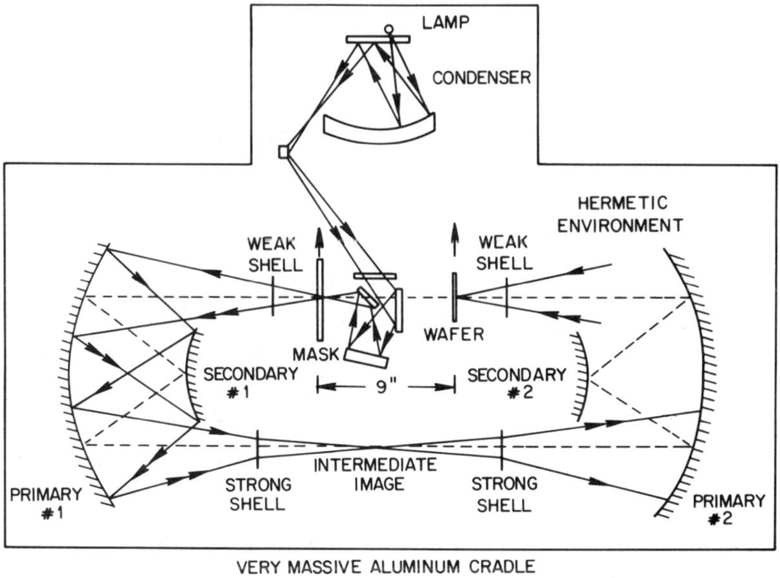

FIG. 10—*Side view of main projection optics in Perkin-Elmer Model 500 projection aligner (courtesy of J. D. Cuthbert, Bell Laboratories).*

Some typical results obtained with this projection aligner are given in Figs. 11 and 12.

The possibility of variable projection magnification is of significance; this feature will permit independent corrections to the magnification in directions perpendicular and parallel to the scan direction. Corrections to the magnification perpendicular to the scan direction (parallel to the slit) would be accomplished by small motions of the strong shells in the optical path. Corrections to the magnification parallel to the scan direction (normal to the slit) would result from the addition of a small differential velocity to the wafer (relative to the mask) by using the microstage fine alignment mechanism. The exact effect of the motion of the shell elements on the machine resolution capability, or the feasibility of introducing a reproducible differential velocity to a wafer, remains to be demonstrated, however. To the extent that in-plane wafer distortion is isotropic, the possibility of a variable, compensating projection magnification may lead to an overlay accuracy comparable with that available on current step-and-repeat systems.

Refractive, Step-and-Repeat Projection Systems

Refractive projection exposure systems require many lenses of the highest quality in order to obtain the needed contrast at high spatial frequencies. Typical refracting optics may contain many individual lens elements which

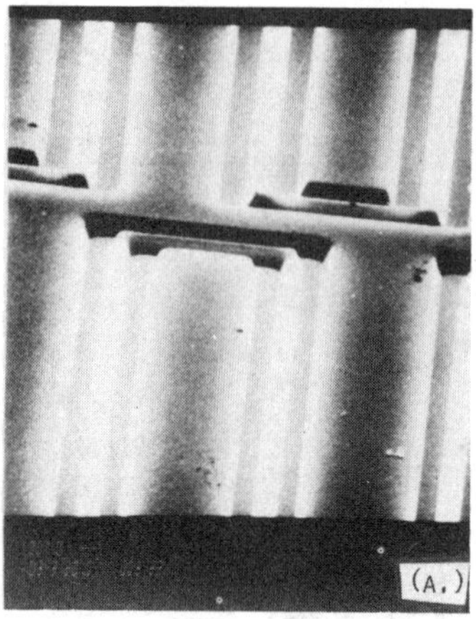

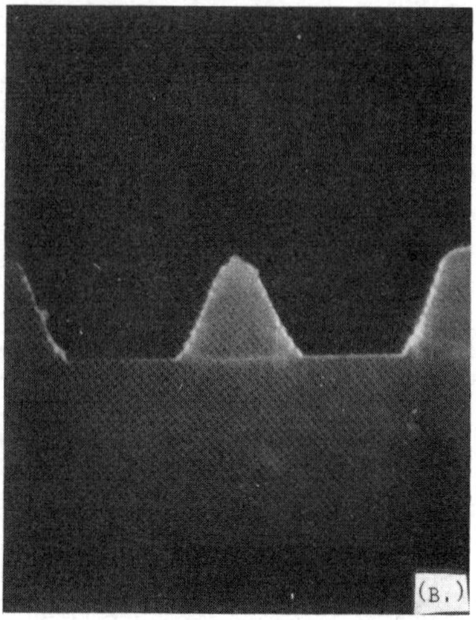

FIG. 11—*Structures defined by Perkin-Elmer Model 500 projection aligner in 1.2-μm AZ-1470 photoresist showing 1.25-μm lines/spaces in resist.* (a) *Top view.* (b) *In profile.* (*Courtesy of M. C. King, Perkin-Elmer Corp.*).

cause the scattering of light, and which subsequently affect the resulting photoresist edge acuity. Even when diffraction-limited lenses are designed, chromatic aberrations result in the system.[7] Narrow bandpass filters are used to block wavelengths in the source other than the wavelengths for exposure and for the alignment optics, thus reducing the effect of these aberrations.

The feature size resolvable with these diffraction-limited lens systems is proportional to the wavelength and to the instrument f-number. Decreasing the wavelength of the exposing radiation as a means of defining smaller features leads to difficulties in lens design (high absorption of optical glasses at shorter wavelengths). Decreasing the f-number by increasing the lens aperture is limited by the physical constraints of the lens system for full-field wafer exposures.[8] The flatness variations of practical substrate surfaces may also exceed the resulting depth of focus. By restricting the area of the image field per exposure (step-and-repeat approach), however, it is possible to effect an improvement in resolution. The step-and-repeat reduction printers having a high-aperture lens system to expose a small-image field form the basis for such an improvement in resolution.

The real advantage of optical step-and-repeat projection systems lies in

[7] Broad-band illumination sources (such as obtained using a typical mercury arc lamp) give rise to chromatic aberrations as a result of the change in lens index of refraction with wavelength. Chromatic aberrations thus produce a dependence of focus on wavelength. These aberrations also affect magnification (line broadening).

[8] For example, a lens designed large enough to project an image over a 3-in.-diameter wafer with adequate resolution for 1-μm dimensional control could not be made with the required mechanical tolerances. See Wilcznski, J. S. and Tibbetts, R., *IBM Journal of Research and Development*, Vol. 13, 1969, p. 192.

FIG. 12—*Deep UV exposure* (λ = *240 to 280 nm*) *of PE Model 500 projection aligner, showing nominally 0.7-μm lines and 1.2-μm spaces in 1.0-μm-thick Hitachi resist over 0.5-μm oxide steps (courtesy of M. C. King, Perkin-Elmer Corp.).*

their ability to print high-resolution features on *deformed* substrate surfaces. Run-out errors can be on the order of 0.5 μm due to process-induced wafer deformations, and can limit the resolution in scanning projection systems. Step-and-repeat projection of a reticle directly on a wafer has made possible the definition of 1.0 to 1.25 μm features on image field sizes in the range of 10 to 20 mm^2 [*41*].

Typical step-and-repeat exposure systems [*16,41*] consist of (*1*) a high-intensity mercury light source; (*2*) a collimating, collecting lens system (such as a fly's eye light integrator) to focus the illumination on a reticle; and (*3*) a reduction lens system to project the image of the reticle onto the surface of a wafer (Fig. 13). Precisely controlled mechanical systems are required to position and step the wafer relative to the mask. Multiple exposures are needed to cover an entire wafer. After each exposure, the wafer is mechanically stepped a given distance, realigned, refocused, etc., until the image of the reticle is projected onto the entire wafer surface.

New designs in the illumination system [*16,25*] have been introduced to reduce the total wafer exposure time to a value comparable with that required in a practical production process. Some of these systems involve more efficient ways to collect the illumination. Other modifications have introduced partially coherent illumination [*29*] to improve resolution and linewidth control.

Step-and-repeat exposure systems use demagnifying optics to project the image of a reticle onto a wafer. Reduction ratios of the lens system in use vary between 2X and 10X [*10,25,41*]. Even though the resolution capability of a lens system depends solely on its numerical aperture, not its reduction ratio, a higher reduction ratio permits a higher NA. Increasing the resolution with

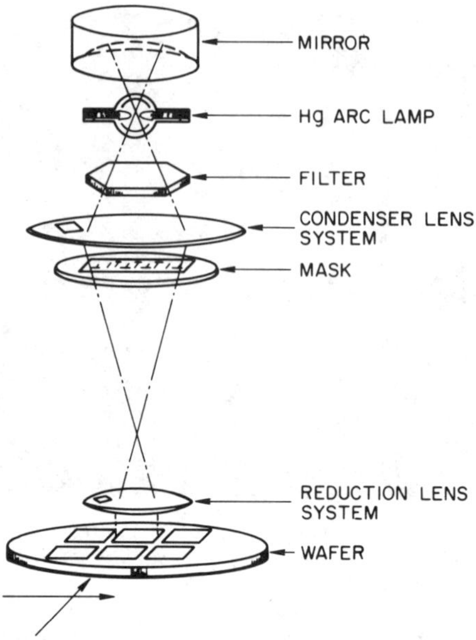

FIG. 13—*Schematic of step-and-repeat reduction projection optical system.*

a large aperture system, however, leads to a reduction in the image field size. Typically, for example, a lens system with a numerical aperture on the order of 0.25 will have a maximum image field diameter of 20 mm when the Rayleigh limit of resolution is 0.5 μm [*41*]. Such a system will have a 1.5-μm linewidth limit in practice.

Two alignment schemes have been developed for the step-and-repeat exposure systems: the so-called off-axis and through-the-lens schemes [*42*]. Both methods of alignment have a typical accuracy of $\pm$0.25 μm. Through-the-lens alignment has the additional capability of individual chip alignment, which is an advantage in view of variable wafer deformation.

Photomask Generation Using e-Beam Lithography

Introduction

In the previous section it was demonstrated that optical exposure systems by themselves have a submicron resolution capability. Full-wafer imaging will be feasible so long as the overall level-to-level registration requirements for device structures do not exceed the limitations set by exposure system overlay accuracy, process-induced wafer deformations, process capability of a multi-level resist system, and registration accuracy inherent in the mask fabrication process. In the design of devices, it is generally required that the level-to-level registration of features be within 20 to 25% of the smallest geometry defined for that device [*43*]. Some of the total $\pm$0.25-μm misregistration allowed in a device with 1-μm design rules then must be apportioned to instrument overlay inaccuracy and to photomask level-to-level misregistration before the effects of wafer processing are even introduced. For an instrument overlay error of ± 0.25 μm, such as specified for the Perkin-Elmer Model 500 aligner (Table 1), the demands on a 1X photomask fabrication processing sequence will be formidable. Using step-and-repeat reduction-imaging systems, registration accuracy is improved significantly, but in these systems sufficiently low defect densities on the reticles become the critical factor determining viability (good yield).

Both of these limiting parameters, level-to-level registration accuracy for full-wafer exposures and defect levels in the reticles needed for step-and-repeat reduction imaging, are dependent on the fabrication process for the photomasks needed for patterning very-large-scale integrated circuit devices. This section considers some aspects of mask fabrication required to support a VLSI device capability.

Photomasks are made currently using electron-beam pattern generating exposure systems [*6,44*]. Systems such as the EBES machine developed by Bell Laboratories [*5*] have a writing scheme that uses electron-beam deflection in one-dimension in a raster-scan fashion. Several excellent reviews describe these systems [*22*]. The discussion in this section will emphasize the definition of patterns on these masks [*45*].

Mask Fabrication Requirements

High-quality photomasks must meet stringent requirements in (*1*) accuracy of pattern placement, (*2*) feature resolution achievable, (*3*) linewidth control over the entire mask substrate area, and (*4*) density of defects on a given mask. Tighter tolerances in all these parameters are needed as the trend towards higher packing densities and smaller features occurs.

Pattern placement accuracy on a master mask is fundamental to the successful superposition of successive mask levels during device fabrication. For a given mask in a set of masks, it must be possible to align each chip pattern with its corresponding pattern on different mask levels.

Temperature variations during the mask fabrication process account for much of the misregistration between different mask levels [*46*]. Differential thermal expansion of the components in a mask-generating system and in the mask substrate itself lead to temperature variations. Environmental control of temperature to a fraction of a degree, not only in the environment where the masks are fabricated but also where they are used, is crucial. A 1°C change in temperature in the pattern generation system, for example, may cause misregistration in two successive mask levels as large as ± 3 ppm (or about ± 0.25 μm over a 76-mm diameter area) in typical glass substrate materials. The use of substrate materials with lower coefficients of thermal expansion (fused quartz, for example) may reduce the corresponding misregistration to about ±1 ppm (or about ±0.10 μm over a 100-mm-diameter area).

The resolution capability obtainable in a mask fabrication process will be determined, in principle, by the smallest feature size that can be written on a given substrate. This minimum dimension is determined by the address structure of the pattern-generating system. Presently available e-beam exposure systems have an operational minimum address structure of 0.25 μm. New developments in e-beam exposure systems emphasize decreasing the address structure size without the undesirable concomitant reduction in throughput, and increase in cost. However, subsequent processing steps involving the electron resist-coated mask substrates which transfer the written features to the chromium films cause a degradation in the actual resolution capability that can be realized. The present resolution level of e-beam generated photomasks of 1.0-μm feature sizes has, until recently, kept ahead of the device design requirements for 2 to 3 μm geometries. As device designs shrink in size, new electron resists and electron-resist processing techniques will be needed to produce the required submicron geometries on photomask substrates.

In the fabrication of IC photomasks, a given level of resolution in feature size is only "realistic" if dimensional uniformity can be maintained across the entire mask substrate. Linewidth control in photomask fabrication is determined by the process control inherent in pattern generation and in pattern transfer to the substrate material. Typical linewidth tolerances presently achievable are usually ±10% of the nominal feature size. For 1-μm features a tolerance control of ±0.1 μm is possible with current processes [*47*]. Further improvements in these processes will be needed as linewidth tolerances are tightened even further.

In addition to feature size definition and proper dimensional control, the density of defects resulting from a given mask fabrication process is of critical importance. Mask defects on the order of 1 μm and above in size are large enough to be transferred to a wafer. Depending on the device design, some of these defects will be fatal with respect to subsequent chip yield. For reticles having 1 to 2 chips per substrate, even more so than for 1X master masks having many chips on a given mask substrate, the production of defect-free masks is vital. Any defects on a reticle used in a step-and-repeat exposure system will be reproduced at each chip site on a wafer, conceivably resulting in zero yield for that device level. The production of perfect reticles is fundamental to the success of a step-and-repeat printing mode.

Once masks have been produced they must be inspected for defects and "repair" of these defects made. Automatic mask inspection techniques are available for detecting defects greater than ~0.2 μm in size [*48*]. These systems operate by using laser beams to compare identical adjacent chip patterns on a given mask. The differences detected are recorded as defects.

The repair of 1X master masks, or of reticles, involves the removal of excess chromium by laser techniques or the reconstruction of areas lacking chromium by photolithographic lift-off techniques. As critical defects decrease in size these techniques will also assume increasing importance.

Summary

Continual improvements in optical lithographic techniques to meet the requirements for fabricating very-large-scale integrated circuit devices have been made possible by (*1*) development of practical exposure systems that eliminate the damaging effects of mask-to-wafer contact, (*2*) dramatic improvements in photomask quality provided by e-beam mask generation systems, and (*3*) use of new multilevel resist structures over profiled surfaces to "fill in" the topography.

Typical projection exposure systems employing scanning reflective optics for full-field wafer imaging are limited by the nonuniformities in exposure that are caused by process-induced wafer deformations and by the instrument overlay accuracy. Step-and-repeat systems having demagnifying refractive optics overcome these limitations. A major disadvantage in their use, on the other hand, is the lower throughput resulting from the required multiplicity of exposures and alignments.

The "practical" resolution capability realized has been maximized in both types of exposure systems through the use of multilevel resist systems. These resist systems effectively planarize the substrate surface, thereby optimizing the linewidth control and uniformity across the substrate.

The use of e-beam-generated photomasks, in the form of 1X master masks for the 1:1 scanning exposure systems and in the form of MX reticles for a step-and-repeat MX reduction system, has produced a dramatic improvement in mask quality. Further improvements in dimensional control across a large-area mask substrate with a concomitant reduction in mask level-to-level misregistration are needed to keep pace with the requirements of VLSI device fabrication using 1X master masks. Nearly perfect, or defect free, reticles are required for step-and-repeat imaging. More sensitive, reliable automatic mask inspection equipment must be developed for submicron defect detection. Precisely controllable mask repair techniques will be needed in the attempts to fabricate a "perfect" mask.

Most likely, optical lithographic techniques will be the dominant pattern-defining technology for the rest of this decade. Both 1:1 scanning projection exposure systems and precision 10:1 step-and-repeat projection exposure systems will be highly competitive for the entire period. The scanning systems are more advantageous from the point of view of throughput, while the stepping systems have an edge with respect to resolution capability and overlay accuracy. As the size of the wafers increases, it is also possible that use of the stepping systems will be required for 1-μm design rules.

Acknowledgments

The author wishes to acknowledge with appreciation helpful discussions with J. D. Cuthbert.

References

[*1*] Keyes, R. W., *IEEE Transactions on Electron Devices*, Vol. ED-26, 1979, p. 271.
[*2*] Dennard, R. H., Gaensslen, F. H., Yu, H. N., Rideout, V. L., Bassous, E., and LeBlanc, A. R., *IEEE Journal of Solid-State Circuits*, Vol. SC-9, 1974, p. 256.
[*3*] Voltmer, F. W., "Manufacturing Process Technology for MOS VLSI," in *VLSI Electronics Microstructure Science*, Vol. 1, N. G. Einspruch, Ed., Academic Press, New York, 1981, pp. 1–40.
[*4*] Noyce, R. N., *Science*, Vol. 195, 1977, p. 1102.
[*5*] Herriott, D. R., Collier, R. J., Alles, D. S., and Stafford, J. W., *IEEE Transactions on Electron Devices*, Vol. ED-22, 1975, p. 385.
[*6*] Skinner, J. G. and Alles, D. S. in *Proceedings*, Kodak Interface '79, 1979, p. 10; see also Skinner, J. G. in *Proceedings*, Kodak Interface '77, 1977.
[*7*] Doane, D. A., *Solid-State Technology*, Vol. 23, No. 8, 1980, p. 101.
[*8*] Markle, D. A., *Solid-State Technology*, Vol. 17, No. 6, 1974, p. 50.
[*9*] Cuthbert, J. D., *Solid-State Technology*, Vol. 20, No. 8, 1977, p. 59.
[*10*] Bruning, J. H., *Journal of Vacuum Science and Technology*, Vol. 17, 1980, p. 1147; see also Roussel, J. M., *Journal of Society of Photo-Optical Instrumentation Engineers*, Vol. 275, 1981, p. 9.
[*11*] King, M. C., "Principles of Optical Lithography," in *VLSI Electronics Microstructure Science*, Vol. 1, N. G. Einspruch, Ed., Academic Press, New York, 1981, pp. 41–81.
[*12*] Shankoff, T. A., Bruning, J. H., and Johnston, R. L., *Polymer Engineering and Science*, Vol. 20, 1980, p. 1102; see also Bruning, J. H., *Journal of Vacuum Science and Technology*, Vol. 16, 1979, p. 1925.
[*13*] Kaneko, T., Umegaki, T., and Kawakami, Y. in *Proceedings*, Kodak Interface '80, 1980, p. 25.
[*14*] Caprari, F. and Doane, D. A., *Electrochemical Society Extended Abstracts*, Vol. 80–2, 1980, p. 858.
[*15*] Massetti, D. O., Hockey, M. A., and McFarland, D. L., *Journal of Society of Photo-Optical Instrumentation Engineers*, Vol. 221, 1980, p. 32.
[*16*] Buzawa, M. J. and Phillips, A. R., Jr., *Journal of Society of Photo-Optical Instrumentation Engineers*, Vol. 135, 1978, p. 77; see also Roussel, J., *Solid-State Technology*, Vol. 21, No. 5, 1978, p. 67.
[*17*] Suwa, K., Nakazawa, K., and Yoshida, S. in *Proceedings*, Kodak Interface '81, 1981, p. 62.
[*18*] Wittekoek, S., *Journal of Society of Photo-Optical Instrumentation Engineers*, Vol. 221, 1980, p. 2; see also Schneider, W. C., *Journal of Society of Photo-Optical Instrumentation Engineers*, Vol. 221, 1980, p. 26.
[*19*] Thornton, P. R., "Electron Physics in Device Microfabrication: I—General Background and Scanning Systems," in *Advances in Electronics and Electron Physics*, Vol. 48, Academic Press, New York, 1979, pp. 271–380.
[*20*] Lepselter, M. P. and Lynch, W. T., "Resolution Limitations for Submicron Lithography," in *VLSI Electronics Microstructure Science*, Vol. 1, N. G. Einspruch, Ed., Academic Press, New York, 1981, pp. 83–127.
[*21*] Watts, R. K., "Advanced Lithography," in *Very Large Scale Integration (VLSI): Fundamentals and Applications*, D. F. Barbe, Ed., Springer-Verlag, Berlin, 1980, pp. 42–88.
[*22*] Herriott, D. W. and Brewer, G. R., "Electron-Beam Lithography Machines," in *Electron-Beam Technology in Microelectronic Fabrication*, Academic Press, New York, 1981, pp. 141–216, and the references contained therein.
[*23*] Special issue on the Hewlett-Packard high-speed electron-beam lithography system: *Hewlett-Packard Journal*, Vol. 32, No. 5, 1981.

[*24*] Moore, R. D., Caccoma, G. A., Pfeiffer, H. C., Weber, E. V., and Woodard, O. C., *Journal of Vacuum Science and Technology*, Vol. 19, 1981, p. 950; see also Ho, C. T., Michail, M. S., Stickel, W., and Woodard, O. C., *Journal of Vacuum Science and Technology*, Vol. 19, 1981, p. 1069.
[*25*] Watts, R. K. and Bruning, J. H., *Solid-State Technology*, Vol. 24, No. 5, 1981, p. 99.
[*26*] Offner, A. and Meiron, J., *Applied Optics*, Vol. 8, 1969, p. 183.
[*27*] Offner, A., *Photographic Science and Engineering*, Vol. 23, 1979, p. 374; see also Ref *11*.
[*28*] King, M. C., *IEEE Transactions on Electron Devices*, Vol. ED-26, 1979, p. 711.
[*29*] Lacombat, M. and Dubroeucq, G. M., *Journal of Society of Photo-Optical Instrumentation Engineers*, Vol. 174, 1979, p. 28.
[*30*] Hatzakis, M., *Solid-State Technology*, Vol. 24, No. 8, 1981, p. 74.
[*31*] Tai, K. L., Sinclair, W. R., Vadimsky, R. G., Moran, J. M., and Rand, M. J., *Journal of Vacuum of Science and Technology*, Vol. 16, 1979, p. 1977.
[*32*] Tai, K. L., Vadimsky, R. G., Kemmerer, C. T., Wagner, J. S., Lamberti, V. E., and Timko, A. G., *Journal of Vacuum Science and Technology*, Vol. 17, 1980, p. 1169.
[*33*] Ong, E., Tai, K., Reichmanis, E., and Wilkins, C. W., Jr., in *Proceedings*, Kodak Interface '81, 1981.
[*34*] Moran J. M. and Maydan, D., *Bell System Technical Journal*, Vol. 58, 1979, p. 1027; see also Moran, J. M. and Maydan, D., *Journal of Vacuum Science and Technology*, Vol. 16, 1979, p. 1620.
[*35*] Watts, R. K., Fichtner, W., Fuls, E. N., Thibault, L. R., and Johnston, R. L., *IEEE Transactions on Electron Devices*, Vol. ED-28, 1981, p. 1338.
[*36*] See, for example, Makita, M., Moriuchi, N., and Kadota K. in *Procedings*, Kodak Interface '80, 1980, p. 104; see also Yau, D. in *Proceedings*, Kodak Interface '81, 1981, p. 124.
[*37*] Gegeworth, R. E. and Laming, F. P., *Journal of Society of Photo-Optical Instrumentation Engineers*, Vol. 100, 1977, p. 66.
[*38*] Cuthbert, J. D., *Proceedings of Microcircuit Engineers*, Aachen, Vol. 79, 1979, p. 300; see also Yau, L. D., *Applied Physics Letters*, Vol. 33, 1978, p. 756, and Yau, L. D., *IEEE Transactions on Electron Devices*, Vol. ED-26, 1979, p. 1299.
[*39*] King, M. C. and Muraski, E. S., *Journal of Society of Photo-Optical Instrumentation Engineers*, Vol. 174, 1979, p. 70.
[*40*] Broers, A. N., *IEEE Transactions on Electron Devices*, Vol. ED-28, 1981, p. 1268.
[*41*] Schneider, W. C., *Journal of Society of Photo-Optical Instrumentation Engineers*, Vol. 174, 1979, p. 6; see also Thomas, G. R. M., Coleman, H. L., and Lanahan, M., *Journal of Society of Photo-Optical Instrumentation Engineers*, Vol. 174, 1979, p. 15.
[*42*] Hershel, R. S., *Journal of Society of Photo-Optical Instrumentation Engineers*, Vol. 174, 1979, p. 54.
[*43*] Sogo, K., Tanaka, Y., and Uchiho, K. in *Proceedings*, Kodak Interface '80, 1980, p. 44.
[*44*] Dan, M., Tanabe, I., Hoga, M., and Torisawa, S. in *Proceedings*, Kodak Interface '79, 1979, p. 103; see also Watakabe, Y., Shigetomi, A., Marimoto, H., and Kato, T., *Journal of Society of Photo-Optical Instrumentation Engineers*, Vol. 275, 1981, p. 62.
[*45*] Ballantyne, J. P., "Mask Fabrication by Electron-Beam Lithography," in *Electron-Beam Technology in Microelectronic Fabrication*, G. R. Brewer, Ed., Academic Press, New York, 1981, pp. 259–307; see also Dey, J. and Harrison, K., *Journal of Society of Photo-Optical Instrumentation Engineers*, Vol. 100, 1977, p. 4.
[*46*] Cowan, M. J., Mattis, D. G., and Chapman, E. W. in *Proceedings*, Kodak Interface '81, 1981, p. 17.
[*47*] Cowan, M. J. and Helm, R. F., *Journal of Society of Photo-Optical Instrumentation Engineers*, Vol. 221, 1980, p. 142.
[*48*] Bracken, R. C., Rizvi, S. A., and Reeves, A. E., *Journal of Society of Photo-Optical Instrumentation Engineers*, Vol. 275, 1981, p. 76; see also, Speck, R. P., *Journal of Society of Photo-Optical Instrumentation Engineers*, Vol. 221, 1980, p. 127, and Schmitzer, A. P., *Journal of Society of Photo-Optical Instrumentation Engineers*, Vol. 221, 1980, p. 132.

Andrew R. Neureuther[1]

Resist Profile Quality and Linewidth Control

REFERENCE: Neureuther, A. R., **"Resist Profile Quality and Linewidth Control,"** *Silicon Processing, ASTM STP 804,* D. C. Gupta, Ed., American Society for Testing and Materials, 1983, pp. 108–124.

ABSTRACT: Key parameters and process tradeoffs in improving resist profile quality and linewidth control are explored for optical projection printing. Simulation with program SAMPLE is used to study the impact of image quality, wafer topography, and resist characteristics in the context of the complete lithographic process. The advantages of the use of a bias and of a "surface induction" resist are illustrated. The tradeoff between quality and throughput is explored as a function of the partial coherence factor. Practical rules of thumb are given for performance with single-layer thick resists. The potential improvements with multilayer and inorganic resists are also discussed.

KEY WORDS: linewidth control, photolithography, resist profile quality, optical projection printing, photoresists

Resist profile quality and linewidth control are important considerations in choosing the set-point of a lithographic process. Both the quality and control are affected by variations in key processing parameters such as defocus, resist thickness change at steps, and development conditions. Systematic studies are needed to explore the tolerance of the process to these parameters and to determine a set-point with suitable process windows. Some improvement in latitude and even resolution can be obtained by tailoring the imaging tools, process conditions, device structures, and resist materials. This tailoring becomes increasingly important as the resolution is pushed into regions where the process latitude degrades rapidly.

An effective way of characterizing a process and understanding limiting tradeoffs is to use simulation together with experimental verification. The computer program for Simulation and Modeling of Profiles in Lithography and Etching (SAMPLE) is a convenient tool for this purpose [*1*].[2] It includes the effects of partial coherence on the aerial image, the standing waves in the

[1] Electronics Research Laboratory, Department of Electrical Engineering and Computer Sciences, University of California, Berkeley, Calif. 94720.

[2] The italic numbers in brackets refer to the list of references appended to this paper.

resist, and the lateral effects in resist development [*2*]. SAMPLE has been used to explore factors affecting linewidth control [*3*], to characterize projection printers [*4*], and to study the trends in developments of lithographic tools [*5,6*]. Similar simulation programs have also been developed and applied to the study of projection printing [*7–11*].

This paper attempts to give an overall perspective on how aerial image quality, wafer topography, and resist characteristics affect resist profile quality and linewidth control in optical lithography. The emphasis is on the use of a single-layer-thick resist, although the potential improvements with multilayer and organic resists are also discussed. The importance of various factors is illustrated by means of scanning electron microscopy (SEM) results from Perkin-Elmer UV-500 with AZ2400 resist. These results were obtained by an exposure primarily at a wavelength of 313 nm at a partial coherence factor of 0.857 on a prototype version of the machine [*12*].[3] The results are used here for illustrative purposes only and are not typical of the current status of the UV-500. Similarly, the model for AZ2400 is a preliminary one [*12*]; a more refined model is forthcoming [*13*]. This resist was chosen because it shows an interesting "surface induction" or delay in development that can improve resist profile quality. The results are summarized in terms of the numerical aperture and wavelength so that they can be applied to any projection printer.

Image Quality

The quality of the aerial image of a projection printing system is an important consideration. We will attempt to identify those features of the aerial image that correlate best with resist performance. It will also be informative to see just how poor the quality of aerial image typically is.

As linewidth decreases the image quality decreases. This makes wafer topography such as steps much more formidable problems. Figure 1 shows SEMs of AZ2400 resist lines crossing a 0.4-μm silicon step. These are the center lines of five lines of equal line and space arrays of pitch 3.0, 4.0, and 5.0 μm. The difference in linewidth control in crossing this step is primarily due to the difference in image quality. For a pitch of 3.0 μm both the profile quality and linewidth control are poor.

The change in the aerial image with linewidth is shown in Fig. 2. The slope of the image is not affected strongly; however, the maximum and minimum are degraded. The development of thick resist proceeds first downward in the highly exposed areas and then laterally into the unexposed areas. The minimum intensity in the unexposed areas plays an important role, since it provides a region of low development rates. This region acts as an etch stop for the lateral development. The increase in the minimum intensity reduces

[3] The exposures were made in the spring of 1981 by M. Hakey for use in Ref *12*.

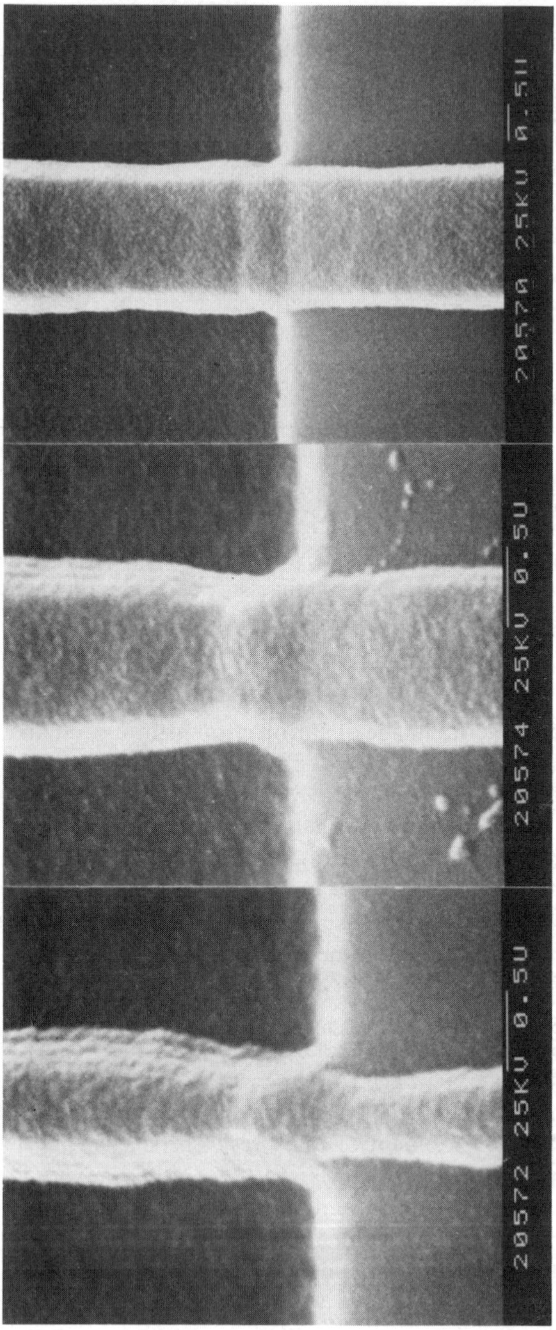

FIG. 1—*AZ2400 resist lines crossing an 0.4 μm step on a silicon substrate. The center line of a five line equal line-and-space array is shown for pitches of 3.0, 4.0, and 5.0 μm.*

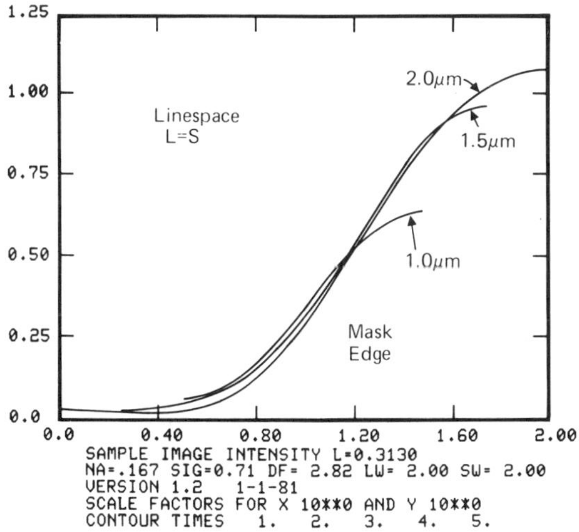

FIG. 2—*Aerial images for the Perkin-Elmer UV-500 at 313 nm. The partial coherence factor is 0.714 and a small amount of defocus is included to approximate possible lens imperfections.*

this etch stop effect and also increases the resist top loss. Thus the change in the minimum intensity is viewed as that property of the aerial image that is primarily responsible for the change in the resist profile.

A potential problem in projection printing is focus tolerance. The classical measure of focal error is the Rayleigh unit (RU) of $\lambda/2(\text{NA})^2$, where λ is the wavelength, and NA is the numerical aperture or sine of the acceptance angle of the lens looking back from the wafer. This distance corresponds to a quarter-wave optical path difference between rays from the center and outer edge of the lens. For the UV-500 (NA $= 0.167$) at 313 nm the RU is 5.64 μm. Figure 3 shows the corresponding aerial image as a function of the focal error. At around one RU the image is degrading very rapidly and a maximum focal error of ⅔ RU is judged to be more appropriate. The effect of defocus on a 4.0 pitch array line is shown in Fig. 4. Both the edge slope and linewidth variation are degraded.

Optical system designers introduce partial coherence to improve the aerial image quality. Partial coherence is described by a factor σ, which is the ratio on the numerical aperture of the illumination system condenser to that of the imaging lens. The numerical aperture of the condenser determines the maximum distance on the mask over which the illumination is coherently related. The imaging lens typically has a larger numerical aperture and hence greater resolving power. It can thus take multiple samples of these coherently related signals as they pass through the mask. The constructive and destructive interference of these image contributions can in turn affect the image quality in a beneficial manner.

The image improvement as a function of the partial coherence factor σ is shown in Fig. 5 for the UV-500 for a 3.0-μm pitch. The partial coherence

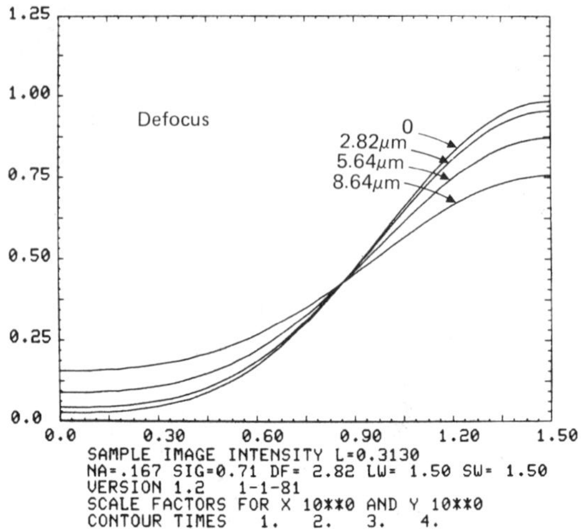

FIG. 3—*Aerial images as a function of defocus for 0, 0.5, 1.0, and 1.5 RU.*

lowers the intensity level at the mask edge (0.75 μm). Incoherent illumination ($\sigma = \infty$) gives a value of 0.5 while coherent illumination ($\sigma = 0$) gives $(0.5)^2$ or 0.25. For the other values of σ shown the intensity decreases from 0.32 to 0.27. In the clear area of the mask constructive interference can yield intensities higher than the intensity in a large clear area (1.0). This tends to improve the intensity at the edge of the clear area. The image slope is also typically increased. In the masked region destructive interference tends to

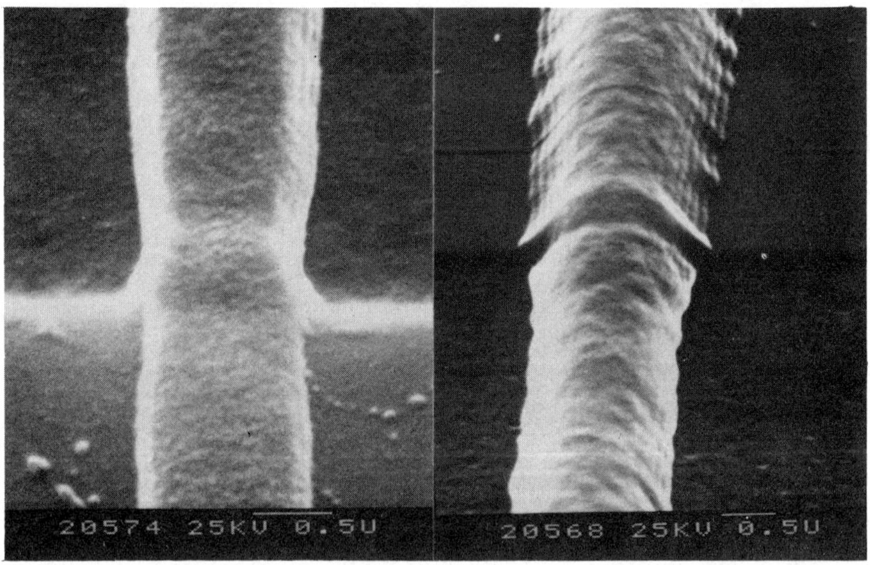

FIG. 4—*The effect of focal error on the resist profile for a pitch of 4.0 μm.*

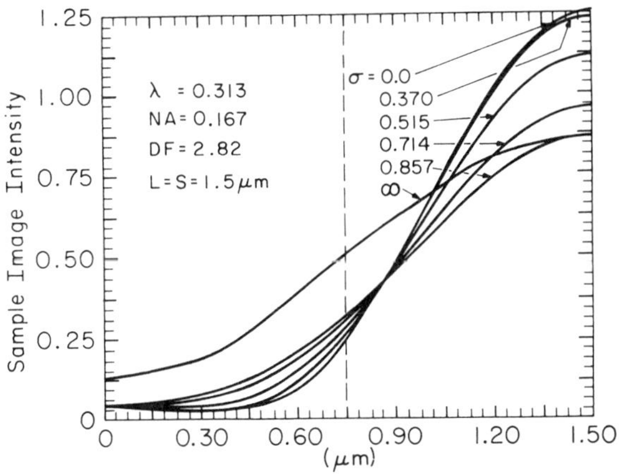

FIG. 5—*Aerial images as a function of the partial coherence parameter σ for a 3.0-μm pitch.*

reduce the intensity, particularly near the mask edge. In agreement with the previous discussion, this reduction of the minimum intensity is by far the most important effect for imaging in single-layer thick resists.

In practice it is convenient to characterize the aerial image by means of the square wave contrast given by $(I_{max} - I_{min})/(I_{max} + I_{min})$ where I_{max} and I_{min} are the maximum and minimum intensities for an equal line-and-space mask pattern. A plot of the square wave contrast of a diffraction-limited lens as a function of the spatial frequency of the patterns is given in Fig. 6 [5]. It is normalized by dividing the period by λ/NA. Note that the square wave contrast for the traditional operating point of 60% modulation transfer function (MTF) is 93% for $\sigma = 0.7$. Decreasing σ to 0.3 gives a square wave contrast of 100%, owing to a perfect phase cancellation in the intensity.

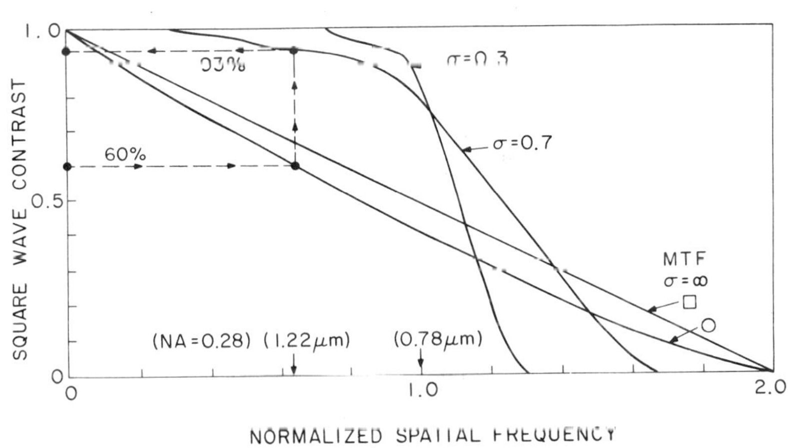

FIG. 6—*Square wave contrast and modulation transfer function for equal line-and-space arrays.*

The amount of partial coherence that can beneficially be used to improve the lithography involves a tradeoff in throughput and eventual excessive image ringing. It is difficult to capture the illumination from the source into a small cone angle. As a result the net throughput decreases roughly as the square of σ. Thus the throughput in decreasing σ from 0.857 to 0.714, 0.515, or 0.370 drops to 69, 36, and 20%, respectively, of the initial throughput. A partial coherence factor as low as 0.3 may introduce excessive ringing at corners [*14*] and may enhance proximity effects [*15*]. Down to a σ of 0.5 the effects appear to be primarily beneficial and a significant improvement in resolution can be obtained. For example the 3.0-μm pitch image in Fig. 5 for a σ of 0.515 is roughly of the same quality as the 4.0-μm pitch image in Fig. 2 for σ of 0.714. This is one way that higher resolution can be obtained with existing equipment at the expense of throughput.

Another consideration is the use of bias. Designers would like the resist linewidth and even the linewidth after pattern transfer to be equal to the mask linewidth. This zero bias case is difficult to control in thick resists due to the poor quality of the optical image near the mask edge. As we had noted previously the image intensity at the mask edge is from 27 to 32% of the clear field intensity. Allowing the resist to develop into the masked region to the point where the intensity drops to 10% will produce steeper profiles that are also less sensitive to development time. Since the intensity of the mask edge is 27 to 32%, the 10% intensity will occur a slight distance into the masked region. A simple rule of thumb can be obtained from the intensity profile of a knife edge pattern at sigma of 0.7 [*2*], which shows that a bias of $0.10\lambda/NA$ into the masked region occurs at each edge of a line.

In practice it is common to operate with a bias. Overexposure and overdevelopment such that large areas clear in half the development time is typically used. This allows small features to clear. An example of bias is shown in Fig. 7 for a 4.0-μm pitch. Here the duty cycle of remaining resist is only about 33% of the pitch. This small duty cycle is due in part to an additional bias that occurred in making the mask itself. For a working resolution of $1.56\lambda/NA$ (corresponding to about the 60% MTF point) a typical bias of $0.10\lambda/NA$ is needed on each line edge. This results in a duty cycle of 37%.

Proximity effects also change the aerial image. The effects are particularly noticeable in the end lines of arrays. For example, in Fig. 8, the 3.0-μm pitch array has end lines that are overexposed. The light from the large clear areas spills over onto the last array lines. At small σ (0.37), where coherent effects are more pronounced, destructive interference can occur and the spillover effects can destructively interact to reduce the intensity as well [*15*].

For inorganic resists the conclusions about typical operating conditions are considerably different. The inorganic resists, such as silver-doped GeSe, appear to respond to image slope and are capable of imaging at much higher resolution. The photodoping in the exposed areas tends to draw additional

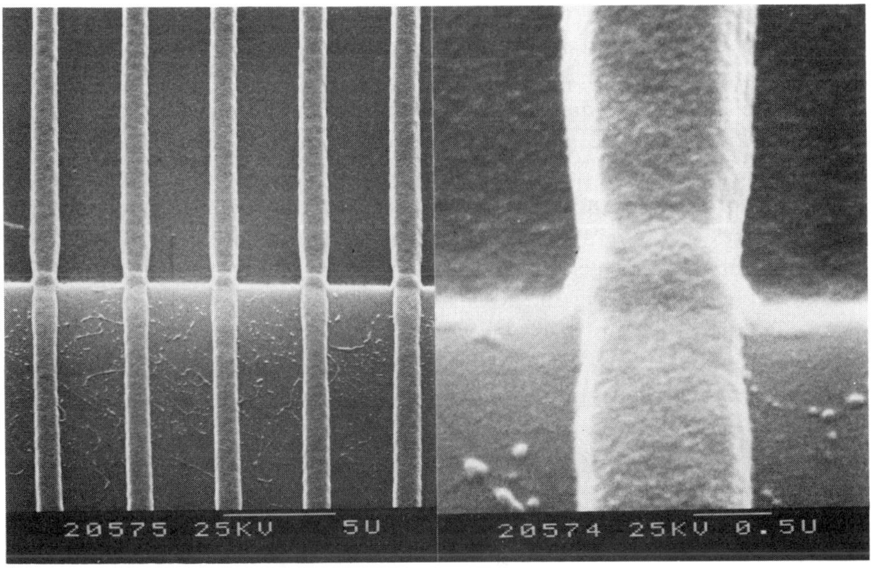

FIG. 7—*Typical bias for an equal line at a pitch of 4.0 μm.*

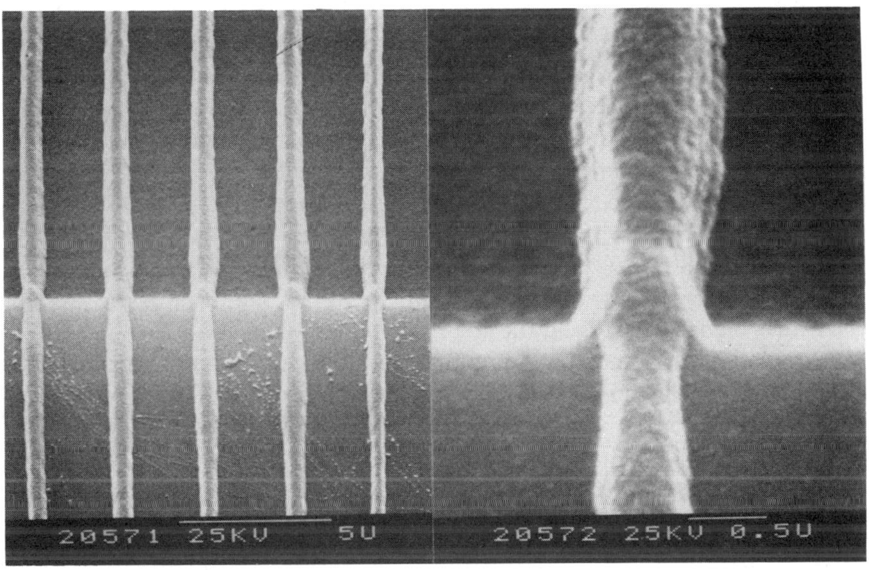

FIG. 8—*Proximity effects on the outside lines in a five line equal line-and-space array at a pitch of 3.0 μm.*

silver from the surrounding unexposed areas, increasing the silver at the image edges. This lateral diffusion of silver sharpens or improves the latent photodoping image. Together with a high-contrast developer a very unusual resolution is produced [*16*]. The edge sharpening is similar to the human eye, which contains interconnected cells that differentiate the signal to recognize edges.

The performance of inorganic resist is very impressive [*16*]. Equal line and space arrays at a pitch of $0.8\lambda/NA$ have been imaged with almost zero bias. A greater latitude between isolated line and space features has also been observed. Potentially, inorganic resists offer as much as a factor of two improvement in resolution. This places the performance in the region where the pitch is less than λ/NA, so that the normalized spatial frequency is larger than 1.0. At this high resolution more incoherent (larger σ) than coherent illumination (smaller σ) is desired, as can be seen in Fig. 5. This may reverse the trend to small σ in present projection printing systems.

Wafer Topography

Steps and features on the wafer greatly complicate the conditions under which the resist must produce a usable profile from the aerial image. Two of the main problems are resist thickness changes at steps and the reflection from materials such as aluminum. The effects of factors such as these have been discussed recently [*3*]. Here the primary considerations will be overviewed.

Topographical features are difficult to planarize with resist and inherently introduce linewidth variations. The resist must be much thicker than the feature height to achieve reasonable resist planarization. A rule of thumb for poly(methyl methacrylate) (PMMA) in electron-beam lithography is to make the resist at least 2.3 times thicker than the feature height [*17*]. The narrowest resist linewidth occurs at the top of the feature and the widest linewidth occurs at the bottom of the step. A local variation in linewidth due to the change in coupling of energy into the resist with resist thickness also occurs. This local linewidth variation is typically less than one third of the global linewidth change in crossing the step.

Highly reflecting substrates require additional aerial image contrast to compensate for the increased contrast in the standing waves in the resist. On silicon substrates, the aerial image contrast of a stepper with $NA = 0.28$, $\lambda = 436$ nm, and $\sigma = 0.7$ for a 2.0-μm pitch is 93%, which easily exceeds the 78% standing wave contrast in the resist. On aluminum, however, the standing wave contrast at the bottom of the resist is 99% and σ must be reduced to 0.3 to get sufficient aerial image contrast (97.5%). Generally it is desirable to reduce the reflectivity seen within the resist to about $(0.8)^2$ or 64% to be comparable to that of silicon.

The use of multiple wavelengths can improve profile quality and linewidth control as well as throughput. To be most effective the resist should be thick enough that a quarter-wave offset in the standing waves of two or more wavelengths can occur. For example, for 436 and 407 nm a thickness of about 0.75 μm is required. At this offset point the contrast in the composite vertical intensity pattern is low and the development can easily be controlled by the high contrast of the aerial image. This results in steep edge profiles around the 0.75 μm height that are also insensitive to development time. This well-controlled aperture 0.75 μm from the silicon interface helps to some extent to improve the control of the linewidth change with development at the silicon interface. The improvement in control is only minor because the standing waves are in phase at the silicon interface. In refracting optical systems that have significant axial chromatic aberration (change of focus position with wavelength) the use of multiple mercury arc lines and even the continuum is generally undesirable [*6*]. The loss on contrast due to part of the energy being out of focus is more detrimental than the gain due to the offset of the vertical standing waves. In reflecting optical systems such as the UV-500 it may also be necessary to limit the spectrum to eliminate the energy that is so strongly absorbed in the resist that it cannot penetrate to the bottom of positive resists [*13*].

Multilayer resists can be an effective way of solving the wafer topography problems. Underlying layers can be used to planarize the surface and thus eliminate problems at steps. By using a planarizing film that is absorbing as well [*18*], the reflections from highly reflecting films can also be eliminated. A lower contrast aerial image could then be utilized. Thus for multilayer resists a ballpark figure for the usable pitch might be 1.2λ/NA.

Resist Characteristics

The resist characteristics have an equally important role in achieving profile quality and linewidth control. We would like the resist to compensate for the poor aerial image quality and difficult wafer topography. The virtues of multilayer and inorganic resists for this have already been discussed. At this point we will primarily focus on the enhancement of positive resist performance, which is possible by means of the "surface induction" effect in thick resists. AZ2400 is well known for showing this effect [*19*]. A first-order model [*11*] will be used here to illustrate the kinds of effects it can have on the lithographic performance.

A convenient starting point for determining a resist development model is a measurement of the thickness remaining during the development versus development time. An example of the thickness remaining versus development time for AZ2400 exposed at 313 nm and developed in 4:1 AZ2401 developer at room temperature is shown in Fig. 9. These curves were recorded by the IBM film thickness analyzer. Note that the slope and hence development rate increases with depth for all levels of exposure. This increase of de-

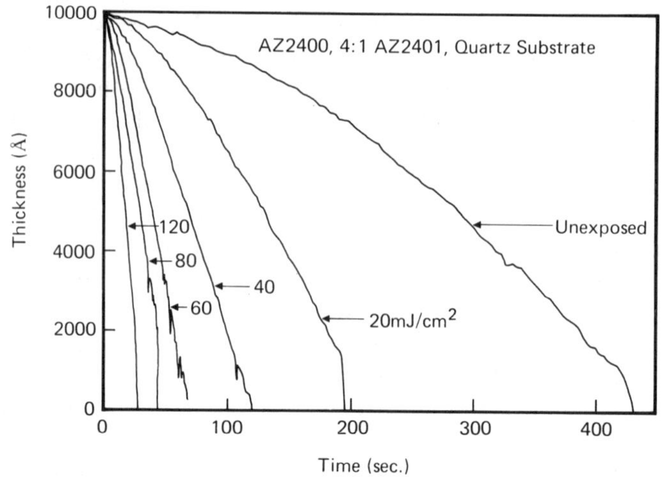

FIG. 9—*Thickness remaining versus development time for AZ2400 resist exposed at 313 nm and developed in 4:1 AZ2401 developer. 1 Å = 0.1 nm.*

velopment rate with depth is shown more explicitly in Fig. 10, where the ratio of the rate at depth z to the rate at a depth of 0.8 μm is plotted as a function of depth. Typically the rate at the top surface is only one third of the bulk rate; this slowness continues a distance from 0.25 to 0.5 μm into the resist. This reduction of the rate near the top surface delays development and hence is termed the "surface induction" effect.

The reduction of the development rate shown in Fig. 10 can be modeled in a manner similar to that used for electron beam resists [*20*]. The net etch rate at any point in the resist can be determined from the product of bulk etch rate $R(M)$ and the surface slowness factor $f(z,M)$ [*12*]. If the slowness factor is assumed to be dose-independent (that is, if it does not depend on the rela-

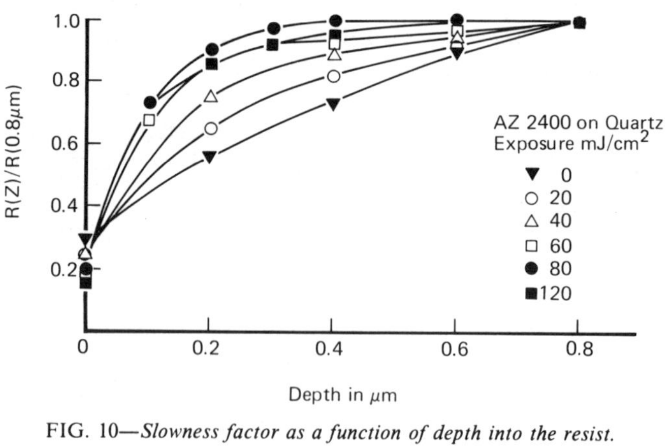

FIG. 10—*Slowness factor as a function of depth into the resist.*

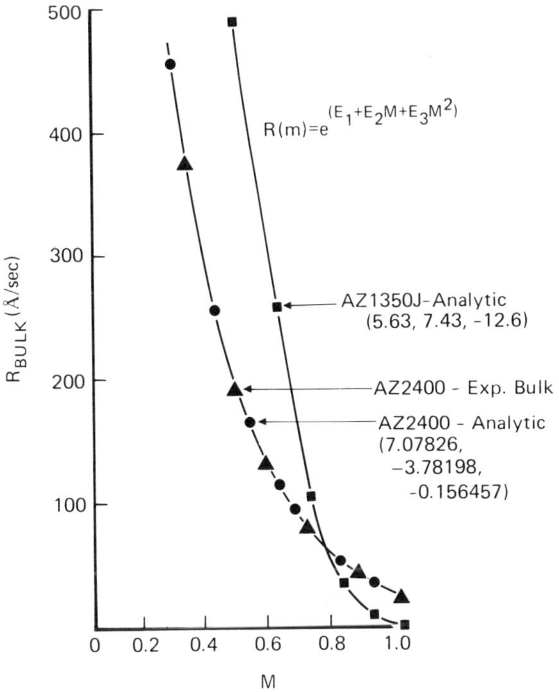

FIG. 11—*Model for the bulk development rate as a function of* M. *1* Å = *0.1 nm.*

tive remaining photoactive compound *M*), a two-parameter model $f(z)$ can be used. The constants *A* and *C* describe this model in Fig. 12. A more accurate dose-dependent model $f(z,M)$ can be made with the four parameters *A*1, *A*2, *C*1, and *C*2 and is also given in Fig. 12. The bulk rate is characterized by the *E*1, *E*2, and *E*3 parameters and the corresponding etch rate curve is shown in Fig. 11. Note that the AZ2400 bulk rate shows less contrast and a higher background rate than AZ1350J.

$$f(z) = 1 - Ae^{-CZ}$$

$$A=0.67 \quad C=4.02\mu m^{-1}$$

$$f(m,z) = 1 - [A_1+A_2(1-m)]\,e^{-[C_1+C_2(1-m)]\,z}$$

$$A_1 = 0.71 \qquad C_1 = 3.68\mu m^{-1}$$
$$A_2 = 0.17 \qquad C_2 = 9.83\mu m^{-1}$$

FIG. 12—*Dose-independent* f(z) *and dose-dependent* f(z,M) *models for the slowness factor. The parameter* M *is the relative remaining photoactive compound.*

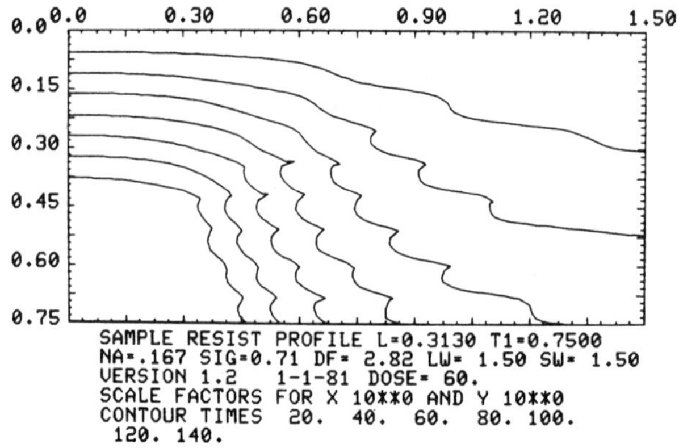

FIG. 13—*Simulated time evolution of the developer resist interface without the "surface induction" effect.*

The above model for AZ2400 will now be used to illustrate the ways in which the "surface induction" effect can influence the profile quality and linewidth control. The simulated time evolution of the developer resist interface is shown in Fig. 13 for AZ2400 without "surface induction." The simulation was done using the data set shown in Fig. 14 in SAMPLE with a modified rate formula. The resist top loss is considerable and the slope of the profile near zero bias is only about 55 deg. The simulation in Fig. 15 includes the "surface induction" effect. The top loss is much smaller and the profile is more vertical, with a slope of 65 deg near zero bias. At a bias of $0.10\lambda/NA$ the edge slope with "surface induction" increases from 63 to 75 deg. The spacing between development contours is similar, indicating that the sensitivity to overdevelopment is basically unchanged. Thus profile quality is improved with the "surface induction" effect with little accompanying improvement in linewidth control at the silicon interface. However, if linewidth is determined by the top of the resist image, as is the case for liftoff or anisotropic dry etching, "surface induction" can improve the linewidth control. The higher quality profile also allows a greater tolerance to defocus. Figure

```
* AZ2400 AT 3130 WITH INDUCTION ON SILICON
LAMBDA  0.3130
PROJ 0.167
LINESPACE 1.5 1.5
TRIAL 2 0 1 0
TRIAL 3 1 0 1 0 0
TRIAL 20 0 0.714 8.46
RUN 1
RESMODEL ((0.3130))  (.162, 0.184, 0.0128) (1.80, ((-0.01)))  (.75)
LAYERS (5.0092,-3.4742)
RUN 2
DOSE 60
RUN 3
ETCHRATE ANALYTIC (7.07826, -3.78198,-0.156457)
DEVTIME 20 TO 140, 7
RUN 4
```

FIG. 14—*SAMPLE input data file.*

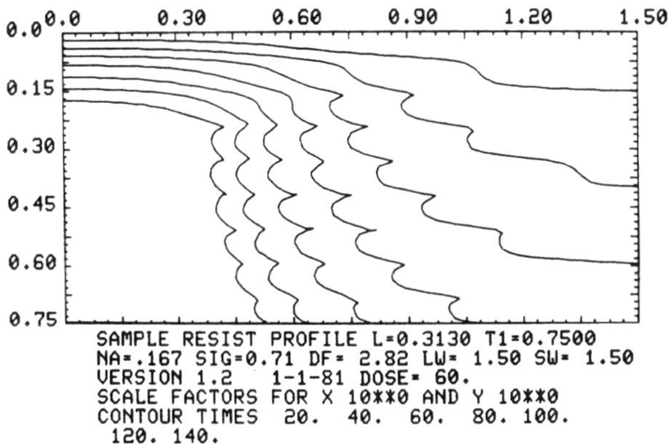

FIG. 15—*Resist profile development with the "surface induction" effect.*

16 shows that the resist profile with "surface induction" for a full RU of defocus is still of usable quality, whereas for convention resist only ⅔ of an RU of defocus is recommended.

Conclusion

Aerial image quality, wafer topography, and resist characteristics are equally important considerations in achieving resist profile quality and linewidth control. Figure 17 summarizes the rules of thumb developed for single-layer thick resists. Specific linewidths are given in micrometres for the GCA DSW4800 and the Perkin-Elmer UV-500 (at a σ of 0.857). The minimum pitch of 1.56λ/NA corresponds to a traditional MTF of 60% and thus may be viewed as somewhat conservative. Bias is recommended for steep profiles and better linewidth control, although it is possible to operate at zero bias if adequate control of the processing can be attained. A maximum

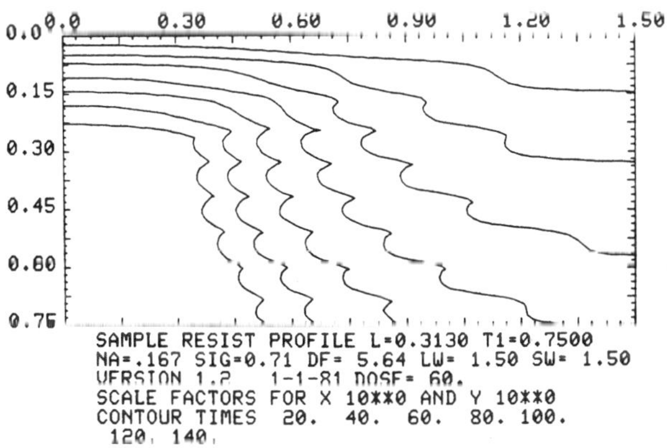

FIG. 16—*Resist profile with the "surface induction" effect for additional defocus.*

Profile Quality and Linewidth Control
in Optical Lithography

Image Quality	GCA	PE500	
min pitch (1.56 λ/NA)	2.43 μm	2.92 μ	
min bias (2 × 0.10 λ/NA)	.31	.37	
resist line	.90	1.09	37%
max defocus $\frac{2}{3}$(λ/2(NA)^2)	1.85	3.74	

Wafer Topography

max steps (h ≤ 2.3t)
max reflectivity ($\rho_R^2 \leq (.8)^2$)
use multiple λ and t > .75 μm
(if axial chromatic aberration not a problem)

Resist Characteristics

low nonbleachable absorption (Bt < .3)
induction effect for quality (.25 μm)
multi-layer for linewidth control (RIE)

FIG. 17—*General parameters developed for single-layer resist.*

focal error of ⅔ RU is recommended to avoid the rapid degradation of the aerial image. Higher resolution could be obtained with existing equipment by operating with the partial coherence parameter σ reduced to 0.5 but at a cost of about a factor of three in throughput. Multilayer resists need less contrast and may operate at a smaller pitch of 1.2λ/NA. Inorganic resists appear to respond to image slope and operation at a pitch of 0.8λ/NA may be feasible. In this small-pitch region it may be more beneficial to have less rather than more coherence in the illumination system.

The primary problems associated with wafer topography are step crossing and highly reflecting materials such as Al. For good step coverage the resist thickness should be as much as 2.3 times the step height. It is desirable to reduce the reflectivity of underlying layers as seen from within the resist to about 64% so that they are comparable to that of silicon. Multiple wavelength exposures can reduce the required contrast if the resist is thick enough that the standing waves are offset. In reflecting optical systems the axial chromatic aberration may degrade the contrast to such an extent that the use of multiple mercury arc wavelengths and even the continuum are generally not desirable. Multilayer resists are an effective means of eliminating both the step coverage and reflectivity problems.

The resist exposure and development characteristics can affect the profile quality and linewidth control. The nonbleachable absorption of the resist should be at a minimum to allow the intensity to penetrate to the bottom of the resist. The "surface induction" or slowing of the etch rate near the top of the resist can be used to reduce the resist top loss and increase the sidewall angle. Multilayer resists offer the advantage of vertical line edge profiles. Inorganic resists show edge sharpening effects similar to the image differentiation of the human eye and may bring almost a factor of two improvement in resolution.

Acknowledgment

The author would like to thank his colleague W. G. Oldham and the SAMPLE group at Berkeley for many discussions from which the rules of thumb for projection printing have evolved. Thanks also to C. G. Willson, D. C. Hofer, and M. Hakey of IBM for collaborative work on AZ2400 resist modeling and the Perkin-Elmer UV-500 projection printer. The resist model and simulations are from this collaborative project. The SAMPLE research at Berkeley was supported in part by National Science Foundation Grant ECS-8106234 and in part by an industrial grant-in-aid.

References

[*1*] Oldham, W. G., Nandgaonkar, S., Neureuther, A. R., and O'Toole, M. M., *IEEE Transactions on Electron Devices,* Vol. ED-26, No. 4, April 1979, pp. 717–722.

[*2*] O'Toole, M. M., "Simulation of Optically Formed Image Profiles in Positive Photoresist," Ph.D. thesis, University of California, Berkeley, June 1979.

[*3*] Neureuther, A. R., Jain, P. K., and Oldham, W. G., *SPIE,* Vol. 275, 1981, pp. 110–116.

[*4*] Ahlquist, C. N., Schoen, P., and Oldham, W. G., "A Study of a High-Performance Stepper Lens," *Kodak Microelectronics Seminar Proceedings,* 1979.

[*5*] Oldham, W. G., Subramanian, S., and Neureuther, A. R., *Solid State Electronics,* Vol. 24, No. 10, Oct. 1981, pp. 975–980.

[*6*] Jain, P. K., Neureuther, A. R., and Oldham, W. G., *IEEE Transactions on Electron Devices,* Vol. ED-28, No. 11, Nov. 1981, pp. 1410–1416.

[*7*] Dill, F. H., *IEEE Transactions on Electron Devices,* Vol. ED-22, No. 7, July 1975, pp. 440–444.

[*8*] Brochet, A., Dubrceucq, G. M., and Lacombat, M., *Revue Technique Thompson-CSF,* Vol. 9, No. 2, June 1977, pp. 287–335.

[*9*] Fujimori, S., *Journal of Applied Physics,* Vol. 50, No. 2, Feb. 1979, pp. 615–623.

[*10*] Bauer, J. and Haferkorn, H., *Feingerätetechnik,* Vol. 28, No. 4, 1979, pp. 166–171.

[*11*] Lin, B. J., *IEEE Transactions on Electron Devices,* Vol. ED-27, No. 5, May 1980, pp. 931–938.

[*12*] Neureuther, A. R., Hofer, D. C., and Willson, C. G., "Characterization of Mid-UV Resist: Models and Simulation," presented at Electronics Materials Conference, Santa Barbara, Calif., 24–26 June 1981.

[*13*] Hofer, D. C., Willson, C. G., Neureuther, A. R., and Hakey, M., "Characterization of the 'Induction Effect' at Mid-UV Exposure: Application to AZ2400 at 313 nm," *SPIE,* Vol. 334, Optical Microlithography, Paper 26, March 1982.

[*14*] Roussel, J. M., *SPIE,* Vol. 275, pp. 9–16, April 1981.
[*15*] Robertson, P. D., Wise, F. W., Nasr, A. N., Neureuther, A. R., and Ting, C. H., "Proximity Effects and Influences of Nonuniform Illumination in Projection Lithography," *SPIE,* Vol. 334, Optical Microlithography, Paper 5, March 1982.
[*16*] Tai, K. L., Vadimsky, R. G., Kemmerer, C. T., Wagner, J. S., Lamberti, V. E., and Timko, A. G., *Journal of Vacuum Science and Technology,* Vol. 17, No. 5, Sept./Oct. 1980, p. 1169.
[*17*] Lin, Y. C. and Neureuther, A. R., *IEEE Transactions on Electron Devices,* Vol. ED-28, No. 11, Nov. 1981, pp. 1397–1405.
[*18*] O'Toole, M. M., Liu, E. D., and Chang, M. S., *IEEE Transactions on Electron Devices,* Vol. ED-28, No. 11, Nov. 1981, pp. 1405–1410.
[*19*] Shaw, J. M. and Hatzakis, M., *IEEE Transactions on Electron Devices,* Vol. ED-25, No. 4, April 1978, pp. 425–430.
[*20*] Kyser, D. F. and Pyle, R., *IBM Journal of Research and Development,* Vol. 24, No. 4, July 1980, pp. 426–437.

Murrae J. Bowden[1] *and John Frackoviak*[1]

Considerations of Resist Processing

REFERENCE: Bowden, M. J. and Frackoviak, J., "**Considerations of Resist Processing,**" *Silicon Processing, ASTM STP 804,* D. C. Gupta, Ed., American Society for Testing and Materials, 1983, pp. 125–143.

ABSTRACT: Resist processing is taken here to refer to the variety of steps performed on the resist pursuant to defect-free pattern delineation. These processing steps include spin coating, pre-baking, exposure, post-exposure curing, development, and post-development treatment (post-baking, plasma de-scumming, etc.). Not all of these processing steps are necessarily performed, since the particular processing steps chosen, together with their operational sequence, are dictated by the chemical and physical properties of the resist. This paper discusses some of the considerations pertinent to process optimization of several electron resists and shows how processing relates to sensitivity, resolution, and defect levels.

KEY WORDS: resist processing, pattern processing, pattern delineation, electron resist, defects in patterns, photo resists

Pattern definition in a resist consists of two steps: (*1*) formation of a latent image in the resist by exposure to some form of radiation, and (*2*) development of that image to produce a relief image in the resist. The fidelity of the latent image is determined by the physical and, to a certain extent, chemical processes involved in latent image formation.[2] The quality of the developed image in the resist is determined by the physical and chemical processes involved in the various processing steps performed on the resist (Fig. 1). The variables associated with each of the resist-processing steps of Fig. 1 are listed in Table 1. Each variable must be optimized in order to achieve the best overall performance. Furthermore, many of the steps are interdependent, which makes the whole process development sequence difficult and time consuming. Nevertheless, the importance of process development cannot be overemphasized. Indeed it is quite probable that many resist materials have been discovered and subsequently rejected because insufficient process engineering was done.

[1] Bell Laboratories, Murray Hill, N.J. 07974.

[2] The italic numbers in brackets refer to the list of references appended to this paper.

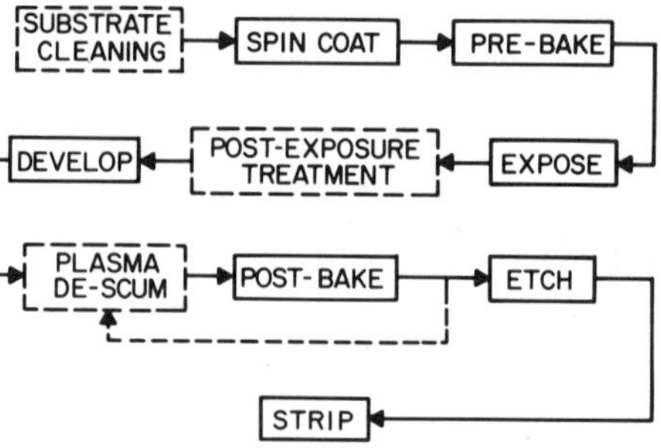

FIG. 1—*Resist-processing flow chart.*

Once the pattern has been defined in the resist, it is then transferred to the underlying substrate. The chemical and physical properties of the resist may well influence the type of etching process that can be used or the etching conditions. However, since these choices are made in order to retain the developed resist pattern quality (and hence ensure fidelity of pattern transfer), they will not be discussed in this paper, whose scope will be limited only to the variety of processing steps pursuant to pattern delineation in the resist. Since much of our processing experience has been primarily concerned with electron resists, the discussion of processing variables as they relate to sensitivity, resolution, and defect levels will mainly refer to these materials, in particular, polybutene-1-sulfone (PBS), novolac-polysulfone resist (NPR), polyglycidyl methacrylate-co-ethyl acrylate (COP), polyglycidyl methacrylate (PGMA), and polyglycidyl methacrylate-co-chlorostyrene (GMC). The conclusions, however, are completely general and should be applicable to any resist system. Although a full discussion of each processing parameter is not

TABLE 1—*Resist-processing variables.*

Step	Processing Variables
Spin coat	molecular weight, solution viscosity (molecular weight, concentration), solvent, speed and acceleration, temperature, ambient atmosphere
Pre-bake	polymer T_g and T_m, polymer stability, spinning solvent composition, time, temperature, atmosphere
Exposure	dose, dose rate, atmosphere
Post-exposure curing	atmosphere, time, temperature
Development	developer composition, rinse composition, spray conditions (pressure, flow rate, droplet size), dipping conditions (agitation, developer age), temperature, atmosphere, time, (developer, overlap in development cycle), drying
Post-bake	same as pre-bake
Plasma de-scum	reactor configurations, time, gas composition, temperature, power frequency

possible here, several of the major variables are discussed with reference to specific examples.

Spin-Coating

The spin-coating process is widely used for depositing the resist film on the surface of a substrate. The process is relatively simple and convenient and produces uniform films which are quite reproducible. Film uniformity depends on several factors including spinning parameters (speed and acceleration, temperature and atmosphere), solution viscosity, and the nature of the spinning solvent. The film uniformity across a single wafer and from wafer to wafer must be within ±150 Å in order to assure reproducible line widths and development times. The major variables affecting film thickness are the molecular weight of the polymeric resist, solution concentration, and spinning speed. The first two are material variables which, together with the type of spinning solvent, determine the solution viscosity. The effect of these variables on thickness (t) has been described by the empirical expression [*2*]

$$t = \frac{KC^{\alpha}}{\omega^{\gamma}} \tag{1}$$

where K, α, and γ are constants that must be determined for a given polymer, solvent, and spinner, respectively; C is the solution concentration, and ω is the spinning speed. The effect of molecular weight can be included as

$$t = \frac{KC^{\alpha}[\eta]^{\beta}}{\omega^{\gamma}} \tag{2}$$

where $[\eta]$ is the polymer intrinsic viscosity which is related to molecular weight via the Mark-Houwink expression $[\eta] = K'M^{a}$. Once the empirical constants have been determined for a given polymer-solvent combination, the equation can be used to predict the film thickness for various concentrations, molecular weights, and spinning speeds.

The choice of solvent is also extremely important. The solvent volatility (vapor pressure) must be such that a uniform film thickness develops during spinning. Solvents that are highly volatile and evaporate rapidly result in too rapid an increase in viscosity, thereby preventing the film from forming with uniform thickness over the whole of the substrate. This problem becomes more severe as substrate size increases.

It is possible for the solvent and spinning conditions to affect the defect density. For example, it was found during studies of acetate-based solvents for PBS resist solutions that chromium masks made using resist spun from such solutions contained an unacceptably high number of circular opaque spots. This problem was usually observed when the temperature of the resist solution during spin coating was above room temperature. When the resist was spun at temperatures 5 to 10°C above room temperature (that is, above ~20°C) and the mask blank subsequently processed, an extremely high opaque spot density was observed (Fig. 2*a*). Masks made using resist coated at intermediate temperatures showed an opaque spot density that decreased with decreasing temperature, eventually becoming zero again for resist which had equilibrated at room temperature prior to spinning (Fig. 2*b*).

This phenomenon was attributed to agglomeration of polymer molecules

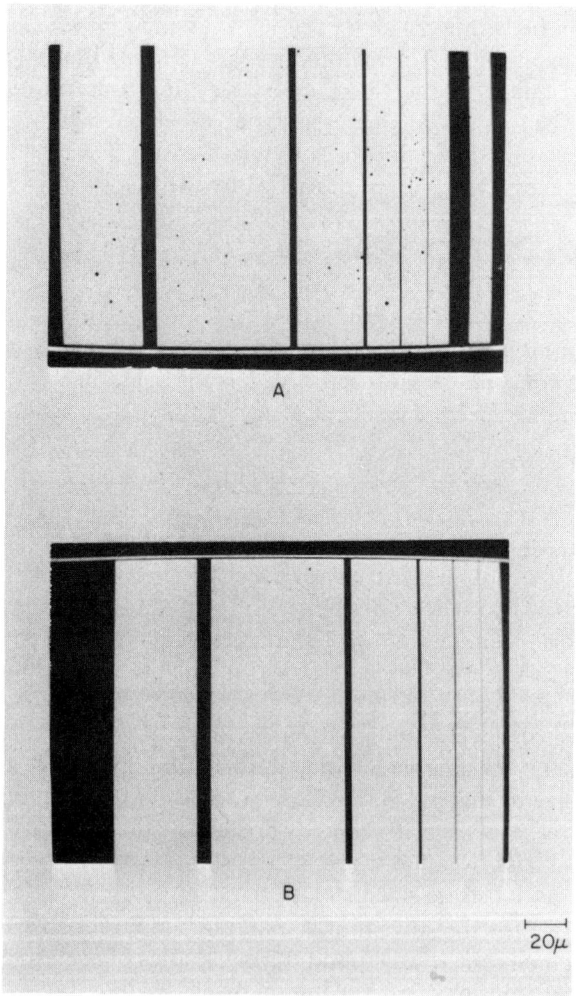

FIG. 2—(a) *Opaque spots in fully processed mask. Pattern was generated in PBS resist spun at slightly elevated temperature.* (b) *Same pattern using PBS resist spun at room temperature.*

at the high molecular weight end of the distribution. Solvents are generally characterized as good or poor, depending on the size of the polymer coil in solution. In a good solvent, the coil is relatively expanded, corresponding to a large value of the mean square end-to-end distance or radius of gyration. As the solvent becomes progressively poorer, the coil shrinks until the polymer and solvent are no longer miscible and the polymer precipitates. Generally, a solvent becomes poorer with decreasing temperature until a point called the θ-temperature is reached below which precipitation of the polymer will occur. The large chains usually precipitate first, with further cooling precipitating out chains of lower molecular weight. Hence this phenomenon forms the basis of a technique for polymer fractionation. There are some solvent systems, however, which display a lower critical solution temperature; that is, the θ-condition is approached as the temperature is increased. Ivin et al [*3*], for example, reported that PBS shows a lower critical solution temperature effect in acetate solvents (23°C for butyl acetate).

Subsequent experiments showed that this was the phenomenon causing opaque spots in PBS. The resist solutions developed a marked turbidity on heating to 40°C, which indicated that the θ-temperature for the particular solvent had been exceeded. No macroscopic precipitate formed; presumably the higher molecular weight chains remained aggregated in colloidal suspension. When this solution was spun onto a mask blank, these regions of aggregation were not broken up by subsequent irradiation. Consequently they stayed behind after development as spots of resist in the developed areas, ultimately giving rise to the opaque spots on subsequent etching. In this case, a simple change of spinning solvent eliminated the problem.

Pre-Baking

After the resist film has been deposited on the substrate it is baked to remove residual solvent and to anneal the polymer, thus promoting adhesion. Typically the polymer is pre-baked above the glass transition temperature (T_g). The upper temperature limit is determined by thermal stability considerations. PBS, for example, begins to decompose at 130°C [*4*]; therefore prebake temperatures must be kept lower than this. Polymethyl methacrylate (PMMA), on the other hand, can be baked as high as 160°C for short periods of time (<1 h) [*5,6*]. Negative resist materials such as COP or GMC are even more constrained by temperature considerations because of thermally induced crosslinking which can lead to a thin insoluble resist layer over the surface of the substrate. This is referred to as resist fogging.

Exposure

In the exposure step, the resist-coated substrate is placed in the exposure system and irradiated with a pre-determined dose of radiation. The exposure dose must be carefully controlled, since this parameter has a marked effect

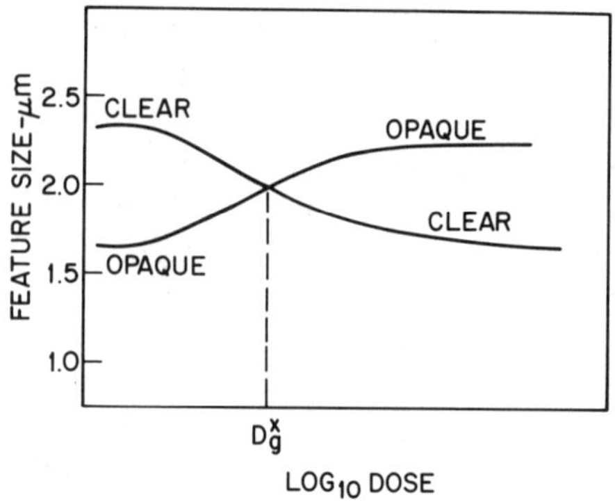

FIG. 3—*Line width of clear and opaque features for COP as a function of exposure dose.*

on the resulting size of the resist image; this effect is shown for COP in Fig. 3. The correct exposure dose in this example is where the coded 2-μm features are of equal size for both opaque and clear lines. Trade-offs are frequently required between dose, initial thickness, development, and post-baking. In negative resists, for example, a final working thickness can be achieved by exposing a thick sample to a low dose or a thin sample to a high dose. This aspect will be discussed later.

It appears that both the crosslinking and scission density are dose-rate independent at the beam currents encountered in electron lithography. Hence one would expect the sensitivity of both negative and positive resists to be dose-rate independent. This conclusion is shown to be true in Figs. 4 and 5

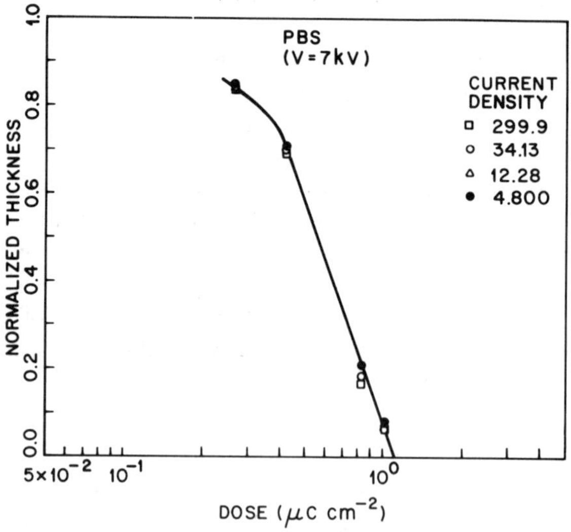

FIG. 4—*Effect of beam current (dose rate) on the lithographic response of PBS.*

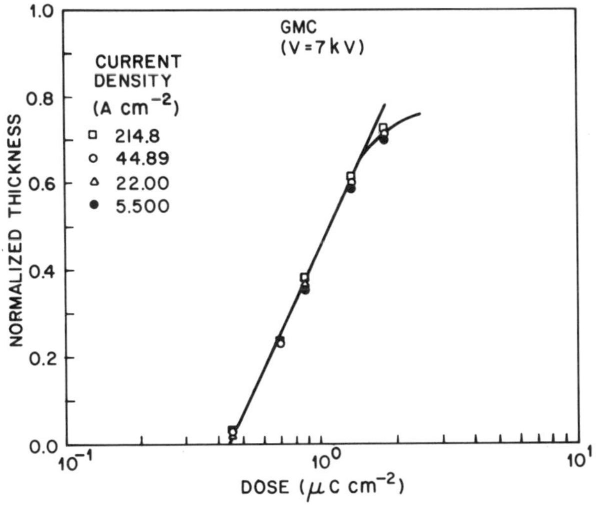

FIG. 5—*Effect of beam current (dose rate) on the lithographic response of GMC.*

for PBS and GMC, respectively. Thus the approach of machine designers of increasing machine through-put by increasing both scanning speed and beam current is practicable. The same is not necessarily true for vapor-developing resists such as poly(2-methyl-1-pentene sulfone) [*7,8*]. In this case the process seems more efficient at lower dose rates (Fig. 6). Further, marked differences in vapor development rate are observed, depending on whether

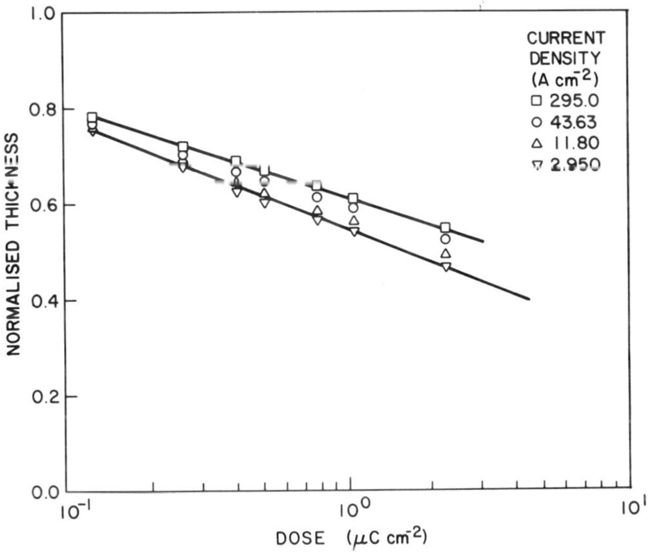

FIG. 6—*Effect of beam current (dose rate) on the lithographic response of PMPS.*

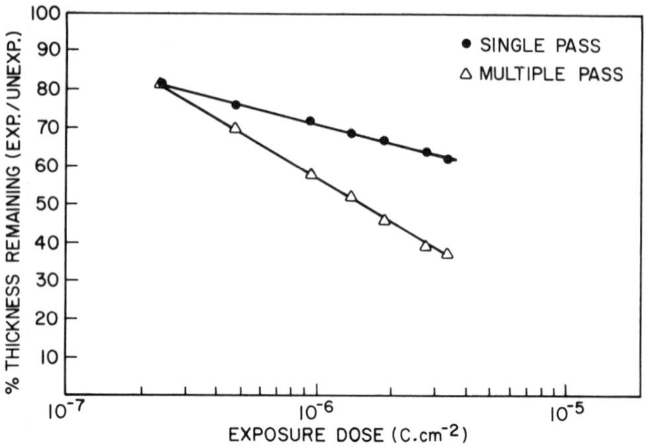

FIG. 7—*Effect of single pass versus multiple pass on vapor development rate of PMPS.*

the dose is received in a single pass (assuming a fixed writing rate) of the beam or cumulated over several passes. The latter method is considerably more efficient (Fig. 7), possibly because the unzip time is finite relative to the dwell time of the beam [*8*].

Post-Exposure Curing

Lithographic exposure systems that operate with the substrate in a vacuum or inert gas often place an additional constraint on the resist process due to the phenomenon of post-exposure curing. This phenomenon is mainly seen with negative resists, particularly those such as COP or PGMA which

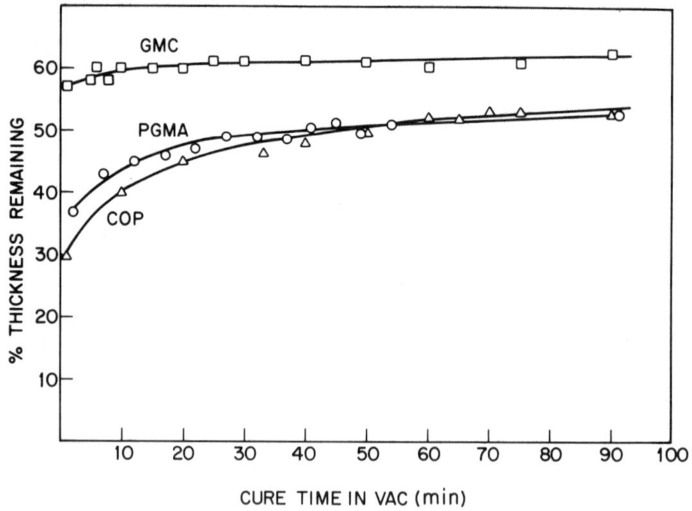

FIG. 8—*Post-curing curves for COP, PGMA, and GMC.*

have either a low T_g and/or a chain reaction mechanism for cross linking [*9,10*]. Post-curing curves for these resists are shown in Fig. 8. The phenomenon is caused by radicals and/or ions produced during irradiation which are trapped in the solid polymer film and which continue to induce the desired reaction after exposure has been completed. Since controlling the effect of the radiation-induced reaction is important if critical dimensions are to be maintained within specified tolerances, it is often necessary to control the length of time each wafer spends in the inert atmosphere. Clearly it would be preferable to eliminate the effect either by using positive resists or through design modification of the negative resist. The post-curing reaction is greatly reduced in GMC (Fig. 8) relative to PGMA and COP, presumably because of the higher T_g of this copolymer over that of COP or PGMA. Similar observations have been made for other high T_g negative resists [*9,11*].

It has been found that post-exposure baking after exposure but prior to development can improve lithographic performance for some systems. NPR, for example, is a positive resist consisting of two components (a novolac resin and a polymeric dissolution inhibitor) in which differential solubility is achieved by removal of the dissolution inhibitor [*12*]. The latter is poly(2-methyl-1-pentene sulfone), which undergoes scission and depolymerization under electron irradiation. The depolymerization process has been termed vapor development [*13*]. Following scission of the main chain the polymer depolymerizes to monomer which subsequently diffuses out of the matrix. A vapor-developing resist thus provides a convenient process for material removal from the matrix, thereby restoring solubility to the exposed area which now consists largely of pure novolac resin. It has been found that the sensitivity of NPR resist increases by a factor of 5 as a result of post-exposure baking at 120 to 130°C (Fig. 9). This is attributed to additional depolymerization resulting from release of trapped radicals at the elevated temperature.

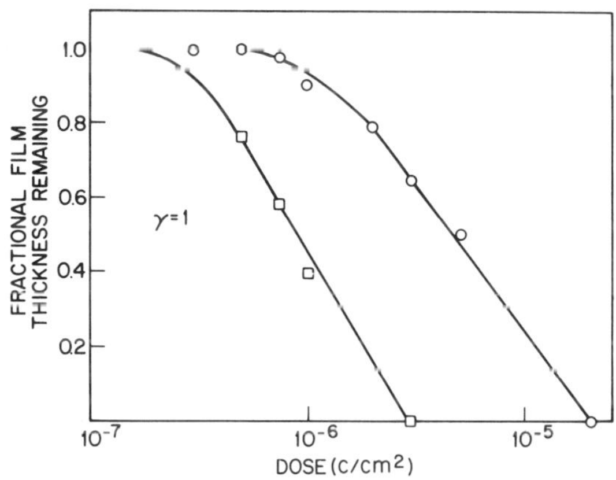

FIG. 9—*Lithographic response for NPR resist* [12] *as a function post-exposure treatment.* (O) *NPR film as exposed.* (□) *NPR film exposed and baked at 120°C for 15 min prior to developing.*

Development

The developing process is the most complex of the processing steps, has the greatest influence on pattern quality, and requires the most process-development time. Both liquid and dry (plasma) development schemes have been used, the former being more widely employed in production. Dry development of resists has been recently reviewed by Taylor [*14*]. Two types of liquid development have been used, spray and immersion. Spray development generally results in better pattern uniformity, process reproducibility, and lower defect densities and is widely used in e-beam lithographic mask-making.

In order to optimize a developing process, it is first necessary to have a set of developing criteria. For example, the following criteria were chosen for the positive resist PBS [*15*]: (*1*) no reduction in original film thickness, (*2*) minimum development time (<1 min), and (*3*) minimum distortion and swelling. The first criterion corresponds to a solubility rate ratio (S_R) (defined as the ratio of the dissolution rate in the irradiated area to that in the unirradiated area) of infinity. In other positive resist systems, particularly PMMA, finite values of S_R as low as 3 have defined the developing criterion. Such a situation is referred to as "force developing".

Of the various developing criteria, the requirement of minimum distortion and swelling is common to all resist systems, since this criterion has a direct bearing on pattern quality and places great importance on the selection of a developer. During normal dissolution, solvent molecules penetrate into the polymer structure by a diffusion process, which produces a swollen surface layer between the solvent and virgin polymer. In the case of positive resists, minimum distortion implies minimum swelling in the regions immediately adjacent to the irradiated area which is removed by dissolution. For negative resists, it implies minimum swelling of the irradiated region during extraction of the soluble (sol) fraction from the gel.

Normally the suitability of a solvent is discussed in thermodynamic and kinetic terms [*16*]. The kinetic influence is reflected in the value of the diffusion coefficient, while the thermodynamic suitability influences the thickness of the swollen layer. However, very little work has been done on quantifying the effect of these parameters on the kinematics of polymer dissolution. Tu and Ouano [*17*] described the dissolution of glassy polymers by a phenomenological model in which they showed that the rate of dissolution and the extent of swelling depend markedly on the concentration dependence of the diffusion coefficient of solvent in the polymer. For most systems the relevant material parameters are not known and the choice of developing solvents remains largely empirical. Figures 10 and 11 show some examples of how solvent choice can influence the quality and appearance of PBS resist images.

Swelling of the polymer during development severely limits the resolution of a negative resist. This is shown in Figs. 12, 13, and 14, which compare resolution in three negative resists (PGMA, COP, and GMC) for the same in-

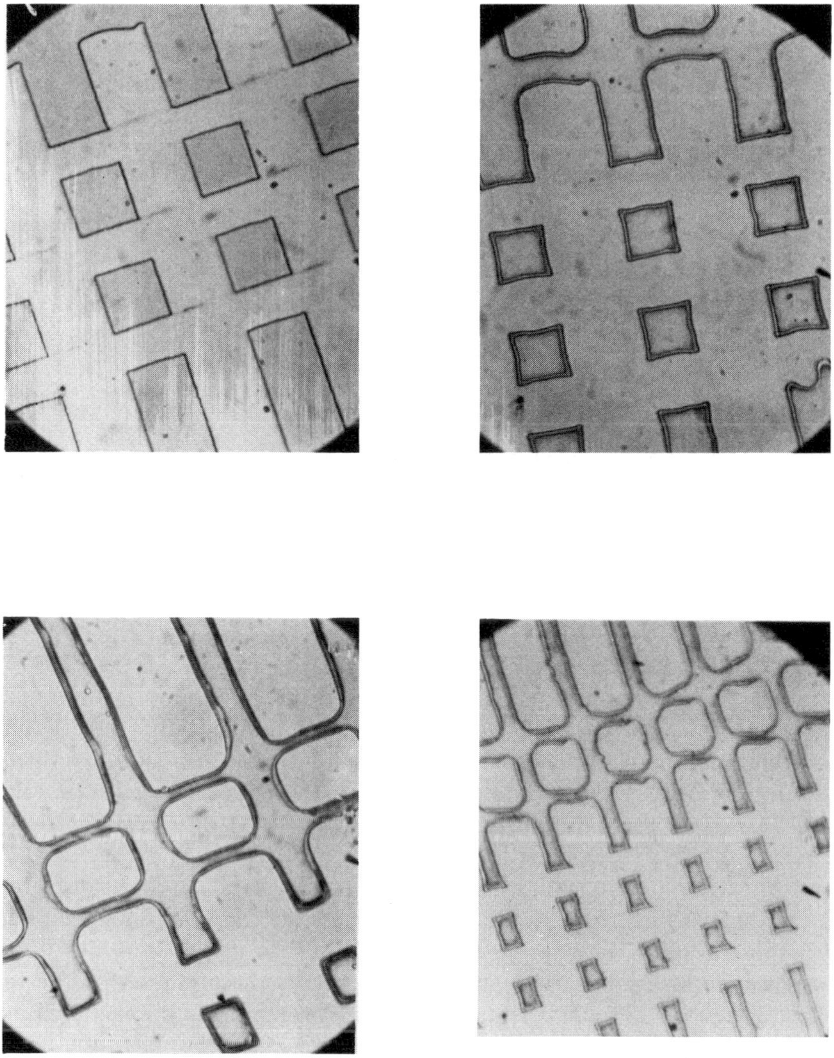

FIG. 10—*Effect of developing solvent on pattern quality of PBS for four different ketonic developers.*

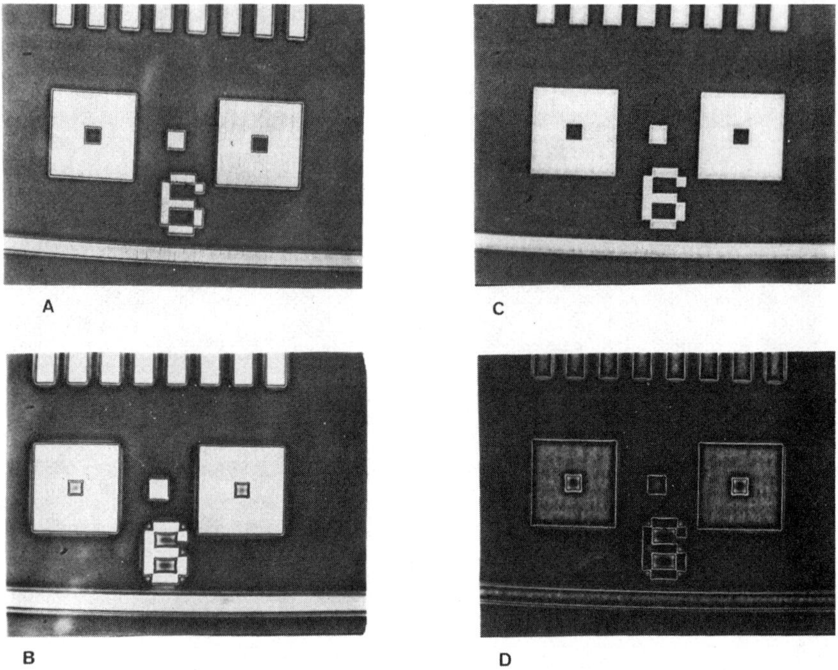

FIG. 11—*Effect of developing solvent on pattern quality of PBS.* (a) *and* (b) *1 and 3 min in Developer* A. (c) *and* (d) *1 and 3 min in Developer* B.

itial resist thickness developed in an MEK/ethanol mixture whose composition has been optimized for each resist. Figures 12 to 14 are for a nominal 1-μm line-space pattern. As the dose is increased, the line width increases but adjacent lines merge well before equal lines and spaces have been obtained. In the case of GMC-1, for example, the "optimum" dose for an initial film thickness of 0.5 μm on a chromium substrate is 4.2×10^{-6} C/cm^2 at 20 kV, which results in 0.7-μm lines and 1.3-μm spaces. The final thickness is ~0.35 μm. At higher doses the lines begin to develop a "skirt", and at still higher doses there is sufficient swelling to cause contact and adhesion between neighboring lines. The "optimum" dose profile for each resist in Figs. 12 to 14 is shown in Fig. 15. Also shown in Fig. 15 is the optimum profile for GMC-2, a more sensitive version of GMC-1. A similar resolution could only be achieved with COP at a final thickness of 0.15 μm owing to the much lower contrast of COP. Clearly additional processing is required to broaden the lines. This may be conveniently accomplished by post-baking.

One problem with very high resolution resists such as GMC is shown in Fig. 16, which shows resolution of the address structure in irradiated areas. The contrast is high enough so that the overlap region for adjacent scans receives a dose insufficient to cause significant gel formation. The "gaps" be-

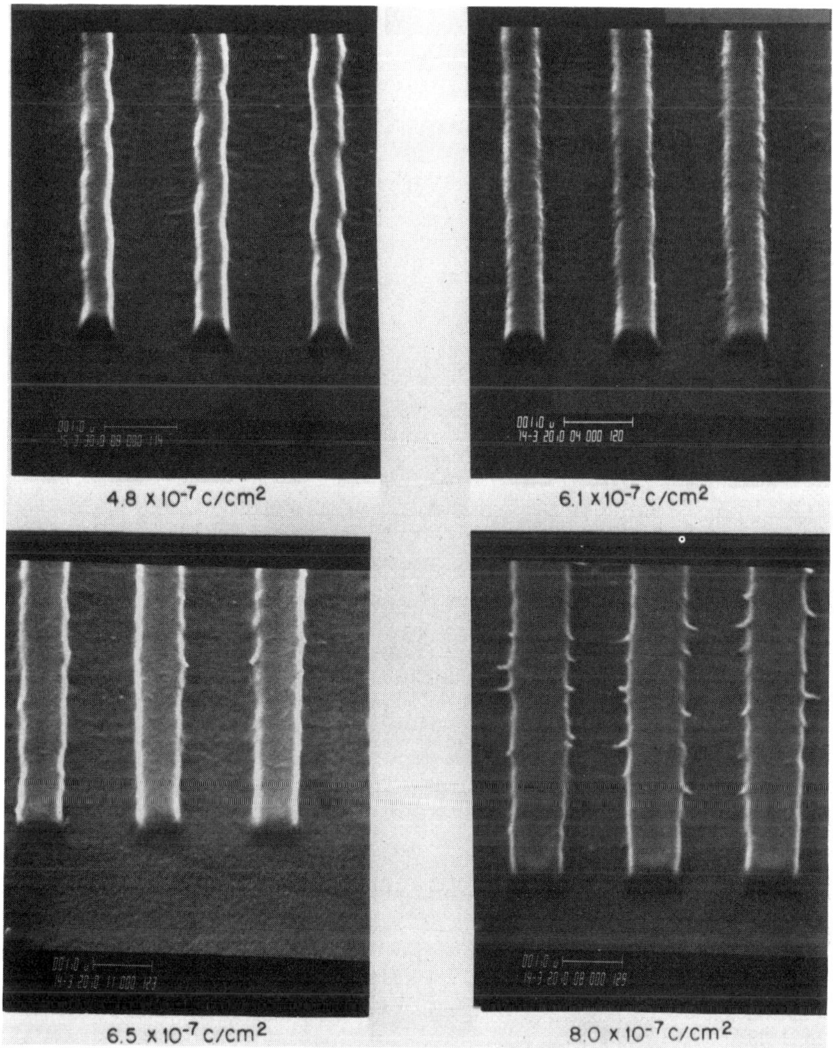

FIG. 12 *COP resist profiles as a function of dose.*

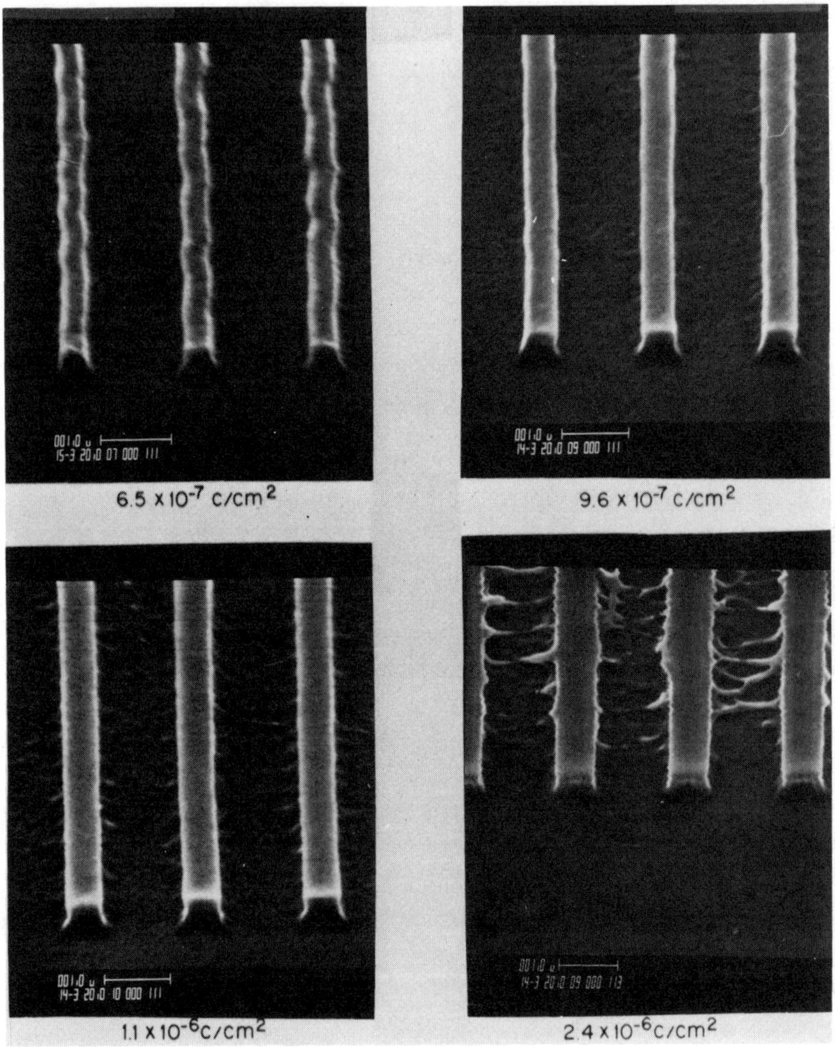

FIG. 13—*PGMA profiles as a function of dose.*

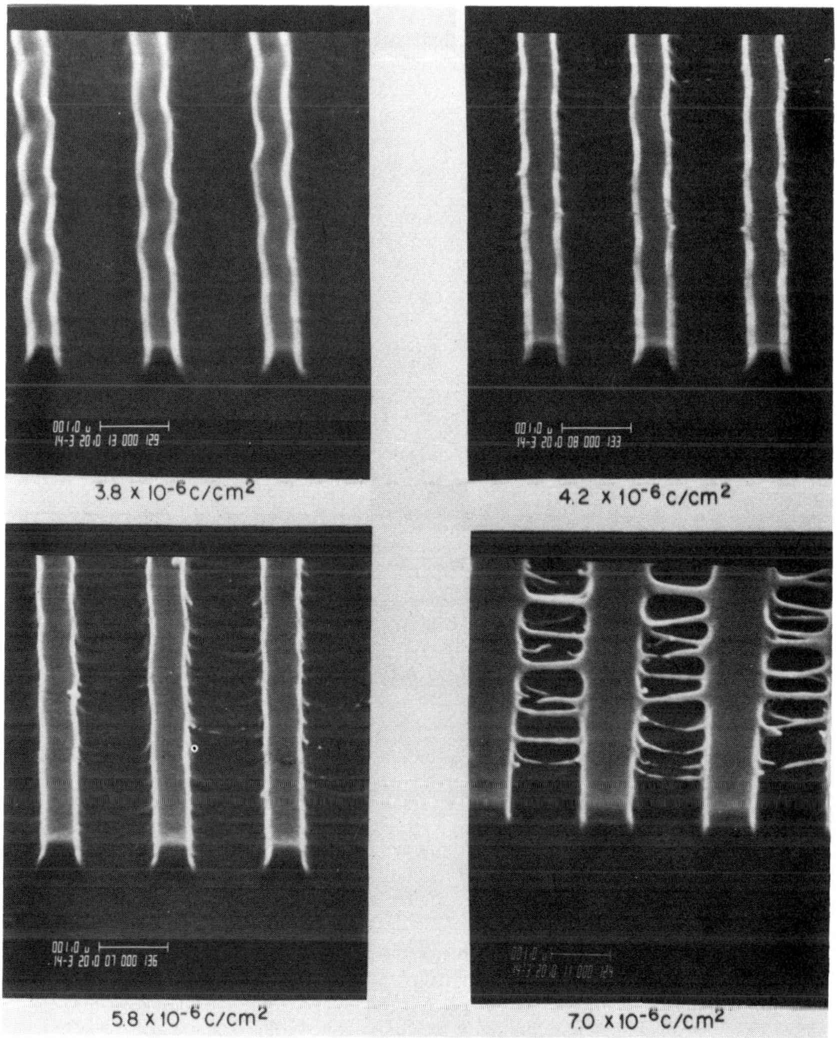

FIG. 14—*GMC-I profiles as a function of dose.*

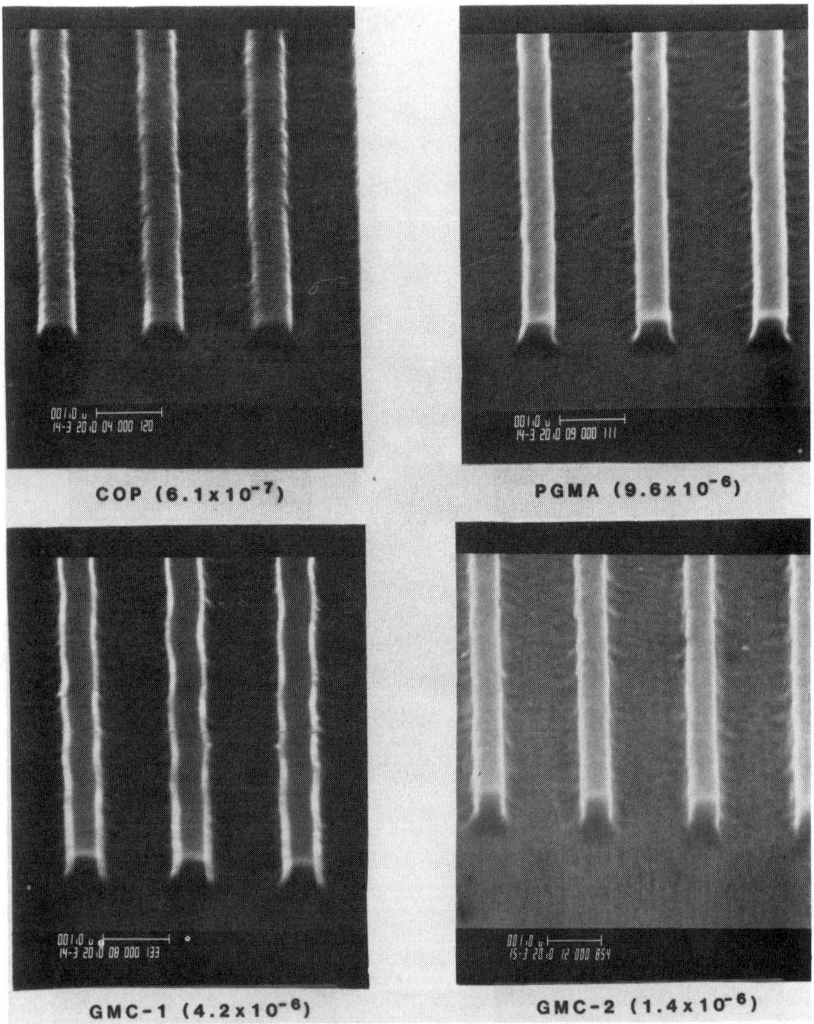

FIG. 15—*"Optimum" dose profile for COP, PGMA, and GMC.*

FIG. 16—*"Swiss cheese" effect in GMC.* (a) *As developed.* (b) *No bake and de-scummed.* (c) *Baked and de-scummed.*

tween the scan lines are not obvious during development, since neighboring regions are considerably swollen. If the sample is immediately quenched in the rinse, however, the lines retract, thereby giving rise to the so-called "Swiss cheese" effect (complete separation is impeded by considerable adhesion between adjacent lines). This effect can be eliminated by the judicious choice of rinsing conditions, by post-baking schedules giving patterns such as shown in Figure 16*c*, or by the use of a smaller address structure. Whichever method is selected, it is obvious that elimination of the "Swiss cheese" effect requires considerable attention to processing details.

Post-Baking

After development, it is necessary to remove residual solvents and anneal the film to promote adhesion. The limitations are again similar to those for pre-baking, with the added requirement that the post-bake temperature must be less than the polymer flow temperature, particularly when the correct pattern dimension has been achieved during development. As discussed earlier for negative resists, this may not be possible and the sample must be deliberately flowed to achieve correct pattern dimensions. This has an added advantage for negative resists, since additional thermally induced crosslinking is not necessarily undesirable.

Conclusions

Resist processing constitutes the most tedious, laborious, and time-consuming aspect of resist research, yet this aspect of the work is vitally important to the overall implementation of a resist in a production environment. Resist discovery contains all the glamour, resist processing contains only hard work. Nonetheless, the success or failure of a resist rests, to a large degree, on being able to sort out the myriad interacting parameters which together constitute processing.

Acknowledgments

The authors wish to thank M. G. R. Thompson for high dose rate exposures and L. R. Thibault for the data presented in Figs. 4 and 5.

References

[*1*] Bowden, M. J., *Journal of the Electrochemical Society,* Vol. 128, No. 5, 1981, p. 195C.
[*2*] Lai, J. H., *Polymer Engineering and Science,* Vol. 19, No. 5, 1979, p. 1117.
[*3*] Ivin, K. J., Ende, H. A., and Meyerhoff, G., *Polymer,* Vol. 3, 1962, p. 129.
[*4*] Bowden, M. J., Thompson, L. F., Robinson, W., and Biolsi, M., *Macromolecules,* in press.
[*5*] Greeneich, J. S., *Journal of the Electrochemical Society,* Vol. 122, 1975, p. 970.
[*6*] Ouano, A. C., *Preprints, ACS Division of Organic Coatings and Plastics Chemistry,* Vol. 45, 1981, p. 431.
[*7*] Bowden, M. J., paper presented at Electronic Materials Conference, Santa Barbara, Calif., June 1981.

[*8*] Bowden, M. J. and Frackoviak, J., to be published.
[*9*] Shiraishi, H., Taniguchi, Y., Horigome, S., and Nonogaki, S., *Polymer Engineering and Science,* Vol. 20, No. 16, 1980, p. 1054.
[*10*] Taniguchi, Y., Hatano, Y., Shiraishi, H., Horigome, S., Nonogaki, S., and Naraoka, K., *Japanese Journal of Applied Physics,* Vol. 18, No. 6, 1979, p. 1143.
[*11*] Ohnishi, Y., *Journal of Vacuum Science and Technology,* Vol. 19, No. 4, 1981, p. 1136.
[*12*] Bowden, M. J., Thompson, L. F., Fahrenholtz, S. R., and Doerries, E. M., *Journal of the Electrochemical Society,* Vol. 128, No. 6, 1981, p. 1304.
[*13*] Bowden, M. J. and Thompson, L. F., *Journal of Applied Polymer Science,* Vol. 17, 1973, p. 3211.
[*14*] Taylor, G. N., *Solid-State Technology,* Vol. 23, No. 5, 1980, p. 73.
[*15*] Bowden, M. J., Thompson, L. F., and Ballantyne, J. P., *Journal of Vacuum Science and Technology,* Vol. 12, No. 6, 1975, p. 1294.
[*16*] Ueberreiter, K. in *Diffusion in Polymers,* J. Crank and G. S. Park, Eds., Academic Press, New York, 1968, p. 218.
[*17*] Tu, Y. O. and Ouano, A. C., *IBM Journal of Research and Development,* Vol. 21, No. 2, 1977, p. 131.

Silicon CVD Techniques

James H. Williams[1]

Epitaxy: An Introspective Review

REFERENCE: Williams, J. H., "**Epitaxy: An Introspective Review,**" *Silicon Processing, ASTM STP 804,* D. C. Gupta, Ed., American Society for Testing and Materials, 1983, pp. 147–150.

ABSTRACT: The evolution of reactor design and gas control techniques for the deposition of silicon epitaxial layers is described. An approach to controlling a large epi factory with digital electronics is discussed. From the viewpoint of ever-growing technology and quality requirements, the present approach is to furnish the options for an automatic operation in an individual reactor.

KEY WORDS: silicon epitaxy, epi facility, DDC system, horizontal reactor, vertical reactor, barrel reactor

The history of silicon epitaxy is basically one of evolution of reactor size and of gas control techniques. An introspective review is one in which the progress is viewed from the standpoint of a single facility.

In our early epi experience, gas control systems were fabricated by using common materials, in contrast to the exotic materials being evaluated by many other investigators. Systems were completely manual and of very low capacity. The mid-1960s ushered in attempts at automation using analog controllers[2] and newly developed miniature valves and flow indicators.

After investigating analog control techniques, the control of a large epi factory with digital electronics was pursued. The functional layout of such a system is shown in Fig. 1. All operations of the reactors are performed in a direct digital control mode (DDC). No routine manual operation of the reactors is provided, and this has proven to be a reliable design criterion due to the system's high reliability.

This DDC system has the capability of:

1. Multiple *P* and *N* layers with programmed impurity grading in each layer.
2. Preprogrammed storage of 500 recipes.

[1] Manager, Materials Development, Motorola Inc., Tempe, Ariz. 85282.

[2] Williams, J. H., "In-Process Control Techniques for Complex Semiconductor Structures," Technical Report AFML-TR-70-235, Air Force Materials Laboratory, Wright-Patterson Air Force Base, Dayton, Ohio, Sept. 1970.

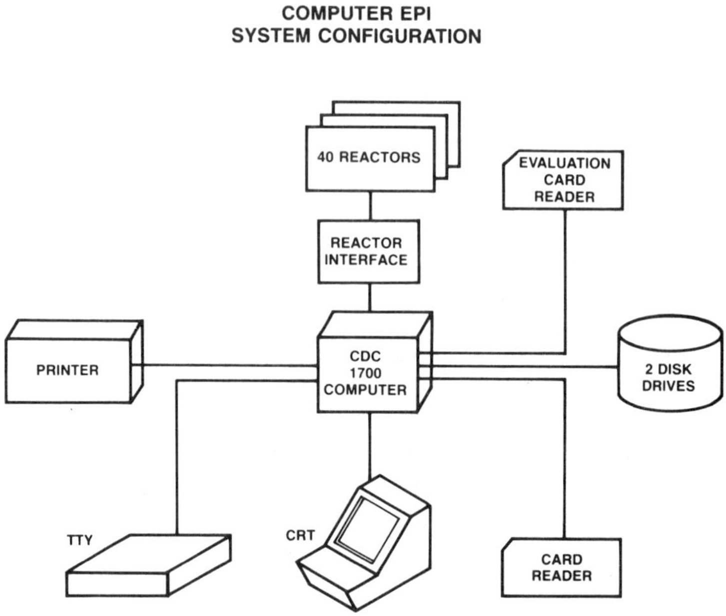

FIG. 1—*Layout of computer-controlled reactors.*

3. Complete real-time production statistics on an hourly, daily, or monthly basis.
4. Automatic correction of dopant flows based on evaluation feedback.
5. Minimal operator interaction.
6. 40-reactor capability.

To further extend the concept of automation, a self-contained epi minifactory was designed and constructed. This was a complete epitaxial system and contained the modules shown in Fig. 2. The work was done in cooperation with Wright-Patterson Air Force Base.[3] All thirteen processing modules were under the control of a central computer operating in a combined DDC and supervisory mode. The system was successfully tested and various ideas from the system were incorporated into our manufacturing operation.

The epitaxial community is gradually phasing out the horizontal reactor in favor of the vertical (pancake) and cylinder reactor types. The advantage of the newer machines is primarily one of epitaxial film quality.

Epitaxial equipment manufacturers today are furnishing the options of automatic operation of the individual reactor. The requirement to incorpo-

[3] Williams, J. H. and van Pul, B., "Manufacturing for the Continuous Epitaxial Processing of Silicon Integrated Circuits," Vol. 1, Technical Report AFML-TR-78-47, Air Force Materials Laboratory, Wright-Patterson Air Force Base, Dayton, Ohio, May 1978.

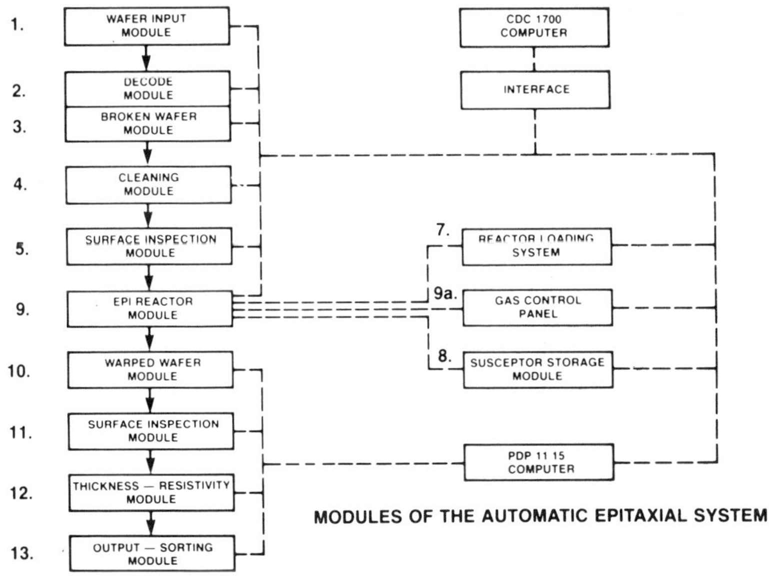

FIG. 2—*Layout of automatic epitaxial system.*

rate individual automatic reactors into a completely supervised system needs to be investigated to provide the following advantages:

1. Extended program storage of large numbers of complex epi structures.
2. More sophisticated monitoring of the gas flows and the temperatures.
3. Methods for automatically correcting for run variations.
4. Automatic data accumulation to optimize reactor productivity.
5. Documentation of maintenance tasks.
6. Collection of production statistics.

The breakdown of labor allocation is roughly shown in Fig. 3 for our factory over a two-decade period. This is a graphic example of one obvious advantage of automation, namely labor reduction. While we have not reached the labor inputs indicated after 1983, such goals are attainable.

Allocations of labor for the purposes of wafer evaluation still remains one of the significant production costs. Resistivity probes and thickness determination techniques have been singularly automated, but are not in general industry use as a system. However, these functions may be combined with a single evaluation system with cassette load and sorting.

The reactor concepts discussed above may be expanded to include the advent of radiant type heating. These cylinder reactors and the larger induction-heated vertical reactors will make up the bulk of reactor sales through the mid-1980s. A chart of comparative reactor capacities is shown in Fig. 4.

PERSONNEL REQUIRED FOR 75,000 SQ. IN. PER WEEK

SYSTEM TYPE	CLEAN	LOAD AND RUN REACTOR	EVALUATION	INSPECTION	TOTAL
ALL MANUAL 1960–1965 (33 REACTORS)	100 OPERATORS 9 SUPERVISORS		48 OPERATORS 3 SUPERVISORS	6 INSPECTORS	166
SEMI-AUTOMATIC REACTOR 1965–1972 (13 REACTORS)	21 OPERATORS 6 SUPERVISORS		21 OPERATORS 3 SUPERVISORS	3 INSPECTORS	54
AUTOMATIC REACTOR 1972–1983 (7 REACTORS)	6 OPERATORS	6 OPERATORS 6 SUPERVISORS	18 OPERATORS 3 SUPERVISORS	3 INSPECTORS	42
AUTOMATIC SYSTEM 1983–?	9 OPERATORS 3 SUPERVISORS				12

FIG. 3—*Labor allocation in a typical epitaxial factory.*

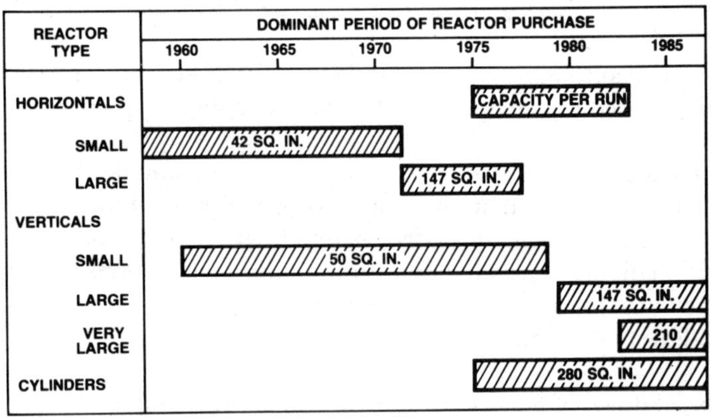

FIG. 4—*Comparison of epi reactor capacities and purchase periods.*

It would appear that the era of reactor design and manufacture by the device manufacturers will soon cease. The independent equipment manufacturing is anticipating and planning for the future, both from the quality and wafer size standpoint.

Gurumakonda R. Srinivasan[1]

Modeling and Applications of Silicon Epitaxy

REFERENCE: Srinivasan, G. R., "**Modeling and Applications of Silicon Epitaxy,**" *Silicon Processing, ASTM STP 804,* D. C. Gupta, Ed., American Society for Testing and Materials, 1983, pp. 151–173.

ABSTRACT: Demands for low-temperature, defect-free, thin silicon epitaxial films have produced significant progress in process modeling and process design. This review paper discusses recent developments with new additional data, with an emphasis on autodoping, defect control, and process-device interactions, as they relate to bipolar applications. A brief discussion of a low-temperature epitaxial process is provided, and calculations are presented to illustrate the advantages of low epitaxial temperatures in terms of profile tolerance and device performance.

KEY WORDS: silicon epitaxy, modeling, defect-free epi, autodoping, impurity distribution, bipolar devices, integrated circuits

The current trend in high-performance silicon bipolar technology is towards shallower junctions and thinner epitaxial films to achieve a reduction in junction capacitances and the charge storage in the emitter and base regions. Films with a thickness of a micrometre or less are being developed for these purposes. These thin films require very sharp dopant profiles and a tight tolerance on the variations in thickness and dopant distribution. In recent years, rapid progress has been made in the understanding of such key areas as autodoping [*1,2*], epitaxial nucleation [*3,4*], and profile modeling [*5*].[2] This review paper is set within the framework which recognizes that any additional enhancement of already mature epitaxial technology requires a deeper understanding of the process basics and a knowledge of the interplay between process and device parameters. Towards this end, we provide a discussion and some new data on the process effects for obtaining abrupt profiles and a discussion of the parameters that affect their control. Defect elimination through proper use of process chemistry is then considered, followed by a discussion of the device modeling with additional data. We end the discussion by following an epitaxial process design that illustrates the application of the process modeling.

[1] Manager, Computer-Aided Device Design Department, General Technology Division, IBM Corporation, East Fishkill Facility, Hopewell Junction, N.Y. 12533.

[2] The italic numbers in brackets refer to the list of references appended to this paper.

Process Modeling

The process of chemical vapor deposition (CVD) epitaxy for bipolar application consists of growing silicon epitaxy over substrates that contain heavily doped buried layers. These buried layers may be of the same conductivity type as the epitaxy, as in subcollectors, or of opposite type, as in subisolation diffusions.

Figure 1 shows a bipolar device structure that uses both the subcollector and subisolation as device elements. For typical large-scale-integration (LSI) applications, these buried layers are separated by a few micrometres, such that the adjacent layers do not overlap. As the epitaxy is grown over these buried layers, some of the dopant that was initially contained in them redistributes into the epitaxy through solid-state outdiffusion and through vapor phase autodoping. Figure 2 shows the vertical profile over the subcollector in which these two components are identified (X_{OD}, X_{AT}). The sharpness of the subcollector profile, together with its variations, determines several of the important characteristics of the device, including the breakdown voltage and the capacitance of the collector-base junction.

For a given species, the extent of outdiffusion (X_{OD}, see Fig. 2) is determined by the epitaxial growth heat cycle and by the initial dopant distribution in the buried layer prior to epitaxy. Figure 3 shows the effect of epitaxial deposition temperature on the outdiffusion of arsenic subcollector for a 2-μm-thick epitaxy that was grown at ~0.1 μm/min. The value of X_{OD} falls off rapidly with a decrease in the deposition temperature (T_{DEP}), from about 1 μm at 1150°C to about 0.1 μm at 1000°C. Below 1000°C, however, the change in X_{OD} with temperature becomes small. Thus, for shallow device applications, epitaxial temperatures of the order of 1000 to 1050°C are most desirable.

The outdiffusion portion of the buried layer profile can be modeled by using numerical solutions to the moving boundary transport equation. On

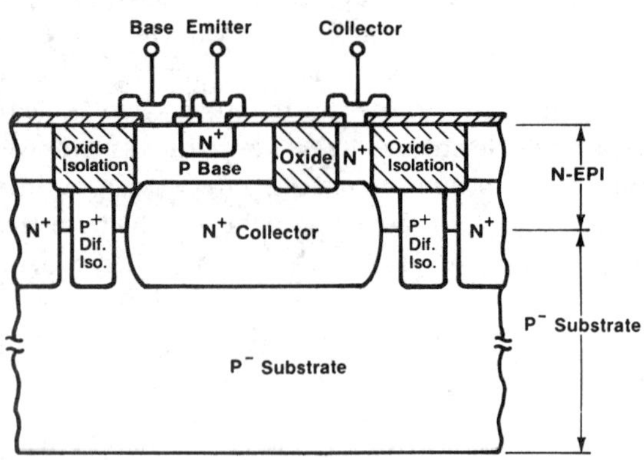

FIG. 1—*A typical NPN bipolar device.*

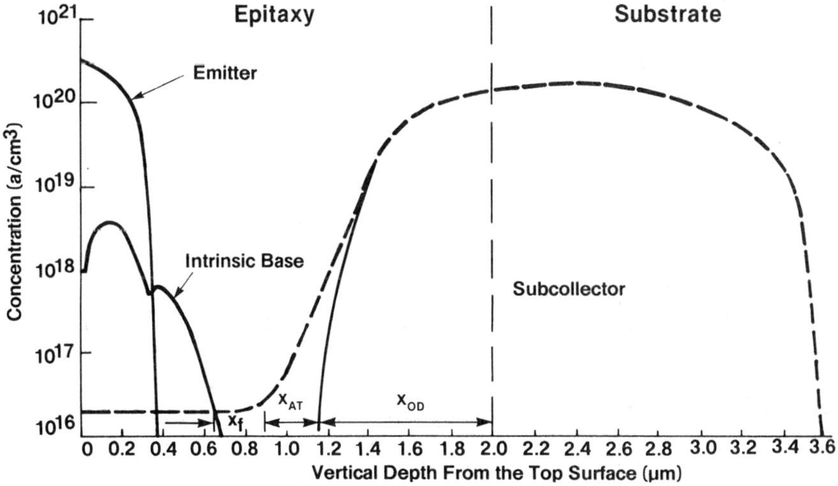

FIG. 2—*Vertical profiles from a high-performance bipolar device.*

the other hand, the autodoping tail (X_{AT}) is somewhat hard to model [*5*]. It depends on several interacting parameters that include both the process as well as substrate characteristics, such as predeposition and deposition heat cycles, growth rate, buried layer density, and number of wafers in the epitaxial run. Nevertheless, since the useful thickness of an epitaxy for device fabrication is determined not only by the actual thickness but also by the extent of the autodoping tail (X_{AT}, see Fig. 2), it is important to understand the parameters that control X_{AT}.

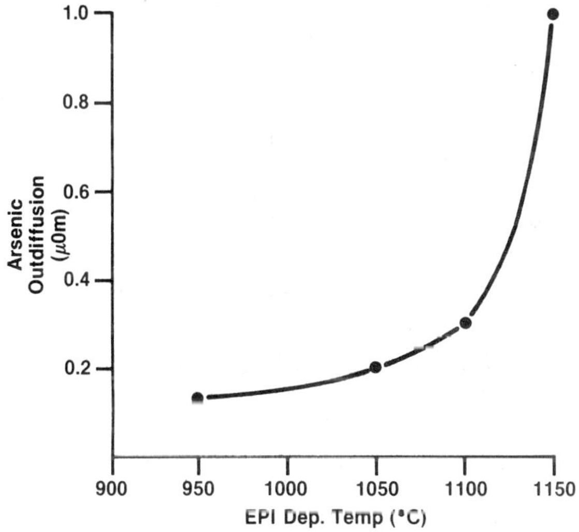

FIG. 3—*Variation of arsenic subcollector outdiffusion with epitaxial deposition temperature.*

Prebake Effects

It has been shown [*1*] that most of the vapor phase autodoping occurs during the thermal cycle prior to the epitaxial deposition (that is, during prebake). The kinetics of the vapor phase autodoping process can be quantitatively modeled [*6*] as expressed by the equation

$$C_A = \text{constant}\ \exp(\beta^2 t_p) \cdot \text{erfc}(\beta t_p^{1/2}) \tag{1}$$

where C_A is the autodoping concentration (maximum concentration in the lateral autodoping peak), t_p is the prebake time, and $\beta = \beta_0 \exp(-\Delta E/kT_p)$ is the ratio of the evaporation velocity to the square root of the dopant diffusivity in silicon in which T_p is the prebake temperature, k is Boltzmann constant, and $\Delta E = E_k - E_D/2$, where E_k and E_D are the activation energies for the evaporation from and diffusion of the dopant in the buried layer. By using this model, it can be shown[3] that about 15% of the arsenic in the buried layer evaporates during a typical prebake cycle of 10 min at 1150°C.

Equation 1 clearly shows that vapor phase autodoping can be reduced by a prolonged prebake cycle at elevated temperatures. Thus it is possible to design a prebake cycle to achieve a given low level of autodoping that is consistent with a given value of the sheet resistance of the subcollector.

Effect of Deposition Temperature

We have already seen how the deposition temperature (T_{DEP}) affects the outdiffusion X_{OD}; that is, X_{OD} decreases with a decrease in T_{DEP}. However, the effect of T_{DEP} on vapor phase autodoping, and hence on the autodoping tail X_{AT}, is entirely different. The vapor phase autodoping *increases* with a decrease in T_{DEP}. The author has analyzed this effect for arsenic by using off-buried layer profiles [*6*], and has shown that this effect can not be explained by an assumption of a thermodynamic equilibrium between the arsenic in the vapor phase and that in the growing epitaxial silicon. The study indicated, instead, a kinetic limiting step to the incorporation process. This effect was later confirmed by Ogirima [*7*], who proposed an adsorption model. However, since in this model the prebake effects and the growth effects were not separated, and since these effects are expected to have different temperature dependencies, accurate modeling of the temperature effect of autodoping is still in a state of development.[3]

The opposing effects of the epitaxial deposition temperature on the outdiffusion and on the autodoping tail parts of the buried layer profile are clearly demonstrated in Fig. 4, which shows arsenic subcollector spreading resistance profiles for a silicon epitaxy that was grown at temperatures in the

[3] G. R. Srinivasan, unpublished results.

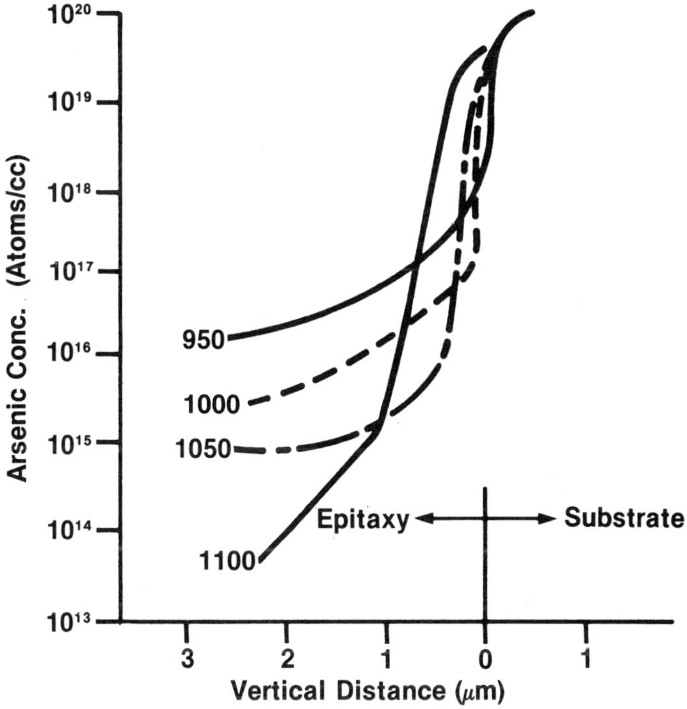

FIG. 4—*Arsenic subcollector profiles in the epitaxy grown at 950, 1000, 1050, and 1100°C (no external dopant added).*

range of 950°C to 1100°C. Depositions in this case were made using $SiCl_4$, with no external doping of the epitaxial layers. We see from this figure that as the deposition temperature is reduced, the subcollector profiles get closer to the substrate-epitaxy interface, while at the same time the level of autodoping increases significantly. It is this increase in vapor phase autodoping that makes the use of low-temperature CVD epitaxy restrictive, unless special techniques are developed to control this increase.

Growth Rate Effects

It has been known for some time that vapor-phase autodoping can be decreased by a decrease in the epitaxial growth rate. In a quantitative formulation of this effect, the author [*1,6*] showed that a model based on dopant trapping is capable of explaining this effect quantitatively. It was shown that the fraction trapped in the growing layers increases exponentially with the growth rate until saturation occurs for large values of the growth rate (see Fig. 5). Also, the trapping is a function of epitaxial growth temperature that defines a characteristic growth rate at which the dopant atoms diffuse to the

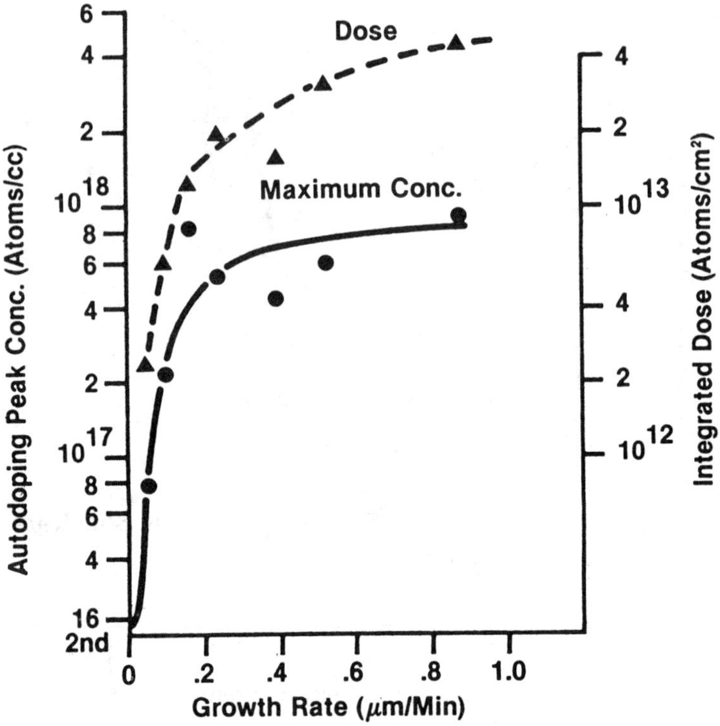

FIG. 5—*Variation of autodoping peak concentration and integrated dose (area under the autodoping peak) with growth rate. Deposition temperature 1100°C. The solid line corresponds to the impurity-trapping model* [1].

surface. This model predicts that at higher deposition temperatures, the dopant trapping would be lower, thus explaining the reduction in autodoping with increasing deposition temperatures as mentioned earlier.

Yet another model that has been proposed is the adsorption model due to Tabe and Nakamura [*8*]. This model, in contrast to the impurity-trapping model mentioned earlier, assumes a thermodynamic equilibrium between the impurity in the absorbed layer and that in the growing epitaxy. However, this model yields a decrease, rather than an increase, in autodoping with increased growth rate, and is contrary to observations in the buried layer studies.

Effect of Pressure

It has been found that arsenic autodoping decreases almost linearly with a decrease in reactor pressure [*9*]. The linearity of this variation suggests an equilibration of arsenic in the vapor phase with that on the growing epitaxial surface at reduced pressures. As discussed earlier, however, this assumption is inconsistent with the observed temperature dependence of autodoping, at least for the 1-atmospheric-pressure case. More work remains to be done to satisfactorily account for this phenomenon.

The effect of reactor pressure on boron autodoping is, however, different from that of arsenic. It has been reported [*10*] that boron autodoping *increases* with a decrease in pressure. The opposite effects of boron and arsenic are not yet completely understood, although when taken together with a similar opposing behavior of boron and arsenic doping with respect to temperature, the phenomena may suggest effects of the electronic states playing a role in the dopant incorporation kinetics.

Effect of Gas Flow

Since the dopant vapor gets redistributed by the carrier gas flow, it is reasonable to expect that the extent of autodoping is determined by the latter. The present author [*11,12*] has developed hydrodynamic models for the vapor transport in the laminar flow epitaxial reactor flow system in which the transport equations are solved for several cases of buried layer geometries. The conclusions arising from these studies are:

1. No strong boundary layer effects exist in the epitaxial reactor when it is operated at 1 atmosphere pressure.
2. Autodoping decreases with an increase in the gas velocity, and this decrease is much more pronounced in the upstream direction than in the downstream direction.
3. Autodoping spreads to considerable distances from each of the buried layer sources, so that for LSI wafers, the autodoping from closely spaced buried layers may be considered to be uniform over the entire wafer.
4. Autodoping increases approximately as the square root of the number of buried layer elements in the wafer.

The conclusion that no stronger boundary layer effects are present in the epitaxial process is a significant departure from previous thinking. Recently, laser interference holography [*13*] has been used for direct experimental observations that confirm the absence of a stagnant boundary layer in the epitaxial reactor flow system.

The mass transport analysis reported in these studies pertains to laminar flow systems such as the horizontal and barrel epitaxial reactors. For nonlaminar flow systems such as the pancake reactors, however, the transport conditions could be different from the ones discussed here.

One of the important concerns for integrated circuits that are built in the epitaxial films for very-large-scale-integration (VLSI) applications is the increased autodoping with increased device density, where the buried layer elements are in close proximity with each other.

A related question concerns the variation in the autodoping between the

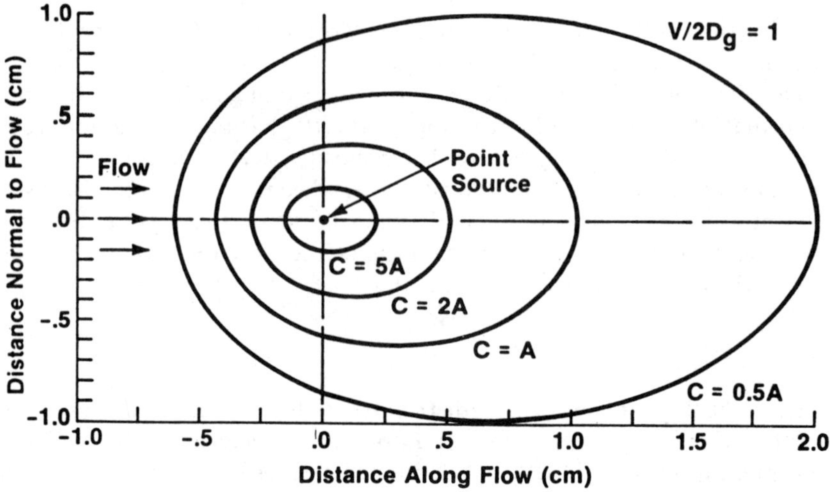

FIG. 6—*Impurity distribution due to a point autodoping source in a moving stream of gas of uniform velocity* V. *Dopant vapor diffusivity in the stream is denoted by* D_g *and* A *is a measure of concentration level.*

closely spaced devices. This autodoping may cause interdevice capacitance, and hence could contribute to device delay. Both these problems of buried layer density and dopant uniformity have been analyzed [*12*], in a study in which each buried layer element was treated as a point autodoping source with a corresponding diffusion field in a carrier gas flow. The impurity distribution as calculated for a point source is shown in Fig. 6. We see from this figure that the diffusion field of each point source reaches distances of the order of several millimetres over the wafer. By summing the autodoping contribution from each point source in a rectangular array, it was shown that the net autodoping increases nearly as the square root of the number of buried layer elements in the wafer. This square-root law has been verified experimentally for LSI wafers [*14,15*].

Also, for a typical closely spaced LSI situation, the overlapping of the diffusion fields from each of the buried layers is so complete as to produce nearly uniform autodoping over the entire wafer. This prediction has also been recently verified experimentally [*15*].

At large distances, on the order of wafer diameters, the autodoping effects are weakly additive, such that autodoping increases somewhat slowly due to an increase in the number of wafers in an epitaxial run until it finally saturates. This saturation effect has been experimentally verified for LSI wafers [*16*].

Profile Tolerance

One of the important aspects of epitaxy for high-performance bipolar applications is the ability to control the buried layer profiles to close tolerances.

The major process parameter that determines this profile tolerance is the variation in the wafer temperature during the epitaxial deposition from run to run and from wafer to wafer in each run. Since the same temperature nonuniformity in the epitaxial reactor causes less profile variation at lower epitaxial temperatures than at higher values, because of the reduced dopant diffusivity at lower temperatures, it would be desirable to choose as low an epitaxial deposition temperature as possible.

To illustrate this point, we make a calculation of the profile tolerances for arsenic buried layers from two epitaxial processes: one using 1150°C as the deposition temperature, and the other using 1050°C. We assume that all the epitaxial parameters other than the deposition temperature and time are held constant for the two processes. This calculation refers to an NPN bipolar device structure that would need 1.5-μm-thick epitaxy if it were grown at 1150°C, but would need only 1.0 μm thickness if grown at 1050°C to provide the same flat region (X_f), because of the reduced outdiffusion of the subcollector at 1050°C. The growth rate used in this calculation is 0.1 μm/min; thus the deposition time is 15 min for the 1150°C epitaxy and 10 min for the 1050°C epitaxy. We also assume a wafer temperature nonuniformity of ±10°C for both cases.

Table 1 shows the results of this calculation. Due to the nonlinear diffusion effects, the experimentally observed outdiffusion value of $\sim 10\sqrt{(Dt)}$ is used in this calculation, where D is the arsenic diffusivity at the epitaxial deposition temperature and t is the deposition time. The calculation shows that the net diffusional tolerance in the arsenic buried layer profile is 2070 Å for the 1150°C epitaxial process versus 530 Å for the 1050°C case.

In addition to the diffusional tolerance one has to take into account the variation in the epitaxial growth rate due to the nonuniformity in the epitaxial temperature from wafer to wafer. Epitaxial growth rate ($\mathring{g}$) is a function of deposition temperature, and is more sensitive at lower temperatures (Fig. 7). Thus nonuniformity in the epitaxial temperature appears as nonuniform-

TABLE 1—*Subcollector profile variations.*

Process	1150°C Epi			High-Low Epi (1050°C)		
Deposition temperature, °C	1140	1150	1160	1040	1050	1060
Deposition time, min	15	15	15	10	10	10
Diffusivity of As in Si, cm^2/s	6.35×10^{-14}	8.09×10^{-14}	1.03×10^{-13}	4.59×10^{-15}	6.08×10^{-15}	8.01×10^{-15}
Diffusion length ($L = \sqrt{(Dt)}$), Å	756	853	963	166	191	219
Subcollector outdiffusion (10L), Å	7560	8530	9630	1660	1910	2190
Total variation		2070 Å			530 Å	

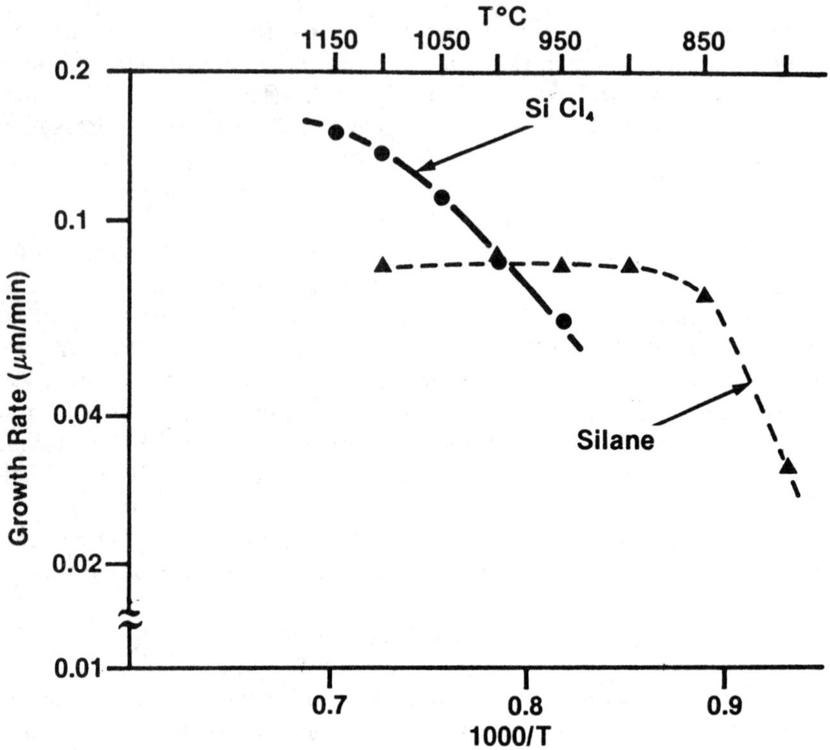

FIG. 7—*Temperature dependence of epitaxial growth rate for silane and $SiCl_4$ depositions at 1 atm pressure. Flow rates are adjusted to give approximately the same ranges of growth rate for the two cases.*

ity in the epitaxial thickness. Table 2 shows the epitaxial thickness variations as calculated for the two processes using the growth rate data ($d\mathring{g}/dT$) of Fig. 7. Since both the epitaxial growth rate and the buried layer outdiffusion increase with increasing deposition temperatures, the variations due to the two effects counteract in terms of the thickness X_f of the epitaxial flat-doped layer. Thus the net effect is the difference between these two variations. Table 3 shows the net calculated variation in X_f due to these two effects. These results show clearly that the total variation in the epitaxial flat region for the 1150°C epitaxy is about three times that for the 1050°C epitaxy, for a wafer temperature nonuniformity of ±10°C. For an NPN transistor built in

TABLE 2—*Epi thickness variations.*

Process	1150°C Epi	High-Low Epi
Deposition temperature, °C	1150 ± 10	1050 ± 10
$d\mathring{g}/dT$ (Å/min/°C)	2	5
Deposition time, min	15	10
Epi thickness variation, Å	±300	±500

TABLE 3—*Epi flat region variations.*

Process	1150°C Epi		High-Low Epi (1050°C)	
Temperature variation	+10°C	−10°C	+10°C	−10°C
Subcollector outdiffusional variation, Å	−1100	+970	−280	+250
Epi thickness variation, Å	+300	−300	+500	−500
Net variation in epi flat region, Å	−800	+670	+220	−250
Total tolerance	1470 Å		470 Å	

these epitaxial layers, the flat region variations occur between the base and the subcollector in the vertical profile, since both the base and the emitter are formed from the top surface of the epitaxial film. We will discuss later the effect of the variation in X_f on the device delay tolerance.

Defect Considerations

Of equal importance to the profile control in the epitaxial film is the control of defects generated during an epitaxial growth, since these defects determine also the device leakage and device yield. The CVD epitaxial process has attained such a state of perfection that it is now possible routinely to make virtually defect-free films. Yet, with the drive towards lower and lower processing temperatures, some of the defect problems need to be re-addressed.

The origin of the epitaxial defects during growth can be traced to two sources: (*1*) the quality of the substrate surface just prior to growth, and (*2*) the epitaxial process parameters, such as deposition chemistry, growth temperature, and growth rate.

Substrate Surface Quality

Modern techniques of silicon substrate preparation have virtually eliminated surface irregularities and defects. (For an editorial review of the current status of semiconductor chemical etching and cleaning, see Ref *17*.) However, there is new evidence that points to the amount of oxygen at or near the substrate surface as a contributing factor to the epitaxial defects. The trend is to design pre-epitaxial treatments to produce an oxygen-denuded layer near the surface, with perhaps precipitated oxygen or SiO_2 in the interior of the substrate acting as the gettering sites for point defects. One can also provide gettering layers in the back surface of the substrate, and this seems to improve the defect quality of the epitaxial layer [*18,19*].

The difficulty encountered in low-temperature silicon epitaxial growth is the inability to efficiently remove the native oxide on the substrate surface just prior to the epitaxial deposition. Sometimes, *in situ* vapor HCl cleaning is employed in order to etch away some of the top silicon surface just prior to deposition; but due to problems of metallic contamination and excessive pattern distortion, this technique is not desirable, especially for VLSI applications.

Process-Induced Defects

With the elimination of thermal slip in modern epitaxial reactors, process chemistry and process temperature have become two important parameters that control epitaxial defects as they relate to processing. It is generally recognized that $SiCl_4$ produces a better quality silicon epitaxial film at elevated temperatures than silane. We can interpret this effect in the following manner. Reduction of $SiCl_4$ is a near-equilibrium reaction that produces nascent Cl_2 or HCl as the reaction product. The reverse etching reaction is likely to take place preferentially at defect sites, due to a high energy state of silicon bonding at these sites. This mechanism allows an *in situ* etching of the defect, while allowing the forward reaction to take place at regular sites. On the other hand, silane deposition is essentially an irreversible reaction with no significant etching component. One would therefore expect a larger number of defects to be trapped in the growing epitaxial layers when silane is used instead of $SiCl_4$.

The preceding mechanism also explains the observation that vapor phase autodoping in $SiCl_4$ epitaxy is lower than that in silane epitaxy under identical deposition conditions [*1*]. The dopant arsenic atoms are likely to be etched away in the $SiCl_4$ epitaxy by the *in situ* etching mechanism described earlier, whereas no such mechanism exists for silane.

Thus, from both the defect and autodoping points of view, $SiCl_4$ epitaxy is preferable to silane epitaxy. Silane is, nevertheless, used for lower temperature depositions instead of the $SiCl_4$. The reason for this stems from observations that the silane deposition reaction reaches temperature-independent, diffusion flow–limited stages at much lower temperatures than $SiCl_4$ (Fig. 7). One should note, however, that the growth rate versus temperature curves are functions of reactor pressure, and that the temperature-independent range tends to become narrower and shift to higher temperatures as the reactor pressure is reduced [*20*]. Also, although the growth kinetics change with increasing deposition temperature from the temperature-dependent to temperature-independent stage, the deposition reactions are still heterogeneous reactions that occur on the substrate surface for both cases. Therefore the epitaxial quality is surface-related at all deposition temperatures.

The effect of deposition temperature on the epitaxial quality for the $SiCl_4$ epitaxy (5 μm thick, 1 Ω-cm arsenic-doped, grown over P^- $\langle 100 \rangle$ substrates) is shown in Fig. 8, in which the metal-oxide semiconductor (MOS) retention time yield (defined as the percentage of the 1.52-mm [60-mil] diameter MOS dots with less than 500 pA leakage [*21*]) is plotted as a function of this temperature. The figure also shows the deterioration of the epitaxial quality at lower temperatures in terms of anodic etch pit density, which measures

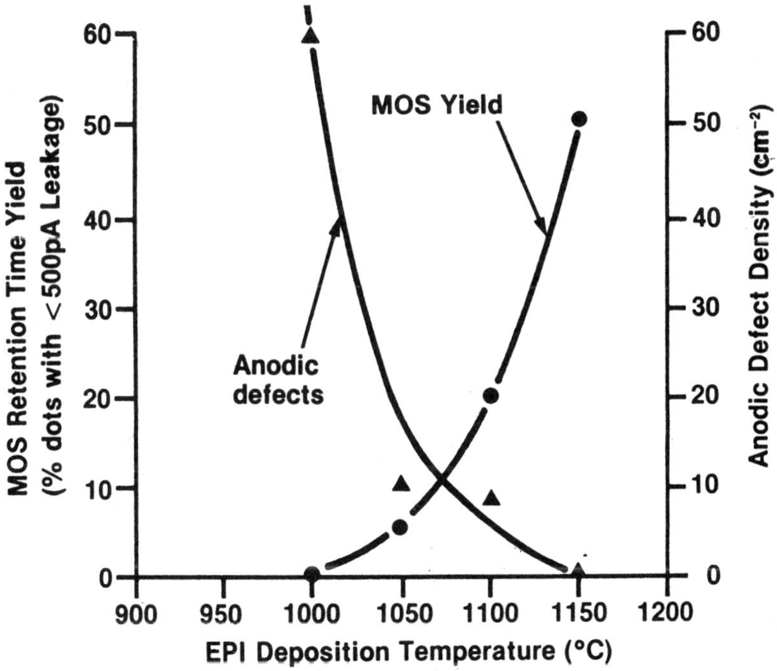

FIG. 8—*Deterioration of epitaxial quality with lowering of the epitaxial deposition temperatures.*

electrically active defects [*22*]. From these data it is clear that the lowering of the epitaxial deposition temperature has disastrous effects on epitaxial quality, unless ways are found to prevent this deterioration. In a later section we will describe a technique that achieves this prevention. The fact that such a defect prevention can be accomplished demonstrates that low epitaxial quality is not inherent to the low deposition temperatures; instead, it is related to the control of the condition of the substrate surface just prior to the epitaxial growth, which has a profound effect on the initial nucleation of the epitaxial layers. The presence of dopant impurity is also an important factor, and it is found[3] that it has a significant effect on the growth kinetics of the epitaxial layer.

On the other hand, the effect of growth rate is simpler to understand in terms of the probability of defect nucleation. The faster the growth, the higher the probability to nucleate and to entrap crystalline defects. At slower growth rates and at high enough thermal activation, the high energy defect sites will have sufficient time and mobility to anneal out.

Device Modeling

We consider now the effects of epitaxial process on the NPN bipolar device characteristics. As stated before, the physical device parameters that are affected by the epitaxial process are the extent of the epitaxial flat region (X_f), the doping level (C_0) of the flat region, the base-collector junction concentration (C_{bc}), and the subcollector sidewall thickness. These parameters determine the collector-base junction breakdown voltage (BV_{CB}), collector-base depletion capacitance (CCB), and collector-to-substrate isolation capacitance (CCS).

One of the simplest ways to analyze the device results is in terms of an abrupt junction approximation. Since the base and the subcollector regions are heavily doped relative to the epitaxial flat doping level C_0, most of the depletion region is in the N region between the base and the subcollector, as shown in Fig. 9. For large values of X_f, BV_{CB} is a function of C_0, which represents the avalanche-limited case. However, as the epitaxial thickness (t_{epi}) is reduced sufficiently, BV_{CB} becomes a function of C_0 as well as X_f; and as the epi thickness is reduced further, BV_{CB} becomes a function of X_f only.

A similar effect holds for the depletion layer capacitance (CCB). For large values of X_f, one-sided abrupt junction theory predicts that $CCB \propto C_0^{1/2}$. As X_f is reduced to very small values, CCB becomes a function of both C_0 and X_f.

Figure 10 shows the variation of BV_{CBO} with the depletion width (W_D) for a transistor with varying epitaxial thickness. The value of W_D was calculated from the capacitance data. The solid lines correspond to the abrupt junction approximation theory for several values of C_0. We see from this figure that the abrupt junction approximation is fairly well obeyed.

Figure 11 shows the variation of $1/CCB^2$ with the applied voltage for two epitaxial thicknesses. According to the one-sided abrupt junction approximation, $1/CCB^2 \propto (V_{bi} + V_{ap})/C_{bc}$, where V_{bi} and V_{ap} are the built-in poten-

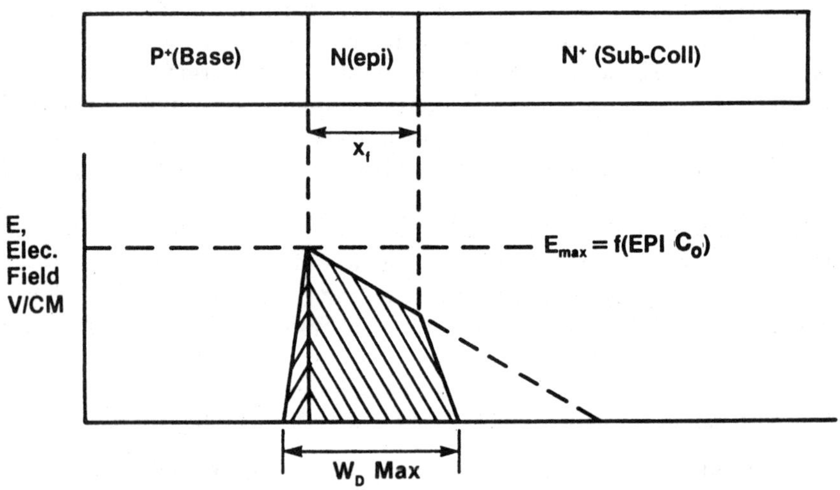

FIG. 9—*Field distribution in a $P^+/N/N^+$ structure with an abrupt P/N junction.*

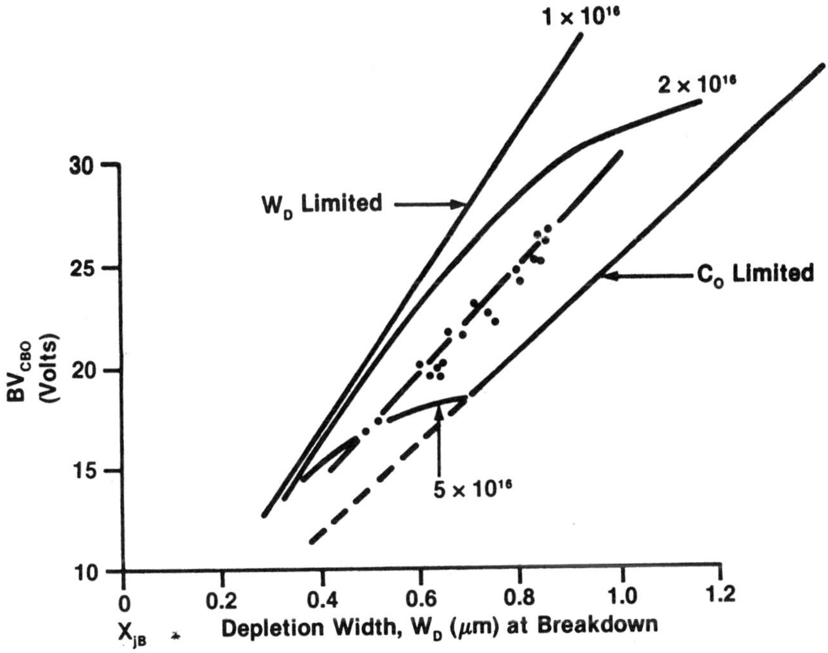

FIG. 10—*Variation of the collector-base breakdown voltage with the depletion width as measured from the capacitance data for transistors built in epitaxial layers of varying thicknesses. Solid lines are calculated from the abrupt junction approximation for several values of* C_0. *The line through the data points is a least-square fit that yields a slope of 25.5 V/μm* (*data courtesy C. A. Berg*).

tial and the applied voltage, respectively. Once again, the linearity in these plots for low values of V_{ap} indicates the near validity of the abrupt junction approximation. The difference in the slope for the two epitaxial thicknesses arises from the two values of the junction concentration C_{bc}, the thinner epitaxy corresponding to a higher crossover of the base and subcollector profiles.

Since both the breakdown voltage and the junction capacitance vary with the epitaxial thickness and with the junction concentration, it is instructive to plot BV_{CBO} versus CCB. This is shown in Fig. 12, which shows that for the thicker epitaxial films, BV_{CBO} is very sensitive to small fluctuations in the epitaxial thickness, while CCB is not. However, as the epitaxial thickness is reduced considerably, the capacitance becomes quite sensitive to the epitaxial thickness fluctuations. This may be seen in Fig. 13, where CCB is plotted as a function of t_{epi}.

The effect of epitaxial temperature nonuniformity on the profile tolerance was discussed earlier for two processes, one using 1150°C deposition temperature and the other 1050°C. These variations in the epitaxial flat region are considerably different for the two temperatures. For example, at

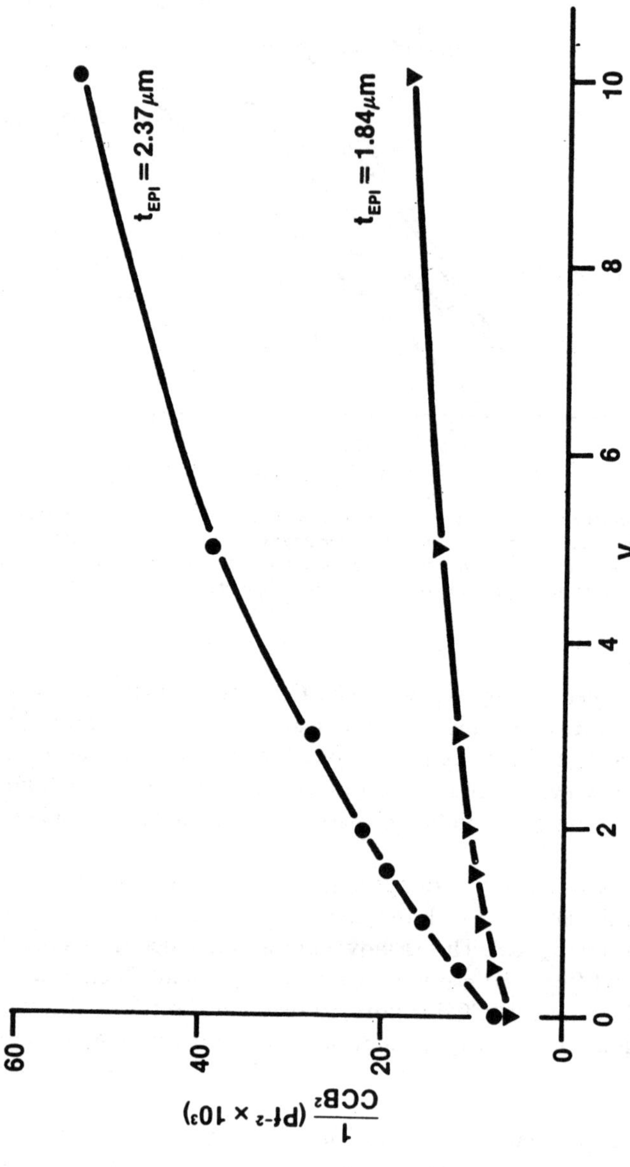

FIG. 11—*Variation of collector-base capacitance with applied voltage for two epitaxial thicknesses (data courtesy C. A. Berg).*

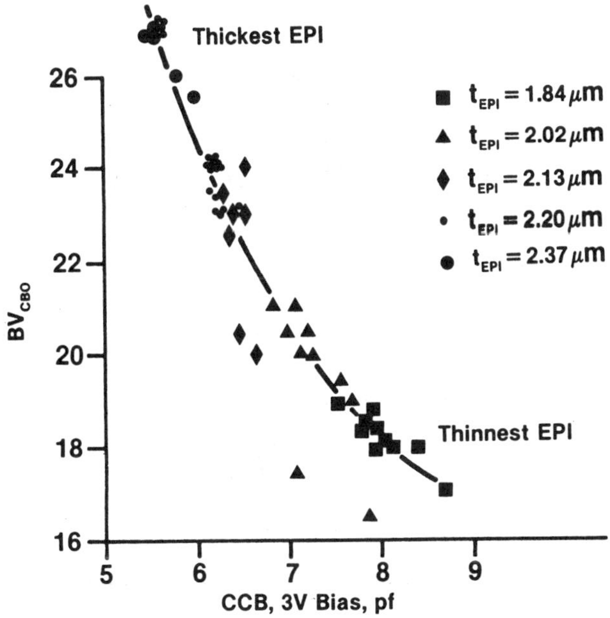

FIG. 12—*Variation of the collector-base breakdown voltage* (BV_{CBO}) *with the collector-base capacitance* (CCB) *for transistors fabricated in different thicknesses of the epitaxial film* (*data courtesy C. A. Berg*).

1150°C, variation in X_f for a temperature variation of ±10°C is 1470 Å as against 470 Å for the 1050°C case (Table 3). These variations have large influence on the device performance. Table 4 shows the result of the performance tolerance calculations for a shallow device structure used in a 4 mW current-switch, emitter-follower circuit using the profile tolerance results of Table 3. It is clear that there is a substantial performance penalty associated with the higher epitaxial temperature used in this example.

In addition to the device considerations discussed in this section, there are other epitaxial process effects that influence device characteristics. For example, the collector-to-substrate isolation capacitance is a function of autodoping and subcollector sidewall thickness, both of which are determined by the epitaxial process parameters. Since these effects are fully discussed elsewhere [*23*], we will not consider them here.

Application to an Epitaxial Process

We have discussed earlier the advantages of low-temperature epitaxy for high speed bipolar applications. We have also seen the difficulties encountered in lowering the epitaxial temperatures for the chemical vapor deposition of silicon in terms of increased autodoping and defects. However, it is possible to design a low-temperature process if we make use of the knowl-

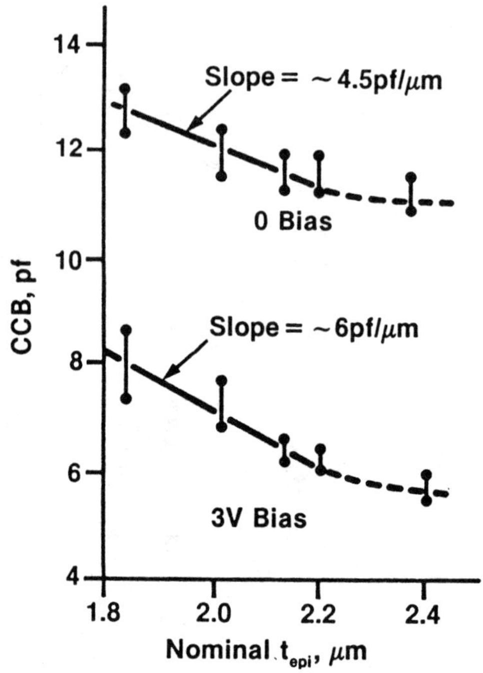

FIG. 13—*Variation of collector-base capacitance with epitaxial thickness (data courtesy C. A. Berg).*

edge of the origin of autodoping and of defects. A process developed by the author [*24,25*], which minimizes the aforementioned problems, will now be briefly described.

Autodoping occurs primarily in the prebake stages [*1*], while outdiffusion of the buried layer depends on the thermal cycle during the deposition stages. Thus one can minimize both outdiffusion and autodoping by decoupling the prebake and growth heat cycles and by optimizing each one separately. For example, a higher prebake temperature produces a lower autodop-

TABLE 4—*Performance parameter variation.*[a]

Device Parameter	1150°C Epi				High-Low Epi (1050°C)			
	1140°C	1150°C	1160°C	Total Δ	1040°C	1050°C	1060°C	Total Δ
Diffusion capacitance (CEB), pf	0.28	0.26	0.21	27.0%	0.27	0.26	0.24	11.5%
CCB, pf	19.7	18.5	16.8	15.7%	19.1	18.5	17.8	7.0%
RDB, kΩ/square	9.9	10.3	11.3	13.6%	10.1	10.3	10.7	5.8%
Delay, ps (0.36 pf load)	245	240	225	8.3%	242	240	236	2.5%

[a] $I_c = 0.8$ mA; 4.0-mW current switch emitter follower circuit (CSEF).

ing, whereas a lower deposition temperature assures a lower buried layer outdiffusion. Thus a high-temperature prebake followed by a low-temperature deposition ("high-low", for short) should satisfy both these requirements.

The next question is one of defects. Previous discussion has shown that it is important to maintain a good quality of the surface just prior to deposition, and that from a consideration of the ability to remove any native oxide on the silicon substrate, a higher prebake temperature should be preferred.

It is also possible [*25*] to relate the effectiveness of the substrate surface preparation for defect prevention to the prebake thermal cycle through Eq 1. This study showed that the defect level in the epitaxial layer decreased with an increase in either the prebake temperature or prebake time. Establishing the defect equivalency through Eq 1, the author showed that a prebake cycle of 10 min at 1150°C was equivalent to a prebake cycle of 7 h at 1050°C or 2 min at 1200°C. However, an extended prebake beyond this equivalent thermal cycle caused excessive buried layer depletion. Thus a high-low epitaxial process consisting of a 1150°C/10 min prebake followed by growth at 1050°C was designed.

Once the thermal cycle was designed, the next consideration was the deposition chemistry. From the previous discussion, it is clear that a deposition with $SiCl_4$ is preferable to that with silane from both the autodoping and the defect points of view. As noted earlier, the epitaxial quality is surface-related at all deposition temperatures, and the high-temperature prebake allows one to lower the $SiCl_4$ deposition temperature by providing a clean, initial nucleating surface.

The last remaining consideration was the choice of an appropriate growth rate. Clearly, from both the thickness controllability and the defect points of view, a low growth rate is to be preferred. In previous studies [*1,6*] it was found that a growth rate beyond 0.1 to 0.2 μm/min led to impurity trapping. Using similar arguments for defect entrapment, we chose a growth rate of a ~0.1 μm/min, which reduced both the epitaxial defects and autodoping.

From all of the preceding considerations, it has been possible to design the high-low process [*24*] for silicon epitaxy. This consisted of a $SiCl_4$ deposition with a 1150°C/10 min prebake in H_2 followed by growth at 1050°C and at a growth rate of ~0.1 μm/min. The results of this process in terms of autodoping, defects and outdiffusion are shown in Figs. 14 to 16. The autodoping increase associated with the lowering of the deposition temperature to 1050°C was reduced by an order of magnitude due to the introduction of the 1150°C prebake (Fig. 14); the outdiffusion was reduced substantially relative to the conventional 1150°C deposition process (Fig. 15); and finally, the 1150°C prebake cycle not only prevented the deterioration of epitaxial quality due to a low temperature deposition, but actually *improved* the MOS retention time epitaxial quality over that of the 1150° deposition process (Fig. 16).

This example serves to illustrate the effective use of the process basics in a successful design of a low-temperature epitaxial process for CVD silicon which has been used in the fabrication of high-performance bipolar inte-

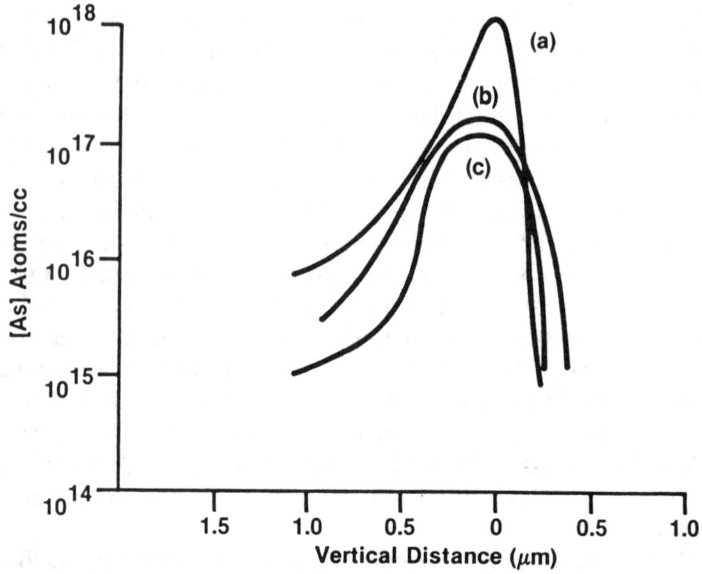

FIG. 14—*Off-subcollector spreading resistance autodoping profiles in the epitaxial layer grown with* (a) *1050°C prebake and deposition,* (b) *1150°C prebake and deposition, and* (c)*1150°C prebake and 1050°C deposition* (*high-low process*).

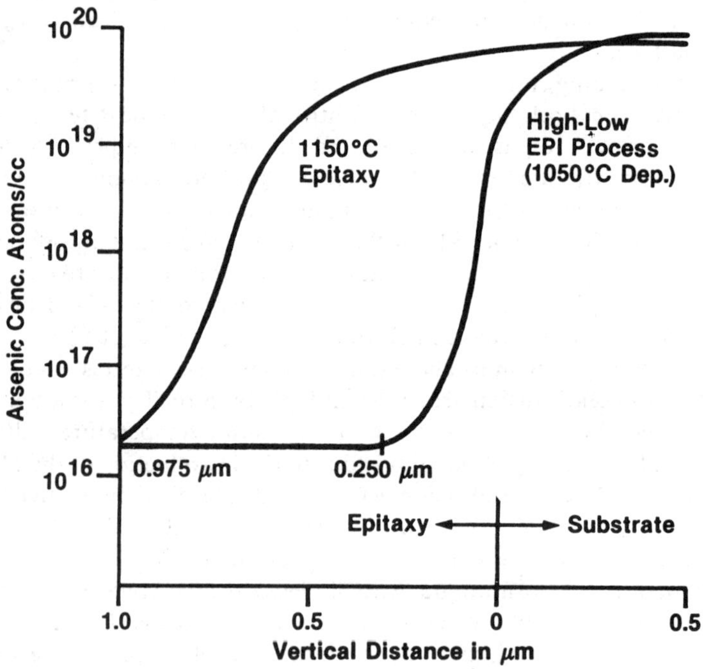

FIG. 15—*Subcollector profiles for 1150 and 1050°C epitaxial depositions.*

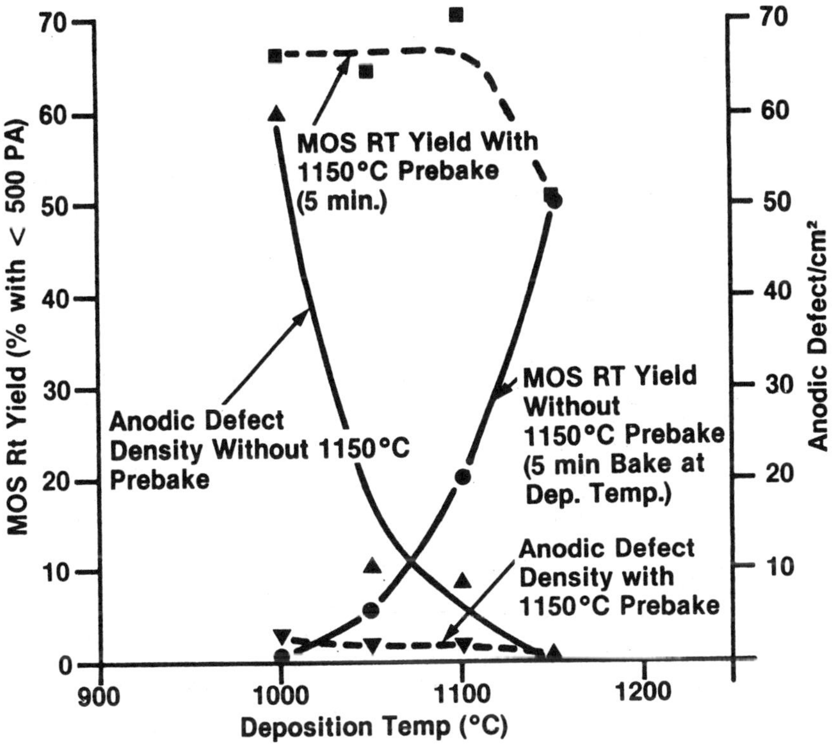

FIG. 16—*Improvement of epitaxial quality due to the 1150°C prebake preceding deposition at lower temperatures. Epi: 1 Ω-cm,* n*-type, 5 μm thick; substrate:* P^-, *⟨100⟩ (data courtesy B. M. Kemlage).*

grated circuits [*26,27*]. In addition to the high-low process, the advantages of lower autodoping at reduced pressures with consequent decrease in CCB and CCS can be exploited by combining the two techniques to obtain a low-temperature and low-pressure process [*28*].

Some Concluding Remarks

The state of the epitaxial art and science has reached stages where it is possible to model and design the process for any particular device application. Further process enhancements can come from such techniques as *in situ* sputtering prior to deposition, plasma-enhanced deposition, and a combination of the high-low process with reduced-pressure epitaxy. Epitaxial temperature can perhaps be reduced to the 900 to 1000°C range with these approaches, thus bringing the epitaxial process step into more conformity with other processing temperatures in the fabrication of silicon integrated circuits. One of the remaining challenges in the silicon deposition technology is to devise schemes for growing good quality silicon single crystal films on insulator surfaces either selectively or over the entire substrate. Exciting work in this field is in progress, and significant developments in this area should occur in the near future.

Acknowledgments

I thank C. A. Berg for the device data used in Figs. 10 to 13, B. M. Kemlage for the MOS retention time data used in Fig. 16, and S. P. Gaur for the two-dimensional simulation of the device performance used in Table 4.

References

[*1*] Srinivasan, G. R., *Journal of the Electrochemical Society,* Vol. 127, No. 6, June 1980, pp. 1334–1342 (also in *Proceedings,* Seventh International Conference on Chemical Vapor Deposition, T. O. Sedgwick and H. Lydtin, Eds., The Electrochemical Society, Pennington, N.J., 1979, pp. 77–101).
[*2*] Langer, P. H., *Journal of the Electrochemical Society,* Vol. 124, No. 4, April 1977, pp. 591–598.
[*3*] Bloem, J., *Journal of Crystal Growth,* Vol. 38, 1977, pp. 364–366.
[*4*] Nishizawa, J. in *Proceedings,* Fifth International Conference on Vapor Growth and Epitaxy, Coronado, Calif., July 1981, in press.
[*5*] Reif, R. and Dutton, R. W., *Journal of the Electrochemical Society,* Vol. 128, No. 4, April 1981, pp. 909–918.
[*6*] Srinivasan, G. R., *Journal of the Electrochemical Society,* Vol. 125, No. 1, Jan. 1978, pp. 146–151 (also in *Semiconductor Silicon, 1977,* H. R. Huff and E. Sirtl, Eds., The Electrochemical Society, Pennington, N.J., 1977, pp. 218–234).
[*7*] Ogirima, M., Saida, H., and Maki, M. in *Extended Abstracts,* The Electrochemical Society, Vol. 78-2, 1978, p. 583.
[*8*] Tabe, M. and Nakamura, H., *Journal of the Electrochemical Society,* Vol. 126, No. 5, May 1979, pp. 822–826.
[*9*] Ogirima, M., Saida, H., Kondo, A., and Shirahata, K., *Journal of the Electrochemical Society,* Vol. 124, No. 6, June 1977, p. 903.
[*10*] Kulakarni, S. B. and Kozul, A. A., *Extended Abstracts,* The Electrochemical Society, Vol. 80-2, 1980, pp. 1351–1352.
[*11*] Srinivasan, G. R., *Journal of Applied Physics,* Vol. 51, No. 9, Sept. 1980, pp. 4824–4829.
[*12*] Srinivasan, G. R., *IEEE Transactions on Electron Devices,* Vol. ED-27, No. 8, Aug. 1980, pp. 1493–1496.
[*13*] Giling, L. H., *Journal of the Electrochemical Society,* Vol. 129, No. 3, March 1982, p. 634.
[*14*] Srinivasan, G. R., *Journal of the Electrochemical Society,* Vol. 127, No. 10, Oct. 1980, pp. 2305–2306.
[*15*] Ackerman, G. K. and Ebert, E., *Journal of the Electrochemical Society,* Vol. 128, No. 11, Nov. 1981, p. 485C (Abstract 624 RNP).
[*16*] Srinivasan, G. R. and Basso, J. E., "Epi Loading Effects on Autodoping and Bipolar Device Characteristics," to be published in the *Journal of the Electrochemical Society.*
[*17*] Marshall, S. L., *Solid State Technology,* Vol. 24, No. 11, Nov. 1981, p. 99.
[*18*] Yang, K. H., Kappert, H. F., and Schwuttke, G. H., *Physica Status Solidi A,* Vol. 50, No. 11, Nov. 1978, pp. 221–235.
[*19*] Chen, M. C. and Silvestri, V. J., *Journal of the Electrochemical Society,* Vol. 128, No. 2, Feb. 1981, pp. 389–395.
[*20*] Duchemin, M. J., Bonnet, M. M., and Koelich, M. F., *Journal of the Electrochemical Society,* Vol. 125, No. 4, April 1978, p. 637.
[*21*] Montillo, F. J., Poponiak, M. R., and Pistor, R. L., "Evaluation of Semiconductor Device Processing Using MOS Charge," *Europhysics Conference Abstracts, Sixth Symposium on Solid State Technology,* Toulouse, France, Vol. 5F, Abstract S6.4, Sept. 1981, p. 242.
[*22*] Deines, J. L., Philbrick, J. W., Poponiak, M. R., and Dove, D. B., *Applied Physics Letters,* Vol. 34, No. 11, June 1979, pp. 746–748.

[*23*] Srinivasan, G. R., *Solid State Technology,* Vol. 24, No. 11, Nov. 1981, pp. 101–110.
[*24*] Srinivasan, G. R., U.S. Patent No. 4,153,486, 8 May 1979.
[*25*] Srinivasan, G. R., "High-Low Silicon Epitaxial Process," to be published in the *IBM Journal of Research and Development.*
[*26*] Chu, S. F., Srinivasan, G. R., Bhatia, H., Kemlage, B. M., Barson, F., Mauer, J., and Riseman, J. in *VLSI Science and Technology/1982,* C. J. Dell'Oca and W. M. Bullis, Eds., The Electrochemical Society, Pennington, N.J., 1982, pp. 306–314.
[*27*] Silvestri, V. J., Ginsberg, B., and Srinivasan, G. R., "Submicron Epitaxial Films," to be presented at 1983 Spring Meeting of the Electrochemical Society, San Francisco, May 1983.
[*28*] Srinivasan, G. R., "High-Low Epitaxial Process under Reduced Pressures," *IBM Technical Disclosure Bulletin,* in press.

Mark R. Boydston,[1] *Gilbert A. Gruber,*[1] *and Dinesh C. Gupta*[1]

Effects of Processing Parameters on Shallow Surface Depressions During Silicon Epitaxial Deposition

REFERENCE: Boydston, M. R., Gruber, G. A., and Gupta, D. C., **"Effects of Processing Parameters on Shallow Surface Depressions During Silicon Epitaxial Deposition,"** *Silicon Processing, ASTM STP 804,* D. C. Gupta, Ed., American Society for Testing and Materials, 1983, pp. 174–189.

ABSTRACT: Buried layer patterns of high dopant concentration are required in semiconductor processing. These patterns are produced by diffusion through patterned oxide and are marked by shallow flat-bottomed depressions bounded by steps of a few hundreds to a few thousands angstroms in height.

The shape and position of the surface depressions relative to the original buried layer patterns are affected by various parameters and processing steps during epitaxial deposition.

Data from a barrel reactor are presented on the lateral displacements of surface depressions, showing the dependence of deposition parameters on symmetrical and asymmetrical distortions. Horizontal and barrel reactor configurations are compared.

KEY WORDS: silicon epitaxy, depressions, buried layers, horizontal reactor, barrel reactor, pattern distortion, pattern shift

Ability to deposit epitaxial layers on shallow surface depressions with controlled pattern shift and distortion becomes important during the fabrication of bipolar devices or wherever there is a requirement for the junction isolation. If not maintained within certain limits, further masking steps in the process may cause misalignment of the pattern and result in device failures. The heavily doped buried layers lower the collector series resistance in devices and are produced by dopant diffusion through patterns cut in oxide by photo-masking methods. After oxide removal and epitaxial deposition, the resulting shape and position of the residual surface depressions relative to the original buried layer patterns are presumed to be affected by several parameters. Deposition variables such as temperature, epi deposition rate, *in situ* etching conditions, and source flow conditions all show some effect, as does the silicon substrate wafer on which buried layers are processed.

Pattern shift and washout are both related to the distortion or obliteration

[1]Siliconix Inc., Santa Clara, Calif. 95054.

of the patterns [*1–7*].[2] The distortion may be an asymmetrical or a symmetrical displacement of the pattern and the size of the pattern may change during epitaxial deposition. Some of the observations made in previous publications are as follows:

1. Pattern shift is approximately proportional to the epitaxial layer thickness [*5*].
2. There is a tendency for pattern shift to occur in a direction opposite to the direction of 3- to 5-deg tilt of the (111) axis on (111)-oriented substrates [*7*], and there is a discontinuity at exactly (111) orientation where the patterns are obliterated [*5*].
3. (100) surfaces behave much differently than (111) surfaces [*3*].
4. Pattern shift is reduced for higher temperatures and lower deposition rates [*2,5*].

Some of these observations conflict among various publications. These conflicts may be due to inaccuracies in the measurement of temperature and pattern shift. Where severe distortions occur, square patterns may become rectangular and the circular patterns may become oval. In pattern washout, one or more edges may disappear, leaving the pattern incomplete. The orientation of the pattern relative to the crystallographic orientation of the wafer plays a dominant role in severe distortion and washout of patterns.

In this paper, experimental data are presented and analyzed for the purpose of optimizing barrel reactor deposition parameters to minimize the pattern shift. The use of the spreading resistance technique for the measurement of pattern shift has been investigated. This measurement is compared with the customarily used bevel measurement. The limitations and the accuracies of the two measurements are discussed.

Experimental Method

Seventy-six-millimetre (3-in.)-diameter silicon substrates, with standard orientations of (100) and (111) off oriented by 4 deg towards (110) were processed through a standard bipolar process. The normal first mask was replaced with one containing grid lines for ease in making shift measurements. Prior to epitaxial deposition, wafer orientations were measured by X-ray diffraction techniques. Silicon depositions were processed on commercially available barrel reactors[3] at atmospheric pressures. The microprocessor-controlled barrel reactor represents the present state of the art in reactor configuration and is shown schematically in Fig. 1. Silicon wafers were placed in surface depressions or pockets on the silicon carbide–coated graphite suscep-

[2] The italic numbers in brackets refer to the list of references appended to this paper.
[3] Applied Materials Models 7600 and 7800.

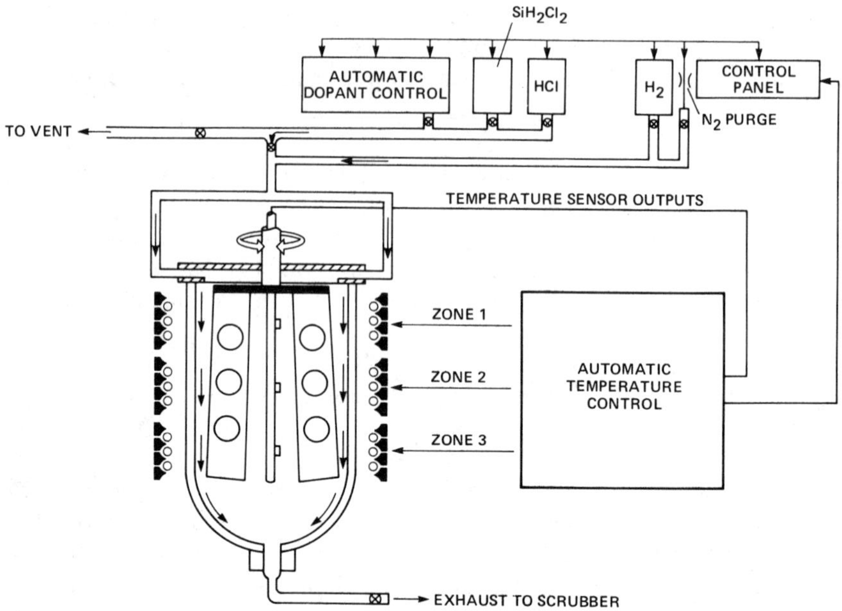

FIG. 1—*Barrel reactor schematic.*

tors tilted at 2.15 deg to the vertical, and lowered into the reaction chamber. The chamber is contained within a cylindrical quartz bell jar and is heated radiantly by tungsten-halogen heating lamps outside the wall of the bell jar.

To achieve the tightest uniformities, a barrel reactor is typically operated at lower Reynolds numbers than a reactor in horizontal configuration. The Reynolds number is directly proportional to the mean velocity of the carrier plus reactant gases. In the horizontal configuration used in the experiment, the Reynolds number was ~300 and in the barrel configuration it was ~100. At the lower gas velocities substrate autodoping effects are maximized. Several experimental runs were completed to observe the effects of temperature, growth rate, slice orientation of substrates, silicon source, and *in situ* HCl etching. During all epitaxial depositions, process parameters were accurately controlled by the microprocessor program, automatic mass flow controllers, and automatic temperature controllers. All temperatures stated are real, and were calibrated by thermocouple prior to experimental runs. Epitaxial layers were deposited to a nominal thickness of 10 μm.

Pattern shift is defined as the displacement of the position of the buried layer pattern after epitaxial growth from its position before growth. If the position of the edge of a buried layer pattern (Fig. 2) prior to and after deposition is at b and a respectively, then the pattern shift (PS) is

$$PS = a - b \tag{1}$$

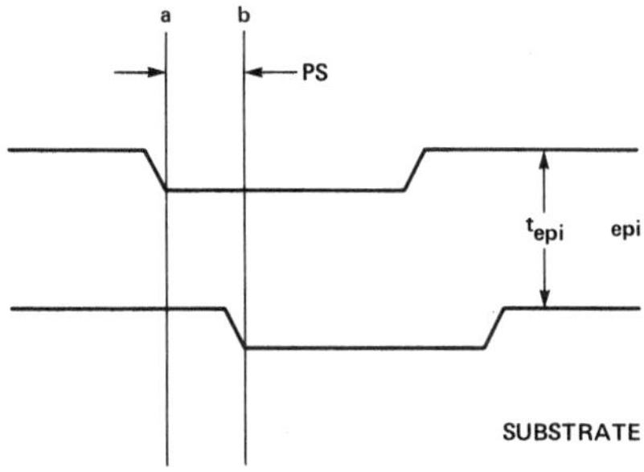

FIG. 2—*Edge displacement.*

With b being zero, the relative shift is

$$PS = a \tag{2}$$

For asymmetrical shift, both sides of the pattern must be measured and averaged. The dimensionless pattern shift ratio (PSR) for shift-normalized to epitaxial thickness t is

$$PSR = PS/t \tag{3}$$

Both parameters necessary for obtaining the pattern shift ratio can be obtained by using the groove-and-stain method. For this work the specimens to be measured were grooved with a Philtec Model 2015 Sectioner and stained with Philtec DSS-3 Stain. Figure 3 is a cross-section sketch of the groove produced by the Philtec Sectioner [8]. Provided $b \ll R$ where b is as shown in Fig. 3 and R is the radius of the sectioning drum, thickness t (epi) by this method is given by

$$t(\text{epi}) = XY/2R = XY/D \tag{4}$$

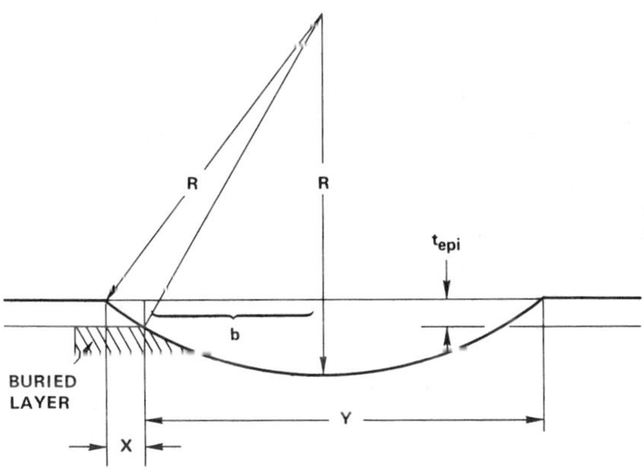

FIG. 3—*Schematic of groove and stain method of measuring layer thickness.*

where X and Y are the dimensions shown in Fig. 3, R is the radius, and D is the diameter of the grooving drum.

The pattern shift was measured (Fig. 2) by using the calibrated Filar eyepiece. The pattern shift ratio was calculated for each measurement. Values of *PSR* were measured in directions both parallel and perpendicular to the major flat on the wafers. A micrograph of a typical grooved and stained surface is shown in Fig. 4.

Results

Experimental pattern shift data collected for epitaxial depositions in typical barrel reactor conditions differed significantly from data reported previously on depositions at higher temperatures on horizontal reactors. Relationships that are established between some processing parameters at low deposition temperatures in the barrel reactor are directly opposed to relationships established at higher temperatures in the horizontal configuration.

For (100)-oriented substrates, at a deposition temperature of 1050°C (Fig. 5) substrate orientation and growth rate seem to be the controlling parameters. As reported by Drum and Clark [*2*], slight variations from the (100) plane may lead to significant pattern shift. However, it was noted that this effect is deposition rate dependent. In fact at low deposition rates (0.5 μm/min), this effect is so dominant that 10 min angular misorientation can lead to shift ratios greater than one. Unless tight controls are specified for wafer orientations, large shifts may result at low deposition rates. Figure 5 also shows that as the deposition rate increases, shift decreases.

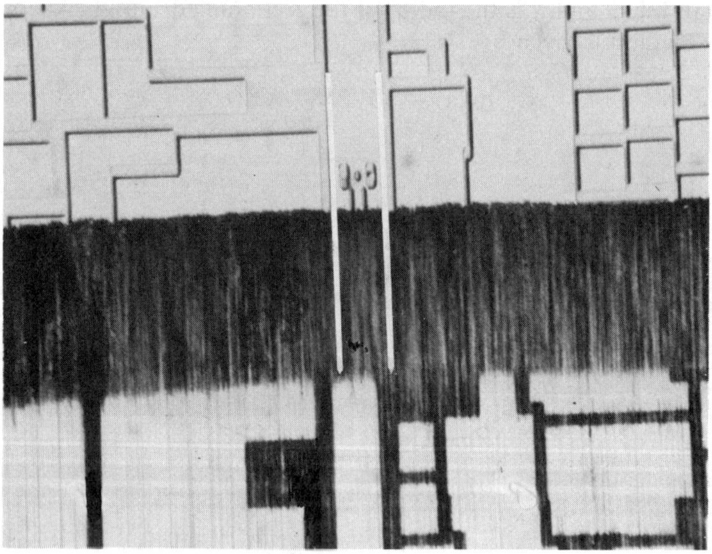

FIG. 4—*Photomicrograph of groove and stained pattern shift sample.*

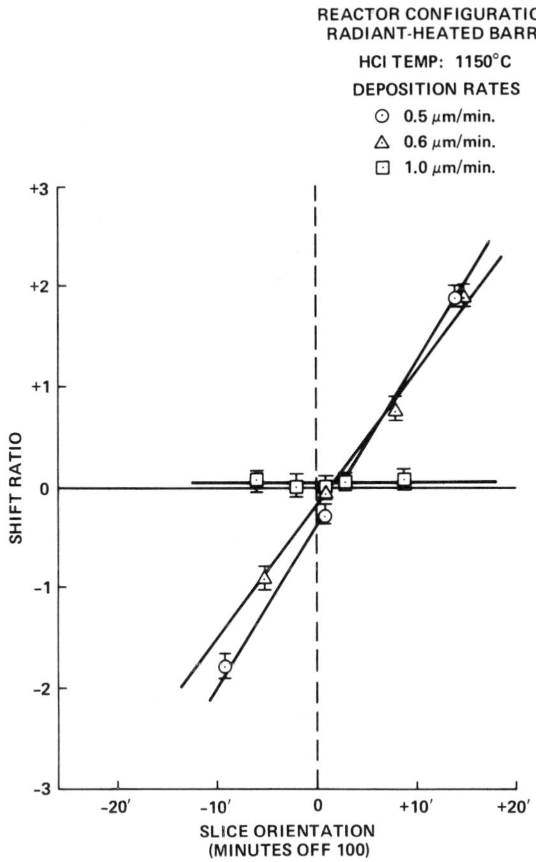

FIG. 5—*(100) shift ratio versus slice orientation, SiH_2Cl_2, 1050°C.*

For similar experiments at higher temperatures, pattern shift ratios were small for all growth rates and misorientations. The significance of these data can be related to process design of silicon epitaxy deposited over highly doped buried layers, particularly where boron is the dopant species. Common techniques used to control autodoping have tended to include variables such as low growth rates and low deposition temperatures. Unfortunately, these same processing parameters appear to be the worst for control of pattern shift on (100) substrates. Obviously, compromises such as the transient control of deposition rates and temperatures must be considered.

Deposition studies of (111) oriented substrates at low deposition temperatures (≤1050°C) also lead to some interesting conclusions. None of the other deposition conditions is found to be critical so long as the (111) slice orientation remains within 4 deg ⊥ 30 min towards (110). This amount of misorientation is recommended by the work of Drum and Clark [*2*], who studied slice misorientation over wide ranges and observed a minimum pattern shift between 3 to 5 deg off orientation toward (110). At 1050°C, pattern shift is constant with deposition rate, although the actual ratios are high (>2.0) (Fig. 6). For depositions at 1200°C, relationships between deposition rate and shift ratio are found to be similar to those reported in literature. For (111) substrates at atmospheric pressures, the least amount of pattern shift can be found by using high deposition temperatures and low deposition rates, as well as when the least chlorinated silane molecules are available. In the production environment, however, these conditions may not be practical. At deposition temperatures of 1150°C, increasing HCl concentration during *in situ* etch reduces pattern shift on (111) (Fig. 7). It is felt that both the concentration of liberated chlorine atoms present during deposition and the deposition temperature have an influence on the thermodynamics of mass transfer and hence the pattern shift. This postulation is based on two sets of data. With a standardized deposition process of 1150°C, 1.0 μm/min deposition rate, the HCl etch concentration was varied for four consecutive runs. Data indicate a steady reduction of pattern shift as a function of increasing HCl concentra-

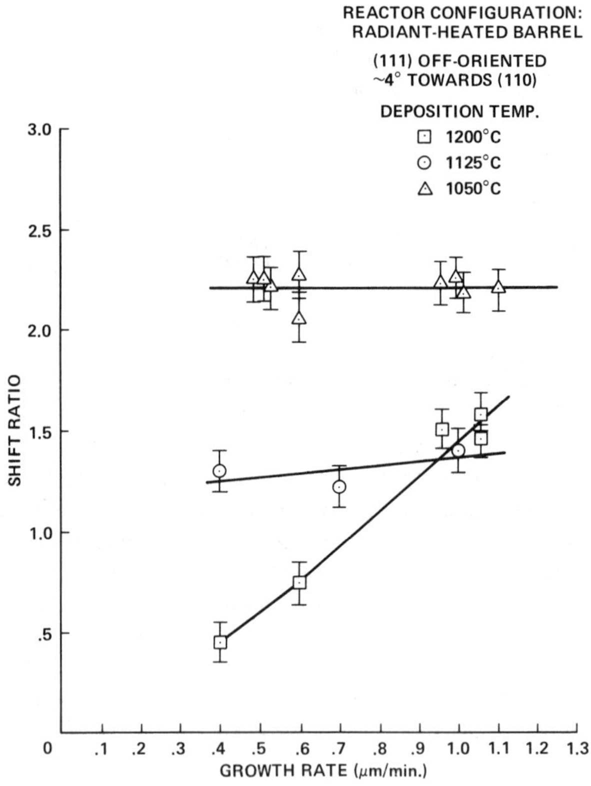

FIG. 6—*(111) shift ratio versus growth rate.*

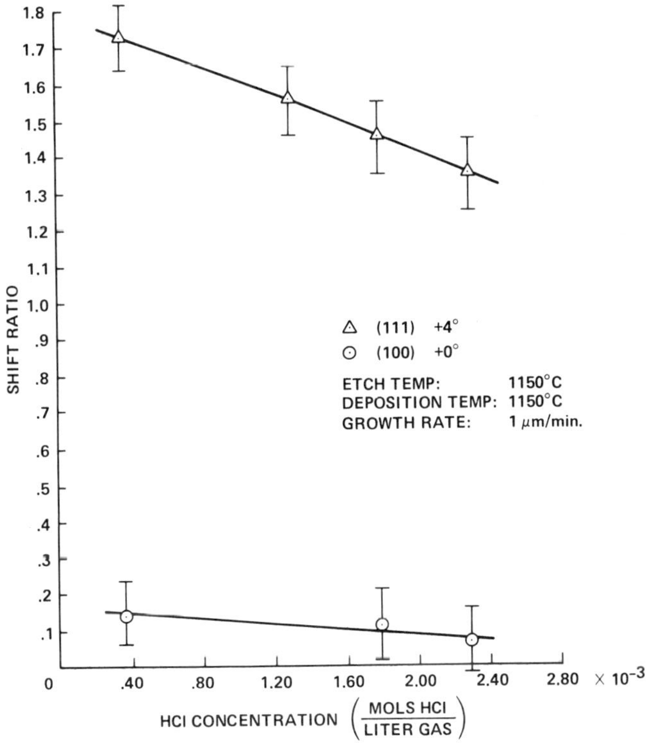

FIG. 7—*Shift ratio versus HCl etch concentration.*

tion (Fig. 7). By comparing the source gas, dichlorosilane and silicon tetrachloride, a similar attempt to study the effect of liberated chlorine atoms was made. Data taken at 1200°C depositions (Fig. 8) indicate a different trend, lower pattern shift for SiH_2Cl_2 than for $SiCl_4$, which confirms data shown by Weeks [7], obtained on a horizontal reactor at high temperatures. At deposition temperatures of 1125°C, however, silicon tetrachloride exhibited less pattern shift than dichlorosilane, which seems to confirm the *in situ* etch concentration data taken at 1150°C. It is important to note that the reduction of chlorinated silanes at deposition temperatures is a complicated multiple-step reaction problem. The complexity of the chemistry makes assumptions quite hazardous in lieu of more detailed information on reaction kinetics and mass transfer.

Finally, an attempt to compare horizontal and barrel configurations was made. For these experiments the barrel reactor was compared with a radio frequency induction-heated horizontal reactor.[4] Reactant concentrations and deposition times were adjusted in the two reactors to achieve the same deposition rates and the layer thicknesses. In addition, processing tempera-

[4] Tempress Model UNICORP 1010.

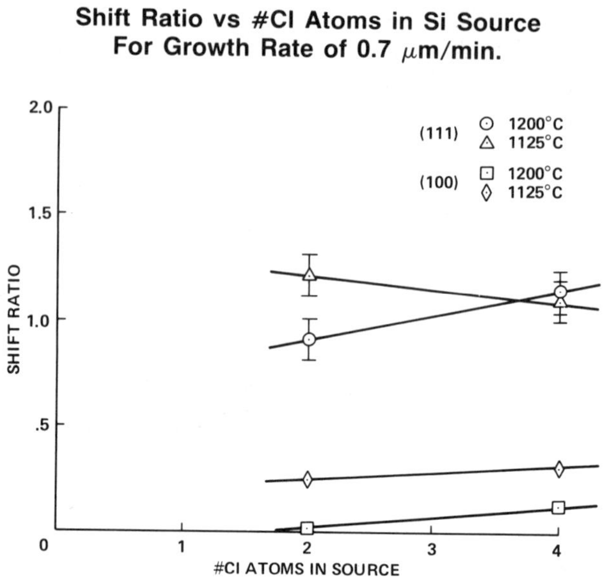

FIG. 8—*Shift ratio versus silicon source molecule.*

tures and the *in situ* etch conditions were maintained the same for the two reactor configurations. Temperature measurement on the horizontal was done optically, and the measured temperature was corrected upwards 75°C to compensate for emissivity, quartz absorption, and mirror reflectance. With these necessary steps taken to normalize the process, pattern shift trends were the same on either of the reactor configurations. As reported by Lee et al [*6*], the major difference in pattern shift between the two systems is the actual process temperature.

Spreading Resistance Technique

A major problem in measuring pattern shift by the groove-and-stain method is that the specimen is destroyed during the measurement. A nondestructive method based on the use of a spreading resistance measurement was therefore investigated. In this method a spreading resistance plot is run perpendicularly across the buried layer pattern by a commercial probe.[5] It has been shown [*9*] that spreading resistances measured on silicon layers which are over layers that are of opposite conductivity type or less conducting than the top layer (for example, epitaxial layer) must be corrected for the proximity of that underlying layer. For silicon of a given resistivity, corrected spreading resistance will vary as follows:

$$R_{corr} = 4a\rho/\text{p}[k(\rho) - 1 + C] \qquad (5)$$

[5] Solid State Measurements Model ASR-100C automatic spreading resistance probe.

where a is the contact radius, p is the number of probes, ρ is the actual epitaxial resistivity, $k(\rho)$ is the correction factor described by Dickey [10], and C is the function of the epitaxial thickness and probe spacing. Therefore

$$R_{\text{corr}} \propto \rho \text{ (epi)}/C \tag{6}$$

As the probe spacing increases by a factor of ten over an insulating boundary, C only changes from 1.5 to 1.85. Figure 9 gives an indication of how C varies with thickness. For a given spacing of 0.1 mm, a change of epi thickness from 8.5 to 2.8 μm changes C from 1.5 to 4. The conducting boundary correction is unaffected by probe separation. It does however vary with the epi thickness, decreasing as thickness decreases. Therefore the position of the buried layer under the epitaxial layer should be determinable, since the measured spreading resistance over the P-type substrate will decrease from what it would be at infinite thickness and over the N^+ buried layer it will increase.

The effect of probe separation is illustrated in Fig. 10. The epi thickness in this case is 14 μm. A minimum resistance over the substrate changes very little

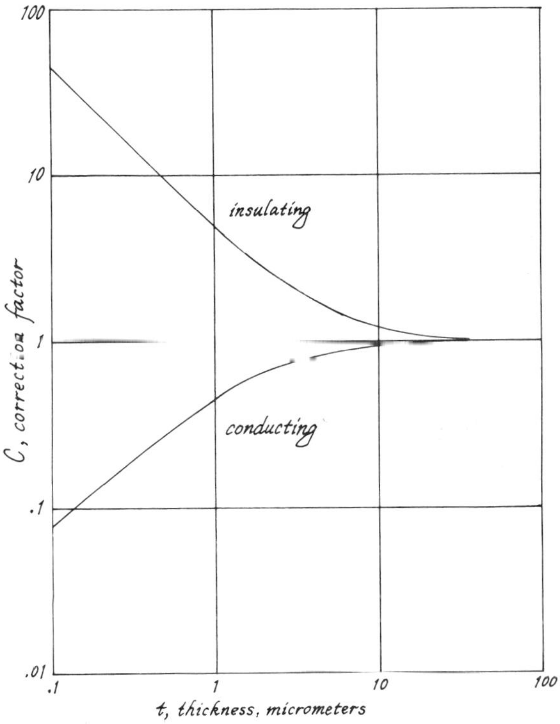

FIG. 9—*Geometric correction factors versus epitaxial layer thickness for both insulating and conducting conditions, 100-μm probe separation, 2-μm contact radius.*

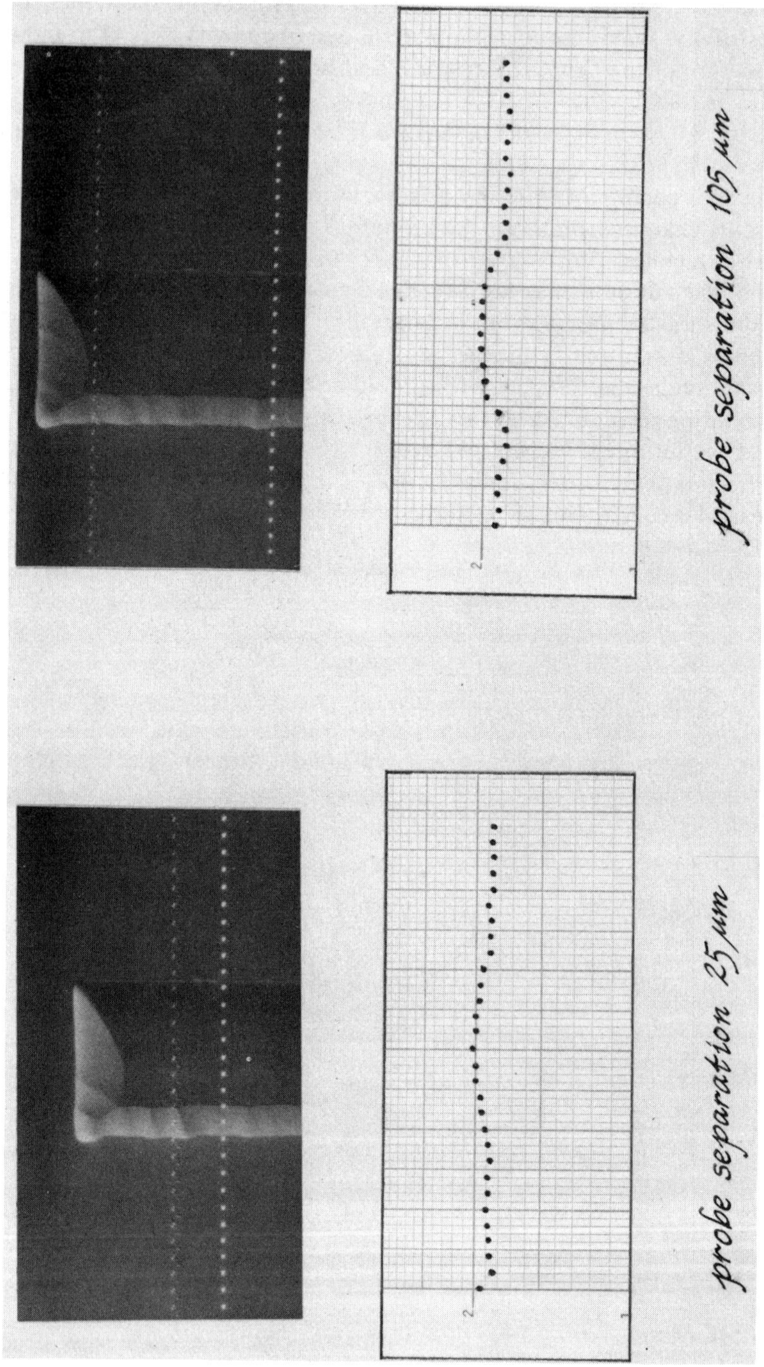

FIG. 10—*Effect of probe separation on buried layer resolution, 2-µm contact radius, 14-µm epi thickness, 5-µm step size.*

as the separation increases from 25 to 105 μm. A small increase is also seen in the resistance measured over the buried layer.

Figures 11*a* to 11*c* shows the result of epi thickness on the measured spreading resistance. Notice that as the epi thickness decreases from 8.5 to 2.8 μm the measured spreading resistance over the substrate decreases and over the buried layer increases as predicted. Measurable change is calculated by taking the decrease of the measured resistance over the substrate together with the increase over the buried layer with an epi thickness of 2.8 μm.

Measuring pattern shift by this method requires that a spreading resistance scan be made across the surface of the epi over the substrate and buried layer. A micrograph of the probe marks on the surface is made under differential interference illumination. The position of the pattern edge on the epi

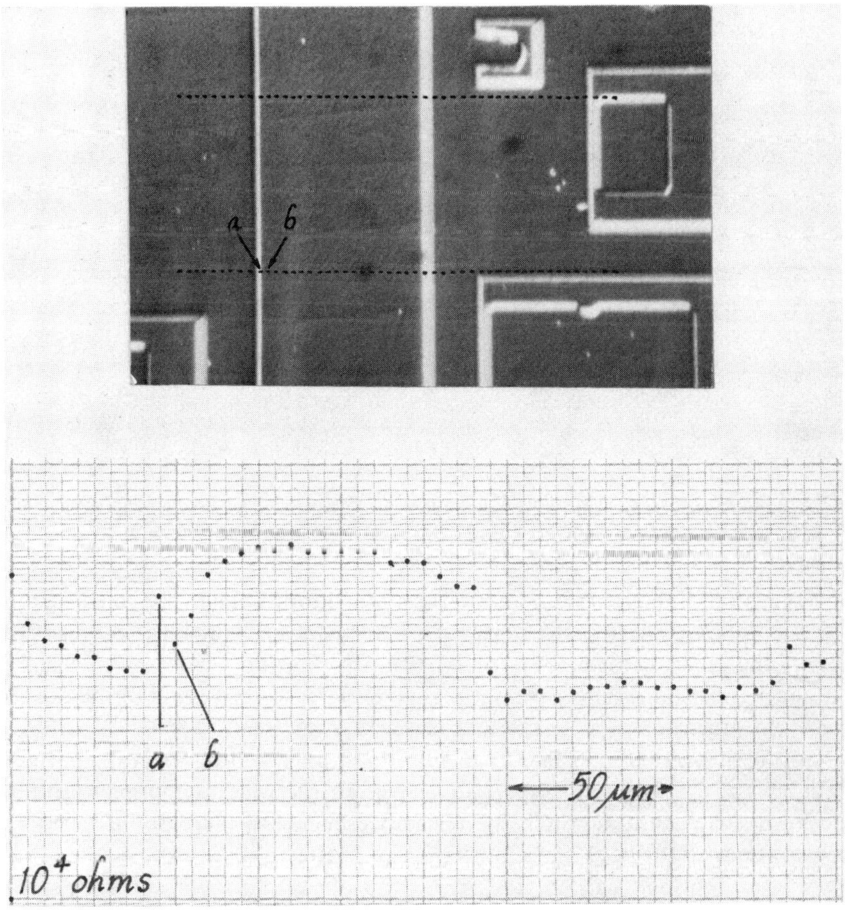

FIG. 11*a*—*Effect of 2.8-μm epitaxial layer thickness on buried layer resolution, 2-μm contact radius, 5-μm step size.*

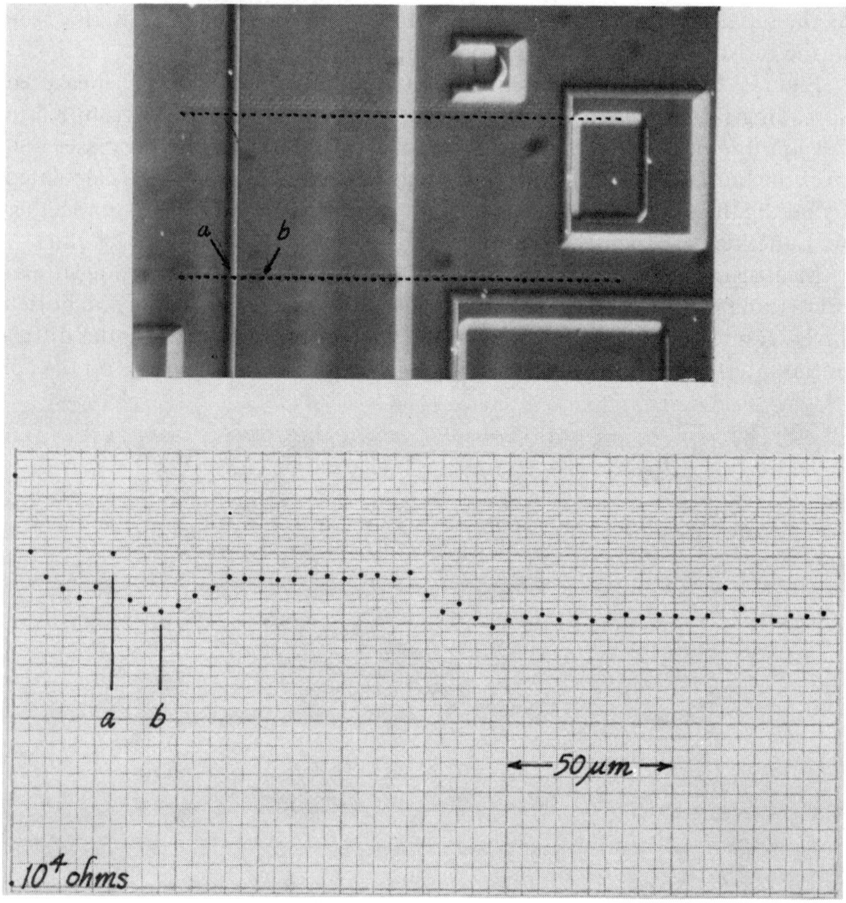

FIG. 11*b*—*Effect of 6.0-μm epitaxial layer thickness on buried layer resolution, 2-μm contact radius, 5-μm step size.*

surface is obtained from the micrograph by counting the probe marks from the start of the scan to the pattern edge. The position of the buried layer edge under the epi is determined from the plot by noting where the measured spreading resistance of the points begins to increase above that of the points over the substrate.

This is illustrated in Figs. 11*a* to 11*c*. Point *a* in each figure is the pattern edge on the epi surface as determined from the accompanying micrograph. Point *b* is the point at which the resistance begins to increase; it therefore corresponds to the edge of the buried layer on the original surface. Note that each point represents a step of 5 μm of the probe on the surface. (Probe separation was 100 μm and probe load was 10 g.) A comparison was made of the pattern shifts with this method and also with the groove-and-stain technique.

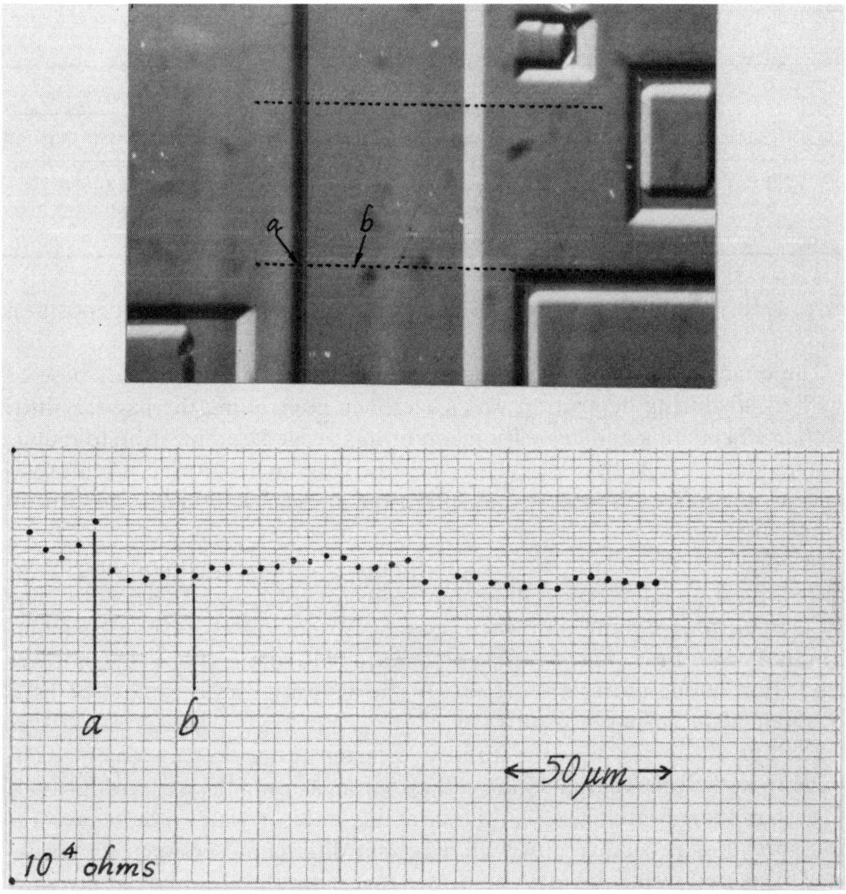

FIG. 11c—*Effect of 8.5-μm epitaxial layer thickness on buried layer resolution, 2-μm contact radius, 5-μm step size.*

Results (Table 1) showed better agreement for thinner epi. For thicker epi, resolution of the spreading resistance technique was limited.

Conclusions

It has been found that reduction of pattern shift requires deposition controls opposite to those required to minimize outdiffusion. In the latter case lower deposition temperatures (≤1100°C) are desired, but for controlling pattern shift higher deposition temperatures (≥1150°C) were found to be better. At 1150°C, lower deposition rates (≤0.6 μm/min.) showed reduced shifts. At lower deposition temperatures, the pattern shift became highly sensitive to the amount of misorientation from the exact (100) plane. At higher temperatures, however, the shift on both (100) and (111) was less sen-

TABLE 1—*Comparison of pattern shifts measured by the groove-and-stain method and by the spreading-resistance technique.*

	Absolute Pattern Shift, μm	
Epi Thickness, μm	Spreading-Resistance Method	Groove-and-Stain Method
2.8	5 (one point)	4.8
6.0	15 (three points)	16.6
8.5	30 (six points)	25.7

sitive to orientation and on (100) it became independent of the deposition rate.

The quantity of liberated chlorine atoms available in the reactor tube or the bell jar during deposition was a factor in controlling the pattern shift. Both *in situ* etching and deposition conditions needed optimization to reduce pattern shift since both these cycles controlled liberated chlorine. Dichlorosilane as the silicon source provided lower shifts at optimum deposition temperatures.

In conclusion, the two reactor configurations (horizontal and barrel) require almost the same controlling conditions for reducing pattern shifts. The shifts in both were minimum when the wafers were exactly on (100) orientation or 4 deg off the (111) orientation toward (110) plane. For (111), if lower deposition temperatures are desired, lower deposition rates should be used to obtain reduced pattern shifts.

The spreading resistance method is a faster and more convenient technique than the groove-and-stain-method for measuring the pattern shift, but it too has some limitations. A thinner epitaxial layer yields a better resolution spreading resistance plot with a greater differential between the resistance measured over substrate and buried layer, but the absolute pattern shift is also smaller for a thinner epi layer.

A second limitation is the resolution of the step size on the spreading resistance probe. Steps as small as 0.25 μm may be taken on the commercial probe used for this study, but with steps below 2.5 μm the probe marks merged into each other, which made it difficult to resolve the individual points.

The spreading resistance method holds promise as a nondestructive technique for monitoring pattern shift on epitaxial layers. However, problems of resolution and optimum conditions require further study.

Acknowledgments

The authors would like to acknowledge P. Butler of NBK Corporation for assisting with the X-ray diffraction topographic work, and J. R. Esparza and W. N. Whitten, Jr., both of Siliconix Inc., for conducting the epitaxial experiments. The guidance of Mr. Whitten during the various phases of the project is also appreciated.

References

[*1*] Boss, D. W. and Doo, V. Y., *Journal of the Electrochemical Society,* Vol. 113, No. 8, Aug. 1966, p. 59.
[*2*] Drum, C. M. and Clarke, C. A., *Journal of the Electrochemical Society,* Vol. 115, No. 6, June 1968, p. 664.
[*3*] Bean, K. E. and Gleim, P. S., *Proceedings of IEEE,* Vol. 57, No. 9, Sept. 1969, p. 1469.
[*4*] Hammond, M. L., *Microelectronics Journal,* Vol. 10, No. 2, Feb. 1979, p. 4.
[*5*] Drum, C. M. and Clark, C. A. *Journal of the Electrochemical Society,* Vol. 117, No. 11, Nov. 1970, p. 1401.
[*6*] Lee, P. H., Wauk, M. T., Rosler, R. S., and Benzing, W. C., *Journal of the Electrochemical Society,* Vol. 124, No. 11, Nov. 1977, p. 1824.
[*7*] Weeks, S. P., *Solid State Technology,* Vol. 24, No. 11, Nov. 1981, p. 111.
[*8*] McDonald, B. and Goetzberger, A., *Journal of the Electrochemical Society,* Vol. 109, No. 2, Feb. 1962, p. 141.
[*9*] Dickey, D. H., "Diffusion Profiles Using a Spreading Resistance Probe," *Extended Abstracts of the Electronics Division, The Electrochemical Society,* Vol. 12, 1963, p. 151.
[*10*] Dickey, D. H., "Two Point Probe Correction Factors," NBS Special Publication 400-10, Semiconductor Measurement Technology: Spreading Resistance Symposium, 1974, p. 45.

Lionel S. White,[1] *G. R. Mohan Rao,*[1] *Peter Linder,*[1] *and Maury Zivitz*[1]

Improvement in MOS VLSI Device Characteristics Built on Epitaxial Silicon

REFERENCE: White, L. S., Mohan Rao, G. R., Linder, P., and Zivitz, M., "**Improvement in MOS VLSI Device Characteristics Built on Epitaxial Silicon,**" *Silicon Processing, ASTM STP 804,* D. C. Gupta, Ed., American Society for Testing and Materials, 1983, pp. 190–205.

ABSTRACT: Two advantages (freedom from ground loops and enhanced performance of diffused guard rings) of P^+/P^- epitaxial versus homogeneous crystal substrates in 64K DRAM are presented. Reduced refresh time and pattern sensitivities of the early TMS 4164, built on homogeneous substrate, are chronicled. To better understand the significant improvements with epitaxy, models are investigated for substrate capacitive/drift current coupling and for guard rings. Substrate-coupling modeling primarily used an instantaneous view of potential distributions via solution of Laplace's equation, followed by time-domain analysis. A comparison is made with a discrete resistor mesh. Guard ring study was empirical. Sets of biased diffused annular rings, surrounding an n^+ dot used for minority carrier injection, were fabricated on both types of substrate. Results of the theoretical and experimental studies are compared with device performance.

KEY WORDS: very large scale integration (VLSI), guard rings, minority carrier injection, DRAMS, substrate noise, epitaxy

A P^+/P^- epitaxial substrate offers two excellent design advantages in the manufacture of high-density NMOS products such as 64K dynamic random access memories (RAMs). The epi structure serves as an excellent ground plane, damping substrate noise/crosstalk. In conjunction with a biased, diffused guard ring it forms an almost impenetrable barrier to injected minority carriers.

Rationale

The primary reason to employ an epitaxial substrate for TMS4164 64K dynamic RAM was the susceptibility of the device to refresh disturb (zero background) algorithms on homogeneous crystal substrates (12 to 15 Ω-cm).

[1] Texas Instruments, Houston, Tex. 77001.

Since the device utilizes a grounded substrate, any node that drops a few tenths of a volt below the substrate potential (or any node where the substrate rises a few tenths above the node voltage) will begin heavily injecting minority carriers into the substrate. The TMS4164 was first attempted on homogeneous crystal substrates. Through internal probing of the device, it was determined that a ground node was being forward-biased by the positive transition of the substrate during the precharge portion of the device cycle (Fig. 1). The positive bias in the substrate was the result of the capacitance coupling of the bit sense lines (BSL) into the highly resistive substrate. The resulting failure was reduced refresh time of the cells closest to the sense amplifier. In addition, reduced refresh time was observed for cells at the ends of the cell array that were closest to the device's peripheral circuitry. A circuit solution was attempted on homogeneous crystal that added a wide guard ring between the injecting ground diffusion and the cell array and the peripheral circuitry and the cell array (Fig. 2). The guard ring between the cell array and periphery consisted of interlaced fingers of N^+ diffusion. Half of the fingers were biased to 5 V and the other half were pulsed from 5 V to ground during the period that the BSLs were being precharged to 5 V. The results were a significant but insufficient reduction in the failure rate of the cells along the sense amplifier portion of the array. Depending on the mode of operation of the interlaced fingers between the periphery and the cell array, the cells of interest either showed improvement similar to that described above (all fingers biased to dc +5.0 V) or showed improvement such that

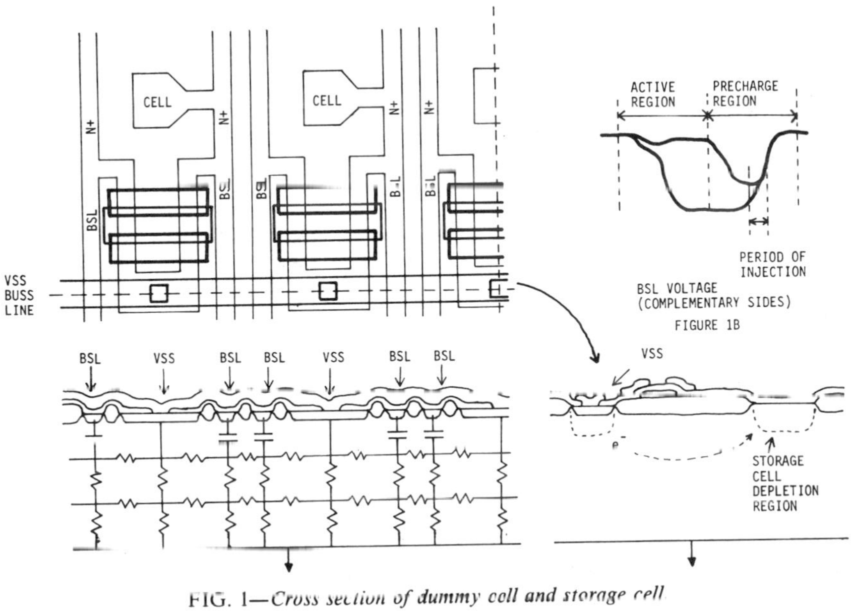

FIG. 1—*Cross section of dummy cell and storage cell.*

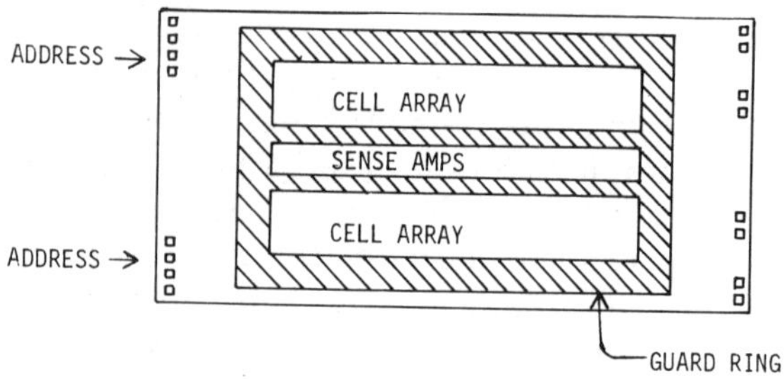

FIG. 2—*Circuit layout on die.*

their refresh times exceeded that of the cells in the center of the array (alternate fingers were pulsed during precharge). It was concluded from this experiment that (*1*) to provide satisfactory performance of the storage cells it would be necessary to eliminate charge injection either by adding a substrate bias or by severely attenuating the substrate transient, (*2*) a more effective guard ring would be necessary to serve as a barrier for minority carriers that appeared in the substrate by means other than junction forward bias injection (for example, impact ionization), and (*3*) intrinsic refresh time of a storage cell could be improved by reducing the substrate drift current that occurred during device precharge.

In addition to reduced refresh times, pattern sensitivity to a walking one/zero algorithm and to the multiplex timing of the address lines was noted on the early devices. The pattern sensitivity to walking one/zero pattern (for instance, background of zeroes/ones with one column out of the 256 trying to sense opposite data, one/zero) was attributed to crosstalk between the transition of the 255 BSLs that formed the background pattern and the test column that was making an opposite transition. This was a purely capacitive coupling effect between the junction capacitances of the parallel diffused bit lines (Fig. 3). Multiplex timing of the addresses could also disturb the sensing sequence due to capacitively coupled noise (Fig. 3). If the address inputs are making a transition from row to column addresses during the sensing period such that the voltages on the one side of the die are making transitions opposite to those on other side of the die, the capacitance coupling into the substrate will bump one side of the array slightly negative while the other side will be bumped slightly positive. This condition is guaranteed to occur under some valid combination of data and address.

The epitaxial substrate has greatly attenuated these effects that were seen on homogeneous crystal substrates. Minority carrier injection and crosstalk have been reduced beyond the capability of observation.

The remainder of this paper will describe the attempt to quantify the substrate phenomena.

Two methods have been employed to determine the a-c transient in the substrate during operation of the device. The first method is a continuum model to extract the static effect of a voltage difference in the substrate. This

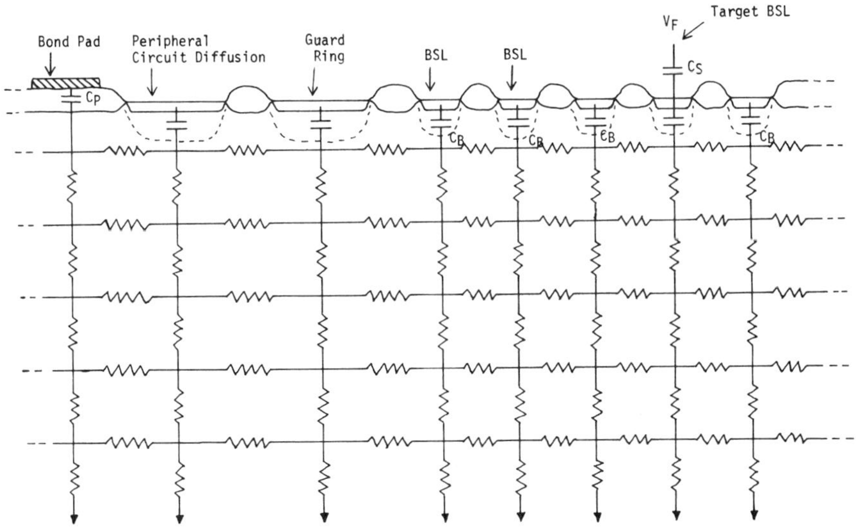

FIG. 3 *Substrate cross-section represented as resistor–capacitor mesh network.* C_P = *capacitance of bond to substrate,* C_B = *BSL junction capacitance,* C_S = *stray capacitance of a BSL, and* V_F = *hypothetical fixed voltage that* C_S *is connected to.*

is then coupled with an a-c model by using an equivalent diode-resistor circuit to determine the voltage response in the substrate. This first method is followed by a study that models the substrate as a mesh of discrete resistors. The a-c response is obtained using a SPICE simulation program. In addition to modeling the substrate a-c transient, the efficiency of a guard ring/epi substrate in collecting minority carriers was studied. The circuit that was used is a set of concentric diffused rings about a center injecting point. The structure is then studied as a lateral NPN transistor. Finally, a theory is proposed that relates degraded refresh time during a refresh disturb algorithm to a drift current in the substrate.

Continuum Model for Bit Line Coupling

Modeling of bit-line crosstalk paths through the substrate can be readily done by use of a mesh of discrete resistors. Even so, such a mesh is only tractable in two-dimensional form. The directions of interest here are the substrate thickness and a direction both perpendicular to the bit lines and parallel to the surface of the slice. We found it more convenient to include the bit-line capacitance components in a continuum model for the substrate resistance. Compare Fig. 4(top) and Fig. 4(bottom) for the relation between cross sections of the actual bit-line structures and the continuum modeling structure. The case depicted is all lines going low (active pull-down), except one "target" line floating high. A substrate voltage dip due to active discharge of background bit lines during sensing will alter the potential of a single bit line (the target), which should remain high.

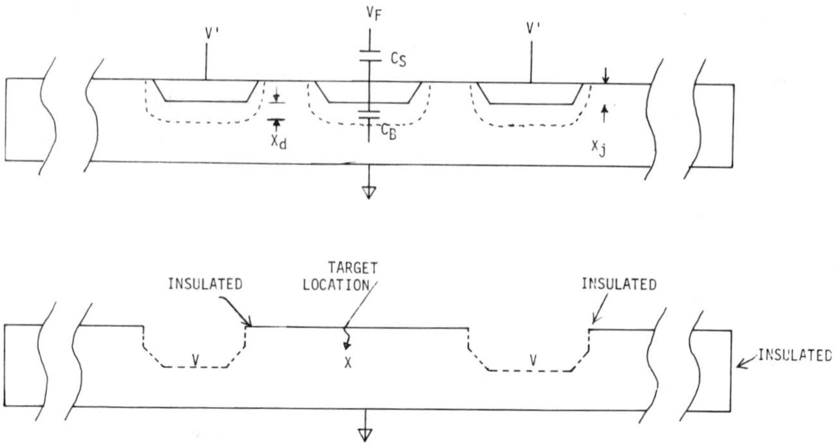

FIG. 4—(*top*) *Actual BSL structure.* (*bottom*) *BSL structure as used in solution of Laplace's equation for continuum model.*

An instantaneous view of the potential distribution within the substrate gives a ratio between V_{xd}, the potential at a driven-line depletion boundary, and V_x, the potential at the site of the floating-high target line. We assume this view to be equivalent to a steady DC conductive picture, that is, we ignore fields induced by changing currents. Further, it can be shown[2] that the potential distribution arising from a steady current is identical to the potential that occurs in an electrostatic problem with the same boundary conditions. If we further neglect changes in depletion layer width as a function of time (only considering results good for small values of elapsed time), we are left with the equivalent circuit shown in Fig. 5, where C_B is the depletion capacitance of the bit-sense line and C_S the stray capacitance. The only difficulty remaining with this model is establishing correlation between external drive V' and depletion boundary potential V, as illustrated in Fig. 4. This correlation will be discussed in the next section.

The initial task is to solve electrostatic boundary value problems, under differing conditions of interest. The major cases studied were (*1*) homogene-

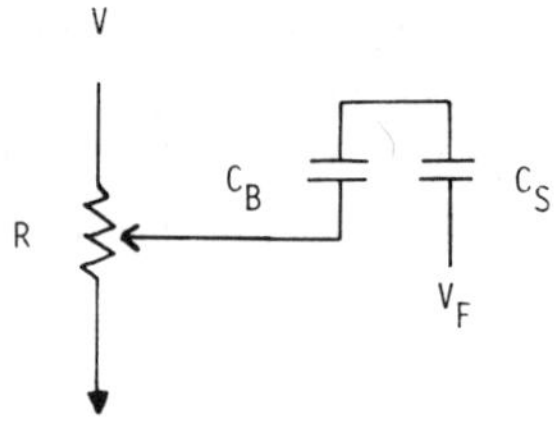

FIG. 5—*Equivalent circuit for background BSL potential coupling to a target BSL.*

[2] Moon, P. and Spencer, D. E. in *Foundations of Electrodynamics*, Van Nostrand, Princeton, N.J., 1980, pp. 86-87.

ous substrate and (2) a substrate with an expitaxial layer. The expitaxial layer introduces only mildly complicating features in the calculations. Since no net charge exists at interior points of the boundaries (we ignore the epi/substrate interface space charge), the task reduces to solution of Laplace's equation in two dimensions:

$$\frac{\partial^2 \phi}{\partial X^2} + \frac{\partial^2 \phi}{\partial Y^2} = 0$$

This continuum model used a square mesh for numerical integration. Mesh densities and other data are provided in Table 1. Kernels used for various types of interior points are given in Table 2. (Kernels II, III, and IV were developed by the authors, although no claim is made of originality.) Cases II and III use symmetry arguments, and Case IV guarantees continuity of current at the substrate-EPI interface.

The approximations used for numerical integration of Laplace's equation can, in one case, be given a direct interpretation in terms of discrete resistors connected at Nodes A, B, C, and D of Fig. 3. These four resistors connect four potentials to a central node. The potential at this central node is exactly the arithmetic mean of the four nearest neighbor nodes. The results of the Laplacian solution were used directly to give the potential ratio V_x/V_{xd}, which is a relative measure of the substrate resistive coupling. Because only ratio information is sought at this point of the present study, an arbitrary potential could be used on the boundary of the depletion region of the driven bit lines. Other parameters (resistance) were of interest. By use of the potential field values, numerical calculation of the electric field along a driven-line depletion boundary was made.[2] By use of Ohm's law in the form

$$\vec{J} = \sigma \vec{E}$$

and the definition of current density

$$J = I/A$$

where $\vec{J}$ is current density, σ is conductivity, $\vec{E}$ is electric field, I is current, and A is area, the total current pumped through a given bit line could be determined. Lastly, the ratio of bit line potential to current through a bit line gives the resistance of any bit line to ground (zero) potential.

Table 1 lists the results of the numerical solution of the continuum model. The static solution was first solved by using a TI 990 computer. The maximum array size feasible was approximately 740 by 250. First the static coupling was studied as a function of epi thickness, assuming that the low-resistivity substrate (1×10^3 lower resistivity) was essentially a ground plane. This allowed numerical integration grid spacing of 0.25 μm. Secondly, the

TABLE 1—*Continuum model coupled with ΔV from discrete diode/resistor simulation. Result is ΔV under target bitline.*[a]

Run	Kernel Distance, μm	BSL	Epi Thickness, μm	Epi ρ, Ω-cm	Substrate ρ, Ω-cm	R Effective, Ω	Coupling Factor	ΔV Under Target BSL, V: ρ = 2.5 Ω-cm	ρ = 5.0 Ω-cm	ρ = 10.0 Ω-cm	ρ = 20.0 Ω-cm	ρ = 40 Ω-cm
1[b]	0.25	8	20	10	0	1.1E5	0.36	0.014	0.028	0.056	0.11	0.22
2[b]	0.25	8	10	10	0	5.5E4	0.11	0.0021	0.0043	0.0085	0.017	0.03
3[b]	0.20	8	5	10	0	2.6E4	0.0072	0.00013	0.0003	0.0005	0.001	0.002
4[c] (homogeneous case)	1.0	37	...	10	10	1.2E6	0.933	...	...	0.135	...	...
5[c]	1.0	37	10	10	1	1.8E5	0.70	...	...	0.018	...	...
6[c]	1.0	37	10	10	0.1	6.9E4	0.24	...	...	0.0024	...	...
7[c]	1.0	37	10	10	0.01	5.9E4	0.13	...	...	0.0001	...	...

[a] Coupling factor is the ratio of voltage from the discharge of the BSLs that appears under the target BSL to the voltage that is generated by the discharging BSL's.

[b] Assuming 10-μm thick epi with ground potential at epi interface.

[c] Assuming 10-μm thick epi with ground potential at back of slice (total thickness is 250 μm).

TABLE 2—*Kernels used for numerical solution of Laplace's equation.*

Notation: U = previous iteration value
V = present iteration value
Indices, where I is in substrate thickness direction:
$V(I, J) = V_{0,0}$
$V(I + 1, J) = V_{+,0}$

CASE I—General interior point, not on interface.

$$V_{0,0} = (U_{0,-} + U_{0,+} + U_{+,0} + U_{-,0})/4$$

CASE II—Boundary point on an insulated wall, (for instance, left-hand wall).

$$V_{0,0} = (2U_{0,+} + U_{+,0} + U_{-,0})/4$$

CASE III—Boundary point at an insulated corner (for instance, top left-hand corner).

$$V_{0,0} = (2U_{0,+} + 2U_{+,0})/4$$

CASE IV—Interior point lying on a row that is a conductivity interface.

Example: Upper rows correspond to σ_1 material; lower rows correspond to σ_2 material.

Definitions: $C1 = \sigma_1/[2(\sigma_1 + \sigma_2)]$
$C2 = \sigma_2/[2(\sigma_1 + \sigma_2)]$

$$V_{0,0} = C2(U_{+,-} + U_{+,+}) + C1(U_{-,+} + U_{-,-})$$

static coupling was studied as a function of substrate resistivity with a constant epi thickness of 10 μm. The spacing was 1 μm. The bit line spacing was assumed to be 20 μm with the depth of the depletion region being 3 μm and the width being 4 μm. The a-c transient on the BSL was taken from SPICE simulations of the TMS4164 and is approximately 0.3 V/ns. Room temperature conditions were assumed.

Resistor Mesh Substrate Model

The substrate is decomposed into a mesh of resistors and simulated by using the SPICE simulation program. Each BSL is simulated as two parallel diodes with the anodes connected to a surface node in the resistor mesh and the cathodes connected to a piece-wise linear voltage source. The two parallel diodes represent the bulk and sidewall diodes of the BSL. The voltage source has the same waveform used in the continuum study. The target BSL is modeled as two parallel diodes connected to a node of the resistor mesh. The cathodes of the diodes are connected to a capacitor with the other side of the capacitor tied to ground. The capacitor represents the stray capacitance C_S of the BSL other than the depletion capacitance and is assumed to be tied to a fixed potential. The common node of the diodes and capacitor is considered to be the BSL node. The resistance values used are derived assuming the resistor to be a block that is square and 1 μm thick in a direction parallel to the BSL. Thus the resistance is given by the equation

$$R = \rho\frac{L}{A} = \rho\frac{L}{Lt} = \rho/t$$

where t = 1 μm and where R is resistance, ρ is resistivity, L is length, A is area, and t is thickness. The simulations were run to obtain two values: the resistance of a BSL to ground and the a-c transient coupling to the target BSL. These can then be compared to the continuum model. The results are shown in Table 3.

Collection Efficiency of Guard Rings

The injection of minority carriers into the substrate constitutes an important failure mechanism in high-density DRAMs, because these carriers are all too often collected by the depletion layers of the memory cells, thereby destroying the stored data. Minority carriers can be injected in a number of ways. The most common mechanism in dynamic circuitry with a grounded substrate is for an N-diffusion to be capacitively coupled to a negative voltage, thereby forward-biasing the P-N junction. Because this is a capacitive effect, the amount of injected charge is usually fixed (as opposed to being a fixed negative voltage with unlimited current). An additional source of injection in DRAMs is through the address-input diffusion when an address input is negative. For this case it is necessary to specify a maximum injection current level. The third source of injection is through the hot electron phenomenon of MOSFETs.

In order to achieve proper DRAM operation in spite of a significant amount of minority carrier injection, it is necessary to keep the carriers away from the memory cells. A cursory analysis would indicate that this could be quite easily accomplished by making the minority carrier extrinsic Debye length much shorter than the distance from an injection source to a memory cell. However, this is exactly contrary to what is needed for good DRAM operation. The substrate doping, P^-, cannot be changed to meet this criterion. The only other alternative is to surround the injection source (or the memory cells) with a guard ring whose function is to absorb the injected carriers. Our goal, then, is to increase the effectiveness of these guard rings.

Epitaxial material can be used to increase the effectiveness of the guard rings by reflecting electrons in the P^- layer into the ring or by enhanced recombination of these that enter the P^+ substrate. There is a built-in downward directed electric field at the P^+/P^- interface. This interface will reflect

TABLE 3—*Results of resistor mesh simulation.*[a]

Run	Substrate ρ, Ω-cm	R Effective, Ω	ΔV BSL, V	ΔV Under BSL, V
8	2.5	...	0.040	0.043
9	5.0	...	0.070	0.076
10	10.0	1.24E6	0.080	0.141
11	20.0	...	0.150	0.257
12	40.0	...	0.260	0.460

[a] Mesh spacing is 5 μm; assumed thickness of slice is 250 μm; BSL = 6.

any minority electrons incident upon it from the P^- region. As sketched in Fig. 6, this will help guide the injected electrons along the surface, thereby reducing the average distance travelled by the electrons in going from the emitter to the collector of the lateral transistor. This will reduce the amount of electrons that recombine and consequently will increase the β (current gain) of the structure. Much more importantly, the collection efficiency of the guard ring will be improved. It might also be feasible to bias the guard ring with a high enough positive voltage so that its depletion region punches all the way through the P^- layer. There would then be no path by which minority carriers injected from the periphery could enter the array.

It should be mentioned that the only mechanism by which electrons might not be reflected from the P^+/P^- interface would occur if injection through an input diode were so heavy as to locally forward-bias the interface junction. The electrons would then be injected into the P^+ substrate where they would quickly recombine with holes, thereby posing no threat to memory cells. When using epitaxial material, we should expect two other effects, both of them stemming from the fact that the P^+ material results in a much lower surface-to-back-side resistance. For a fixed negative voltage imposed on an input diode, higher currents will be injected with epi than with homogeneous crystal material. While not valid for capacitively coupled nodes, this should be kept in mind for address inputs. It is for this reason that address input currents should be limited. The other effect involves rapidly changing diffusion voltages. With the reduced "base" resistance of the epi material, we should expect a shorter dielectric relaxation time and therefore better high-frequency performance of the guard rings.

There is one last aspect of guard ring operation that has not yet been mentioned. The situation often arises in which a guard ring is desired around a known injector of minority carriers, and a V_{SS} (ground) but no V_{DD} (positive supply) diffusion is conveniently available. In such a situation it would be nice to be able to use the V_{SS} diffusion.

Measurements were performed on a structure like that sketched in Fig. 7, consisting of concentric N-type diffusion rings surrounding an N-type central "dot." In operation, the substrate was grounded, electrons were injected from the center dot, and the rings were monitored. The surface P-layer of the

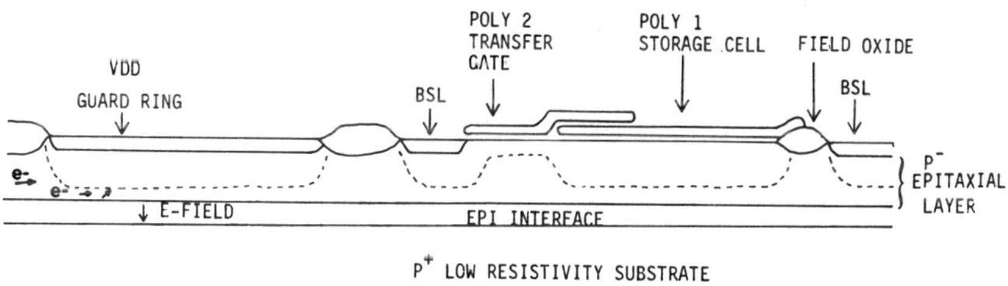

FIG. 6—*Cross section of epitaxial interface.*

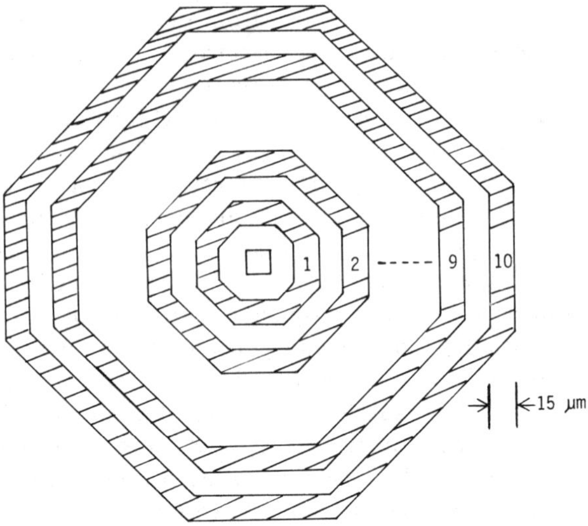

FIG. 7—*Fabricated guard ring structures, consisting of a central diffused dot plus rings.*

epitaxial wafer had the same doping as the homogeneous crystal wafer. Although the ring structures on the homogeneous crystal and the epi wafers were not strictly identical, they were very similar. The static tests were performed over a wide range of currents. The low injection conditions would appropriately model capacitively coupled diffused nodes and hot-electron effects, while the high-current conditions would model the negative address input problem. Care had to be taken to properly shield the wafers from ambient light.

The "middle dot" electron injector was connected to a current source and all tests were performed at 10 mA, 1 mA, 100 μA, 10 μA, and 1 μA of emitter current. The first ring (R1) was connected to a +5-V supply; the currents into R1 and into the substrate were measured with a digital multimeter (DMM) with all the other rings floating. The DMM had a resolution of 10 nA. Next, R1 was connected to ground with all the other rings floating and again the current into R1 was measured. The same two measuremens were repeated for Rings R3 and R10. In order to test the effectiveness of guard rings, a set of measurements were subsequently made for which one ring, the collector ring, was held at 5 V and its current monitored while other rings between the collector ring and the middle dot were biased to 0 V or 5 V. Measurements were made with R3 as the collector ring and with R1 at 5 V, with R1 and R2 at 5 V, with R1 at 0 V, and with R1 and R2 at 0 V. Similarly, R10 was used as the collector ring with R1 at 5 V, R1 and R4 at 5 V, R1 and R4 and R7 at 5 V, R1 at 0 V, R1 and R4 at 0 V, and finally with R1 and R4 and R7 at 0 V.

The results of the measurements are presented in Tables 4 and 5. Table 4 indicates the current gain, or β = (ring current/substrate current), for each of the three collector rings for each of the five emitter currents. As expected,

TABLE 4—*Comparison of guard-ring transistor actions.*

	Homogeneous Crystal		EPI	
Current	β	Current Gain[a]	β	Current Gain[a]
10 mA	R1 → 1.7 R3 → 0.2 R10 → 0.04	87% 81% 77%	R1 → 8.2 R3 → 1.3 R10 → 0.2	96% 94% 97%
1 mA	R1 → 4.2 R3 → 0.5 R10 → 0.1	84% 78% 74%	R1 → 19.5 R3 → 2.8 R10 → 0.5	97% 96% 98%
100 μA	R1 → 5.6 R3 → 0.6 R10 → 0.2	80% 72% 63%	R1 → 16.6 R3 → 3.0 R10 → 0.7	97% 96% 98%
10 μA	R1 → 6.7 R3 → 0.8 R10 → 1.1	82% 59% 58%	R1 → 9.0 R3 → 2.0 R10 → 0.4	96% 95% 97%
1 μA	R1 → 8.8 R3 → 1.0 R10 → 3.9	81% 50% 68%	R1 → 4.0 R3 → 1.3 R10 → 0.2	94% 94% 95%

[a] Ring current at 0 V as percentage of ring current at 5 V.

β falls off rapidly with increasing distance from the collector to the emitter. The variations of β with emitter current are rather more difficult to explain, but a few simple models are appropriate. It should be noted first, however, that increases in emitter current were accomplished (by the current source) by increasing the emitter-to-base voltage. There was no increase in base-to-collector voltage, however. Hence, we would not expect any base-width modulation effect, which in normal vertical NPN structures accounts for increasing β with increasing voltage drop-collector to emitter (V_{CE}). For the single-crystal wafer, we would expect that as the emitter voltage becomes increasingly negative with respect to the substrate (i.e. larger emitter currents), the vertically directed component of the emitter current would become larger in proportion to the horizontally directed component. This would result in an increase in effective base width and hence a decrease in β for increasing emitter currents, as indicated by the data in Table 4.

For epitaxial material the vertical component would also increase in proportion to the horizontal component for increasing emitter current, but this would have little effect on the base current, since the emitter current is simply reflected and redirected at the P^+/P^- interface. At large enough emitter currents the P^+/P^- junction might however become forward-biased as discussed earlier, causing a rapid decrease in β, as indicated in Table 4 for 10 mA of emitter current. Table 4 also indicates that for currents below about 100 μA for the epi material, β decreases with decreasing current. The reason for this is unknown, but a possible explanation might be the existence of a fixed number of generation-recombination centers at the P^+/P^- interface that place

TABLE 5—*Guard ring efficiencies under varying bias conditions.*

Emitter (Center Dot) Current	I_{R3} expressed as % of I_{R3} Without Guard Rings								I_{RIO} expressed as % of I_{RIO} Without Guard Rings											
	R1 = 5 V		R1 = R2 = 5 V		R1 = 0 V		R1 = R2 = 0 V		R1 = 5 V		R1 = R4 = 5 V		R1 = R4 = R7 = 5 V		R1 = 0 V		R1 = R4 = 0 V		R1 = R4 = R7 = 0 V	
	Single Crystal	Epi	Single Crystal	Epi	Single Crystal	Epi	Single Crystal	Epi	Single Crystal	Epi	Single Crystal	Epi	Single Crystal	Epi	Single Crystal	Epi	Single Crystal	Epi	Single Crystal	Epi
10 mA	37.7	0.5	29.0	0.0	49.4	5.3	43.8	0.4	37.2	0.4	32.6	0.0	30.2	0.0	44.2	0.6	39.5	0.2	37.2	0.2
1 mA	24.9	0.0	15.0	0.0	42.0	2.9	28.8	0.2	21.8	0.0	10.9	0.0	7.9	0.0	38.6	1.8	26.7	0.1	20.8	0.0
100 μA	23.9	0.0	14.1	0.0	45.5	2.5	31.3	0.1	4.8	0.0	1.7	0.0	1.4	0.0	29.3	1.1	15.6	0.0	12.8	0.0
10 μA	17.4	0.0	10.3	0.0	38.7	2.5	25.2	0.1	0.4	0.0	0.4	0.0	0.4	0.0	2.7	1.3	1.5	0.0	1.5	0.0
1 μA	13.5	0.0	7.7	0.0	28.8	1.9	19.2	0.0	1.3	0.0	1.3	0.0	1.3	0.0	1.3	0.0	1.3	0.0	1.3	0.0

a lower limit on the base defect current and therefore reduce β for small emitter currents.

Indicated on Table 4 is also the ring current with the ring at 0 V, expressed as a percentage of the ring current with the ring at 5 V. It should be pointed out that these numbers are consistently and significantly higher for the epi wafer than for the homogeneous crystal wafer. This is doubtlessly caused by the fact that the vertical component of emitter current is larger for the homogeneous crystal wafer when the ring voltage is 0 V than when the ring voltage is 5 V. Because of P^+/P^- interface reflection, this has little effect on epitaxial material.

Table 5 shows the results of the measurements made to evaluate guarding effectiveness. The table is divided into two sections, one for the case where Ring 3 served as the collector, and the other for Ring 10 as the collector. In each instance the collector (ring) current with the indicated guard rings at the indicated voltages is expressed as a percentage of the collector current in the absence of guard rings. The worst case for both single crystal and epitaxial materials occurs at 10 mA of emitter current, with R3 as the collector and R1 at 0 V as the only guard ring. For this case, the collector current is 49.4% of its guard-ring-free value for homogeneous crystal material and 5.3% for epitaxial material.

The enhanced effectiveness of guard rings on epitaxial material is clear throughout Table 5. In fact, in many cases the collector current was reduced to less than 10 nA, which resulted in a 0.0% entry in the table. It should also be pointed out that the guard ring effectiveness on epi material tends to increase with decreasing emitter current in spite of the β roll-off with decreasing current. This helps substantiate the lossy interface explanation for β roll-off, since the guard rings tend to pull the current away from the P^+/P^- interface. For homogeneous crystal material, the fact that a larger portion of the emitter current is near the surface for small emitter currents than for large ones explains the increase in guard ring effectiveness with decreasing emitter current.

Drift Current During Precharge

An often ignored aspect of refresh degradation is a drift current component that results from the voltage transient in the substrate during the precharge portion of the cycle. This drift component will exist regardless of substrate bias or lifetime degradation mechanisms incorporated in the substrate material. It will be a very strong function of substrate resistivity.

During a substrate voltage transient an electric field is generated that originates at the surface of the chip and is terminated at the backside of the chip. The electric field is directed such that it repels holes from the surface of the chip and attracts electrons to the surface. This transient will exist until sufficient hole current can flow to bring the substrate back to equilibrium potential. (This phenomenon is identical to the action that occurs as a PN junction depletion region expands due to a change in voltage. If a step function voltage is applied to a PN junction such that it expands the space charge region, the holes are swept out of the space charge region and must then flow out of the diode to bring the bulk of the diode back to equilibrium potential. The speed with which this occurs is directly proportional to the bulk resistivity of the *p*-type material). As the resistivity is increased to reduce bit line capacitance, both the concentration of minority carriers and the voltage transient in the substrate increase. The worst case pattern will be a background of zeroes with a single column trying to retain ones. The following equation approximately expresses what the theoretical number of electrons N_e collected should be

$$N_e \approx \frac{A_C}{A_C + A_{BSL}} \cdot A_T \cdot n_p \cdot \mu_e \cdot \frac{T_{REF}}{T_{CYCLE}} \int E_S(t)dt$$

where

T_C = area of the depletion region under the storage cell,
A_{BSL} = area of the BSL depletion region per cell,
A_T = total area of the cell,
T_{CYCLE} = cycle time at which the device is operated,
E_S = electric field under the storage cell due to the substrate precharge transient, and perpendicular to the slice surface,
T_{REF} = refresh period specified for the device,
n_p = minority carriers (electrons) in the P-type region, and
μ_e = mobility of an electron in silicon.

No effort was made to precisely correlate this equation to an observed phenomenon. However, both the experiment on homogeneous crystal using the interlaced guard band and observations of pattern sensitivity of 16K DRAMs built on high-resistivity material tend to support this theory.

Conclusions

The rationale for converting a VLSI NMOS product (64K DRAM) from homogeneous crystal substrates to epitaxial substrates has been described. The two primary advantages of a P^+/P^- substrate are damping of substrate voltage transients and efficient collection of minority carriers injected into the substrate. Substrate voltage transients can couple noise into circuit nodes, causing a malfunction of the circuit. It has been shown that noise attenuation up to two orders of magnitude is possible with P^+/P^- substrate. In addition, voltage transients in the substrate can create an electric field such that minority carriers in the substrate are channeled into the depletion region of the storage cell, thus reducing the refresh time of the device. This is a drift current component of cell leakage that has generally been ignored. The epitaxial substrate in conjunction with a V_{DD} guard is an extremely effective barrier to minority carriers, due to the lateral NPN transistor action of the guard ring. Protection of up to six orders of magnitude was demonstrated.

Acknowledgments

We express our sincere gratitude to Carol Sirianni, Bart Holland, Jeanne Hughes, and Dinh Dung for assistance with simulations, preparation of figures, and monitoring progress of the numerical solution of Laplace's equation.

M. L. Hammond[1]

Reduced-Pressure Chemical Vapor Deposition of Polycrystalline Silicon

REFERENCE: Hammond, M. L., "**Reduced-Pressure Chemical Vapor Deposition of Polycrystalline Silicon,**" *Silicon Processing, ASTM STP 804,* D. C. Gupta, Ed., American Society for Testing and Materials, 1983, pp. 206–217.

ABSTRACT: The practice of depositing polycrystalline silicon by low-pressure chemical vapor deposition (LPCVD) is discussed in terms of effect of deposition parameters on structural property relationships. Electrical and oxidation properties are also discussed.

KEY WORDS: polycrystalline silicon, integrated circuits, interconnects, chemical vapor deposition, low-pressure chemical vapor deposition, reduced-pressure chemical vapor deposition

Thin polycrystalline silicon (poly Si) films are vital to MOS integrated circuits for gate electrodes and interconnects. Poly Si also has use as silicon resistors, emitter diffusion sources, and as a lossy dielectric, when doped with oxygen (O_2). Future applications of poly Si include solar cells and laser-annealed films for device structures.

Chemical vapor deposition (CVD) poly Si films were initially prepared in silicon epitaxy reactors; however, the development on low-pressure chemical vapor deposition (LPCVD), with wafers standing vertically in a heated tube, revolutionized the process by reducing the cost of depositing a poly Si film by more than 90%. This reduction in cost, combined with the significant improvements in film uniformity, allowed tremendous growth in poly Si gate MOS devices [*1–4*].[2]

Commercial Reactor Design

In silicon epitaxy reactors, the wafers lay against a heated surface and reactive gases flow across the wafer surface. In LPCVD systems, the wafers are standing up parallel to each other and coaxial with the process tube. To prevent deposition on the wafer backside and to double production, wafers

[1] Vice President, Technical Director, TETRON Inc., Cupertino, Calif. 95014.
[2] The italic numbers in brackets refer to the list of references appended to this paper.

are sometimes loaded back-to-back in the same slot. The reactive gas flows in the annular region between the wafer edge and the tube wall. Typical flow conditions are noted in Table 1. Gaseous HCl is sometimes used before or during deposition as a desiccant and purifying agent. SiH_4 is the chemistry of choice for MOS poly Si interconnects. SiH_2Cl_2, $SiHCl_3$, and $SiCl_4$ react at temperatures above 800°C where chlorine-based chemistry will etch exposed single-crystal silicon and also attack weak areas in the oxide. With a reaction temperature range of 550 to 650°C, SiH_4 provides good control of grain structure with no attack of the oxide or exposed single crystal.

Two different vacuum pumping systems are used for LPCVD of polycrystalline silicon:

- Mechanical Pump Only (Non-Dilute Process).
- Mechanical Pump with Roots Blower (Dilute Process).

Mechanical pump only systems utilize 100% SiH_4 with no carrier gas; therefore they are called nondilute systems. Adding a roots blower with a larger mechanical pump permits a carrier gas to dilute the SiH_4 to 20 to 30%. Such systems are termed dilute systems.

In the LPCVD process, the gases flow rapidly along the annular space between the wafer and the tube wall while reactants diffuse between the wafers to react on the heated surfaces. Reacted by-products diffuse back into the annular flow stream to be carried to the vacuum pump. A high gas velocity down the tube reduces the effects of depletion. Increasing the temperature from 10 to 30°C at the downstream end of the deposit zone also reduces the effect of depletion. This temperature ramp is an important factor in achieving uniformity over the deposition zone.

Most commercial LPCVD poly Si systems use mechanical pump only systems and a 10 to 30°C temperature ramp over the deposition zone. Increas-

TABLE 1—*Typical polycrystalline silicon deposition conditions (100-mm wafers).*

Parameter	Value
Wafer spacing	4.76 mm (3/16-in.) on center
Temperature	600 to 650°C
Temperature ramp	
Non-dilute process	10 to 30°C
Dilute process	0 to 5°C
Pressure	0.2 to 0.4 torr
SiH_4	
Non-dilute process	60 SCCM
Dilute process	160 SCCM, 650 SCCM N_2
Deposition zone	61 cm (24 in.)
No. of Wafers	100 to 200
Deposition rate	
Non-dilute process	100 Å/min
Dilute process	150 Å/min

ing the pumping speed with a roots blower and adding N_2 to the flow stream allows this temperature ramp to be reduced to 0 to 5°C while significantly increasing the deposition rate. Distributing the SiH_4 over the deposition zone by mechanical ducting can also allow a lower temperature gradient.

System maintenance is a major factor in LPCVD processing. Poly Si deposits on the quartz tube wall at the same rate as on the wafers. The tube may either be removed and wet etched every 10 to 20 μm of deposition to minimize particles or it may remain in place for up to 50 μm and not be re-used. Poly Si tube deposits greater than 50 to 100 μm thick may crack the process tube when it is cooled. Pump oil needs to be filtered and changed as required to remove abrasive particles and corrosives. Quartz wafer boats are commonly used; however, it has been suggested that poly Si boats would generate fewer particles because of a better match of thermal expansion coefficients.

Safety in handling SiH_4 is a very important aspect of LPCVD poly Si. SiH_4 is a pyrophoric gas. Cylinders must be stored in a safe place; adequate purge systems and cycle purging are essential, and N_2 purge of the various pump oil reservoir chambers as well as other such areas is required for safe operation [*5*].

Flow Conditions

Flow conditions in an LPCVD system can be characterized by the Pe'clet Number (Pe), which is a ratio of the dominant transport mechanisms [*6,7*]. Transport by forced convection is characterized by the velocity (v) of a gas times the distance (d) to be traveled. Transport by diffusion mixing can be characterized by the diffusion coefficient (D), which is a function of pressure, temperature, and molecular weight. The Pe'clet Number is the ratio of these parameters, and three different flow regimes are recognized (Table 2).

Flow conditions in LPCVD have been characterized as follows:

Between Wafers

- Stagnant diffusion with deposition on wafers
- Stagnant flow bordering on slip flow

Annular Space Around Wafers

- Plug flow with some mixing at wafer edges
- Slip flow bordering on laminar flow

TABLE 2—*Flow characterization.*

Type	Pe[a]	Dominant Transport Mechanism
Laminar	$\gg 1$	forced convection, minimum mixing
Slip	1	moderate mixing
Molecular	$\ll 1$	diffusion, high mixing

[a] Pe = Pe'clet No. = vd/D = (velocity) (distance)/Diffusion coefficient.

Slip to laminar flow in the annular region means that if the annular space is too large, reactants at the outer wall will not diffuse across the annular space to react with the wafers before the reactants are swept out of the reaction zone. If the annulus is too small, reaction with the wall and the wafers will cause rapid depletion and result in a large thickness gradient down the deposition zone.

Stagnant flow between the wafers means the reactant arrival rate will be proportional to:

- (Molecular weight)$^{1/2}$
- (Temperature)^{constant}
- (Pressure)$^{-1}$
- Distance between plates
- Concentration of SiH_4

For a fixed gas composition, temperature, and distance between the wafers, the deposition rate over the wafers will be optimum for one pressure and thicker at the edge for a higher pressure.

To achieve the goal of a uniform reaction rate down the deposition zone, in spite of depletion, the corrective action is to:

1. Increase the pumping speed and the reactant mass flow rate.
2. Increase temperature at downstream end of deposition zone.
3. Add inert gas to reduce gas phase reaction and increase gas velocity.
4. Optimize temperature, pressure, and spacing for fixed chemistry.
5. Increase gas velocity by mechanical ducting.

Nucleation and Growth

Two nucleation modes for gas source CVD are recognized: (*1*) heterogeneous or foreign surface nucleation and (*2*) homogeneous or gas phase nucleation. The purpose of any gas source CVD system is to bring the reactants near the conditions for homogeneous reaction and provide a surface on which the desired reaction is nucleated.

The exact nature of the surface will determine the initial density of nuclei and that, in turn, will determine the growth structure on the film [*8*]. For a fixed set of conditions, CVD-deposited Si_3N_4 provides a higher initial nucleation density than dry thermally grown SiO_2, and dry SiO_2 provides a higher initial nucleation density than SiO_2 grown from steam. Because high nucleation density promotes many points to start growth, it assures a film of more uniform properties and less haze.

Once the film has started growing, the grains will grow with a columnar structure or with a random orientation of new grains [*9*]. In a columnar structure, most of the grains have the same favorite orientation, and an X-ray diffraction pattern will show strong preferential texture. At lower deposition temperatures, the arriving atoms do not have sufficient energy to create long-range order and a micro-crystalline (grains less than 100 Å) or amorphous structure is formed.

Film properties are a function of the nature and size of the grain structure, and the grain structure can be altered by annealing as well as by doping.

Process Observations

The goal of poly Si CVD is to provide a uniform film with reproducible structure and properties. Some process observations for poly Si deposition are summarized below [*10,11*]:[3]

Deposition Temperature

1. *Less than 580°C*—Amorphous or microcrystalline (grain size <100 Å). Structure and properties subject to strong change by annealing at 800 to 1200°C.
2. *600°C*—Polycrystalline, fine grained, equiax grain structure. Structure and properties not strongly changed by 800 to 1200°C anneal.
3. *620 to 640°C*—Polycrystalline, fine grain, tending to columnar structure with strong (110) texture. Structure and properties independent of annealing temperature up to 1000°C.
4. *>650°C*—Gas phase reaction begins to cause high particle density.

Deposition Pressure

1. *<0.4 torr*—Shiny, uniform deposit.
2. *>0.4 torr*—Becoming thicker at edge, then hazy, then brown, then black as film becomes less dense and more gas phase reaction occurs.

Device interconnect quality poly Si should be shiny to indicate a smooth surface and should be transparent to visible light to about 6000 Å to show uniform structure. Haze is a term used to signifiy an unsmooth surface or an optically absorbing film. Haze is detected by observing light from a collimated source scattered by the film. Some origins of haze in poly Si are [*3*]:

- Pressure over-shoot during SiH_4 turn-on
- Air leaks
- Dirty wafers
- Oil backstreaming
- Temperature too high/low
- Pressure too high
- SiH_4 purity
- SiH_4 flow rate too low
- High moisture content in oxide underfilm

[3] See Bibliography: Structure.

Some observations regarding tuning for thickness over deposition zone are [*3,4*]:

1. Increasing pressure or average temperature preferentially increases deposition rate upstream.
2. Increasing SiH_4 flow rate at fixed temperature and pressure increases deposition rate downstream, if depletion is the dominant factor.
3. Increasing temperature locally will increase growth rate locally.
4. Nonreproducible thickness over deposition zone is usually caused by nonreproducible temperature profiles due to lack of thermal recovery following wafer insertion.

Doped Poly Si

Undoped poly Si, as-deposited, has an electrical resistivity of several hundred thousand ohm-cm. Poly Si can be doped post-deposition by gaseous sources such as $POCl_3$ and ion-implantation. Poly Si can also be doped during film deposition; however, there are several problems with CVD doping:

1. The addition of PH_3 to SiH_4 lowers the growth rate dramatically.
2. CVD doping at 600°C has not achieved the same low resistivities obtained by 1000°C $POCl_3$ or high-dose ion-implantation.
3. CVD-doped films must be annealed above 1000°C to activate the charge carriers.

CVD-doped poly Si will continue to be an area of research because of the possibility of process simplication.

Structure/Property Relationships

Electrical resistivity and oxidation rate are two important use parameters for poly Si. Both parameters are functions of structure.

To understand its electrical conductivity, poly Si can be considered to be an array of crystallites whose surfaces or boundaries can reversibly trap and release both dopant atoms and charge carriers. Several observations generate this model [*10–13*]:

1. Low dopant concentrations give unexpectedly high electrical resistivities (charge carriers are trapped in boundaries).
2. Resistivity will decrease sharply with small increase in dopant as crystallite boundary traps are saturated.
3. At high dopant concentration, the resistivity will approach that of single-crystal silicon.

4. Large crystallite poly Si will have lower resistivity than smaller crysallite material.
5. At intermediate doping levels, poly Si deposited below 600°C (very small grains) and annealed after doping (large grains due to strong grain growth) will have lower resistivity than poly Si deposited above 600°C (medium grain growth) and annealed after doping (limited grain growth).
6. At the highest doping levels, there is little difference in electrical resistivity between poly Si deposited at 580°C (very small grains) and at 620°C (larger grain size) because grain boundaries are saturated with dopant.
7. Grain boundary segregation of dopant is reversible. At intermediate doping levels, the electrical resistivity can be varied cyclically between 0.07 and 0.10 ohm-cm by cyclically annealing between 800 and 900°C.

For arsenic and phosphorus, there is appreciable segregation of dopant into the grain boundaries or the body of the grain, depending upon the annealing sequence. Boron does not show appreciable grain boundary segregation [*12,13*].

The oxidation rate of poly Si is directly dependent upon doping level and therefore indirectly dependent upon structure [*14*]. For a 620°C LPCVD poly Si doped to solid solubility at 1000°C with $POCl_3$ and oxidized at 1000°C, the oxidation rate is:

1. Reduced by an 800°C anneal after doping and prior to oxidation due to phosphorus atoms segregating to the grain boundary and no longer being chemically or electrically active.
2. Enhanced more for thinner poly Si films (0.5 μm) than for thicker films (1 μm), because the phosphorus rejected from the oxide into the remaining poly Si is at a higher concentration in the thinner poly Si.

For higher temperature oxidation (1100°C), the difference in oxidation rates due to thickness is less because less dopant is segregated to the grain boundaries.

The oxidation of poly Si thin films is not perfectly uniform because of asperities in the wafer surface and because different grain orientations oxidize at different rates. The result is that oxide films grown on poly Si have a higher apparent leakage current than oxide films grown over single-crystal silicon. Laser annealing of poly Si provides a smoother structure and a larger grain size, thereby significantly reducing the leakage current of oxide grown over poly Si [*15–17*].

Resistivity versus Doping Concentration

For single-crystal silicon, the conduction is controlled simply by the interaction of the active charge carriers with the periodic structure of the crystal.

For poly Si, the periodic structure of the trapping grain boundaries is added to that of the crystal. The result is that there is a critical concentration of dopant below which the charge carriers near the grain boundaries are primarily trapped in the grain boundaries, and the resulting depletion region produces a potential barrier to be overcome at each grain boundary.[4] At higher concentrations, there are sufficient carriers to saturate the traps and to fill in the depletion regions. At intermediate concentrations (approximately 10^{18} phos/cm^3) the mobility is at a deep minimum. For high dopant concentrations, the mobility increases to "normal" levels and the higher carrier concentration allows the resistivity to approach that of single crystal silicon. Arsenic-doped films show lower mobility than phosphorus-doped films in all reports.

The critical doping concentration for minimum mobility is proportional to the reciprocal of the grain size. For saturation doping, the concept of critical doping concentation has no effect. At lower doping levels, the effect is quite strong and devices that utilize intermediate doping levels could be strongly affected by this concept.

Effect of Dopant on Structure

Dopants at high concentration can have a dramatic effect on poly Si grain structure for a given deposition and annealing sequence [*9*,*13*,*18*]. Below 1×10^{20} phos/cm^3 there is little effect. Between 1 and 10×10^{20} phos/cm^3, the effect is superlinear; that is, the grain size changes by more than a factor of 10 for a ten-fold increase in phosphorus concentration. This observation is contrary to the usual metallurgical observation that impurities pin grain boundaries and reduce grain growth.

In many thin films, the maximum grain size developed by annealing is expected to be comparable to the film thickness. For phosphorus-doped poly Si, grains of more than ten times the film thickness have been obtained. Such large final grain diameters indicate the possibility of creating device-quality, thin, single-crystal films from poly Si. Unfortunately, these large grains also show that delineating heavily doped and annealed poly Si for interconnects could be a major problem for very-large-scale integration circuit (VLSI) processing.

Poly Si Limitations

For 0.5 μm film thickness, the sheet resistance of poly Si apparently cannot be reduced below about 8 to 10 Ω/square regardless of the deposition/annealing technique used. Extreme limits of processing produce a large grain size which can be difficult to pattern. This limitation must be overcome to permit lower resistance interconnects and faster integrated circuits. In the fu-

[4] See Bibliography: Electrical Properties, Dopant Segregation, Carrier Trapping.

ture, poly Si interconnects will be augmented by metals and/or silicides [19–22].

Summary

Poly Si deposited by LPCVD is a commercial process of high productivity. Deposition conditions for good results are fixed in a relatively narrow range of temperature, pressure, and wafer spacing. Deposition kinetics, nucleation and growth, doping, process requirements, and structure/property relationships are reasonably well understood.

Electrical and oxidation properties can be explained by considering poly Si to be an array of crystallites whose boundaries can reversibly absorb and release both dopant atoms and charge carriers. Grain size can be increased dramatically by annealing phosphorus-doped poly Si films. The large grain size permits lower electrical resistivity, but may cause difficulties in delineating poly Si for fine line geometries. Soon, poly Si for integrated cirucit (IC) interconnects must be augmented with lower resistance metals or silicides or both.

A selected Bibliography on various aspects of the properties of polycrystalline silicon is provided after the Reference section.

Acknowledgments

I wish to thank Ted Kamins of Hewlett-Packard for many helpful discussions. The bibliography is largely based on his summary [23].

References

[1] Chruma, J. L. and Hilton, P. G., U.S. Patent 3900597, 19 Aug. 1975.

[2] Hammond, M. L., "Silicon CVD Technology: A Review," Technical Note GEN-003, Applied Materials Inc., Santa Clara, Calif., 1977.

[3] Hammond, M. L., "Polycrystalline Silicon and Refractory Silicide Films for Integrated Circuits," Boston, Mass.; University Extension and College of Engineering, University of California, Berkeley, Calif., Nov. 1980.

[4] Rosler, R. S., *Solid-State Technology,* Vol. 20, 1977, p. 63.

[5] Hammond, M. L., *Solid-State Technology,* Vol. 23, 1980, p. 104.

[6] Hitchman, M. L., Kane, J., and Widmer, A. E., *Thin Solid Films,* Vol. 59, 1979, p. 231.

[7] Gieske, J., McMullen, J., and Donaghey, L. F. in *Proceedings,* Sixth Chemical Vapor Deposition Symposium, Los Angeles, The Electrochemical Society, 1977, p. 182.

[8] Bloem, J. in *Proceedings,* Seventh Chemical Vapor Deposition Symposium, Atlanta, The Electrochemical Society, 1979, p. 41.

[9] Kamins, T. I., Mandurah, M. M., and Saraswat, K. L., *Journal of the Electrochemical Society,* Vol. 125, 1978, p. 927.

[10] Kamins, T. I., *Journal of the Electrochemical Society,* Vol. 127, 1980, p. 686.

[11] Mandurah, M. M. and Saraswat, K. L., *Journal of the Electrochemical Society,* Vol. 126, 1979, p. 1019.

[12] Mandurah, M. M. et al, *Journal of Applied Physics,* Vol. 51, 1980, p. 5755.

[13] Sharma, P. D., Wear, T. G., and Seto, J. Y. W., *Extended Abstracts,* Vol. 81-2, Abstract 402, The Electrochemical Society, 1981.

[14] Singh, H. and Saraswat, K. L., *Extended Abstracts,* Vol. 81-1, Abstract 282, The Electrochemical Society, 1981.

[*15*] DiMaria, D. J. and Kerr, D. R., *Applied Physics Letters*, Vol. 27, 1975, p. 505.
[*16*] Anderson, R. M. and Kett, D. R., *Journal of Applied Physics*, Vol. 48, 1977, p. 4834.
[*17*] Yaron, G., Hess, L. D., and Kokorowski, S., *IEEE Transactions*, Vol. ED-27, 1980, p. 964.
[*18*] Wada, Y. and Nishimatsu, S., *Journal of the Electrochemical Society*, Vol. 125, 1978, p. 1499.
[*19*] Muraka, S. P., "Polycrystalline Silicon and Refractory Silicide Films for Integrated Circuits," Boston, Mass.; University Extension and College of Engineering, Berkeley, Calif., Nov. 1980.
[*20*] Miller, N. E. and Beinglass, I., *Solid-State Technology*, Vol. 23, 1980, p. 79.
[*21*] Beinglass, I. et al, *Extended Abstracts*, Vol. 81-2, Abstract 380, The Electrochemical Society, 1982.
[*22*] Gargini, P. A. and Beinglass, I., *Extended Abstracts*, Vol. 81-2, Abstract 381, The Electrochemical Society, 1982.
[*23*] Kamins, T. I., "Polycrystalline Silicon and Refractory Silicide Films for Integrated Circuits," Boston, Mass.; University Extension and College of Engineering, University of California, Berkeley, Calif. Nov. 1980.

Selected Bibliography—Properties of Polycrystalline Silicon

For a more extensive reference list up to 1974, see review paper by T. I. Kamins, *IEEE Transactions on Parts, Hybrids, and Packaging*, Vol. PHP-10, 1974, p. 221.

Structure

Feldman, C. et al, *Journal of Electronic Materials*, Vol. 7, 1978, p. 309.
Gabilli, E. et al, *Solid-State Electronics*, Vol. 20, 1977, p. 925.
Gat, A. et al, *Applied Physics Letters*, Vol. 33, 1978, p. 775.
Haberle, K. and Froschle, E., *Thin Solid Films*, Vol. 61, 1979, p. 105.
Horiuchi, S., *Solid-State Electronics*, Vol. 18, 1975, p. 1111.
Jain, G. C. et al, *Applied Physics Letters*, Vol. 33, 1978, p. 445.
Kamins, T. I. and Cass, T. R., *Thin Solid Films*, Vol. 16, 1973, p. 147.
Kamins, T. I. et al, *Journal of the Electrochemical Society*, Vol. 125, 1978, p. 927.
Nagasima, N. and Kubota, N., *Japanese Journal of Applied Physics*, Vol. 14, 1975, p. 1105.
Nagasima, N. and Kubota, N., *Journal of Vacuum Science and Technology*, Vol. 14, 1977, p. 54.
Nakamura, K. et al, *Journal of Applied Physics*, Vol. 46, 1975, p. 4678.
Ouwens, C. D., and Heijligers, H., *Applied Physics Letters*, Vol. 26, 1975, pp. 569–71.
Schmidt, C., *Materials Research Bulletin*, Vol. 13, 1978, p. 873.
Wada, Y. and Nishimatsu, S., *Journal of the Electrochemical Society*, Vol. 125, 1978, p. 1499.

Oxidation

Anderson, R. M. and Kerr, D. R., *Journal of Applied Physics*, Vol. 48, 1977, p. 4834.
Barnes, J. J. et al, *Journal of the Electrochemical Society*, Vol. 126, 1979, p. 1779.
Hu, C. et al, *Applied Physics Letters*, Vol. 35, 1979, p. 189.
Irene, E. A. et al, *Journal of the Electrochemical Society*, Vol. 127, 1980, p. 705.
Kamins, T. I., *Journal of the Electrochemical Society*, Vol. 126, 1979, p. 838.
Sunami, H., *Journal of the Electrochemical Society*, Vol. 125, 1978, p. 892.

Diffusion

Coe, D. J., *Solid-State Electronics*, Vol. 20, 1977, p. 985.
Kamins, T. I. et al, *Journal of Applied Physics*, Vol. 43, 1972, p. 83.
Nakamura, K. and Kamoshida, M., *Journal of Applied Physics*, Vol. 48, 1977, p. 5349.
Shimakura, K. et al, *Solid-State Electronics*, Vol. 18, 1975, p. 991.

Electrical Properties

Angelucci, R., Severi, M., and Solmi, S., *Extended Abstracts*, Vol. 81-2, Abstract 410, The Electrochemical Society, 1981.
Briska, M. and Kiofsky, A., *Journal of the Electrochemical Society*, Vol. 121, 1974, p. 972.
Joseph, J. D. and Kamins, T. I., *Solid-State Electronics*, Vol. 15, 1972, p. 355.
Kamins, T. I., *Journal of the Electrochemical Society*, Vol. 126, 1979, p. 833.
Mandurah, M. M. et al, *Journal of the Electrochemical Society*, Vol. 126, 1979, p. 1019.
Robinson, M., "Impurity Doping Process in Silicon," in *Materials Processing Theory and Practices*, Vol. II, F. F. Y. Wang, Ed., North-Holland, New York, 1981.
Wada, Y. and Nishimatsu, S., *Denki Kagaku*, Vol. 47, 1979, p. 118.

Dopant Segregation

Cowher, M. E. and Sedgwick, T. O., *Journal of the Electrochemical Society*, Vol. 119, 1972, p. 1565.
Fripp, A. L., *Journal of Applied Physics*, Vol. 46, 1975, p. 1240.
Kazmerski, L. L. et al, *Applied Physics Letters*, Vol. 36, 1980, p. 323.
Mandurah, M. M. et al, *Applied Physics Letters*, Vol. 36, 1980, p. 683.
Mandurah, M. M. et al, *Journal of Applied Physics*, Vol. 51, 1980, p. 5755.

Carrier Trapping

Baccarani, G. et al, *Journal of Applied Physics*, Vol. 49, 1978, p. 5565.
DiStefano, T. H. and Cuomo, J. J., *Applied Physics Letters*, Vol. 30, 1977, p. 351.
Ghosh, A. K. et al, *Journal of Applied Physics*, Vol. 51, 1980, p. 446.
Hasegawa, S. et al, *Journal of Applied Physics*, Vol. 50, 1979, p. 7256.
Hirose, M. et al, *Journal of Applied Physics*, Vol. 50, 1979, p. 377.
Kamins, T. I., *Journal of Applied Physics*, Vol. 42, 1971, p. 4357.
Korsh, G. J. and Muller, R. S., *Solid-State Electronics*, Vol. 21, 1978, p. 1045.
Lu, N. C. C. et al, *IEEE Electron Device Letters*, Vol. EDL-1, 1980, p. 38.
Pike, G. E. and Seager, C. H., *Journal of Applied Physics*, Vol. 50, 1979, p. 3414.
Seager, C. H. and Pike, G. E., *Applied Physics Letters*, Vol. 35, 1979, p. 709.
Seager, C. H. and Castner, T. G., *Journal of Applied Physics*, Vol. 49, 1978, p. 3879.
Seto, J. Y. W., *Journal of Applied Physics*, Vol. 46, 1975, p. 5247.
Snejdar, V. and Jerhot, J., *Thin Solid Films*, Vol. 37, 1976, p. 303.
Young, R. T. et al in *Proceedings*, Twelfth IEEE Photovoltaic Specialists Conference, Baton Rouge, La., 1976, pp. 65-66.

Hydrogenation

Makino, T. and Nakamura, H., *Applied Physics Letters*, Vol. 35, 1979, p. 551.
Seager, C. H. and Ginley, D. S., *Applied Physics Letters*, Vol. 34, 1979, p. 337.

Miscellaneous Properties

Bean, K. E. et al, *Journal of Applied Physics*, Vol. 40, 1969, p. 2358.
Brown, W. A. and Kamins, T. I., *Solid-State Technology*, Vol. 22, July 1979, p. 51.
Kamins, T. I., *Journal of the Electrochemical Society*, Vol. 127, 1980, p. 686.
Kuhl, C. et al, *Journal of the Electrochemical Society*, Vol. 121, 1974, p. 1496.
Muraka, S. P., Sinha, A. K., and Levenstein, H. J., *Journal of the Electrochemical Society*, Vol. 127, 1980, p. 2446.
Onuma, Y. and Sekiya, K., *Japanese Journal of Applied Physics*, Vol. 11, 1972, p. 20.
Seto, J. Y. W., *Journal of Applied Physics*, Vol. 47, 1976, p. 4780.

MOSFETs in Polycrystalline Silicon

DeGrauff, H. C. and DeGroot, J. G., *IEPM Technical Digest*, IEEE, Vol. 46–9, 1980, p. 46.
Fa, C. H. and Jew, T. T., *IEEE Transactions on Electron Devices*, Vol. ED-13, 1966, p. 290.
Gibbons, J. F. and Lee, K. F., *Applied Physics Letters*, Vol. 34, 1979, p. 831.
Kamins, T. I., *Solid-State Electronics*, Vol. 15, 1972, p. 789.
Kamins, T. I. et al, *IEEE Transactions on Electron Devices*, Vol. ED-27, 1980, p. 290.
Lee, K. F. et al, *Applied Physics Letters*, Vol. 35, 1979, p. 173.

Deposition

Brown, W. A. and Kamins, T. I., *Solid-State Technology*, Vol. 22, 1979, p. 51.
Gieske, R. J. and McMullen, J. J., *Proceedings of the Electrochemical Society*, Sixth CVD Symposium, Vol. 77-5, pp. 183-4.
Hitchman, M. L., Kane, J., and Widmer, A. E., *Thin Solid Films*, Vol. 59, 1979, p. 231.
Rosler, R. S., *Solid-State Technology*, Vol. 20, 1977, p. 63.

Oxygen-Doped Polycrystalline Silicon

Adachi, T. et al, *Extended Abstracts*, Vol. 79-1, Abstract 444, The Electrochemical Society, 1979.
Knowlle, W. R. and Maxwell, H. R., Jr., *Journal of the Electrochemical Society*, Vol. 127, 1980, p. 2254.
Matsushita, T. et al, *Proceedings of 7th Conference on Solid-State Devices*, Tokyo, 1975, supplement to *Japanese Journal of Applied Physics*, Vol. 15, 1976, p. 35.
Matsushita, T. et al, *Applied Physics Letters*, Vol. 35, 1979, p. 549.
Matsushita, T. et al, *IEEE Transactions on Electron Devices*, Vol. ED-33, 1976, p. 826.
Maxwell, H. R., Jr., and Knowlle, W. R., *Journal of the Electrochemical Society*, Vol. 128, 1981, p. 576.
Mochizuki, H. et al, *Proceedings of 7th Conference on Solid-State Devices*, Tokyo, 1975, supplement to *Japanese Journal of Applied Physics*, Vol. 15, 1976, p. 41.
Yamoto, H. et al, *Extended Abstracts*, Vol. 79-1, Abstract 447, The Electrochemical Society, 1979.

Dennis W. Hess[1]

Plasma-Enhanced Chemical Vapor Deposition of Silicon and Silicon-Containing Films

REFERENCE: Hess, D. W., "**Plasma-Enhanced Chemical Vapor Deposition of Silicon and Silicon-Containing Films**," *Silicon Processing, ASTM STP 804*, D. C. Gupta, Ed., American Society for Testing and Materials, 1983, pp. 218–225.

ABSTRACT: The use of a radio frequency (rf) glow discharge or plasma has recently come into favor for the deposition of thin films. In plasma-enhanced chemical vapor deposition (PECVD), chemical reactions can be carried out at low (<300°C) temperatures, since the energy needed to break chemical bonds is supplied by the high-energy electrons generated by the applied rf field. Further, due to the complex chemistry occurring within the glow discharge, materials with unique properties can be fabricated. This paper reviews the characteristics of nonequilibrium glow discharges as used for PECVD of silicon and silicon-containing films. The glow-discharge-modified chemistry of film deposition is discussed as it relates to the physical and chemical properties of the resulting films. Examples include amorphous and single-crystal silicon films, as well as silicon dioxide and silicon nitride films, which are widely used as intermetallic and passivation layers in integrated circuit manufacture.

KEY WORDS: chemical vapor deposition, plasma-enhanced chemical vapor deposition, silicon film, silicon-containing film, silicon dioxide, silicon nitride, passivation layer

Chemical vapor deposition (CVD) has been used for approximately 50 years to form metallic, semiconducting, and insulating thin films for a variety of applications. A serious limitation in the applicability of CVD is the generally high (>500°C) temperature required to obtain film deposition at a reasonable rate. These temperatures preclude the use of substrates or underlying film materials such as polymers or low-melting metals (that is, aluminum or gold). Of course, physical methods of film deposition (evaporation or sputtering) can be used when temperature is a concern, but a wide range of material properties is often difficult to achieve with these techniques.

Recently, the use of a radio frequency (rf) glow discharge has become an attractive method for carrying out low-temperature CVD [*1*].[2] In plasma-

[1] Associate Professor, Department of Chemical Engineering, University of California, and Materials and Molecular Research Division, Lawrence Berkeley Laboratory, Berkeley, Calif. 94720.

[2] The italic numbers in brackets refer to the list of references appended to this paper.

enhanced CVD (PECVD), high-energy electrons break chemical bonds, thereby promoting chemical reactions at reduced temperature and allowing temperature to be used as a variable to tailor film properties.

Properties of rf Glow Discharges

As utilized in PECVD, an rf glow discharge or plasma is a partially ionized gas containing ions, electrons, and a variety of neutral species. Typical pressures are in the 50 m torr to 5 torr range, while frequencies are generally between 20 kHz and 40 MHz. Under these conditions, the plasma properties are rather unique (Table 1). Since the plasma is electrically neutral, the concentrations of electrons and ions are the same. However, the concentration of neutral species (n_N) is several orders of magnitude higher than that of the electrons. The electron temperature (T_e) of 1 to 10 eV corresponds to temperatures of 10^4 to 10^5 K, whereas the ion (or thermal) temperature (T_i) is ~300 K. Because of the large difference between T_e and T_i, such glow discharges are referred to as nonequilibrium plasmas, and it is this characteristic that is responsible for their unique properties, namely that high-energy electrons can promote chemical reactions while the thermal temperature remains low.

Due to the difference in mobility between positive ions and electrons, all surfaces within the rf glow discharge assume a potential that is negative with respect to the plasma [*2*]. Thus, positive ion bombardment of all surfaces, including the depositing film, occurs. The ion energies are sufficiently high (>100 eV) that bond breaking generally takes place in the film material. Further, depending upon the gas pressure and the geometrical configuration of the reactor [*2*], ions with energies above 500 eV may strike the surfaces, resulting in implantation damage or sputtering. Naturally, film properties such as composition and stress are greatly affected by such processes. It should be mentioned that electron (and photon) bombardment of surfaces immersed in the plasma also occurs, but due to the negative potential imposed by the discharge, ion bombardment is the dominant process affecting film properties.

Precise control and reproducibility of film properties has proven to be difficult in PECVD. The reason for this observation is the large number of process parameters available. The plasma variables may be broken down into two general groups (Table 2). The microscopic parameters are the ones that we would *like* to control. If this were possible, the plasma could be precisely defined in every case. Macroscopic parameters are parameters that can be

TABLE 1—*Properties of rf glow discharges used for plasma processing.*

$n_e = 10^9$ to 10^{12} cm^{-3}
$n_N \simeq 10^{15}$ cm^{-3}
$T_e = 1$ to 10 eV
$T_i \simeq 0.04$ eV

TABLE 2—*Parameter control problem in plasma processing.*

Macroscopic Parameters	Microscopic Parameters
rf power	electron concentration
rf frequency	electron energy distribution
gas flow rate	gas density
pump speed	residence time
pressure	
etchant gas	
electrode temperature	
electrode material	
electrode potential	
reactor geometry	

varied by equipment settings, machine design, or materials of construction. However, a change in one macroscopic parameter results in a change in two or more microscopic parameters, and usually the change is not predictable from first principles.

Types of Plasma Reactors

For the most part, PECVD for thin film applications is carried out in four basic types of reactors. The most common, from the standpoint of films used in integrated circuit (IC) manufacture, are the capacitively coupled systems shown in Figs. 1*a* and 1*b*. In both cases, the substrates and thus the growing films are exposed to the glow discharge and therefore to particle and photon bombardment.

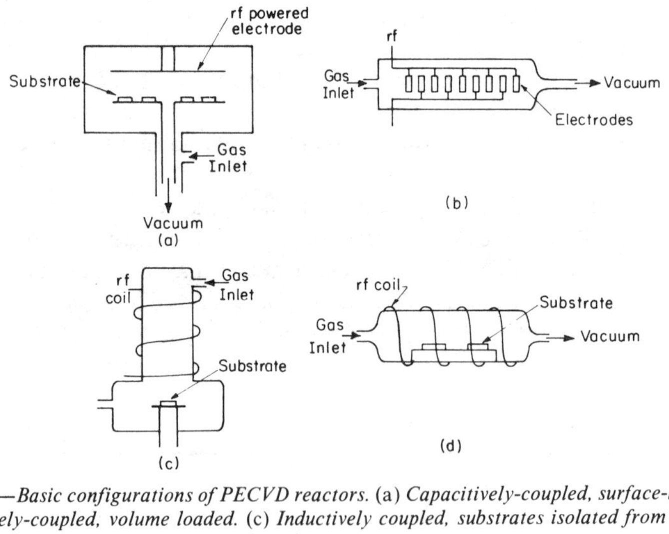

FIG. 1—*Basic configurations of PECVD reactors.* (a) *Capacitively-coupled, surface-loaded.* (b) *Capacitively-coupled, volume loaded.* (c) *Inductively coupled, substrates isolated from discharge.* (d) *Inductively coupled, substrates within glow region.*

Chronologically, the first reactor used extensively in IC processes (Fig. 1*a*) was the parallel plate system of Reinberg [*3*]. In this configuration, the substrates sit on the grounded electrode, which can be heated. Production processes dictate that large [>610-mm (24-in.)]-(diameter) electrodes be used to increase throughput of 102-mm (4-in.)-diameter wafers.

Recently, a PECVD system that is positioned within a furnace tube [*4*] and is capable of large throughput has been introduced (Fig. 1*b*). This system has a series of parallel electrodes that are alternately powered and grounded with substrates positioned on each plate.

For higher temperature processes, such as those used to grow epitaxial layers by PECVD, inductively coupled systems are generally used [*5*], although capacitively coupled reactors with electrodes positioned outside the quartz wall have been reported [*6*]. In some instances, the glow is upstream from the heated substrates (Fig. 1*c*), while in other cases, the substrates again reside within the glow region (Fig. 1*d*).

Silicon-Containing Films Deposited by PECVD

As a result of their applications in IC technology and in solar cell fabrication, PECVD of silicon-containing films has received considerable attention [*1,7,8*]. In the following sections, specific film materials will be briefly discussed, and some of their properties related to plasma deposition conditions. It should be pointed out that because of the complex nature of the chemistry occurring in the plasma, it is critical that one measurement technique, such as refractive index measurements in the case of dielectric films, *not* be used as the sole criterion for process control. For instance, a change in composition could result in the same refractive index for two films, but widely different film properties (etch rate, density, conductivity, etc.).

Silicon Nitride

Owing to its potential as a passivation coating, silicon nitride is the film material that has been most intensely studied in the IC area. This material can be deposited from either $SiH_4 + NH_3$ or from $SiH_4 + N_2$. The former reactants are preferred because of the high bond energy of N_2, which results in silicon-rich films unless large (>95/1) ratios of N_2/SiH_4 are used.

The exact properties of the films, such as density, composition, stress, etc., can of course be varied by changing the macroscopic parameters listed in Table 2, as discussed in Refs *9* to *11*. With the proper choice of parameters, adherent, conformal, crack-resistant, pinhole-free films can be formed [*9,10*]. Typical properties of silicon nitride films deposited at low frequencies are presented in Table 3. Low frequencies are generally preferred for such depositions, because the enhanced ion bombardment at low frequencies results in small compressive stresses under nearly all deposition conditions [*10*].

Owing to the reactive atmosphere in a glow discharge, any chemical spe-

TABLE 3—*Selected properties of CVD SiO_2, plasma oxide, and plasma nitride.*[a]

Property	CVD SiO_2	Plasma Oxide	Plasma Nitride
Pinhole density	1 to 10 cm^{-2}	≤ 1 cm^{-2}	≤ 0.5 cm^{-2}
Particulate density	10 to 100 cm^{-2}	≤ 0.2 cm^{-2}	≤ 0.2 cm^{-2}
Step coverage	poor	conformal	conformal
Adhesion to aluminum	good	good	good
Adhesion to gold	bad	good	good
Crack resistance	poor	good	excellent
Intrinsic film stress	tensile 6 to 8 $\times 10^8$ dynes/cm^2	compressive 2 $\times 10^9$ dynes/cm^2	compressive 4 $\times 10^9$ dynes/cm^2
Breakdown voltage	8 $\times 10^6$ V/cm	8 $\times 10^6$ V/cm	2 $\times 10^6$ V/cm
Film resistivity	$>10^{17}$ Ω-cm	$>10^{17}$ Ω-cm	2 $\times 10^6$ Ω-cm
Dielectric constant (1 MHz)	3.8	4.6	7.0

[a] From van de Ven [*10*]. Plasma deposition conditions are frequency = 50 kHz, pressure = ~300 m torr, power = ~0.1 W/cm^2, and temperature = 300°C.

cies present in the reactant gases or as contaminants will be incorporated into the films. In particular, large concentrations ($\sim 10^{21}$ cm^{-3}) of hydrogen originating from SiH_4 or from NH_3 are present in silicon nitride, for the most part bonded to either silicon or nitrogen [*12,13*]. Clearly these bonding structures greatly affect film properties such as density and stress. Further, out-diffusion of hydrogen can cause instabilities in devices passivated by plasma-deposited silicon nitride [*14*].

Silicon Dioxide

The need for intermetallic dielectric layers has led to the investigation of plasma-deposited silicon dioxide films [*10,13,15*]. Generally, the films are deposited from SiH_4, plus either oxygen (O_2), carbon dioxide (CO_2), or nitrous oxide (N_2O), with the last favored due to control considerations. Phosphorous-doped films can also be obtained by addition of phosphine to the reacting gas mixture.

Typical properties of plasma-deposited SiO_2 films are shown in Table 4, where the films were deposited from SiH_4 and N_2O [*15*]. Note that a small amount of nitrogen is incorporated during deposition. In addition, hydrogen is present in these films ($\sim 10^{21}$ cm^{-3}) in the form of SiH, SiOH, and H_2O [*16*]. Comparison of PECVD film properties with those of normal CVD SiO_2 is given in Table 3, where the effects of the plasma environment are evident [*10*].

Amorphous Silicon

Amorphous silicon (a-Si), prepared by the rf glow discharge decomposition of silane, has been found to be suitable as a low-cost solar cell material

TABLE 4—*Plasma-deposited silicon oxide films.*[a]

$$SiH_4 + 2N_2O \rightarrow SiO_2 + 2N_2 + 2H_2$$

Properties of Plasma-Deposited Films

Flow Ratio, cm^3/min, STP	Deposition Temperature, °C	Density,[b] g/cm^3	Refractive Index[c]	Stress × 10^9, dynes/cm^2[d]
N_2O/SiH_4	250	2.44	1.5	<0.1
(900/60)	350	2.23	1.45	1.2
$N_2O/SiH_4/Ar$	250	2.39	1.4	4
(900/60/300)	350	2.18	1.48	3

Sample	Flow Ratio N_2O/SiH_4, cm^3/min	Electron Spectroscopy for Chemical Analysis (ESCA) Empirical Formula	Atom Percent Si	Atom Percent O	Atom Percent N
As-received	900/200	$SiO_{1.48}N_{0.15}$	52.06	44.03	3.90
	900/60	$SiO_{1.96}N_{0.14}$	45.66	51.14	3.20
After ion etching	900/200	$SiO_{1.13}N_{0.17}$	57.77	37.31	4.91
A few hundred Å[e]	900/60	$SiO_{1.89}N_{0.09}$	48.23	49.60	2.00
Theoretical (pure material)	0	SiO_2	46.67	53.33	0

[a] From Hollahan [*15*]. Plasma deposition conditions are 250 W, 660-mm (26-in.)-diameter electrode, 350 m torr, 51-mm (2-in.) electrode spacing, 50-kHz rf supply.

[b] Pure $SiO_2 \Rightarrow 2.27$.

[c] Normal $\Rightarrow$ 1.45 to 1.47.

[d] All compressive on silicon substrates.

[e] 1 Å = 0.1 nm.

[*17,18*]. Compared with sputtered a-Si, the plasma-deposited material has few structural defects, and thus a relatively low density of states in the band gap. The resulting properties (high photoconductivity, reasonable carrier lifetimes, etc.) are principally due to the large concentrations of hydrogen ($\sim 10^{21}$ cm^{-3}) bonded into the structure. In effect, this material is a silicon-hydrogen alloy [*19*], or "polymer" material composed of SiH_x units [*20*], as indicated by the fact that films of similar properties to those deposited from SiH_4 can be formed from discharges of Si_2H_6 and Si_3H_8 [*21*]. Care should be exercised, however, in labeling the aforementioned materials *amorphous*. Under certain glow discharge conditions and substrate temperatures, polycrystalline films result [*22*].

Hydrogen is not the only additive that will generate photoconductive a-Si films. Films modified by the addition of fluorine [*23*] and chlorine, boron, nitrogen, carbon, and germanium [*24*] have been reported to possess certain advantages over the solely hydrogenated material.

Epitaxial Silicon

Very large scale integration (VLSI) requires reduced film thicknesses and in many cases lowered processing temperatures. In the growth of epitaxial

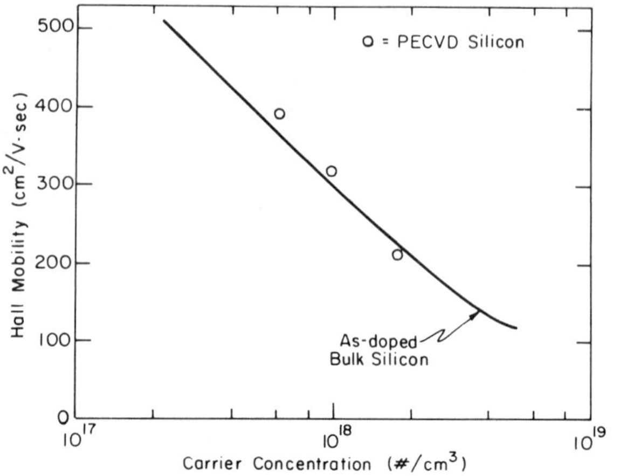

FIG. 2—*Hall mobility versus carrier concentration (open circles) for epitaxial PECVD phosphorus-doped silicon deposited at 750°C. The solid line represents the relationship between Hall mobility and carrier concentration for single-crystal, arsenic-doped silicon (after Ref* 5).

silicon layers, PECVD appears to offer the potential for reduced temperature epitaxy, ostensibly because of ion and electron bombardment effects along with the resulting improvement in surface cleanliness. Indeed, the substrate temperature necessary for epitaxial growth of evaporated metal and semiconductor films onto sodium chloride substrates is substantially reduced when charged particles impinge on the growing surface [*25*]. With regard to silicon, epitaxial film growth has been reported in systems similar to those described in Figs. 1*c* and 1*d*, at temperatures between 650 and 800°C [*5,6*]. The films display a high degree of crystalline perfection with few stacking faults [*6*]. Further, as shown in Fig. 2 [*5*], phosphorous-doped films that have the same electron mobility as arsenic-doped bulk single crystal silicon can be deposited from a glow discharge at a temperature of 750°C.

Summary and Conclusions

Because of the unique properties of rf glow discharges, plasma-enhanced chemical vapor deposition offers the capability for depositing a wide variety of silicon-based materials at low temperature. Unfortunately, the technique suffers from control problems that arise from the complex chemistry occurring in the reactive atmosphere. Despite these problems, PECVD is currently used in IC fabrication, and its use will increase as improved equipment and processes are developed. Toward this end, additional efforts are needed to establish the fundamental relationships between macroscopic plasma variables and the microscopic parameters that determine plasma and ultimately film properties. When these relationships are defined, extensive application of PECVD to all types of film materials will occur.

References

[1] Hollahan, J. R. and Rosler, R. S. in *Thin Film Processes*, J. L. Vossen and W. Kern, Eds., Academic Press, New York, 1978, p. 335.

[2] Vossen, J. L., *Journal of the Electrochemical Society*, Vol. 126, 1979, p. 319.

[3] Reinberg, A. R., U.S. Patent No. 3,757,733, 1973.

[4] Rosler, R. S. and Engle, G. M., *Solid State Technology*, Vol. 22, No. 12, Dec. 1979, p. 88, and Vol. 24, No. 4, April 1981, p. 172.

[5] Suzuki, S., Takai, H., Okuda, H., and Itoh, T. in *Proceedings*, Eleventh Conference on Solid State Devices, Tokyo, 1979.

[6] Townsend, W. G. and Uddin, M. E., *Solid State Electronics*, Vol. 16, 1973, p. 39.

[7] Reinberg, A. R., *Annual Review of Material Science*, Vol. 9, 1979, p. 341.

[8] Rand, M. J., *Journal of Vacuum Science and Technology*, Vol. 16, 1979, p. 420.

[9] Sinha, A. K., Levinstein, H. J., Smith, T. E., Quintana, G., and Haszako, S. E., *Journal of the Electrochemical Society*, Vol. 125, 1978, p. 601.

[10] van de Ven, E. P. G. T., *Solid State Technology*, Vol. 24, No. 4, April 1981, p. 167.

[11] Dun, H., Pan, P., White, F. R., and Douse, R. W., *Journal of the Electrochemical Society*, Vol. 128, 1981, p. 1555.

[12] Lanford, W. A. and Rand, M. J., *Journal of Applied Physics*, Vol. 49, 1978, p. 2473.

[13] Stein, H. J., Wells, V. A., and Hampy, R. E., *Journal of the Electrochemical Society*, Vol. 126, 1979, p. 1750.

[14] Sun, R. C., Clemens, J. T., and Nelson, J. T. in *Proceedings*, IEEE Reliability Physics Symposium, New York, 1980, p. 244.

[15] Hollahan, J. R., *Journal of the Electrochemical Society*, Vol. 126, 1979, p. 930.

[16] Adams, A. C., Alexander, F. B., Capio, C. D., and Smith, T. E., *Journal of the Electrochemical Society*, Vol. 128, 1981, p. 1545.

[17] Spear, W. E., LeComber, P. G., Kinmond, S., and Brodsky, M. H., *Applied Physics Letters*, Vol. 28, 1976, p. 105.

[18] Carlson, D. E., *IEEE Transactions on Electron Devices*, Vol. ED-24, 1977, p. 449.

[19] Fritzsche, H., Tsai, C. C., and Persans, P., *Solid State Technology*, Vol. 21, No. 1, Jan. 1978, p. 55.

[20] Revesz, A. G., *Thin Solid Films*, Vol. 50, 1978, p. L29.

[21] Scott, B. A., Brodsky, M. H., Green, D. C., Kirby, P. B., Plecenik, R. M., and Simonyi, E E., *Applied Physics Letters*, Vol. 37, 1980, p. 725.

[22] Messana, C., De Angelis, B. A., Conte, G., and Gramaccioni, C., *Journal of Physics D*, Vol. 14, 1981, p. L91.

[23] Madan, A., Ovshinsky, S. R., and Benn, E., *Philosophical Magazine*, Vol. 40, 1979, p. 259.

[24] Tanner, D. P., Johnson, G. R., and Sefcik, M. D., *Solar Energy Materials*, Vol. 3, 1980, p. 533.

[25] Shimaoka, G. in *Proceedings*, Second Symposium on Ion Sources and Application Technology, Tokyo, IEEE, 1978, p. 93.

Dielectrics and Junction Formation

Rakesh Kumar,[1] *Matthew M. Nowak,*[1] *Ernie H. Tyler,*[1] *and Mark A. Vinson*

Process Technology Requirements for VLSIC Fabrication

REFERENCE: Kumar, R., Nowak, M. M., Tyler, E. H., and Vinson, M. A., **"Process Technology Requirements for VLSIC Fabrication,"** *Silicon Processing, ASTM STP 804,* D. C. Gupta, Ed., American Society for Testing and Materials, 1983, pp. 229–237.

ABSTRACT: The fabrication of high-density, high-performance integrated circuits to provide a lower cost per function requires N-channel metal-oxide-semiconductor (NMOS) circuits with smaller minimum dimensions, thinner gate oxides, and shallower junction depths. This paper provides an overview of process technology constraints and requirements for the fabrication of scaled NMOS very-large-scale integrated circuits (VLSICs). Devices with shallow junction depths are required to minimize short channel effects in such VLSICs; this places numerous constraints on processing. These constraints require low-temperature processing and cause several deviations from present production processing techniques. We shall review some solutions to these constraints, especially in the junction formation and low pressure chemical vapor deposition areas.

KEY WORDS: fabrication requirements, N-channel metal-oxide-semiconductors (NMOS), very large scale integrated circuits, low-pressure chemical vapor deposition, junction formation, scaling, gates

Metal-oxide-semiconductor (MOS) circuit performance improvement by the extension of semiconductor technology to smaller dimensions is a prevailing trend in very-large-scale integrated (VLSI) circuit development. The main driving force behind this trend has been the capacity to obtain improved performance, lower cost per die and cost per function, and, hopefully, greater yield as the area of a particular circuit is reduced.

Background

Patterning Requirements

Improving the lithographic and etch capabilities in semiconductor processing is the most apparent requirement in the shrinking of circuits. Enhanced resolution with tighter dimensional control and more precise alignment accuracy are pushing the lithographic process to the use of direct wafer

[1] Burroughs Corporation, San Diego, Calif. 92127.

step and repeat projection alignment for current high-performance circuit fabrication. Electron-beam and X-ray lithography are being developed for future circuit manufacturing where minimum feature sizes less than 1 μm will be required.

Extensive effort is being devoted to the development of dry etching processes and equipment for the accurate delineation of small features. The importance of the planar plasma etch technology is based on its ability to accurately replicate photoresist images into underlying material without undercutting the mask. This eliminates the need for appreciable mask compensation when film thicknesses comparable to feature linewidth are to be etched. Increased packing density capability with minimum dimensional distortion results.

Device Structural Requirements

In addition to the requirement of improved patterning of surface features, the realization of high-performance MOS VLSI circuits demands that strict attention be given to the vertical components of the device structure. These demands on VLSI device structure place restrictions on the fabrication technology, and the realization of high-performance circuits within the confines of these restrictions is the main concern of this paper.

Device Scaling

Guidelines for the definition of a basic device structure can be founded on the application of the "constant field" scaling principle [*1*].[2] Scaling theory predicts that a K-fold decrease in critical linewidth can result in a reduction of circuit delay by K and a density increase by K^2, provided that gate oxide thickness and source-drain junction depth are also scaled down K-fold while channel doping is increased by K.

Failure to properly scale the device structure leads to enhanced short-channel and narrow-channel effects and poor VLSI device characteristics. Figure 1 shows that the control over the value of the MOSFET device threshold voltage is lost as the gate length L of the transistor is reduced. Measured electrical data for several device structures are plotted. As oxide thickness and junction depth are scaled down, smaller gate length devices can be fabricated with good control over threshold voltage.

Figure 2 shows the drain to source breakdown voltage BV_{DSS} versus gate length, indicating that source to drain punch-through becomes a problem when gate lengths less than the scale-predicted value are used. The use of surface channel implants at appropriate doping levels helps to reduce this susceptibility to punch-through. This also reduces threshold voltage sensitivity to substrate resistivity variation. The use of such implants also allows the use

[2] The italic numbers in brackets refer to the list of references appended to this paper.

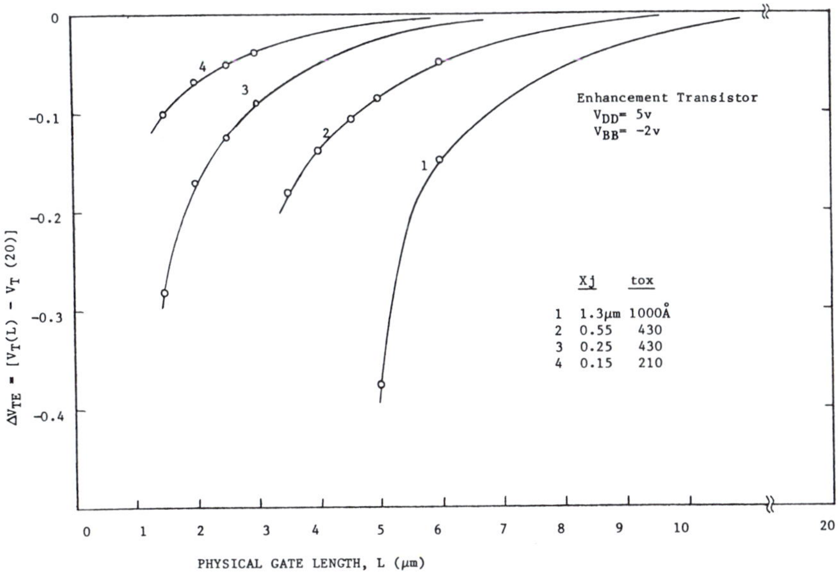

FIG. 1—*Threshold voltage variation for various device structures.*

of higher resistivity substrates in order to reduce parasitic capacitances and hence enhance the device speed performance.

Junction Formation

The requirements to achieve shallower source-drain junction depths as device structures are scaled down to minimize short channel effects places limits on the extent of high temperature processing allowable. Some of the constraints and possible solutions will be discussed in this section. The N^+ source drain regions of typical self-aligned silicon-gate NMOS transistors have traditionally been formed using phosphorus deposition from a dopant source such as $POCl_3$.

These present-day MOS manufacturing processes require temperatures in excess of 1000°C to "reflow" the phosphorus-doped glass used to insulate conductor lines. Reflow is utilized for planarization because the atmospheric, chemically vapor deposited (CVD) vapox films deposit leaving cusps and sharp edges at underlying steps. Minimum reflow temperature is dependent on phosphorus concentration, but phosphorus concentration in excess of approximately 5% creates potential metallization reliability problems, thus confining reflow temperatures to near 1050°C.

The final junction depth of the source drain regions is determined primarily by the diffusion time at the high reflow temperature. Junction depths around 1.5 μm are typical for phosphorus-doped junctions driven in at temperatures around 1050°C. For such diffused junctions there is a very sig-

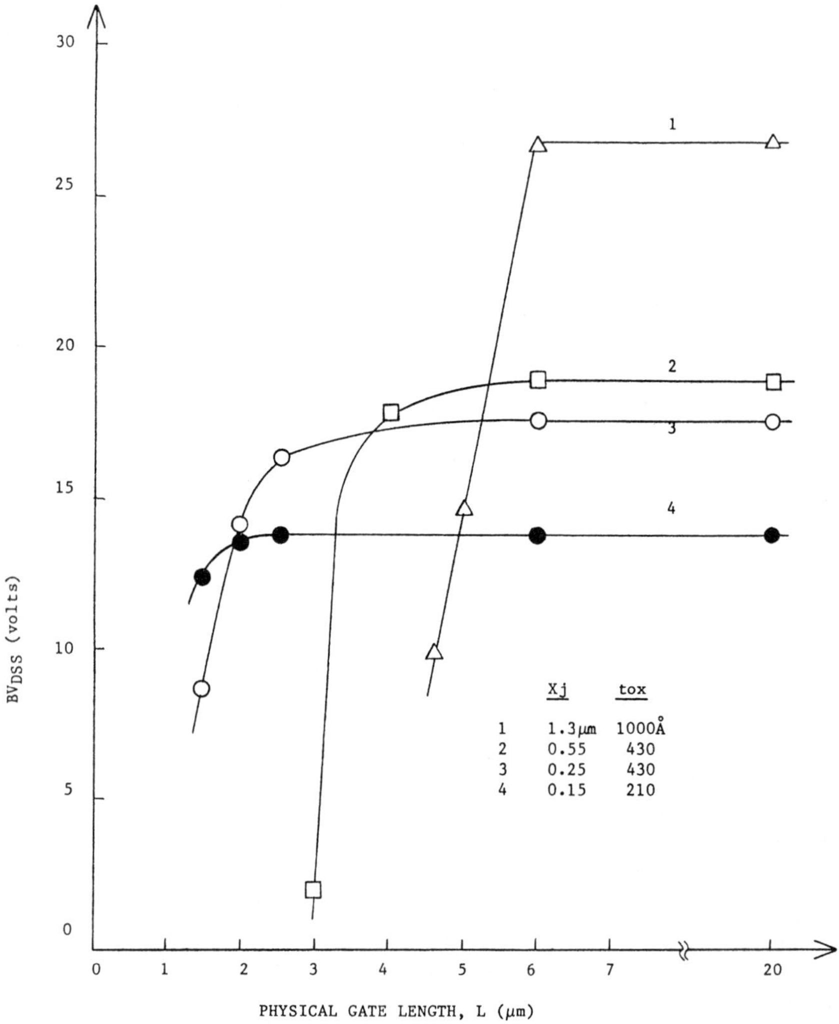

FIG. 2—*Drain-to-source punch-through voltage* BV_{DSS} *for various device structures.*

nificant lateral diffusion on the order of 70 to 80% of junction depth x_j. The electrical channel length L_{eff} is $\leqq L - (2{*}0.8x_j)$ due to this encroachment. Clearly this encroachment must be minimized for higher density devices.

The use of arsenic as the dopant source allows shallower junction depths at the same diffusion temperature because of its lower diffusivity in silicon. Whereas other dopant sources are possible, ion implantation is the most common method for incorporating arsenic into the silicon. Figure 3 is a plot of junction depth versus anneal temperature for a typical 100 keV As^+ implant annealed during furnace ramp up and down cycles at various peak tempera-

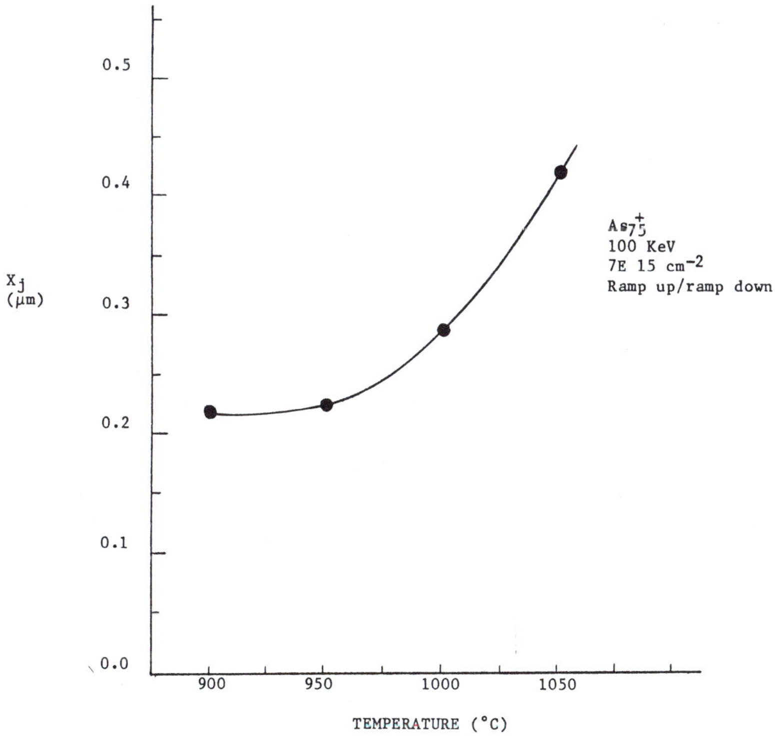

FIG. 3—*Junction depth versus anneal temperature.*

tures. These are results of process modeling using the program SUPREM. It is evident that with delay times of 10 to 20 min at peak temperatures, junction depths of ~0.5 μm are possible under practical implant and reflow conditions. Further reductions in junction depth are possible but are limited by at least two major factors.

1. Temperatures below ~1000°C inherently render the reflow process, and therefore P-Vapox, unusable. An alternative dielectric not requiring reflow is required; such a dielectric and the system we have used is described in the next section.

2. Thermal annealing of high-dose ion implants at temperatures below 1000°C requires special attention. Inadequate annealing could lead to leaky *n-p* junction characteristics due to the presence of microdefects. We have successfully fabricated arsenic-implanted junctions implanted in the dose range of 4 to 7 *E* 15 cm^{-2} and annealed at maximum temperatures of 900 to 925°C. Junction depths are 0.15 to 0.25 μm. Particular attention has been given to the requirements of epitaxial regrowth of the amorphous layers at temperatures between 500 to 600°C, after Csepregi et al [*2*] and Tamminga et

al [*3*]. The fabricated junctions have excellent electrical characteristics; reverse leakage currents of ~1 nA/cm^2 at 7-V reverse bias are achieved routinely.

Incomplete electrical activation and mobility effects result in higher resistivities for the N$^+$ source drain regions annealed at lower temperatures. This effect is shown in Fig. 4, which is a plot of sheet resistivity versus implant dose for different anneal temperatures. Sheet resistivities of ~40 Ω/square are achievable under practical conditions.

Special profiling techniques to minimize the lateral N$^+$ encroachment under the gate have been attempted for several years. The concept of using a lightly doped N-region in the vicinity of the gate has recently been reported [*4,5*]. Since the diffusivity of arsenic is concentration-dependent, the more lightly doped region has a shallower junction depth and hence a reduced lateral encroachment. The increased resistivity of the N$^+$ region that results, though, must be compensated either by a heavier N$^+$ region removed from the gate or by the use of a high-conductivity refractory metal [*6*] or silicide [*5*].

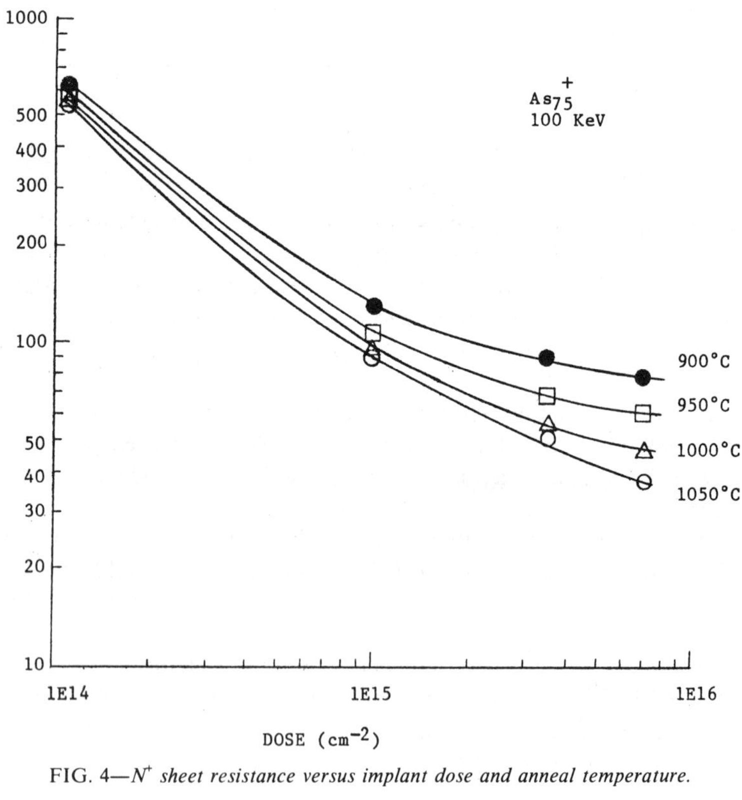

FIG. 4—*N$^+$ sheet resistance versus implant dose and anneal temperature.*

Low-Pressure CVD

As discussed in the previous section, the fabrication of high-density NMOS integrated circuits requires elimination of the high-temperature reflow step. The replacement of the atmospheric CVD *p*-vapox with a dielectric deposited conformally, without cusps, etc., at low temperatures is a key to VLSIC fabrication. Whereas plasma-assisted and sputtered silicon nitride and oxide films are possible options, we shall report on a low-pressure, low-temperature, chemically vapor deposited oxide (LPLTO) film. This oxide conformally covers underlying features, allowing acceptable step coverage without high temperature reflow, as shown in Fig. 5.

A simplified cross-section of the low-pressure CVD system that has been used is shown in Fig. 6. The system was first reported by Hochberg [7] and has been fabricated at our facility. The oxide is deposited via the use of silane and oxygen. The gases are injected into the quartz tube at either end through plenums. Holes in the plenums restrict gas injection into the tube only underneath the enclosed quartz boat that carries the silicon substrates. The system design constrains the gas flow to be in a radial pattern across the wafer surface. The system has been operated typically at 400 to 450°C and deposition conditions have been optimized to provide deposition rates ~15 nm/min

FIG. 5 *LPLTO film deposited over polysilicon line.*

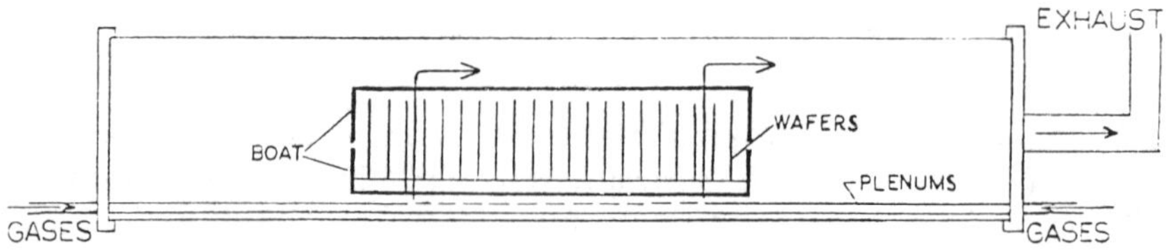

FIG. 6—*Schematic of reactor.*

with good uniformity and step coverage. Similar systems are now becoming available commercially.

High-Conductivity Gate

Reduction of processing temperatures requires other changes to processing technology. In current MOS processes, undoped polysilicon is deposited and subsequently doped with phosphorus by either a diffusion or ion-implanted predeposition step. The final polysilicon sheet resistance is dependent on the maximum process temperature, generally the "reflow" step. Higher sheet resistance results when drive temperature is reduced. Increases in the gate interconnect resistance also result from the reduced linewidths.

The use of better doping methods for the polysilicon, such as *in-situ* doping during deposition, can provide an interim solution. Resistivities of ~10 Ω/square are achievable. Over the past few years considerable work has been reported utilizing refractory metal silicides such as $MoSi_2$, $TaSi_2$, WSi_2, etc. These high-conductivity silicides allow resistivities to be ~1 to 5 Ω/square; they are used either directly or as a silicide/polysilicon sandwich on the gate dielectric.

The approaches outlined in Ref *5* (platinum silicide) and Ref *6* (tungsten) offer unique solutions to maximizing the conductivity of the gate interconnect and the source drain regions simultaneously.

Discussion

The fabrication of VLSICs places numerous constraints on the process techniques required. Some of these constraints and their solutions in the context of low-temperature processing have been reviewed. It should be pointed out, however, that considerable attention must be given to other areas in optimizing any VLSIC process. In particular, the use of optimum gettering techniques and contact structures is a must. Process and two-dimensional device modeling can provide very useful insights and can be used to optimize the device structures.

References

[*1*] Dennard, R. H., Gaensslen, F. H., Yu, H. N., Rideout, V. L., Bassous, E., and Le Blanc, A. R., *IEEE Journal of Solid State Circuits*, Vol. SC-9, No. 5, Oct. 1974, pp. 256–267.

[*2*] Csepregi, L., Kennedy, E. F., Gallagher, T. G., Mayer, J. W., and Sigmon, T. W., *Journal of Applied Physics*, Vol. 48, No. 10, Oct. 1977, pp. 4234–4240.

[3] Tamminga, Y. and Josquin, W. J. M. J., *Applied Physics Letters*, Vol. 32, No. 1, 1 Jan. 1978, pp. 13–15.
[4] Ogura, S., and Tsang, P., Walker, W., Shepard, J., and Critchlow, D. in *Technical Digest*, 1981 International Electron Devices Meeting, Washington, D.C., Institute of Electrical and Electronics Engineers, New York, 1981, Abstract 28.3, pp. 651–654.
[5] Shibata, T., Hieda, K., Sato, M., Konaka, M., Dang, R., and Iizuka, H. in *Technical Digest*, 1981 International Electron Devices Meeting, Washington, D.C., Institute of Electrical and Electronics Engineers, New York, 1981, Abstract 28.2, pp. 647–650.
[6] Gargini, P. A. and Beinglass, I. in *Technical Digest*, 1981 International Electron Devices Meeting, Washington, D.C., Institute of Electrical and Electronics Engineers, New York, 1981, Abstract 3.2, pp. 54–57.
[7] Hochberg, A. and Dozier, A. in *Proceedings*, Electrochemical Society Meeting, Pittsburgh, No. 78-2, The Electrochemical Society, Pennington, N.J., Abstract 221, pp. 592–594.

Lewis E. Katz[1]

Defects and High-Pressure Steam Oxidation of Silicon

REFERENCE: Katz, L. E., "**Defects and High-Pressure Steam Oxidation of Silicon,**" *Silicon Processing, ASTM STP 804,* D. C. Gupta, Ed., American Society for Testing and Materials, 1983, pp. 238–244.

ABSTRACT: Recently high-pressure oxidation has been successfully applied to both bipolar and metal-oxide-semiconductor (MOS) technologies. This paper reviews those aspects of high-pressure oxidation, from a perfection standpoint, which are necessary to qualify this technique for incorporation into device processing.

Oxidation-induced stacking faults, which are known to degrade junctions, are shown to be reduced with high-pressure oxidation when compared with conventional wet oxidation at atmospheric pressure. Models are discussed which correlate well with this observation.

Impurity levels, as measured by defect state concentration measurements and neutron activation, have values equal to or lower than those for wafers oxidized by conventional techniques. A review of oxide charges present, as observed by several authors on various equipment including commercial pyrogenic and water-injected systems, shows values comparable to one atmosphere control. Additionally, process parameters affecting stress in the growing oxide film are discussed.

Overall, the high-pressure oxidation technique appears to grow high-quality oxides comparable to those grown in qualified conventional systems.

KEY WORDS: steam oxidation, high-pressure oxidation, defects in oxidation, silicon oxide on silicon, oxide quality

High-pressure oxidation of silicon is a technology making the transition from an experimental process to one that can be incorporated into very-long-scale-integration (VLSI) processing. Early work [*1–4*] studied the kinetics and showed the feasibility of such a technique.[2] Recently both experimental systems and production-oriented equipment have become available. Along with these advances device applications have begun to appear. For example, a high-speed, high-density oxide-isolated bipolar process using this technology has recently been described [*5–7*]. Additionally, applications to metal-oxide-semiconductor (MOS) technology have been successfully made to the growth of a thick-field oxide layer in a dynamic random access memory (RAM) [*8*]. In this paper those aspects of high-pressure steam oxidation

[1] Supervisor and Member of Techical Staff, Bell Laboratories, Allentown, Pa. 18103.

[2] The italic numbers in brackets refer to the list of references appended to this paper.

of silicon, from a perfection standpoint, which are necessary to qualify this technology for incorporation into device processing will be reviewed.

Oxidation-Induced Stacking Faults

The formation of oxidation-induced stacking faults (OSF) typically associated with thermal oxidation of silicon are observed to be suppressed at high pressure [*9, 10*]. This is a beneficial effect, since it is well known that such defects have deleterious effects on device performance. Tsubouchi et al [*9*], working in the 950 to 1100°C range and at 6.4 atm pressure, conclude that for the same grown oxide thickness, the high-pressure steam oxidation results in a significant reduction in the length and density of OSF when compared with conventional wet oxidation at atmospheric pressure. This effect is shown in Fig. 1. Katz and Kimerling [*10*], working at lower temperature (700°C) and higher pressure (20 atm), observed complete suppression of OSF for oxides of all thickness studied (up to 5 μm) and for all substrates that included (100) or (111) *n*-type and *p*-type lightly doped wafers. Preexisting OSF tended to grow during a high-pressure steam oxidation treatment; however, the net length of the OSF was reduced by the consumption of silicon. This is shown in Fig. 2. Lin et al [*11*] have presented a model where ox-

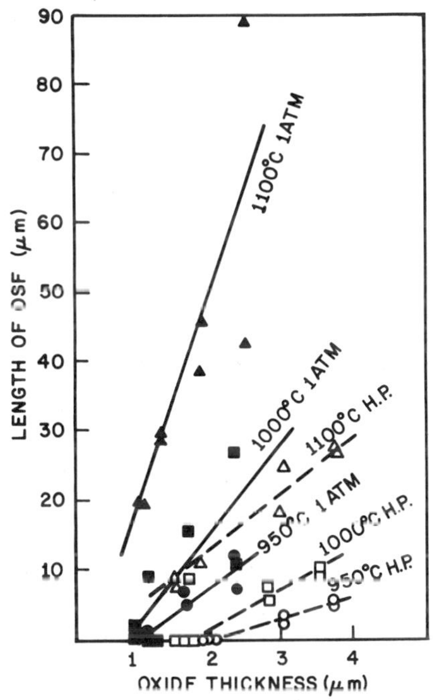

FIG. 1—*Length of OSF versus oxide thickness for conventional and high-pressure* (*6.6* kgf/cm^2) *steam oxidations* [9].

INITIAL SIZE	H.P. OXIDE THICKNESS (700°C)	FINAL SIZE
34μm	1.6μm TO 4.2μm	SHRINKAGE UP TO 20%
5μm	5.5μm	2.5μm

FIG. 2—*Size change in pre-existing stacking faults* [10].

idation rate is the controlling parameter in OSF length. The equation describing this is [*11*]

$$\frac{dr}{dt} = K_1 \left(\frac{dT_{ox}}{dt}\right)^n - K_2 \quad (1)$$

where

r = stacking fault length,
t = time,
K_1 = is related to the OSF growth mechanism and defect generation at the Si-SiO_2 interface,
$\frac{dT_{ox}}{dt}$ = oxidation rate,
n = power dependence, and
K_2 = is related to the retrogrowth mechanism.

Application of this equation to OSF experimental data in the literature gives values for n, K_1, and K_2. A 0.4th power dependence is observed [*11*]. This less-than-linear dependence of the OSF growth rate dr/dt on oxidation rate means that a smaller OSF will result at a higher oxidation rate at the same temperature for the same oxide thickness. This, of course, is the case with high pressure oxidation where the oxidation rate is increased. Figure 3 shows the good agreement obtained when the model in Eq 1 is fitted to the high-pressure data of Tsubouchi et al [*8*].

Additionally, Murarka [*12*] has described the OSF length L in (100) silicon as

$$L = K^1 p_{O_2}{}^m t^n \exp(-Q/kT) \quad (2)$$

where K^1 is a constant, P_{O_2} is the partial pressure of oxygen in the oxidizing ambient, m and n are number exponents, t is the oxidation time, Q the activation energy, and T the temperature of oxidation. Use of this equation gives values comparable to that observed experimentally [*10*].

Impurities

One aspect of qualifying the high-pressure technique is to examine both the oxide and the silicon from a contamination standpoint. This has been

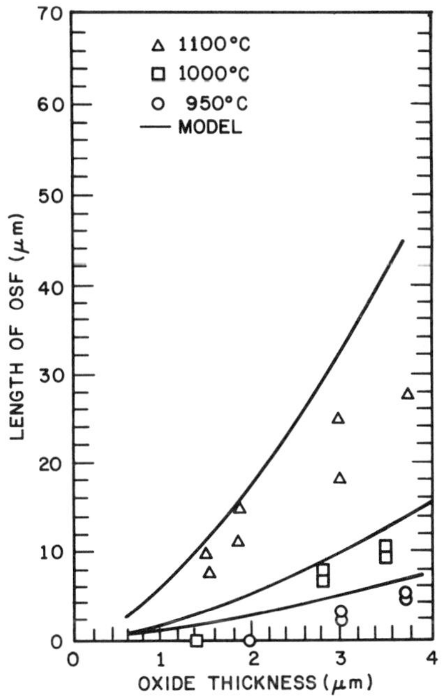

FIG. 3—*Length of OSF versus oxide thickness for high-pressure (6.6 kgf/cm^2) steam oxidation. Experimental data from Ref* 9; *model from Ref* 11.

done in the silicon by using capacitance transient measurements to look at the defect state spectrum following oxidation and subsequent removal of the oxide [*10*]. Results, shown in Fig. 4, indicate that both wet and dry 1-atm oxidation introduce deep states in the 10^{11} to 10^{12} cm^{-3} concentration range which can act as effective generation-recombination centers. The low-temperature (700°C, 20-atm) oxidation shows no new states other than the pair of shallow states in the 10^{11} cm^{-3} concentration range, which are related to oxygen associates. The important feature is that no states representative of contamination are observed. Additional data at higher temperature and pressure (950°C and 25 atm) give similar results.[3]

To examine for contamination in the oxidized silicon wafer, neutron activation analysis [*13*] has been used.[4] This technique is particularly important, since the concentration of transition group metals, which are known to degrade junction performance, can be measured directly. Samples used in this study were oxidized at 950°C and 25 atm pressure to grow ~2.0 μm thick oxides in a water-injected system. The neutron activation analysis results, presented in Table 1, integrate the impurity concentration over the entire volume of the sample, including both oxide and silicon [76-mm (3-in.) diameter, 0.5 mm (0.020 in.) thick]. These values are equal to or lower than values mea-

[3] L. E. Katz and L. C. Kimerling, Bell Laboraties, Allentown, Pa., unpublished data.
[4] P. F. Schmidt, Bell Laboratories, Allentown, Pa., private communication, 1980.

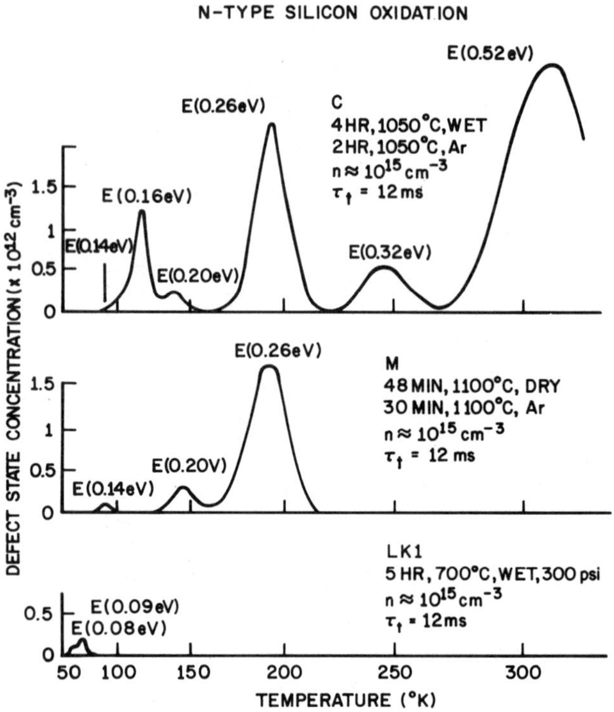

FIG. 4—*Typical defect state spectrum following oxidation of* n-*type silicon* [10].

sured on wafers oxidized in a qualified MOS facility,[3,4] even though the oxide thickness is greater.

Oxide Charges

Perhaps the question most frequently raised about the accelerated oxidation technique concerns the qualification of the oxide from a MOS standpoint. The answer determines the suitability for use in device processing. There are several reports of such an evaluation on oxides grown at high pressure in various systems and under various conditions. Miyoshi and coworkers [*14,15*], using a pyrogenic high-pressure system at 6.4 atm and 800

TABLE 1—*Impurity concentration from neutron activation analysis.*

Impurity	Concentration per Wafer, atoms
Copper	1.0E14
Tantalum	1.5E11
Cobalt	9.5E11
Chromium	1.1E13
Nickel	9.5E13
Iron	4.5E14

to 1000°C, report positive mobile charge levels of 4×10^{10} per cm^2 and an average fixed charge level of 3×10^{11} per cm^2 for (100) silicon. Katz and Howells [*16*] report a value of 3.2×10^{10} per cm^2 for the fixed charge, also on (100) silicon, for 20 atm and 725°C oxidations following appropriate annealing and baking. These authors used a water-injection system. Deal et al [*17*] have evaluated oxides grown in a commercial pyrogenic high-pressure system and report oxide charge densities (oxide fixed, interface trap, and mobile ionic charge densities) to be approximately the same for high-pressure oxides (up to 25 atm) as for 1-atm oxides. Specifically, a value of 4×10^{10} per cm^2 for the mobile ionic charge density is reported. Additionally, a commercial water injection high-pressure system has been evaluated for oxide quality at 25 atm pressure in the 900°C range.[4] Values obtained for oxide fixed charge and mobile ionic charge on samples oxidized at high pressure were comparable to 1-atm controls.

The overall conclusion with respect to oxide quality is that high-pressure oxidation equipment is capable of growing oxides which are comparable in quality to those grown under atmospheric conditions.

Stress

Selective oxidation of silicon represents one application of the high-pressure technique. It is possible that stress along the edge of the oxidized area, especially in recessed oxides, may produce severe damage in the silicon in the form of dense dislocation networks [*18*]. It is necessary that the stress generated by the growing oxide, whose volume is over twice that of the consumed silicon, be relieved without damaging the silicon. Katz and co-workers [*18*] have examined various parameters related to the isolation process for recessed structures and conclude that the oxidation temperature must be sufficiently high to allow the oxide stress to be relieved in the oxide by viscous flow. Their process parameter results [*18*] are summarized in Table 2. The critical temperature to prevent stress-induced defect formation in the recessed structures is in the 950°C range. This temperature correlates well with that found by Eer Nisse [*19,20*] for stress-free growth in his study of stress in oxide films grown at 1 atm.

TABLE 2—*Process parameters for recessed structures.*

Parameter	Value to Achieve Fewer Defects
Masking nitride thickness	thinner
Suboxide thickness	thicker
Trench depth	shallower
Trench shape	shallow slope
Isolation oxide thickness	thinner
Bulk versus epi	bulk
Oxidation temperature	higher

Summary

In summary, high-pressure oxidation is being introduced into both MOS and bipolar technologies. From a defect standpoint for comparable temperature and oxide thickness, oxidation stacking fault length is reduced for high-pressure oxidations. If high pressure and low temperature are used, stacking fault formation can be completely suppressed. From a contamination standpoint, no measurable defect states are introduced into the silicon as a result of high-pressure oxidation. Additionally, neutron activation analysis of the oxide and silicon show impurity levels at least as good as qualified MOS oxides. Charge levels in the oxide are comparable to 1-atm oxides. Finally, in the growth of thick recessed oxides, the stress of the expanding oxide must be controlled by varying the process parameters, particularly the temperature, to prevent stress-induced dislocation formation.

References

[*1*] Ligenza, J. R., *Journal of Physical Chemistry,* Vol. 65, No. 11, Nov. 1961, p. 2011.

[*2*] Ligenza, J. R., *Journal of the Electrochemical Society,* Vol. 109, No. 2, 1962, p. 73.

[*3*] Panousis, P. T. and Schneider, M., *Extended Abstracts,* Electrochemical Society Abstract 53, 1973 Spring Meeting, The Electrochemical Society, Pennington, N. J., 1973, p. 137.

[*4*] Powell, R. J., Ligenza J. R., and Schneider, M., *IEEE Transactions on Electron Devices,* Vol. ED21, No. 10, Oct. 1974, p. 636.

[*5*] Tsukamoto, K., Akasaka, Y., Miyoshi, Y., Tsubouchi, N., Horiba, Y., Kijima, K., and Nakata, H. in *IEDM Technical Digest,* IEEE, 1979, pp. 340-343.

[*6*] Agraz-Guerena, J., Pritchett, R. L., and Panousis, P. T. in *IEDM Technical Digest,* IEEE, 1978, pp. 209–212.

[*7*] Agraz-Guerena, J., Panousis, P. T., and Morris, B. L., *IEEE Transactions on Electron Devices,* Vol. 27, No. 8, Aug. 1980, p. 1397.

[*8*] Tsubouchi, N., Miyoshi, H., Abe, H., and Enomoto, T., *IEEE Transactions on Electron Devices,* Vol. 26, No. 4, April 1979, p. 618.

[*9*] Tsubouchi, N., Miyoshi, H., and Abe, H., *Proceedings,* Ninth Conference on Solid State Devices, Tokyo, 1977; *Japanese Journal of Applied Physics,* Vol. 17, Supplement 17-1, 1978, p. 223.

[*10*] Katz, L. E. and Kimerling, L. C., *Journal of the Electrochemical Society,* Vol. 125, No. 10, 1978, p. 1680.

[*11*] Lin, A. M., Antoniadis, D. A., Dutton, R. W., and Tiller, W. A., *Extended Abstracts,* Abstract 539, 1979 Fall Meeting, The Electrochemical Society, Pennington, N. J., 1979, p. 1349 (accepted for publication in the *Journal of the Electrochemical Society*).

[*12*] Murarka, S. P., *Physics Review B,* Vol. 16, No. 6, 15 Sept. 1977, p. 2849.

[*13*] Schmidt, P. F., Riley, J. E., and McMillian, D. J., *Analytic Chemistry,* Vol. 51, No. 2, Feb. 1979, p. 189.

[*14*] Miyoshi, H., Tsubouchi, N., and Nishimoto, A., *Journal of the Electrochemical Society,* Vol. 125, No. 11, 1978, p. 1824.

[*15*] Miyoshi, H., Tsubouchi, N., Nishimoto, A., and Abe, H., *Extended Abstracts,* Abstract 359, 1977 Fall Meeting, The Electrochemical Society, Pennington, N. J., 1977, pp. 943–945.

[*16*] Katz, L. E. and Howells, B. F., *Journal of the Electrochemical Society,* Vol. 126, No. 10, 1979, p. 1822.

[*17*] Deal, B. E., Rayouk, R. R., and Lie, L. N., *Extended Abstracts,* Abstract 308, 1980 Fall Meeting, The Electrochemical Society, Pennington, N. J., 1980, pp. 802–804.

[*18*] Katz, L. E., Agraz-Guerena, J., and Morris, B. L., *Extended Abstracts,* Abstract 306, 1980 Fall Meeting, The Electrochemical Society, Pennington, N. J., 1980, pp. 797–799.

[*19*] Eer Nisse, E. P., *Applied Physics Letters,* Vol. 30, No. 6, 15 March 1977, p. 290.

[*20*] Eer Nisse, E. P., *Applied Physics Letters,* Vol. 35, No. 1, July 1979, p. 8.

Joseph R. Monkowski,[1] *Michael D. Monkowski,*[2]
Ignatius S. T. Tsong,[3] *and Joseph Stach*[1]

Hydrogen/Chlorine Distributions in Silicon Dioxide: Detection and Model

REFERENCE: Monkowski, J. R., Monkowski, M. D., Tsong, I. S. T., and Stach, J., **"Hydrogen/Chlorine Distributions in Silicon Dioxide: Detection and Model,"** *Silicon Processing, ASTM STP 804*, D. C. Gupta, Ed., American Society for Testing and Materials, 1983, pp. 245–259.

ABSTRACT: Secondary ion mass spectrometry (SIMS) and electron microscopy have been used to investigate the distribution of hydrogen and chlorine in SiO_2 films thermally grown on silicon. These films have been grown in various ambients, including pure O_2, steam, HCl/O_2, and Cl_2/O_2, as well as HCl/O_2 mixtures diluted with N_2. The data suggest that there exists a strong interaction between the chlorine and the hydrogen, with the chlorine being a very effective gettering agent for hydrogen. The results of this investigation, as well as a model explaining the incorporation and distribution of the hydrogen and chlorine, are presented.

KEY WORDS: silicon dioxide, modeling, hydrogen/chlorine distributions, SIMS, electron microscopy

In order to obtain reasonable levels of yield and reliability for very-large-scale-integration (VLSI) devices, the maintenance of very low contamination and defect levels in both the silicon wafers and the overlying SiO_2 films is mandatory. Since the requirements for impurity and defect levels are continually becoming more stringent, the mere prevention of impurities and defects is becoming increasingly difficult, and manufacturers are looking more and more at gettering techniques that actively remove or annihilate defects and impurities from the device structures. One of the most prevalent of these techniques is the oxidation of silicon in ambients containing chlorine.

The beneficial properties derived from the use of chlorine include the passivation of mobile alkali ions in the oxide [*1,2*],[4] an improvement in the di-

[1] Department of Electrical Engineering, Pennsylvania State University, University Park, Pa. 16802.

[2] Department of Materials Science and Engineering, Pennsylvania State University, University Park, Pa. 16802.

[3] Department of Physics, Arizona State University, Tempe, Ariz. 85287. Formerly, Department of Materials Science and Engineering, Pennsylvania State University.

[4] The italic numbers in brackets refer to the list of references appended to this paper.

electric strength of the oxide [3], and the removal of metallic impurities [4] and crystalline imperfections [5] from the silicon. However, if the processing is done inappropriately, the use of chlorine can lead to catastrophic results, some of the more severe including separation of the oxide from the underlying silicon and etching of the silicon surface. Furthermore, in what could prove to be either beneficial or deleterious, the chlorine appears to play a central role in the incorporation of relatively large amounts of hydrogen into the oxide.

In order to understand these various effects, and to use the chlorine most effectively in its many roles as a gettering agent, a quantitative model of its incorporation and its various properties is required. This investigation dealt with the development of such a model. Secondary ion mass spectrometry (SIMS) and scanning transmission electron microscopy (STEM) were the primary tools used for the detection of the chlorine and hydrogen, and thermodynamic and kinetic calculations took on the central role in explaining the distributions of these elements in the oxide and their attendant properties.

Experimental Procedure

Specimen Preparation

All oxidations were performed on 50-mm, (100)-oriented, phosphorus-doped, 2 to 3 Ω-cm silicon wafers. Before oxidation, the wafers were processed through a standard cleaning operation consisting of a solution containing ammonium hydroxide and hydrogen peroxide, followed by a solution containing hydrochloric acid and hydrogen peroxide, and finally a dilute hydrofluoric acid dip. After every solution, the wafers were rinsed in deionized water. The oxidations were done in a resistance-heated furnace containing a 75-mm diameter fused silica tube and a mullite liner.

To be certain that the gas mixture reached equilibrium by the time it contacted the wafers, the mixture first flowed through the hot zone of the furnace in a 9-mm inside diameter fused silica tube that was located within the larger oxidation tube. At the far end of the hot zone, the gases exited the smaller tube and flowed back through the hot zone in the oxidation tube, where the silicon wafers were oxidized.

High-purity gases were used for the various oxidations. The chlorine and HCl were of 99.99% purity, while the oxygen was of 99.999% purity. The nitrogen was obtained from a liquid source and was dried and purified before use.

The oxidations were carried out in ambients consisting of either pure O_2, a mixture of O_2 and HCl, or a mixture of O_2 and Cl_2. Some of the wafers were processed in ambients containing N_2. The wafers were pushed into the furnace with the desired gas mixture flowing. At the end of the oxidation period, which ranged from 4 min to 81 h, the oxidizing ambient was turned off and the wafers were annealed in N_2 for approximately 5 min. The total gas flow for each of the oxidations was approximately 1 L/min. Volume fractions for HCl/O_2 ranged from 3% to 50%, while for Cl_2/O_2 they ranged from 0.7 to 20%. In no case was a deliberate addition of hydrogen or water vapor carried out. Temperatures of oxidation ranged from 900 to 1300°C.

Specimens to be observed in the STEM were prepared by first etching the back side of the wafers with a fine jet (approximately 0.5-mm diameter) of a 9/1 solution of HNO_3/HF. The etching, which formed a bowl-shaped depression in the silicon, was stopped when only several micrometres of silicon remained in the center of the depression. An ion beam thinner was then used to remove the few micrometres of silicon that remained. The thin layer of silicon left after chemical etching was necessary to prevent preferential etching of additional phases present at the oxide/silicon interface.

Equipment

The only piece of equipment used in this investigation that is unique and of particular interest is the SIMS. In order to allow hydrogen detection via SIMS, the hydrogen and H_2O background pressures needed to be kept very low. For this reason, during analysis the specimen chamber was evacuated to 0.67 μPa (5×10^{-9} torr) by a combination of a turbomolecular pump and a titanium sublimation pump. Furthermore, proton contamination of the primary ion beam needed to be prevented, and this was done by passing the Ar^+ primary ion beam through a mass analyzer. The 7-keV Ar^+ ion beam was focussed to 100 μm in diameter with a current of 0.4 μA. The beam was rastered in a 500-μm by 500-μm pattern, and the secondary ion signal was electronically gated so that the analysis was only over the central portion of this larger area. This gating prevented any crater effects, which give rise to poor depth resolution. The sputtering rate was approximately 0.5 nm/s.

To overcome charging of the SiO_2 surface due to Ar^+ ion bombardment, a 1.6-keV electron beam focussed to a 2-mm diameter spot was directed onto the specimen surface. Charge neutralization was usually obtained with an electron beam current of about 40 μA, although this value varied somewhat from one specimen to another. In all cases, complete neutralization was considered to be achieved when the $^{16}O^+$ and $^{30}Si^+$ depth profiles appeared undistorted.

Calibration of the SIMS for hydrogen concentrations was done by recourse to resonant nuclear reaction experiments. Several specimens were analyzed via both techniques, and the calibration factor between counts per second and hydrogen concentration was determined. In a similar manner, the chlorine profiles were calibrated by comparison with Rutherford backscattering data.

Results and Discussion

Oxide Microstructure

Figure 1 shows an STEM micrograph of an oxide grown in 6 volume-percent HCl at 1150°C for 1 h. The micrograph clearly shows the existence of

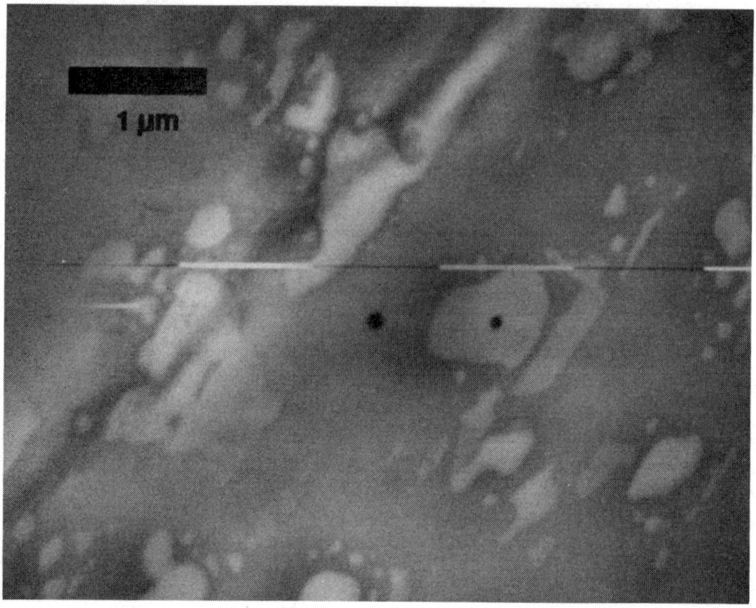

FIG. 1—*STEM micrograph of oxide grown in 6% HCl at 1150°C for 1 h.*

isolated additional phase regions. By means of electron diffraction, these additional phase regions were found to be amorphous. Two X-ray microanalyses were done on this specimen, and the contamination spots from the electron beam can be seen in the micrograph. For the spot on the left, which is in the SiO_2 matrix, no chlorine was detected down to a limit of 10^{15} Cl/cm^2. For the spot on the right, which is inside an additional phase region, a significant amount of chlorine was found. Similar analyses were done on other oxides grown under conditions ranging from 2 volume-percent HCl at 1150°C to 10 volume-percent HCl at 1300°C, and in each case the chlorine was found only in the additional phase regions.

Other investigations have shown that this additional phase, which is believed to be a chlorosiloxane, is responsible for the passivation of alkali ions; in fact, the appearance of the phase in these 1-μm isolated regions always occurs at approximately the same time that complete passivation occurs [*2,6*]. Cross-sectional scanning electron microscope (SEM) micrographs of oxides with this additional phase have shown that at this stage of growth the oxide is contiguous with the silicon, except where the phase is sandwiched between the two. The isolated phase regions are only a few tens of nanometres thick, and in this form are primarily beneficial to the electrical properties of the oxide. At this point, removal of the oxide in HF and subsequent viewing via SEM shows no etching of the silicon surface.

For longer oxidation times, the microstructure in Fig. 2 results. As shown

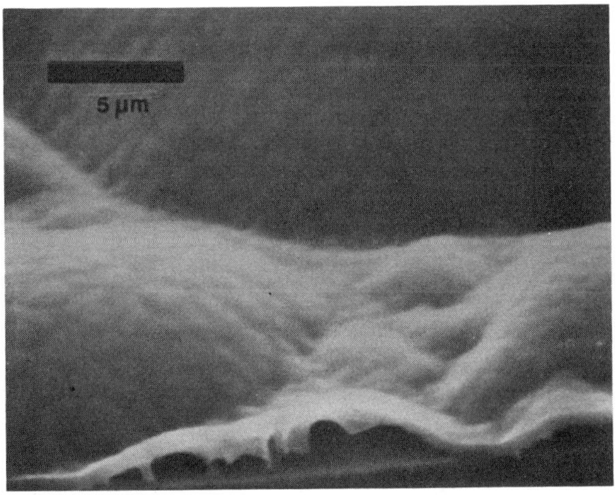

FIG. 2—*Cross-sectional SEM micrograph of oxide grown in 10% HCl at 1200°C for 6 h.*

in this cross-sectional SEM micrograph of an oxide grown in 10 volume-percent HCl at 1200°C for 6 h, if the oxidation is carried out for too long a time with too much chlorine, the oxide can bubble up from the silicon. Under extreme conditions the oxide separates completely from the silicon.

This bubbling up of the oxide is believed to be a result of a build-up of a gas at the interface. Mass spectrometric analyses have shown this gas to be some form of silicon chloride [7]; this is further substantiated by the etching of the silicon surface that occurs concurrently with the separation of the oxide. Significantly, the formation of this interfacial gas is not limited only to extended oxidations; it can occur under a number of different situations that are discussed in detail below.

Chlorine Distributions

Figure 3 shows the density of chlorine per unit area as a function of time for oxides grown in 6 volume-percent HCl at temperatures of 900°C to 1200°C. These curves were obtained by integrating the chlorine profiles such as those shown in Fig. 4 for the oxides grown at 1200°C. The most significant aspect of Fig. 3 is the similarity in the forms of the five curves. They all rise quite rapidly in the initial stages of oxidation. After this relatively rapid incorporation of chlorine, there exists a period of time when the curves tend to level out, with most of the curves even showing a dip at this point. The 1100°C curve does not show a dip, but it is difficult to conclude whether there is actually no dip or whether insufficient data points were taken. For example, in three of the curves, the dip is shown by only one point in each case.

The 1200°C curve is most striking, showing an extraordinarily large de-

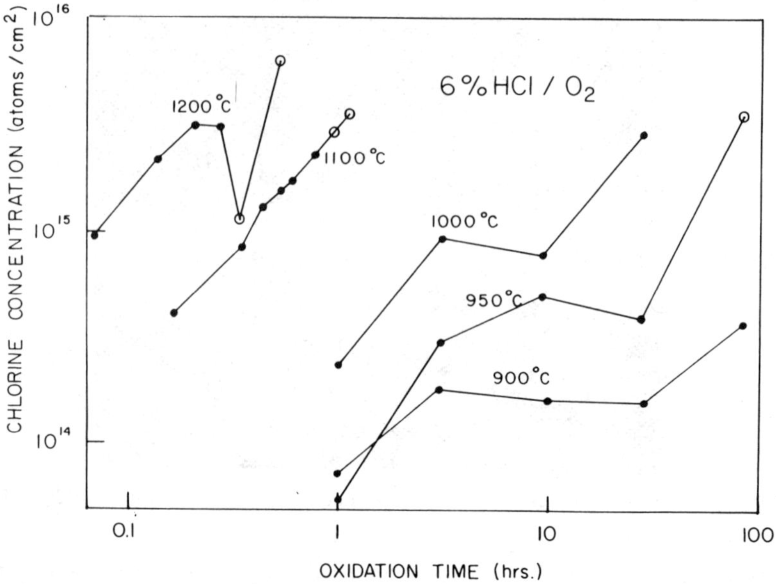

FIG. 3—*Density of chlorine per unit area as a function of oxidation time for oxides grown in 6% HCl. Open circles indicate oxides showing phase separation.*

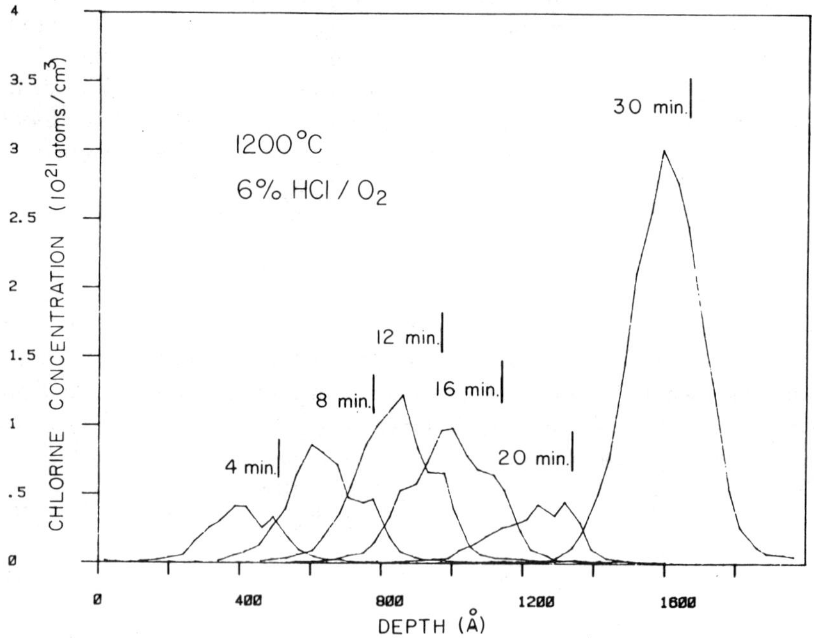

FIG. 4—*Chlorine profiles determined by SIMS for oxides grown in 6% HCl at 1200°C. Lines indicate oxide/silicon interface for each growth time.*

crease in the chlorine content at just the time that the phase separation occurs. Moreover, this large decrease appears to be real and far outside of experimental error. The SIMS data is reproducible to within several percent, whereas the data differ by approximately a factor of two. Also, the temperature would need to be off more than 100°C, or the time would need to be 5 min instead of 20 min, or the fraction of HCl in the ambient would need to be approximately 2% instead of 6% if this data point were to be rationalized as experimental error. Consequently, the loss of chlorine in the oxide appears to be real, although a systematic artifact in the SIMS measurements cannot be ruled out.

Although the decrease in chlorine content as a function of time appears to be real for most of the temperatures investigated, the reasons for this loss are not clear. Interestingly, the loss occurs just before the rapid incorporation region associated with the phase separation. In the work of Rohatgi et al [*8*] and Stagg et al [*2*], the amount of chlorine in the oxide was plotted as a function of the HCl content of the ambient for a particular oxidation time; in both cases the researchers found a sudden increase in the chlorine content of the oxide once they got to a sufficient HCl content to yield phase separation. Apparently, there is a change in the manner of chlorine incorporation associated with the formation of the micrometre-sized additional phase regions, although at this time the reasons for this change are not clear.

Another important similarity among at least several of the curves is that after sufficient time, the phase separation does eventually occur. Although there was much evidence that at temperatures of 1100°C and above, one could obtain the additional phase [*1,2*], there were no reports of the appearance of this phase below these temperatures. As pointed out in Fig. 3, phase separation has been observed in this investigation at 950°C, and would have undoubtedly occurred at 1000°C had the oxidation been carried out longer. Although the trend appears to be the same at 900°C, the eventual occurrence of phase separation at this temperature is not so conclusive.

The significance of the phase separation is that at least at the higher temperatures, the phase separation has been found to be necessary for complete passivation of mobile alkali ions in the oxide [*1,2*]. Stagg and Boudry [*2*] pointed out that the amount of chlorine in the oxide required for phase separation and the attendant passivation is approximately 3×10^{15} per cm^2. The data shown in Fig. 3 agree well with their data, and furthermore show that this same chlorine density required for phase separation is valid at lower temperatures as well.

Although the exact mechanism of the chlorine incorporation is unclear, several observations can be made. The transmission electron microscopy (TEM) work of Monkowski et al [*1*] has shown that even prior to the rapid increase leading to phase separation, there is some trace of an additional phase present, albeit on hardly more than the scale of a monolayer. During the rapid increase, most of the chlorine is incorporated as the additional phase, and when the amount of chlorine becomes sufficiently high the phase coalesces into the 1-μm regions seen in Fig. 1. As pointed out above, regardless of the temperature, the amount of chlorine in the oxide at the time of coalescence into the large isolated regions appears to be a constant, namely 3×10^{15} per cm^2. Assuming that the additional phase is a chlorosiloxane with approximately 2×10^{22} Cl/cm^3, the uniform interfacial layer breaks up when it is approximately 1.5 nm thick. Apparently this is because the balance between the energies of the oxide/silicon interface, the phase/silicon interface, and the phase/oxide interface allow a stable, uniform film only until it reaches this critical thickness. After attaining this thickness, coalescence occurs, and subsequent chlorine incorporation feeds these isolated regions.

Hydrogen Distributions

Figure 5 shows the chlorine profiles for oxides grown in 6 volume-percent HCl at 1100°C for times of 10, 35, and 55 min, as well as a profile of an oxide grown in 5 volume-percent Cl_2 at 1100°C for 10 min. The original data showed very little noise or scatter, and the double peaks in the 35- and 55-min curves are reproducible.

Although the chlorine profile for the oxide grown in a Cl_2/O_2 mixture appears strikingly different from those grown in HCl/O_2 mixtures, this difference results not from an inherent difference between Cl_2 and HCl, but rather from the difference in the amount of Cl_2 present within the furnace after the

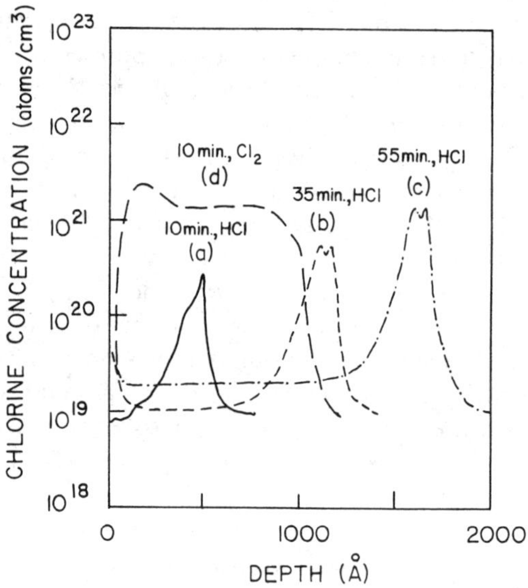

FIG. 5—*Chlorine profiles determined by SIMS for oxides grown in either 6% HCl or 5% Cl_2 at 1100°C.*

high-temperature gaseous reactions occur. As pointed out by Tressler et al [*9*], regardless of the gases put into the furnace, the imporant parameter is the equilibrium partial pressure of Cl_2 within the furnace. In Fig. 5, the Cl_2/O_2 oxidation was carried out in approximately ten times the amount of Cl_2 as those oxidations carried out in HCl/O_2 mixtures.

Figure 6 shows hydrogen profiles for these same oxides. In each of the oxides grown in an HCl ambient, the hydrogen content is approximately two times that which was observed for a similar oxide grown in pure O_2. Conceivably, this hydrogen could have been introduced from the HCl, but this is unlikely. Curve *d* on this same figure shows the hydrogen content for the oxide grown in an ambient into which there was no deliberate addition of hydrogen.

The large amount of hydrogen, which appears to be related to the amount of chlorine in the oxide, is very likely a result of gettering of hydrogen in the form of water vapor by the chlorine in the oxide. This behavior is consistent with the contention that the additional phase is a chlorosiloxane, since it is well known that chlorosiloxanes are very hygroscopic. In view of the important role that hydrogen is known to play in MOS (metal-oxide-semiconductor) device characteristics, this hygroscopic nature of the chlorine could be quite significant.

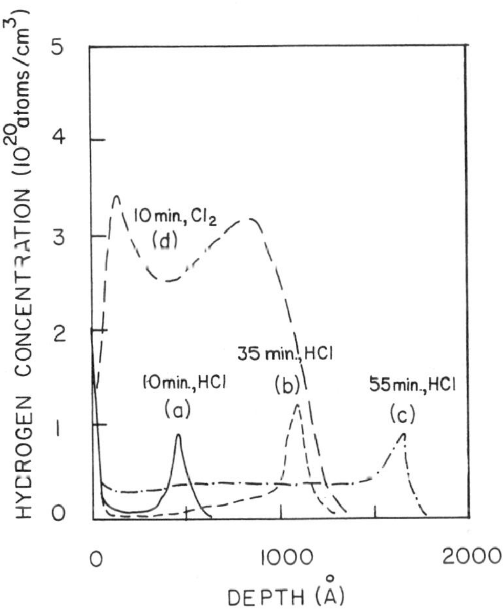

FIG. 6—*Hydrogen profiles determined by SIMS for oxides grown in either 6% HCl or 5% Cl_2 at 1100°C.*

Model

The incorporation of chlorine at the oxide/silicon interface and the consequent formation of the additional phase can be explained with a model that accounts for the changing thermodynamic situation at the interface as the oxidation or subsequent processing is carried out. Figure 7 shows a series of chemical activity gradients for a growing oxide. In order to make the model somewhat clearer, the activities have been normalized such that the phase formation, discussed below, occurs when the normalized activities are equal. At early times, such as those shown for time = T, the oxidation is in the linear regime, and the activity of oxygen is high with respect to that of chlorine. At this time only SiO_2 is formed, since the difference in activities makes an oxide much more stable than a chloride. However, an interesting situation occurs as time progresses. As is well known, the diffusion of oxygen through the already grown oxide is rate-limiting, thus indicating that the activity of

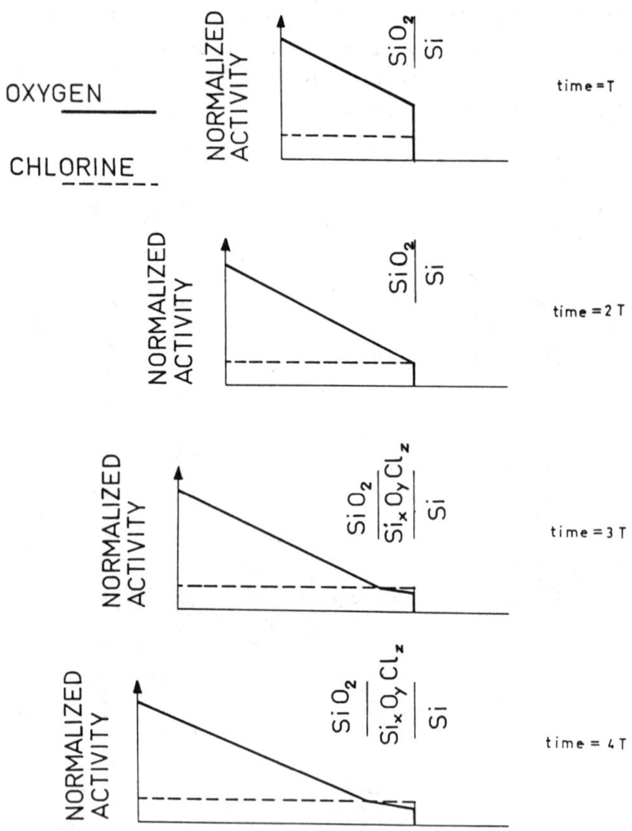

FIG. 7—*Normalized activity gradients of oxygen and chlorine in the oxide as a function of oxidation time.*

oxygen must decrease as one proceeds into the oxide. In contrast to this, the chlorine has been shown to possess a much greater diffusivity, allowing only a negligible drop in activity across the oxide [*10*]. Because the activity of oxygen is decreasing through the depth of the oxide, while that of the chlorine is not, a point is reached, represented by time $= 2T$, where the ratio of chlorine activity to oxygen activity at the interface allows the formation of a reaction product containing chlorine. In Fig. 7, the activity ratio at the onset of the formation of this reaction product is shown as unity; this is only to make the figure clearer—the actual ratio is discussed below. As time progresses, the amount of this reaction product increases, but it must follow the interface because this is the only region where it is stable. This reaction product, of course, is the additional phase, which is believed to be a chlorosiloxane.

At even longer oxidation times, the activity ratio of oxygen to chlorine decreases even further. Eventually, a regime may be reached where yet another additional phase is stable. Very likely, this second chlorine-rich phase contains more chlorine than oxygen, and could perhaps contain only chlorine and silicon. Since most silicon chlorides are gaseous at oxidation temperatures, this is believed to be the origin of the chlorine-rich interfacial gas.

Overall, four phases are being discussed. At the outset of the oxidation, SiO_2 and silicon are present. The next phase to form is the chlorosiloxane, which is believed to be a viscous liquid at oxidation temperatures. Cross-sectional SEM micrographs have shown this layer to be only several tens of nanometres thick, and with the high density of chlorine present the phase must necessarily be a condensed material. Furthermore, TEM micrographs taken at room temperature show the frozen-in patterns that occur during coalescence, which look very much like those of a viscous liquid [*1*]. This chlorosiloxane is the additional phase that is required for complete passivation.

The last phase to form is the chlorine-rich gas that is responsible for separating the oxide from the silicon (Fig. 2). In contrast to the condensed phase, this gaseous phase vigorously attacks the silicon surface, leaving anisotropic etch pits up to several micrometres deep.

In order to test this thermodynamic model, and to make it quantitative, the interfacial activities of O_2 and Cl_2 were calculated for a number of oxides grown for various times at temperatures of 900°C to 1150°C in HCl concentrations ranging from 3 to 10 volume-percent. The activity of oxygen at the interface was calculated by setting the flux of oxygen through the oxide equal to the flux of oxygen reacting with the silicon, where these fluxes were obtained through use of the parabolic and linear growth rate constants. The details of these calculations have been reported in an earlier paper [*7*]. The activity of chlorine at the interface was taken to be equal to its activity in the ambient, since there is little change in its activity through the oxide. The abscissa is put in terms of the Cl_2 partial pressure because, as mentioned above, Cl_2 has been shown to be the active species [*9*].

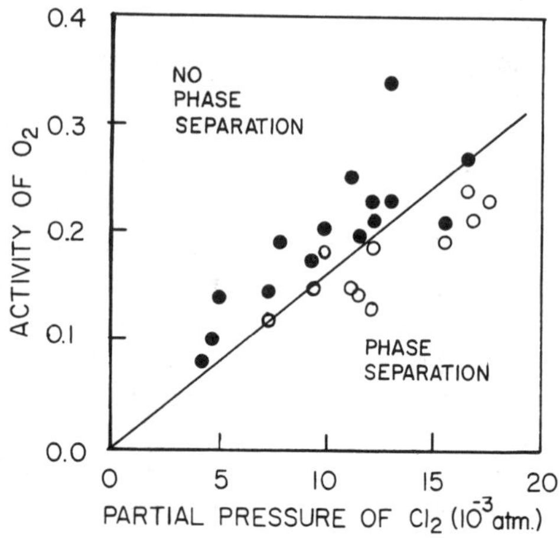

FIG. 8—*Activity fields showing where isolated phase regions can and cannot form. Open circles indicate oxides showing phase separation.*

The open circles in Fig. 8 represent the interfacial activities of oxides grown for a sufficient time to just produce the onset of phase separation. For all times longer than those represented by the open circles, the data points would fall lower in the figure. For each set of oxides represented by an open circle, where one set corresponds to one particular temperature and HCl content, the oxide grown for just a slightly shorter period of time is shown represented by a closed circle. In other words, the closed circles represent the interfacial activities of those oxides just short of showing phase separation. Data points for oxides grown for yet shorter periods of time would be located higher in the figure.

As can be seen in Fig. 8, despite that a large range of experimental conditions are considered, the onset of phase separation does occur at a relatively fixed ratio of activities.

Use of the Model in Device Processing

The most apparent role for the proposed model in device processing is in choosing the appropriate oxidation conditions to match a particular oxide parameter, such as thickness, with one or more of the benefits of chlorine, such as alkali ion passivation. The oxidation conditions that can be varied include time, temperature, and chlorine concentration; this model, along with previously published passivation data [*1*] should greatly aid in determining the proper conditions.

The model has other uses, however, that may not be as apparent. Figure 8, for example, indicates that there is a particular region of activity ratios that will yield the additional phase and the attendant passivation. On one side of this region the additional phase is unable to form, and on the other side there is danger of obtaining the gaseous chlorine-rich phase. As explained above, the usual way to move from one region to another is by oxidizing for longer times and obtaining different oxide thicknesses. Other methods that are less obvious are also possible.

Consider an oxide grown in 10 volume-percent HCl for 1 h at 1150°C. Calculation of the activity ratios for this oxide indicates that it is very near the ideal ratio, and indeed, micrographs show a very nice passivated oxide film. After being annealed for 5 h at 1150°C in N_2, however, the film develops large bubbles, as shown in Fig. 9. The reason for the bubble formation is that when the oxide is put into the N_2 ambient, the oxygen activity is lowered to zero and the interfacial gas becomes stable. The beneficial condensed phase consequently reacts to form the gaseous phase.

This same type of behavior can also occur for oxides deposited by chemical vapor deposition (CVD) using chlorinated source gases. During CVD, chlorine is incorporated into the oxide. Upon subsequent annealing, the chlorine moves to the interface, and forms the interfacial gas.

The model is also important for multiple step oxidations. Kriegler has shown, for example, that an oxide grown in pure O_2 can be passivated by subsequently oxidizing in an HCl/O_2 ambient [*11*]. Both Kriegler and Hashimoto et al [*12*] have found, however, that an unpassivated oxide cannot be passivated in an ambient containing only HCl and an inert gas; at least a fraction of a percent of O_2 must be present. These observations can be readily explained since the model shows that regardless of when or how the chlorine is introduced, it will go to the interface, and if there is enough, it will

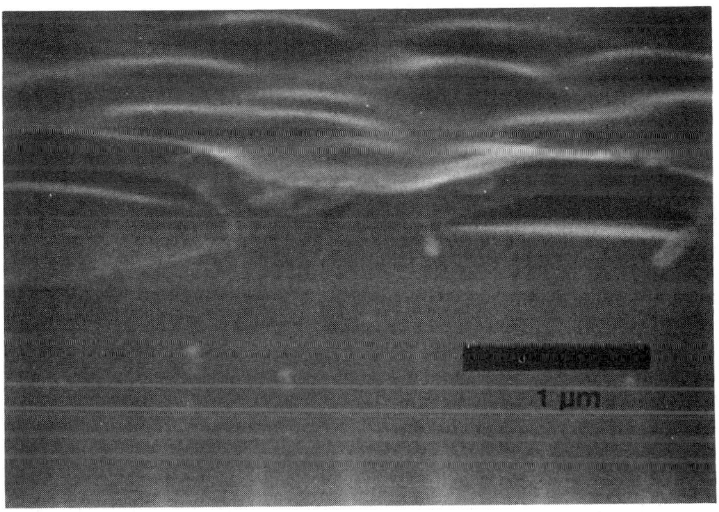

FIG. 9—*Cross-sectional SEM micrograph of oxide grown in 10% HCl at 1150°C for 1 h, and subsequently annealed in N_2 at 1150°C for 5 h.*

form the additional phase. In addition, there appears to be a kinetic barrier to the chlorination of SiO_2. That is, given the appropriate thermodynamic situation, O_2 can oxidize the chlorosiloxane, but chlorine appears not to be able to react with SiO_2 to form a chlorosiloxane. Consequently, since the chlorosiloxane contains both oxygen and chlorine, both constituents must be present in the ambient for the additional phase to form.

Some very recent work [*13*] indicates that regardless of whether an HCl oxidation follows, percedes, or is put in between pure O_2 oxidations, the chlorine stays at the interface as expected. The chlorine can be introduced even after the growth of an oxide in a wet ambient. One case, however, cannot be done, and that is to introduce a chlorinated oxide to a wet ambient. The chlorosiloxane, which is very hygroscopic, readily reacts with the H_2O forming HCl gas at the oxide/silicon interface. This gas formation results in an oxide which looks very much like that in Fig. 9.

These examples are an indication of how the model has been used to explain several observations concerning chlorinated oxides; very likely other examples exist. An important point which is meant to be illustrated by these examples, is that when using chlorine, one needs to pay close attention to the entire processing sequence, and not only to the particular step in which chlorine is directly involved. Hopefully, the use of the model will help to make the use of chlorine somewhat more straightforward.

Acknowledgments

This work was supported in part by the U.S. Army Research Office under Contract DAAG29-80-K-0085.

References

[*1*] Monkowski, J. R., Stach, J., and Tressler, R. E., *Journal of the Electrochemical Society*, Vol. 126, No. 7, July 1979, pp. 1129–1134.

[*2*] Stagg, J. P., and Boudry, M. R., *Journal of Applied Physics*, Vol. 52, No. 2, Feb. 1981, pp. 885–899.

[*3*] Osburn, C. M., *Journal of the Electrochemical Society*, Vol. 121, No. 6, June 1974, pp. 809–815.

[*4*] Katz, L. E., Schmidt, P. F., and Pearce, C. W., *Journal of the Electrochemical Society*, Vol. 128, No. 3, March 1981, pp. 620–624.

[*5*] Hattori, T., *Solid State Technology*, Vol. 22, No. 11, Nov. 1979, pp. 85–89.

[*6*] Monkowski, J. R., Tressler, R. E., and Stach, J., *Journal of the Electrochemical Society*, Vol. 125, No. 11, Nov. 1978, pp. 1867–1873.

[*7*] Monkowski, M. D., Monkowski, J. R., Tressler, R. E., and Stach, J. in *Surfaces and Interfaces in Ceramic and Ceramic-Metal Systems*, Plenum Press, New York, 1980, pp. 361–366.

[*8*] Rohatgi, A., Butler, S. R., Feigl, F. J., Kraner, H. W., and Jones, K. W., *Journal of the Electrochemical Society*, Vol. 126, No. 1, Jan. 1979, pp. 143–149.

[*9*] Tressler, R. E., Stach, J., and Metz, D. M., *Journal of the Electrochemical Society*, Vol. 124, No. 4, April 1977, pp. 607–609.

[*10*] Butler, S. R., Feigl, F. J., Titcomb, S. L., Tsai, H., Jones, K. H., and Kraner, H. W. in *Surfaces and Interfaces in Ceramic and Ceramic-Metal Systems*, Plenum Press, New York, 1980, pp. 367–377.

[11] Kriegler, R. J., *Denki Kagaku* (*Journal of the Electrochemical Society of Japan*), Vol. 41, No. 7, July 1973, pp. 466–475.
[12] Hashimoto, C., Muramoto, S., Shiono, N., and Nakajima, O., *Journal of the Electrochemical Society*, Vol. 127, No. 1, Jan. 1980, pp. 129–135.
[13] Monkowski, M. D., Monkowski, J. R., Tsong, I. S. T., and Tressler, R. E., Abstract No. 370, *Proceedings*, 160th Meeting of the Electrochemical Society, Denver, Oct. 1981, pp. 887–888.

Stefan K. Lai[1]

Dependence of Thin-Gate Oxide Properties on Processing

REFERENCE: Lai, S. K., "**Dependence of Thin-Gate Oxide Properties on Processing,**" *Silicon Processing, ASTM STP 804,* D. C. Gupta, Ed., American Society for Testing and Materials, 1983, pp. 260–272.

ABSTRACT: Thin-gate oxides (23 nm or less) will be very important for future very-large-scale-integration (VLSI) circuits. Oxides in this thickness range can be grown in different environments and temperatures. Breakdown, interface properties, and electron and hole trapping are affected by the growth conditions. The effect of growth and anneal temperature on insulator and interface properties will be examined. Such information is useful in optimizing process conditions for the long-term reliability of thin-gate oxides.

KEY WORDS: silicon dioxide, electrical breakdown, electron traps, hole traps, radiation damage, interface traps

The principle of scaling [*1*] plays a very important part in the development of very-large-scale-integration (VLSI) circuits.[2] It gives a mathematical formalism for the design of devices with smaller dimensions based on present knowledge of larger dimension devices. Even in the case where simple scaling laws do not apply, it provides the basis for fine tuning of the design. Of course, there are parameters that cannot be scaled. One example is the resistance of interconnecting aluminum and polysilicon [*1*]. There are also other possible problems associated with the smaller dimensions used [*2*]. One example is the fluctuation in the number of dopants in the channel region for very small dimension devices [*3*]. Another potential problem is the gate insulator. It has been projected that for devices with half-micrometre dimensions, 10-nm oxides will be used for the gate insulator [*1*]. This oxide has to be prepared defect-free, and should be able to withstand additional high-temperature processing and repeated electric field stress in varied environments for the projected life of the device.

Most of the earlier work on silicon dioxide has been on oxides that are 50 nm or thicker [*4,5*]. In this thickness range, oxides prepared are very repro-

[1] Intel Corporation, Santa Clara, Calif. 95051; formerly at Thomas J. Watson Research Center, IBM Corporation, Yorktown Heights, N.Y. 10598.

[2] The italic numbers in brackets refer to the list of references appended to this paper.

ducible and of very high quality. As a result, present day integrated circuits can be fabricated with high yield and are very reliable in actual use. For oxides thinner than 25 nm, however, there is not much work that has been reported in the literature. Most of the work had been on the growth kinetics of oxides in the thin regime [*6–9*]. Such work demonstrates one of the problems in the study of thin oxides. The oxide growth temperatures discussed have been in the range of 600°C [*6*] to 1100°C [*8*]. The gas environment was from dry oxygen [*7*] to steam [*6*], from one tenth of one percent oxygen in nitrogen [*8*] to pure oxygen, and from 33.3 Pa (0.25 torr) low pressure [*9*] to atmospheric pressure. The growth kinetics are different for the different growth conditions, from linear growth kinetics showing reaction-limited growth to parabolic kinetics showing diffusion-limited growth.

The growth condition is not the only determining factor of the properties of the gate oxide. In modern polysilicon gate technology, the gate oxide is grown in the early stages of processing. Subsequently, it has to go through additional high-temperature processing, subjected to implantation of ions [*10*] and high-energy radiation from electron beam [*11*] or X-ray lithography [*12*] or reactive ion etching (RIE) [*13*]. All these can change the properties of the oxide drastically [*14–16*]. To fully characterize the properties of oxides after all the different process conditions involves studying variations due to a large number of parameters. It is almost impossible to determine the effect of all the process conditions. The purpose of this paper is to identify some of the more important properties of thin-gate oxides and to try to understand the effect of growth temperatures and subsequent heat cycles.

Most of the experiments have been performed on oxides with a thickness of approximately 23 nm, which is the projected oxide thickness for field effect transistors (FETs) with gate lengths slightly more than 1 μm [*17*]. Table 1

TABLE 1—*Processing matrix.*[a]

Runs 1, 2, and 3: Dry, Dry-Wet-Dry, and HCl Oxides	
A 800°C no anneal	**B** 800°C 5 min 800°C anneal
C 900°C no anneal	**D** 900°C 5 min 900°C anneal
E 900°C 5 min 800°C anneal	**F** 1000°C no anneal
G 1000°C 5 min 1000°C anneal	**H** 1000°C 5 min 800°C anneal
Run 4: Dry Oxides, High-Temperature Anneal	
A 800°C 5 min 800°C anneal	**B** 800°C reinsert 5 min 800°C anneal
C 800°C 5 min 1000°C anneal	**D** 800°C 60 min 1000°C anneal
E 800°C no anneal	
Run 5: Dry Oxides with Polysilicon Gates	
A 800°C $POCl_3$-doped poly	**B** 800°C *in situ* P-doped poly
C 800°C no anneal, Al gate	**D** 1000°C $POCl_3$-doped poly
E 1000°C *in situ* P-doped poly	**F** 1000°C no anneal, Al gate

[a] The oxides are approximately 23 nm thick. The wafers are N-type, ⟨100⟩ in orientation, phosphorous-doped with resistivity of 0.2 Ω-cm. Postoxidation anneal was in argon, and drive-in for polysilicon was in nitrogen.

gives a summary of the different process conditions. In other experiments, thinner oxides were examined. The experiments were not designed to study all the parameters in detail but to shed some light on the effect of processing. The results were obtained with a limited data set, but they are sufficient to indicate trends in the change in properties. A lot of useful information is obtained from the results.

Breakdown in Silicon Dioxides

Electrical breakdown in silicon dioxides has been studied as a function of oxide thickness [*18–20*]. It was shown that for intrinsic breakdown, the breakdown field increased for decreasing oxide thickness, going from less than 10 MV/cm for oxides thicker than 50 nm to larger than 14 MV/cm for oxides thinner than 15 nm. On the other hand, the defect density increased with decreasing thickness [*9*], giving a high probability for low field breakdown in very thin oxides. In actual devices, except for special transient conditions, the electric field on the oxide never approaches the values for intrinsic breakdown. It is the low field breakdown that causes problems in yield and infant mortality in devices. Experimentally, very good breakdown statistics were obtained for the oxides listed in Table 1 with aluminum top electrodes. The wafers were N-type so that the breakdown testing was under positive bias, the normal bias condition for FET operation. Figure 1 shows a typical breakdown histogram. In other experiments, very good breakdown statistics were also obtained for oxides 10 nm thick.[3] The results show that under clean control conditions, oxides with very good breakdown statistics can be obtained down to thicknesses of 10 nm. It must be noted, however, that these oxides have not seen any additional processing and, in many cases, the additional processing may degrade the oxide.

The details of the breakdown process were examined by the Ramp I–V technique [*21*], which involved ramping the gate voltage at a fixed rate (0.5 V/s in the present experiments) and measuring the current with a logarithmic picoammeter. The technique is very useful for studying electron trapping in the oxide. When the Ramp I–V curve for the device is examined (Fig. 1), it can be seen that there is a distinct trapping ledge in the high field region. Similar curves were obtained in other oxides. From the ramp-rate and the current at which the ledges appeared, it was concluded that they were due to electron traps with capture cross-sections in the 10^{-17} to 10^{-18} cm^2 range [*21*]. The densities of the traps were difficult to determine because the oxides were destroyed before the end of the ledges; also, at the very high field, detrapping of the trapped electrons occurred. From previous work on electron trapping in oxides, it can be concluded that the trapping was due to water-related traps in the oxides [*22*]. In the present experiments, even the dry oxides have a lot

[3] Private communication from C. M. Osburn, IBM T. J. Watson Research Center, Yorktown Heights, N.Y., 1981.

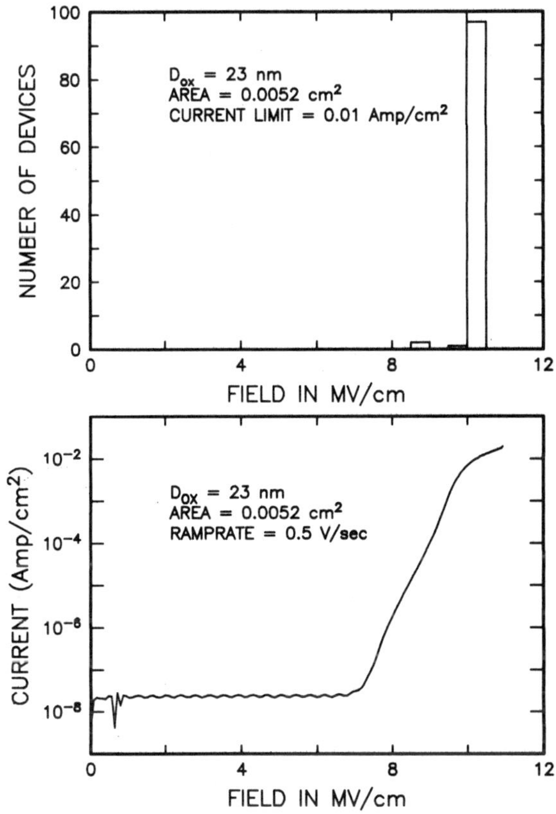

FIG. 1—*Breakdown histogram and Ramp I–V curves for a dry-wet-dry oxide grown and annealed for 5 min at 900°C.*

of water-related electron traps. This probably explains the very good breakdown statistics observed [23]. Low field breakdown in silicon dioxide is initiated in nonuniform spots in the contact that give a local high field region with high-current injection. In an oxide with a high density of electron traps, local high-current injection will cause trapping in the local region. The build-up of negative charge acts as a negative feedback to lower the field locally, quenching the local injection. Breakdown will now occur only in the high field region and the current is more uniform in the oxide. Of course, this mechanism works only when there is no gross defect in the oxide. For oxides prepared in a clean, controlled environment, the electron traps probably play a very important part in controlling breakdown.

The foregoing results were for capacitors with aluminum gate. For wafers with polysilicon gate, the breakdown results were not as consistent when compared with aluminum gate. The results are preliminary and a more detailed study has to be carried out to understand the detail of the effect. The breakdown histogram for one of the oxides is shown in Fig. 2. The same oxide with aluminum gate gives good breakdown statistics. Since the silicon–silicon dioxide interface is the injecting interface in the present experiment, it is unlikely that the top electrode contact causes the change in breakdown properties. More probably, the change is due to additional processing in preparing the polysilicon gate. (The additional processes are the deposition of silicon by low-pressure chemical vapor deposition (LPCVD), the deposition of $POCl_3$ at 870°C and the drive-in of phosphorus at 1000°C.) Again the details of the breakdown were examined by the Ramp I–V technique. When the Ramp I–V curve was examined (Fig. 2), the trapping ledge was not observed. The absence of the trapping ledge has been verified by lowering the ramp rate so that the ledge came in at a lower current level [*21*]. This absence or low density of electron traps in polysilicon gate capacitors has been reported before [*22*] and is believed to be due to the driving out of water by the high-temperature process of polysilicon deposition and the capping of the

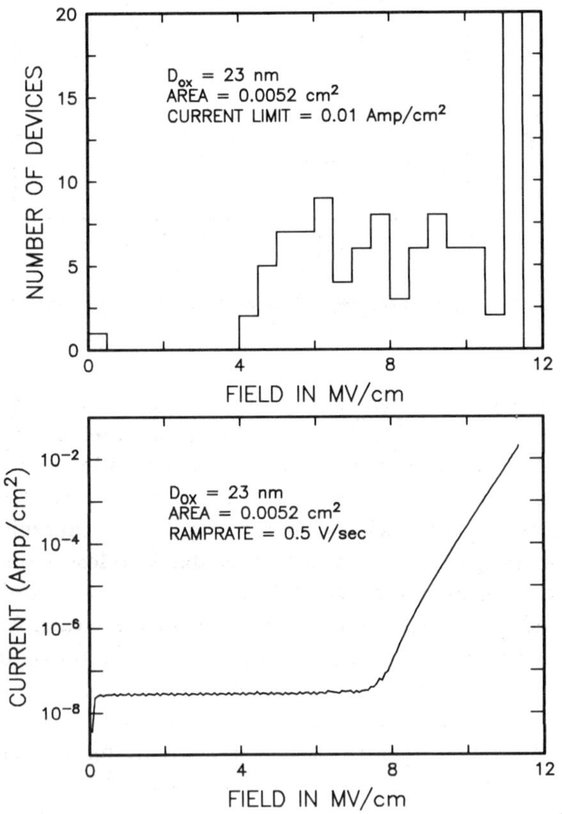

FIG. 2—*Breakdown histogram and Ramp I–V curves for a dry oxide grown at 800°C. The top contact is polysilicon doped by $POCl_3$.*

oxide by the polysilicon to prevent further in-diffusion of water. This is certainly one possible explanation for the difference between the aluminum gate and the polysilicon gate. The present result seems to contradict earlier work [*24*] that showed that there was no difference in breakdown between aluminum gate and polysilicon gate capacitors. One possible reason for the discrepancy will be discussed in the next section on electron trapping. In conclusion, it is possible to grow 22.5-nm oxides with high yield and good breakdown statistics. Such oxides seem to be degraded when polysilicon gates are used. It seems as though electron traps play a very important part in controlling breakdown in the thin oxide regime.

Electron Trapping in Silicon Dioxide

The dependence of electron trapping on the thickness of silicon dioxide has been studied [*22*]. The intrinsic electron traps in silicon dioxide were shown to be due to water-related centers. They were shown to be distributed in the bulk of the oxide; as the thickness of oxide was reduced, the total number of traps were reduced with the thickness reduction. The flatband voltage shift for a given injected charge increases with the square of the thickness. Therefore the effect of electron trapping on the shift of the threshold voltage for thin oxides is expected to be small. In terms of processing, the general trend is that the higher the temperature and the more postoxidation processing, the lower the density of electron traps. This is consistent with the picture that the intrinsic electron traps in silicon dioxide are due to water-related centers, and that the effects of higher temperatures and more processing are to reduce the amount of water in the oxide.

As the charge due to bulk trapping is reduced, the interface effects become more and more important [*25*]. Interface traps generated due to hot electron injection and trapping are mostly acceptors. Therefore, even though the flatband voltage shift may be small, the acceptors give a larger change in the inversion voltage (Fig. 3). The threshold voltage shift in FETs will be large and the transconductance of the device will be degraded. To evaluate the effect of electron trapping on FETs, it is not sufficient to determine flatband voltage shift alone. Finally, the interface effects are reduced when the density of bulk traps is reduced [*26*].

The foregoing results were for capacitors with aluminum gates. As mentioned in the last section, the use of the polysilicon gate changes the density of water-related traps in the oxide due to the high-temperature processing. When the effects of oxide thickness on flatband voltage shift in polysilicon gate capacitors were examined, a very different thickness dependence was observed. Instead of changing as the square of oxide thickness, the flatband voltage shift is more like a logarithmic function of oxide thickness (Fig. 4). The result is consistent with the picture that the high-temperature process drives water out of the oxide. This can also explain the discrepancy mentioned

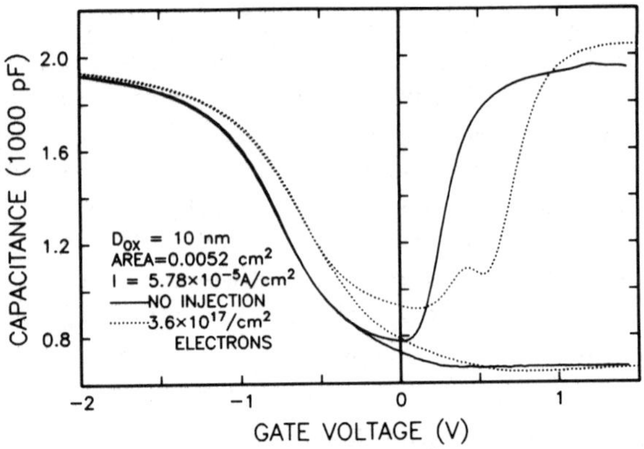

FIG. 3—*High frequency and quasistatic capacitance curves for a 10-nm oxide before and after injection of 3.6×10^{17} per cm^2 electrons.*

in the previous section. Thicker oxides were used in earlier work [*24*], and there may still be a high density of electron traps in the oxide. When thinner oxides were used in the present experiments, the density of electron traps was lower and thus a larger difference in breakdown properties was found. The density of water-related traps depends critically on the processing conditions, from oxidation to annealing, and even on the way the wafers are cooled to room temperature [*26*]. Unless there is close control on all the process steps, very different results may be obtained. Oxidation in dry oxygen does not give

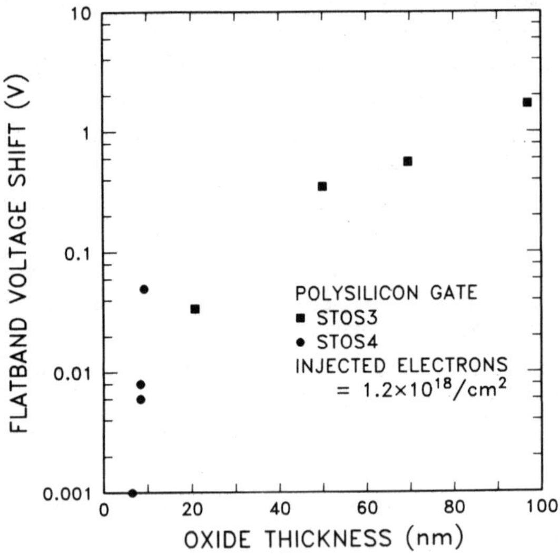

FIG. 4—*Flatband voltage shift as a function of oxide thickness after injection of 1.2×10^{18} per cm^2 electrons. The results were for two different runs in different thickness ranges.*

a dry oxide automatically. The dry oxide prepared in one laboratory may be very wet when compared with oxides prepared in another laboratory.

There is one more effect observed in oxides after the injection of hot electrons: the oxide becomes slightly leaky [*25*]. The effect was observed even in 50-nm oxides but is more pronounced in 5-nm oxides. In this thickness range, no permanent electron trapping is observed. It is possible that the trapped electrons may tunnel out of the trap under the electric field used for the hot electron injection, so that electrons are not trapped in the oxide. However, the passage of the current may cause some irreversible change in the oxide to give rise to the leakage current. This effect has to be studied in more detail.

Radiation Effects and Hole Trapping

The dependence of radiation damage on oxide thickness has been studied for thicker oxides. The flatband or threshold voltage shifts were reported to be proportional to either d_{ox}^2 [*27*] or d_{ox}^3 [*28*], where d_{ox} is oxide thickness. When extrapolated to below 25 nm, a very small flatband voltage shift is expected. It was even reported that for oxides less than 15 nm, there was no change in flatband voltage after irradiation [*29*]. At the same time, other experiments showed that the recovery time of hardened oxides to radiation was significantly shorter for thinner oxides [*30*]. From electron-beam damage experiments on very thin oxides, only a small generation of interface traps was observed [*31*]. All these experiments indicate that very thin oxides are highly resistant to radiation damage.

There are, however, still a number of unanswered questions. First of all, it has been demonstrated that radiation damage is related to hole trapping at the interface [*32,33*]. Even though the oxide thickness is reduced, the interface is still present and one would expect the hole traps to be still present. This has been confirmed by hole-trapping experiments in which hole traps were measured for oxides as thin as 8 nm (Fig. 5). Therefore the advantage of the thin oxides may be due to the fact that the volume for absorption of radiation is reduced. Once a hole gets into the oxide, there is still a high probability for it to be trapped to give positive charge and interface trap. Another area of concern is the fact that most of the reported results on thickness dependence of the radiation damage have been with aluminum gate structures. It is known that when polysilicon gate structures are used, they tend to be very susceptible to radiation damage.[4] This has been explained by the additional high temperature processing which has been known to degrade radiation hardness [*28,32,34*]. Since polysilicon gates are used in most VLSI circuits, and high energy radiation like electron beams [*17*] or X-rays [*12*] will be used for future fine line lithography, it is important to understand the radiation effect

[4] Private communication from J. M. Aitken, IBM T. J. Watson Research Center, Yorktown Heights, N.Y., 1981.

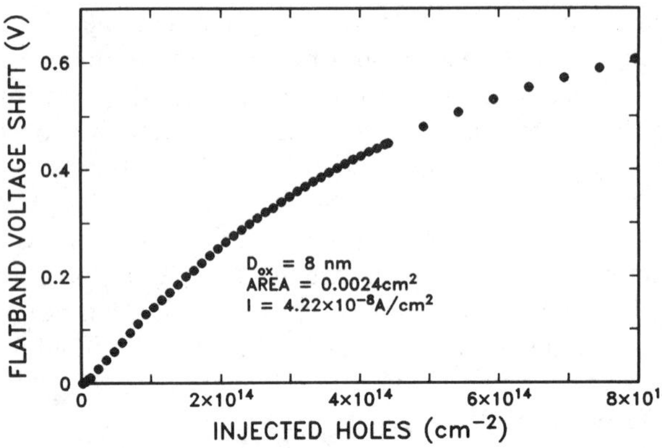

FIG. 5—*Flatband voltage shift as a function of injected holes for an 8-nm oxide with polysilicon gate.*

of polysilicon gate structures. The advantage of going to thinner oxides observed in aluminum gates may not carry over directly to polysilicon gate.

Hole-trapping experiments have been carried out on the wafers listed in Table 1. Hole injection by avalanche in silicon is a very convenient way of studying hole trapping in silicon dioxide [*32*]. A direct correlation between hole trapping and radiation damage has been established [*33*]. In using hole injection, one does not have to deal with dangerous radiation in special systems. Instead, individual devices can be tested and the trapping kinetics obtained. Some typical results are shown in Fig. 6. In general, the results confirmed previous observations [*28,32,34*]. Higher temperature and longer postoxidation anneal treatments give a higher density of hole traps. There is, however, one exception. While oxidation with HCl gives a low density of hole traps at 900 and 1000°C, the trend is reversed at lower temperatures. HCl ox-

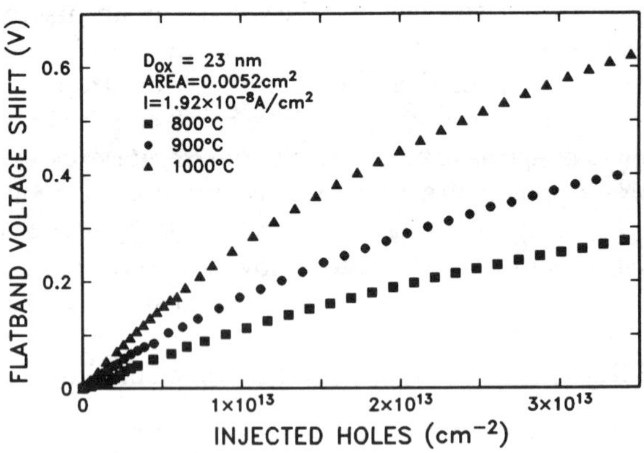

FIG. 6—*Flatband voltage shift as a function of injected holes for three 23-nm dry oxides grown and annealed for 5 min at three different temperatures.*

ides grown at 800°C give a very high density of hole traps (Fig. 7). The results have been confirmed by using direct electron-beam damage in P-type wafers processed under similar conditions. Dry-wet-dry oxides at 800°C also give a high density of hole traps, though not as much as HCl oxides. Dry oxides at 800°C give a low density of hole traps. This result is significant because in order to have sufficient control of the growth of thin oxides, the trend has been to go to lower temperatures. Low temperatures are also desirable for VLSI processing. Given the present results, one must be very careful in studying and understanding new process conditions so they do not degrade the oxide properties unnecessarily. Hole trapping in polysilicon gate capacitors was also studied. The results were consistent with the fact that the additional processing increased the density of hole traps. Better results were obtained for *in situ* doped polysilicon. This suggests that with low-temperature processing, it may be possible to fabricate radiation hard devices with polysilicon gates so long as one is careful not to introduce other effects such as were observed in the HCl oxides described previously.

Fixed Charge and Interface Trap Density

There have been limited reports on the density of fixed charge and interface traps for very thin oxides. In thin oxide systems, the oxide capacitance is high. A small difference in flatband voltage gives a large calculated charge density. Any uncertainty in the metal silicon work function difference gives a large error in fixed charge density. Similarly, the high oxide capacitance can give a large error when the interface trap density is determined. From the literature, it has been reported that the detrimental effect of nitrogen anneal is enhanced for thin oxides [*35*]. Low fixed charge density could not be obtained and an increase is observed after a short anneal at 1000°C. No such effect is observed in argon anneal. As for interface traps, the reported trap density

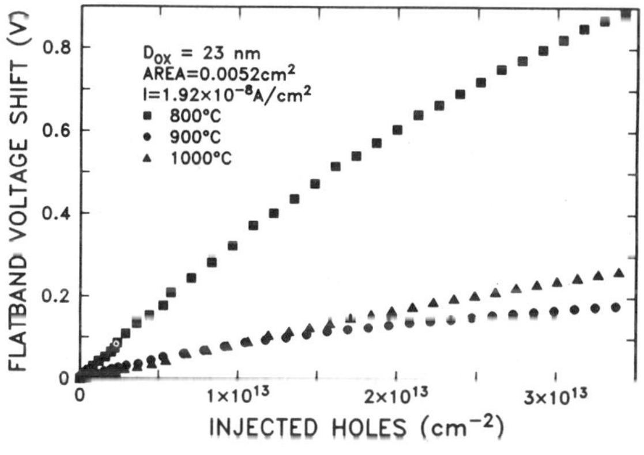

FIG. 7—*Flatband voltage shift as a function of injected holes for three 23-nm HCl oxides grown and annealed for 5 min at three different temperatures.*

for thin oxides have always been higher (in the 10^{11} per cm^2 range) compared to thicker oxides [*36*]. Ogata et al reported that the interface trap densities were increased with decreasing oxide thickness [*37*]. This may be due to an intrinsic limitation in thinner oxides or it may simply be the result of the lower processing temperature required to prepare the oxide.

The flatband voltages and interface trap densities for the oxides listed in Table 1 were examined. Basically, the fixed charge density followed the Deal triangle. In the worst case with no anneal, the fixed charge density was approximately 4×10^{11} per cm^2, whereas the lowest fixed charge density observed was in the low 10^{10} per cm^2 range. The interface trap density varies from the mid 10^{10} to high 10^{10} per cm^2 range. This is slightly higher than the density observed for thicker oxides.

Conclusion

The breakdown, electron trapping, radiation damage and hole trapping, fixed charge, and interface trap properties for thin oxides (25 nm or less) were discussed and some experimental results for 23-nm oxides were presented. For breakdown, very good breakdown statistics can be obtained for aluminum gate capacitors, whereas preliminary results on polysilicon gate capacitors show some degradation compared with aluminum gate. The observed results can be partly accounted for by the beneficial role of electron traps. For electron traps, the total number is expected to be low in thin oxides, especially for the case of polysilicon gate capacitors. This is because they are water-related traps. The density of hole traps can be very high in thin oxides when HCl is used for oxidation at 800°C. Otherwise, the density of hole traps increases with process temperature and process time. Fixed charge density follows the expected temperature behavior if argon is used for the anneal. Interface trap densities were in the mid to high 10^{10} range for 22.5-nm oxides. From the present results, it can be concluded that it is possible to fabricate oxides in the 20 nm thickness range with high yield and high reliability.

Acknowledgments

The experiments on hole trapping were performed in collaboration with J. M. Aitken and C. M. Osburn, who designed the processing matrix shown in Table 1. The Silicon Processing Facility was responsible for preparing most of the oxides in the present study. The technical support of J. A. Calise and F. L. Pesavento is greatly appreciated. The author would also like to thank D. R. Young, D. J. DiMaria, J. M. Aitken, C. M. Osburn, and E. A. Irene for valuable discussions on various aspects of thin gate oxides, and D. R. Young and M. H. Brodsky for a critical reading of the manuscript. The work was supported by the Defense Advance Research Projects Agency, Washington, D.C., and monitored by the Rome Air Development Center, Deputy for

Electronic Technology, Solid State Science Division, Hanscom AFB, Massachusetts, under Contract MDA903-81-C-0100.

References

[1] Dennard, R. H., Gaensslen, F. H., Yu, H. N., Rideout, V. L., Bassous, E., and LeBlanc, A. R., *IEEE Journal of Solid State Circuits,* Vol. SC-9, 1974, p. 256.
[2] Dennard, R. H., *Journal of Vacuum Science and Technology,* Vol. 19, 1981, p. 537.
[3] Keyes, R. W., *IEEE Journal of Solid State Circuits,* Vol. SC-10, 1975, p. 245.
[4] Sze, S. M., *Physics of Semiconductor Devices,* Wiley, New York, 1969, p. 425.
[5] Deal, B. E. and Grove, A. S., *Journal of Applied Physics,* Vol. 36, 1965, p. 3770.
[6] Goodman, A. M. and Breece, J. M., *Journal of the Electrochemical Society,* Vol. 117, 1970, p. 982.
[7] Irene, E. A., *Journal of the Electrochemical Society,* Vol. 121, 1974, p. 1613; Vol. 125, 1978, p. 1708.
[8] Kamigaki, Y. and Itoh, Y., *Journal of Applied Physics,* Vol. 48, 1977, p. 2891.
[9] Adams, A. C., Smith, T. E., and Chang, C. C., *Journal of the Electrochemical Society,* Vol. 127, 1980, p. 1787.
[10] Fang, F. F. and Rupprecht, H. S., *IEEE Journal of Solid State Circuits,* Vol. SC-10, 1975, p. 205.
[11] Fang, F. F., Hatzakis, M., and Ting, C. H., *Journal of Vacuum Science and Technology,* Vol. 10, 1973, p. 1082.
[12] Lepselter, M. P., *IEDM Technical Digest,* International Electron Devices Meeting, Institute of Electrical and Electronics Engineers, New York, 1980, p. 42.
[13] Ephrath, L. M., *IEEE Transactions on Electron Devices,* Vol. ED-28, 1981, p. 1315.
[14] DeKeersmaecker, R. F. and DiMaria, D. J., *Journal of Applied Physics,* Vol. 51, 1980, p. 1085.
[15] Aitken, J. M., *IEEE Transactions on Electron Devices,* Vol. ED-26, 1979, p. 372.
[16] Ephrath, L. M. and DiMaria, D. J., *Solid State Technology,* Vol. 24, 1981, p. 182.
[17] Yu, H. N., Reisman, A., Osburn, C. M., and Critchlow, D. L., *IEEE Transactions on Electron Devices,* Vol. ED-26, 1979, p. 318.
[18] Osburn, C. M. and Ormond, D. W., *Journal of the Electrochemical Society,* Vol. 119, 1972, p. 597.
[19] DiStefano, T. H. and Shatzkes, M., *Journal of Vacuum Science and Technology,* Vol. 13, 1976, p. 50.
[20] Harari, E., *Journal of Applied Physics,* Vol. 49, 1978, p. 2478.
[21] DiMaria, D. J., Ghez, R. and Dong, D. W., *Journal of Applied Physics,* Vol. 51, 1980, p. 4830.
[22] Young, D. R., Irene, E. A., DiMaria, D. J., DeKeersmaecker, R. F., and Massoud, H. Z., *Journal of Applied Physics,* Vol. 50, 1979, p. 6366.
[23] DiMaria, D. J., Young, D. R., and Ormond, D. W., *Applied Physics Letters,* Vol. 31, 1977, p. 680.
[24] Ormond, D. W., *Journal of the Electrochemical Society,* Vol. 126, 1979, p. 162.
[25] Lai, S. K. and Young, D. R., *Journal of Applied Physics,* Vol. 52, 1981, p. 6231.
[26] Lai, S. K., Young, D. R., Calise, J. A., and Feigl, F. J., *Journal of Applied Physics,* Vol. 52, 1981, p. 5691.
[27] Huges, G. W., Powell, R. J., and Woods, M. H., *Applied Physics Letters,* Vol. 29, 1976, p. 377.
[28] Derbenwick, G. F. and Gregory, B. L., *IEEE Transactions on Nuclear Science,* Vol. NS-22, 1975, p. 2151.
[29] Share, S., Epstein, A. S., Kumar, V., Dahlke, W. E. and Haller, W., *Journal of Applied Physics,* Vol. 45, 1974, p. 4894.
[30] Boesch, H. E., McLean, F. B., McGarity, J. M., and Winokur, P. S., *IEEE Transactions on Nuclear Science,* Vol. NS-25, 1978, p. 1239.
[31] Ma, T. P., *Applied Physics Letters,* Vol. 27, 1975, p. 615.

[32] Aitken, J. M. and Young, D. R., *IEEE Transactions on Nuclear Science,* Vol. NS-24, 1977, p. 2129.
[33] Deal, B. E. and Razouk, R. R., "Process Variable Dependence and Interrelationship Between Avalanche-Injected and Radiation-Induced Carrier Trapping in Thermal Oxides," Technical Report, Contract N00014-79-C-0297, Office of Naval Research, Arlington, Va., July 1980, p. 23.
[34] EerNisse, E. P. and Derbenwick, G. F., *IEEE Transactions on Nuclear Science,* Vol. NS-23, 1976, p. 1534.
[35] Revitz, M., Raider, S. I., and Gdula, R. A., *Journal of Vacuum Science and Technology,* Vol. 16, 1979, p. 345.
[36] Dressendorfer, P. V., Lai, S. K., Barker, R. C., and Ma, T. P., *Applied Physics Letters,* Vol. 36, 1980, p. 850.
[37] Ogata, T. and Okamoto, K., *Report of the University of Electrocommunications* (*Japan*), Vol. 28, 1978, p. 207.

Daniel J. Belton,[1] *Pieter van Pelt,*[1] *and Alan E. Morgan*[1]

Polyimide-Substrate Effects During Wet Chemical Processing of Polyimide Films in VLSI

REFERENCE: Belton, D. J., van Pelt, P., and Morgan, A. E., **"Polyimide-Substrate Effects During Wet Chemical Processing of Polyimide Films in VLSI,"** *Silicon Processing, ASTM STP 804,* D. C. Gupta, Ed., American Society for Testing and Materials, 1983, pp. 273–289.

ABSTRACT: The application of polyimides as insulators in multilevel interconnect systems is receiving considerable attention for very-large-scale integration (VLSI) production. For a tight control of wet chemical pattern formation, as well as for the ultimate adhesion in the finished device, a fundamental understanding of polyimide-substrate interactions is important.

We have examined the effects of molecular weight, chemical composition, and interphase structure on the dissolution behavior of polyimides on silicon substrates. The dissolution rate of partly cured polyimides was found to depend strongly on the above parameters, and showed distinct regions of behavior. The overall-etching characteristics of the materials examined were dominated by surface adhesive interactions.

A method was developed to quantify the chemical adhesion during etching of polyimides. A phenomenological model of the dissolution behavior will be presented and its implications for VLSI processing will be discussed.

KEY WORDS: polyimide, integrated circuit, adhesion, interphase structure, interface dissolution behavior, etching behavior, adhesion testing, silane, organosilane

The continuing trend in integrated circuit manufacturing towards higher packing densities, and smaller cell sizes requires the use of multilevel interconnect systems. Conformable dielectrics, such as chemically vapor deposited (CVD) oxide or plasma nitride, become problematic as geometries decrease. This is due in large part to the development of stress concentrations and cracks at sharp features. As a result, polyimide materials are being investigated as alternatives to existing intermetallic dielectrics [*1*].[2] Figure 1 depicts the basic structural unit of three proprietary materials. The obvious differences in chemical composition lead to differences in process conditions, as well as process limitations. In general, however, these materials exhibit properties

[1] Philips Research Laboratories Sunnyvale, Signetics Corporation, Sunnyvale, Calif. 94086.
[2] The italic numbers in brackets refer to the list of references appended to this paper.

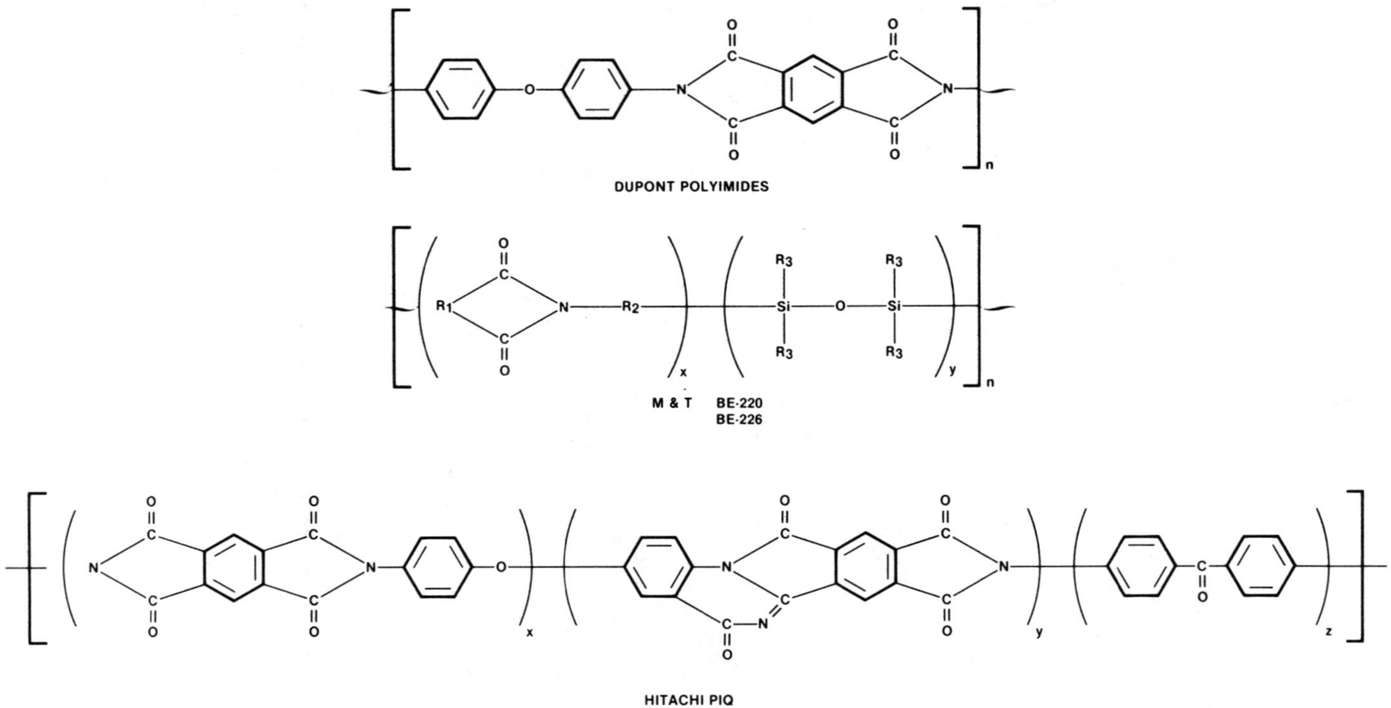

FIG. 1—*Schematic representations of DuPont, M&T, and Hitachi polyimides.*

(planarizability, defect densities, chemical resistance, electrical behavior, etc.) that make them suitable candidates for intermetallic applications.

During wet chemical pattern formation in poly(amic acid-imide) systems a number of factors govern the dissolution behavior. These include polymer composition, molecular weight distribution, thermal history, solvent ionic strength, solvent temperature, mode of etching (spray, puddle, dip, etc.), and polymer-substrate adhesive interactions. In this paper we will be concerned with the importance of adhesive interactions upon the dissolution behavior of polyimides of differing molecular weight distributions, and chemical compositions.

The general use of organosilanes as adhesion promotors in composite systems is well known [*2*], and the necessity of their use during many photolithographic processes is recognized [*3*]. The importance, however, of the organosilane interphase structure upon polyimide processing has not received attention. Schrader et al [*4*] studied ^{14}C labeled γ-aminopropyltriethoxysilane (γ-APS) deposited on polished Pyrex,® and described a three-phase layer structure based upon desorption data. This structure was proposed to consist of a thick layer physisorbed upon the surface. Underlying this layer were two chemisorbed layers, the lower of which was an irreversibly adsorbed monolayer. Ishida and Koenig [*5*] examined via Fourier transform infrared spectroscopy (FTS) the hydrothermal degradation of γ-methacryloxypropyltrimethoxysilane (γ-MPS) and proposed an analagous three-layer structure. In their model, the outermost layer was believed to consist of small oligomers that were physisorbed upon the substrate. Closer to the glass surface a layer of oligomers, similar to those in the outermost layer, was proposed to be connected by a few bonds. Finally, close to the glass surface existed a highly cross-linked three-dimensional network. Thus the uniformity and cross-link density increased as the surface was approached. Sung et al [*6*] examined the structure of γ-APS films of varying thickness adsorbed upon aluminum oxide single crystals by FTS. They concluded that an ionic type of bonding exists between the sapphire and the aminosilane film involving NH_3^+ species. They also demonstrated that the crosslink density increased as the film thickness decreased.

We have examined the polysiloxane layer structure using both ellipsometric and Auger electron spectroscopic techniques. The silane film structure has a large effect upon the overall dissolution behavior of polyimides applied to silicon substrates, and a possible technique of monitoring chemical adhesion is described, based on this phenomenon.

Sample Preparation

Clean silicon wafers with a native oxide on the order of a few tenths of a nanometre were used in all experiments. Solutions of DuPont VM-651 γ-APS coupler of concentrations from 5 μL to 150 μL per 100 mL of methanol:deionized water (195 mL:5 mL) were prepared. Solutions containing 10 μL/100 mL were designated as standard solutions. All solutions were applied to the substrate in one of three fashions: dispensing followed by spinning at 5000 rpm for 30 s (STD 5000) or puddle application followed by 5 min quiescent coupler adsorption prior to either 5000 rpm spin for 30 s (IMM 5000) or 1000 rpm spin for 45 s (IMM 1000). These application techniques provided a wide range of silane film thicknesses.

DuPont PI-2545 and PI 2555 and M&T Chemicals BE-220 were used in film thicknesses ranging from 1.5 to 2.5 μm. DuPont PI-2555 has a $M_n \approx 4.5$ times that of DuPont PI-2545.

Data Collection and Analysis

Kinetic Experiments

The dissolution behavior of polyimide films is determined in part by the precuring conditions. Therefore, the thermal ring closing imidization reaction of poly(amic acid) films was monitored by measuring changes in the integrated intensities of the imide carbonyl absorptions (1780 cm^{-1} [ν_s] and 1727 cm^{-1} [ν_{as}]) with respect to a fully cured standard, using a Digilab FTS-20C spectrophotometer. A digital compensation of the sample spectrum by the fully

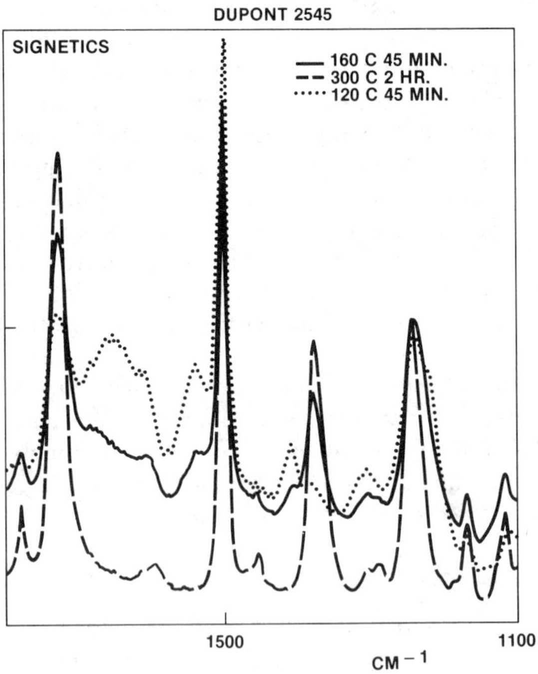

FIG. 2—*FTIR spectra of fully cured (300°C) and partially cured (120°C, 160°C) polyimides.*

cured reference spectrum yielded the percent imidization through a scaling factor. The data were of course corrected for film thickness variations. An example of the spectral data obtained is illustrated in Fig. 2, and the percent imidization versus precure temperature (45 min) is given in Fig. 3.

Preimidization was followed by the application and exposure of HPR-204 photoresist. The kinetics of the dissolution process was monitored by developing samples for a given time in either Shipley AZ351 (1:3 dilution) or Shipley Microposit 351 (1:2 dilution), and by measuring the remaining film thickness with a Nanospec film thickness measuring system.

Film Thickness Measurements

Ellipsometry—The ratio of the complex reflection coefficients ρ for the parallel component r_p and the perpendicular component r_s of an incident beam of electromagnetic radiation at the boundary between two dielectrics is an experimentally measurable quantity. These reflection coefficients are given by

$$r_p = E_p''(t)/E_p(t) \tag{1}$$

and

$$r_s = E_s''(t)/E_s(t) \tag{2}$$

where E_p'' and E_s'' are the instantaneous complex amplitudes of the reflected p and s components, and E_p and E_s are the instantaneous complex amplitudes of the incident p and s components. The reflectivity ratio is therefore given as

$$\rho = |E_p''|\,|E_s''| \exp i[(\phi_p'' - \phi_s'') - (\phi_p - \phi_s)]/|E_p|\,|E_s| \tag{3}$$

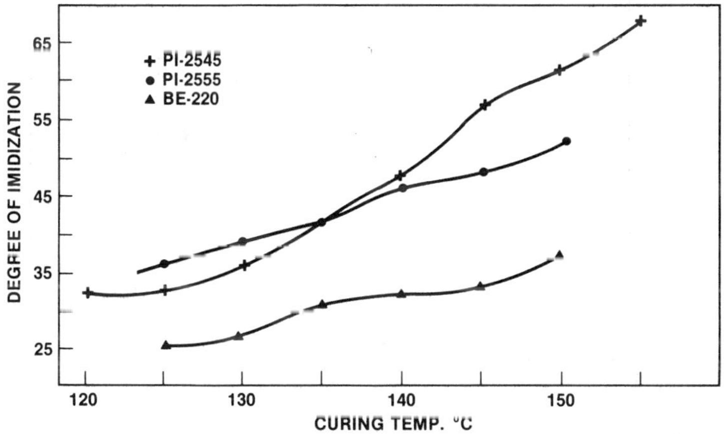

FIG. 3—*Degree of imidization versus curing temperature.*

where ϕ denotes the time-independent phase. The amplitudes and phases allow the definition of four angles in the complex plane. These are

$$\tan \psi_i = E_p/E_s, \ \tan \psi_r = E_p''/E_s'' \tag{4}$$

$$\Delta_i = \phi_p - \phi_s \text{ and } \Delta_r = \phi_p'' - \phi_s'' \tag{5}$$

The basic equation of ellipsometry follows:

$$\rho = \tan \psi_r \exp i(\Delta_r - \Delta_i)/\tan \psi_i \tag{6}$$

The angles ψ and Δ are functions of incident angle, wavelength, complex refractive index of the substrate and the ambient, the real index of a transparent film, and the film thickness. In this work calculations were based upon a model of a single transparent film on a reflecting substrate. The thickness calculations, therefore, neglect the effect of a refractive index variation normal to the reflecting surface, as well as the presence of an underlying native oxide film. The effect of refractive index on thickness is shown in Fig. 4. Calculations involving the presence of a native oxide layer and taking into account the anisotropy of the refractive index normal to the reflecting surface would lead to values of thickness smaller than those presently calculated.

All measurements were conducted with a Rudolph Research Auto-el system. The index of refraction of the silane film was determined to be 1.528, and measurements were conducted at a 70-deg incident angle.

Auger Electron Spectroscopy—The basic phenomenon underlying the use of Auger electron spectroscopy for film thickness determination is the exponential decrease of the Auger electron escape probability with distance from the surface [*7,8*]. Consider a substrate *s* covered by a film *f* of thickness *d*. If

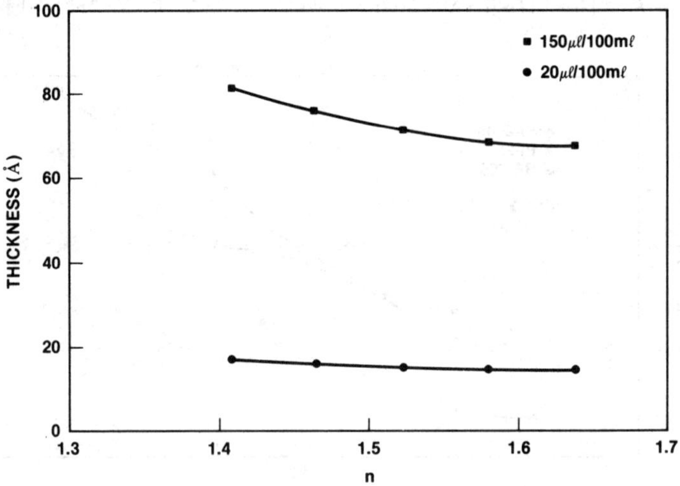

FIG. 4—*Ellipsometric thickness versus index of refraction.*

the film is uniform in both thickness and composition, the Auger electron signal from an element A in the substrate is given by

$$I_A^s = I_{A,\infty}^s \exp\,[-d/\lambda(E_A)\cos\theta] \tag{7}$$

and from an element B in the film by

$$I_B^f = I_{B,\infty}^f\,\{1 - \exp\,[-d/\lambda\,(E_B)\cos\theta]\} \tag{8}$$

where I_∞ is the signal from a sample of infinite thickness, λ is the inelastic mean free path of Auger electrons of energy E in the film, and $\cos\theta$ is a geometrical correction for the analyzer collection angle. Equations 7 and 8 are valid only if factors such as primary electron beam attenuation, backscattering, matrix dependence of λ, surface roughness effects, etc., can be considered negligible. The film thickness can be derived from either equation if $\lambda(E_A)$ or $\lambda(E_B)$ is known. If both λ's are known, d may be calculated by using the ratio of film and substrate Auger signals:

$$I_B^f/I_A^s = I_{B,\infty}^f/I_{A,\infty}^s\,\{\exp\,[d/\lambda(E_A)\cos\theta] - \exp \times (d[\lambda(E_D) - \lambda(E_A)]/\lambda(E_A)\lambda(E_B)\cos\theta)\} \tag{9}$$

Consider an element common to both film and substrate, with an Auger peak at E_A in the substrate chemically shifted to E_A' in the film, then $E_A \approx E_A'$ and Eq 9 reduces to

$$I_A^f/I_A^s = I_{A,\infty}^f/I_{A,\infty}^s\,\{\exp[d/\lambda(E_A)\cos\theta] - 1\} \tag{10}$$

Immediately following silane film deposition and ellipsometric measurements the samples were placed in a Physical Electronics 590 scanning auger microscope containing a cylindrical mirror energy analyzer equipped with an integral electron gun. They were analyzed using a 5-keV, 30-deg incident, 1.6-μA electron beam rastered over 0.5 mm^2. Auger data were taken in the $d[E \cdot N(E)]/dE$ mode by using a 2-V peak-to-peak a-c modulation and a 0.3% energy analyzer resolution setting. Peak-to-peak heights were used as a measure of Auger intensities. Ellipsometric readings were unchanged after sample removal from the Auger apparatus, indicating that no vacuum evaporation of the films had occurred.

Results and Discussion

The normalized thickness (film thickness ÷ original thickness) versus etching time in 1:3 AZ351:deionized water is illustrated in Figs. 5 to 7 for the DuPont materials PI-2545 and PI-2555 and M&T BE220, respectively. The DuPont materials exhibit three regions of differing etch rates: (I) an inhibition period characterized by solvent invasion of the polymeric matrix; (II) a region that could be attributed to gel formation and molecular dissolution behavior; and (III) a region characterized by rapid film removal. The

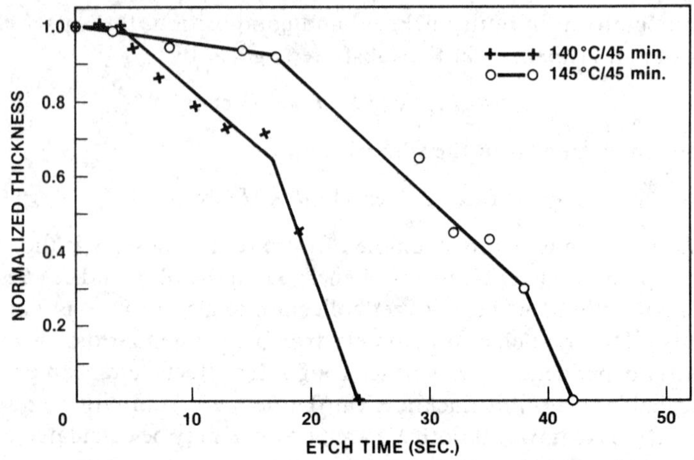

FIG. 5—*Normalized thickness versus etching time for DuPont 2545. Etchant is AZ351:deionized water, 1:3. Silicon substrate exposed to the standard γ-APS solution.*

BE-220 exhibits only two regions of behavior (Fig. 7); no rapid film removal toward the end of the etching process was observed. The structural differences between the DuPont and M&T systems lead one to assume that the differences in dissolution behavior may be related to an interfacial effect; that is, one would expect a more pronounced adhesive interaction of the silicone copolymer with the silicon substrate. Indeed, if one observes the spincoating behavior of BE-220 on unprimed substrates, smooth adhering films are obtained free from retraction or "fish-eye" effects such as occur with DuPont polyimide materials.

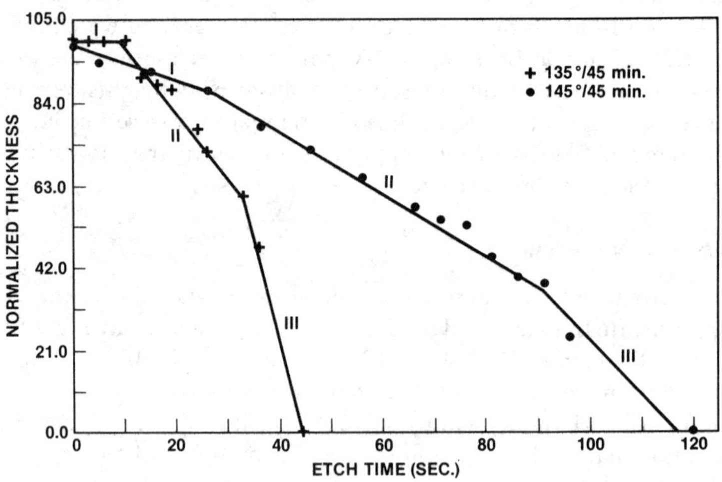

FIG. 6—*Normalized thickness versus etching time for DuPont 2555. Etchant is AZ351:deionized water, 1:3. Silicon substrate exposed to the standard γ-APS solution.*

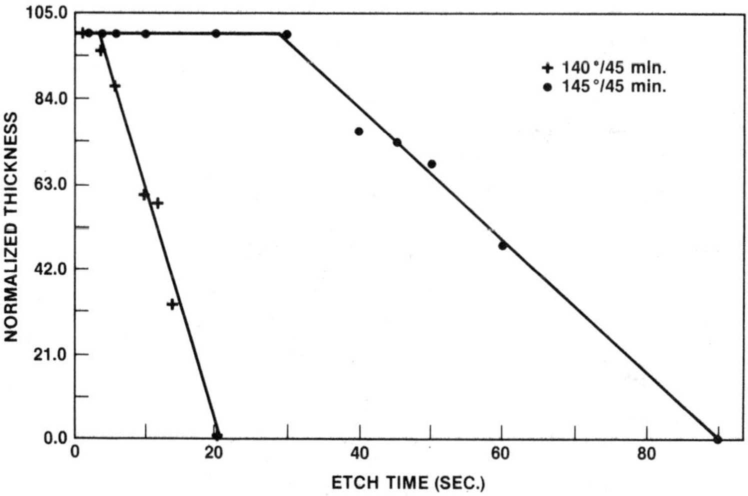

FIG. 7—*Normalized thickness versus etching time for M&T BE-220. Etchant is AZ351:deionized water, 1:3. Silicon substrate exposed to the standard γ-APS solution.*

The effect of the precure temperature of the PI-2545 and PI-2555 materials is reflected in two fashions: (*1*) the onset of Region II behavior is delayed, and (*2*) the slope of the Region II curve is decreased. Both effects can obviously be ascribed to a reduction in solubility resulting from both intermolecular and intramolecular imidization reactions (Fig. 3). The reduced etch rate in Region III, as well as the increased time for its onset, appear again to be related to an improvement of the interfacial adhesive interactions. This improvement is likely if one considers intermolecular interactions occurring between amine groups associated with the γ-APS and the carboxylic acid group of the polymer as temperature is increased.

The effect of the surface adhesive interaction, dominating the overall etch time of a polyimide film, is more dramatically illustrated in Figs. 8 and 9. Polyimide films interacting with a substrate devoid of organosilane, a surface treated by spincoating a standard silane solution, and surfaces treated by puddle techniques in a standard solution and one of 100 μL/100 mL (×10) are illustrated. In both cases the unprimed surface shows a dramatic loss of film thickness over a very short time period. It is speculated that during this time period rather large agglomerates are being removed. This speculation is based upon microscopic examination of samples just prior to complete film removal, which shows that dissolution in Region III is a very nonuniform process.

The etch rate can be decreased in Region III and the overall time for film removal increased by varying the organosilane solution concentration or application technique. The decreasing etch rate of Region III is particularly apparent for the ×10 immersion-coupled PI-2555 sample. In order to exam-

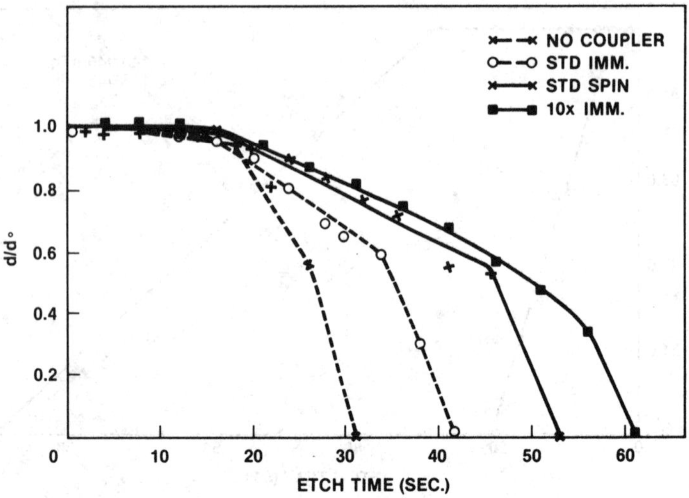

FIG. 8—*Normalized thickness versus etching time for DuPont 2545 with various surface treatments. Curing temperature is 150°C/45 min.*

ine this behavior in more detail, the film thickness and desorption characteristics of the organosilane films were studied. The film thickness as measured ellipsometrically is given in Fig. 10 as a function of both the concentration of the organosilane solution and the application technique. Thickness values increase continuously as solution concentration increases for all modes of application. Adsorption does not follow the classical Langmuir adsorption isotherm characteristic of chemisorption only. This is particularly true of the

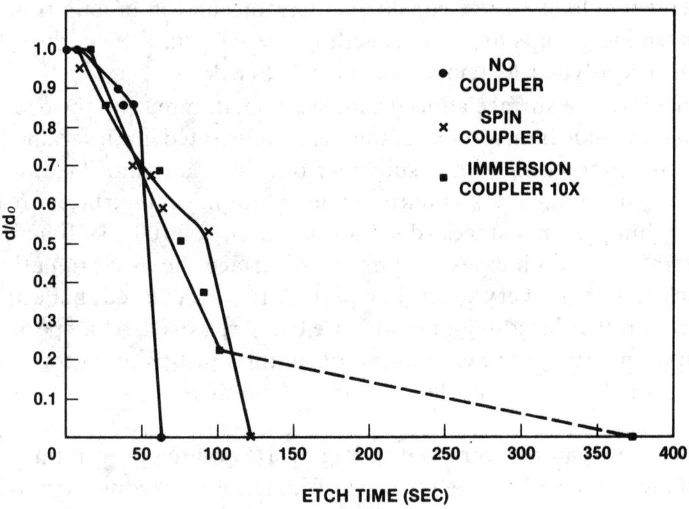

FIG. 9—*Normalized thickness versus etching time for DuPont 2555 with various surface treatments. Curing temperature is 150°C/45 min.*

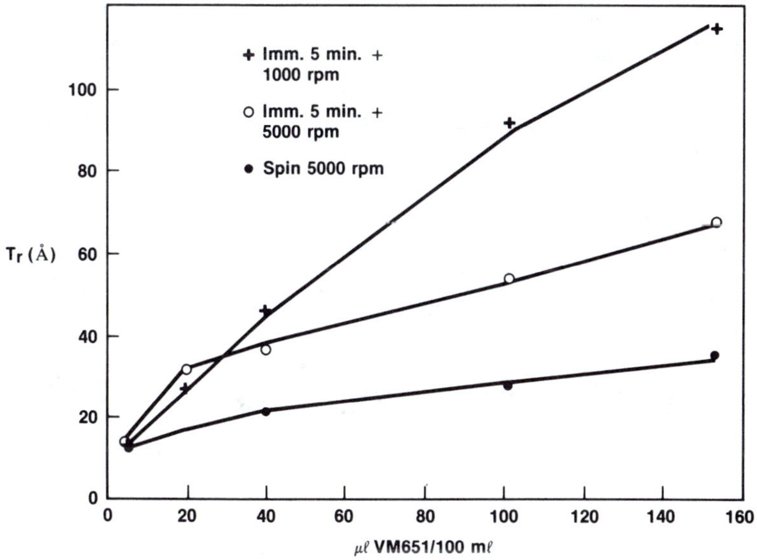

FIG. 10—*Organosilane film thickness versus solution concentration for various application techniques (ellipsometric measurements).*

immersion techniques. On the other hand, if physisorption plays a role due to favorable intermolecular interactions, then deviations of the type illustrated in Fig. 10 can be expected. Therefore the application mode has a large influence on the thickness of the physisorbed layer, the thickest films being obtained with the IMM 1000 technique.

As previously mentioned, the ellipsometric values are an overestimation. With this in mind an Auger electron spectroscopic technique was employed to try to determine more precisely the thickness of the silane layers that were being produced. Auger spectra from a sputter-cleaned Si sample and from thin and thick organosilane films are shown in Fig. 11. One sees a shift to lower energies in the film for both the Si_{LVV} and Si_{KLL} Auger transitions when compared to elemental silicon. Thus, a signal from a thin enough film contains both the chemically shifted and unshifted peaks. Peaks for C and O were visible in all of the fims with an N peak becoming apparent in the thicker films.

The inelastic mean free paths were derived from the general formula for organic compounds [*9*]:

$$\lambda = 49\, E^{-2} + 0.11\, E^{1/2}\ \mathrm{mg \cdot m^{-2}} \tag{11}$$

by assuming a density of 1 $g \cdot cm^{-3}$. Values of 1 nm for Si_{LVV}, 1.8 nm for C, 2.5 nm for O, and 4.4 nm for Si_{KLL} Auger electrons resulted. With the present experimental setup cos $\theta = 0.74$[10].

Table 1 contains the thicknesses derived from various computational ap

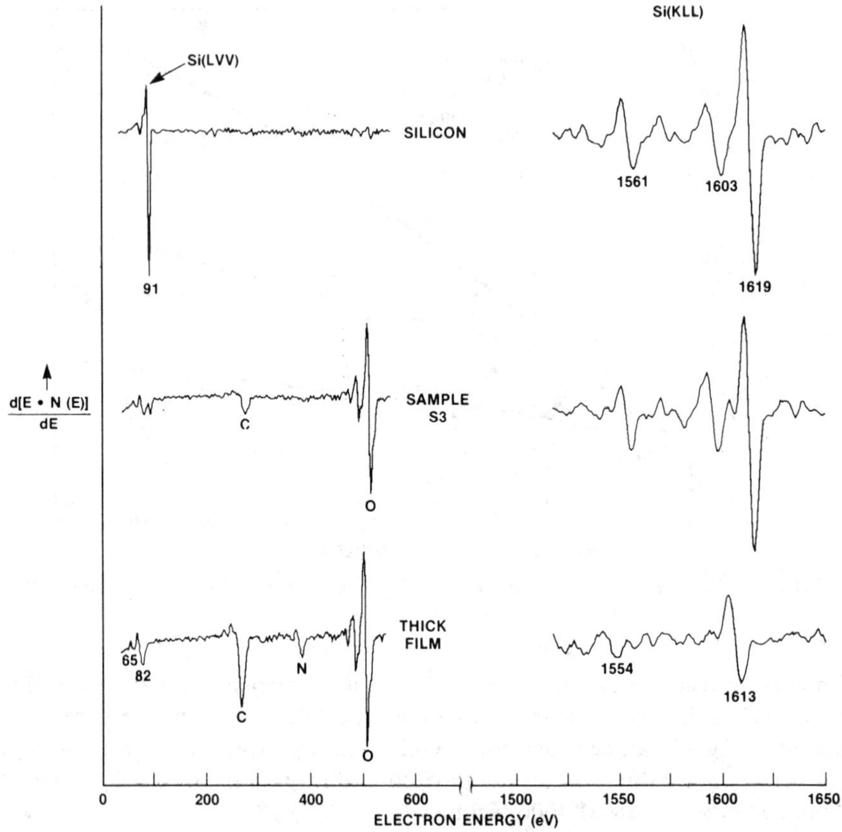

FIG. 11—*Auger derivative spectra for bare silicon and two silane films.*

proaches. The Si_{LVV}^f/Si_{LVV}^s and Si_{KLL}^f/Si_{KLL}^s Auger signal ratios were substituted in Eq 10 to obtain Columns A and B respectively, $Si_{LVV}^s/Si_{LVV,\infty}^s$ and $Si_{KLL}^s/Si_{KLL,\infty}^s$ were utilized via Eq 7 for Columns C and D respectively, and Columns E, F, G, and H were derived from C^f/Si_{LVV}^s, C^f/Si_{KLL}^s, O^f/Si_{LVV}^s and O^f/Si_{KLL}^s respectively by using Eq 9. The last column lists the average values, which can be compared with the ellipsometric data. The main source of uncertainty in the Auger data is the lack of a precise value for λ in the actual film material. Judging from the literature [*9*], the presently assumed values appear reasonable, and it is based upon this that the Auger thickness values are considered to be the more realistic ones.

The observation that the organosilane layer thickness is a strong function of application technique and original coupler solution concentration leads to the question of whether structural differences occur concomittantly with thickness differences. In Table 2, the effect of 1-h solvent exposure upon the film thickness is given for silane films prepared by both the spin coating and immersion techniques. The thin layers prepared by the STD 5000 method

TABLE 1—*Auger and ellipsometric thickness data and comparisons.*[a]

Sample	Ellipsometry Data	A	B	C	D	E	F	G	H	Auger
S1	17	...	...	2.4	...	1.9	...	4.9	4.7	3.5
S2	27	13	17	14	...	8.3	...	18	16	14
S3	32	15	17	18	...	12	...	20	17	17
S4	66	...	34	...	22	...	18	...	28	26
S5	80	...	38	...	33	...	21	...	34	32

[a] Data in angstroms.

appear to have a negligible amount of physisorbed oligomers, corroborating the Langmuir-type appearance of the adsorption isotherm. The IMM 5000 technique shows a measureable difference (this is based upon the amount of room temperature water-soluble silane present). The films prepared from the 100 μL/100 mL solution exhibit significant losses when exposed to deionized water, which once again is attributed to physisorbed silane oligomers.[3] These results, as well as those for the other solvents used, suggest that the crosslink density decreases as the film thickness increases. This is in accord with Sung's spectroscopic data [*6*]. Additional structural verification is given by the desorption data presented in Fig. 12. The solvent in this case was Shipley Microposit 351, diluted 1:2 with deionized water. Since this solution is a better solvent than deionized water we were able to observe complete reversibility of adsorption, and therefore the times for complete film removal were accessible. The data illustrate that as film thickness increases, the time for complete film removal decreases. This again may be interpreted in terms of increasing crosslink density as the film thickness decreases.

TABLE 2—*Film thickness remaining after 1-h solvent exposure. STD 5000 and IMM 5000 VM-651 application techniques for standard and ×10 solutions.*

Coupler Solution Concentration	Application Technique	Solvent Exposure (1 h)	Original Thickness, Å	Final Thickness, Å
10 μL/100 mL solvent	spin coat 5000 rpm	deionized H_2O	17.3	17.3
		toluene	17.3	...
		boiling H_2O	17.3	12.4
10 μL/100 mL solvent	immersion 5 min, 5k spin	deionized H_2O	22.5	18.2
		toluene	22.5	17.7
		boiling H_2O	22.5	14.8
100 μL/100 mL solvent	spin coat 5000 rpm	deionized H_2O	41.6	22.2
		toluene	41.6	23.7
		boiling H_2O	41.6	25.7
100 μL/100 mL solvent	immersion 5 min, 5k spin	deionized H_2O	54.7	23.1
		toluene	54.7	25.5
		boiling H_2O	54.7	24.6

[3] Strictly speaking, the use of *physisorbed* is in error due to water hydrolysis of Si—O—Si linkages. *Physisorbed* should be replaced by *loosely bound.*

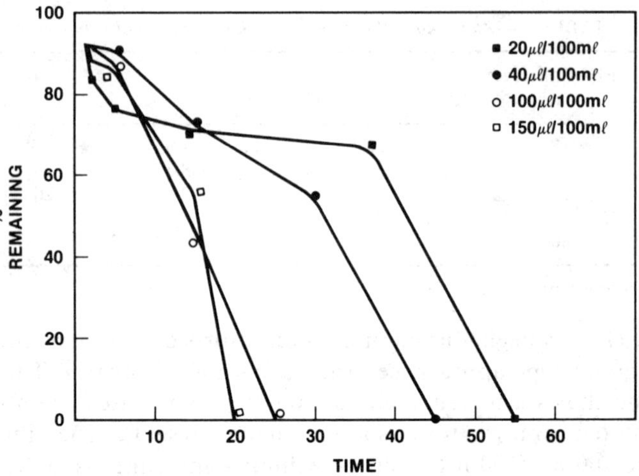

FIG. 12—*Percent film thickness remaining versus solvent exposure time for films applied from VM-651 solutions of concentrations 20 μL, 100 μL, and 150 μL/100 mL solvent by using IMM 5000 technique.*

Plots of time for complete polyimide film removal versus coupler solution concentration and application technique are given in Figs. 13 and 14 for DuPont PI-2545 and PI-2555, respectively. The complete etch time for polyimide films on substrates treated by the IMM 1000 technique increases continuously with concentration for both polymer systems. Those substrates treated by the STD 5000 technique have only a slight variation in overall etch time for polyimide removal. These results can be correlated to the silane layer structure resulting from these application techniques, and the ability of the poly(amic acid) molecules to interact with that structure. Since it appears that the crosslink density of the silane layer decreases as film thickness increases, one expects a more extensive interphase interaction of polyamic acid with the silane film produced by IMM 1000—that is, both the depth of the interaction, as well as the amount of intermolecular NH_2—COOH associations. The exposure of the interphase resulting from either application technique to an aqueous base could result in not only disentanglement or dissolution of poly(amic acid-imide) molecules, but also the removal of a portion of the oligomeric phase of the silane film. It appears through examination of Figs. 12 to 14 that polyimide film removal is governed by both polymeric disentanglement and dissolution, as well as removal of oligomeric silane for the IMM 1000 case. For STD 5000, this cannot be true since the observed etch times are considerably shorter than those times required for a measurable silane removal. Thus it appears that the rather small interpenetration distances likely in this case cause polyimide film removal to be dominated by polymeric dissolution alone. It is of interest to note that the Auger thicknesses are calculated to be between 3.5 and 32 Å. Assuming a value of 10 to

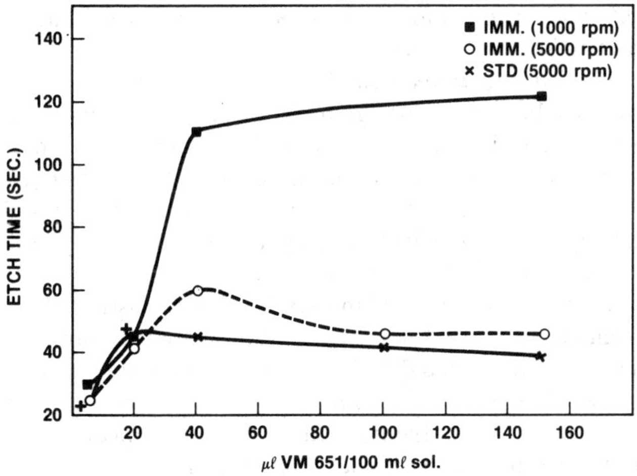

FIG. 13—*Time for complete film removal versus VM-651 solution concentration and application technique. DuPont 2545 cured 150°C/45 min. Solvent was Microposit 351 1:2 deionized water.*

12 Å for the end-to-end length of a γ-APS molecule it appears that our thinnest layer has not only Si-O interactions with the surface but also NH_2 interactions, thus rendering those groups unavailable for interaction with the polyimides. This would certainly provide fewer sites for subsequent polymeric interaction than would a three-dimensional structure, and hence a lower value of chemical adhesion.

Whenever adhesive interactions are considered, care must be taken to properly correlate the experimental observations with the type of adhesive

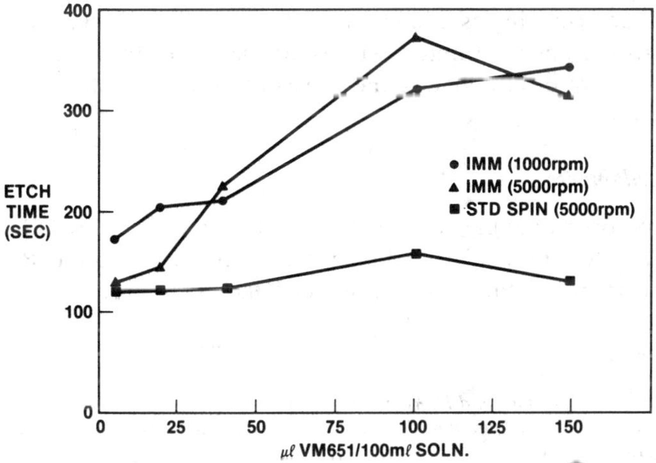

FIG. 14—*Time for complete film removal versus VM-651 solution concentration and application technique. DuPont 2555 cured 150°C/45 min. Solvent was Microposit 351 1:2 deionized water.*

interaction that is being monitored. Adhesive interactions that are of interest with respect to solvent penetration or ionic migration at an interface have been termed *photoresist adhesion*. Mittal [*10*] reviewed a number of techniques that have been described in the literature. These include undercutting ratio, water penetration, and the ion migration technique. In addition, the statistical analysis of the removal of photoresist islands has been mentioned as a technique to examine adhesion [*11*]. In order to maximize the adhesive interaction for a given application it would be of advantage to detect even subtle changes brought about by the process conditions. It is proposed here that the kinetic curves illustrated in Figs. 5 to 9 demonstrate the capability to resolve chemical adhesive differences through both the shape of the curve as well as the endpoints detected, at least for polyimide-substrate systems. This type of analysis has proved useful in defining the process conditions for wet chemical processing of polyimide films. Also, it appears to be a suitable means of monitoring the interfacial interactions that are of importance for reliability of the manufactured device.

Conclusions

The kinetics of polyimide film dissolution are characterized by distinct regions of etching behavior. The appearance of these curves can be manipulated by preimidization conditions, chemical composition, and interphase structure. The interphase structure can be monitored by ellipsometric and Auger electron spectroscopic techniques, and has a profound effect upon the overall etching behavior of polyimide films. The ability of a polyimide to interact with an organosilane film and subsequently display good chemical adhesion depends upon its ability to interpenetrate and interact with the silane. This interpenetration and interaction is highly dependent upon the level of crosslinking occurring as a function of depth within the silane film. The kinetic curves can be interpreted in terms of chemical adhesion, and provide a useful means of discriminating the effects of process conditions upon interfacial behavior.

Acknowledgments

The authors would like to acknowledge the assistance of S. DiIorio in preparing samples and performing kinetic measurements, and T. R. Quackenbush in performing the Auger measurements.

References

[*1*] Wilson, A. M., *Thin Solid Films*, Vol. 18, No. 2, 1981, p. 19.
[*2*] Plueddeman, E. P., *Journal of the Electrochemical Society, Extended Abstracts*, Vol. 81-1, May 1981, p. 340.
[*3*] Deckert, C. A. and Peters, D. A., *Journal of Electrochemical Society, Extended Abstracts*, Vol. 81-1, May 1981, p. 301.
[*4*] Schrader, M. E., Lerner, I., and D'Oria, F. J., *Modern Plastics*, Vol. 45, 1967, p. 195.

[5] Ishida, H. and Koenig, J. L., *Journal of Polymer Science,* Vol. 18, 1980, p. 1931.
[6] Paik Sung, C. S., Lee, S. H., and Sung, N. H., "Adhesion and Adsorption of Polymers," *Polymer Science and Technology,* Vol. 12B, L. H. Lee, Ed., Plenum Press, New York, 1980, p. 757.
[7] Holloway, P. H., *Journal of Vacuum Science and Technology,* Vol. 12, 1975, p. 1418.
[8] Chang, C. C. and Boulin, D. M., *Surface Science,* Vol. 69, 1977, p. 385.
[9] Seah, M. P. and Dench, W. A., *Surface and Interface Analysis,* Vol. 1, No. 1, Feb. 1979.
[10] Mittal, K. L. and Lussow, R. O., "Adhesion and Adsorption of Polymers," *Polymer Science and Technology,* Vol. 12B, L. H. Lee, Ed., Plenum Press, New York, 1980, p. 503.
[11] Yanazawa, H., Matsuzawa, T., and Hashimoto, N. in *Proceedings,* Kodak Interface Seminar, Oct. 1977, p. 153.

Richard B. Fair[1] *and William G. Meyer*[2]

Modeling Anomalous Junction Formation in Silicon by the Codiffusion of Implanted Arsenic with Phosphorus

REFERENCE: Fair, R. B. and Meyer, W. G., **"Modeling Anomalous Junction Formation in Silicon by the Codiffusion of Implanted Arsenic with Phosphorus,"** *Silicon Processing, ASTM STP 804,* D. C. Gupta, Ed., American Society for Testing and Materials, 1983, pp. 290–305.

ABSTRACT: Cross-contamination of phosphorus in arsenic-implanted layers has been found to occur in mechanically scanned ion implantation systems. Phosphorus introduced at low levels exhibits a diffusivity that depends upon n^2, where n is the free electron concentration of the highly doped arsenic background layer. If the arsenic layer is shallow and the phosphorus diffuses through the arsenic junction, an enhanced phosphorus tail region is formed. Tail diffusion coefficients at 900°C were ~20 times greater than the intrinsic phosphorus diffusion coefficient. Also, when a phosphorus tail is formed, the mechanism that enhances the tail diffusion also enhances the high concentration arsenic diffusion. These results are remarkable, since arsenic by itself will not produce an enhanced tail during diffusion in silicon, nor will it cause any emitter push effect.

These results have been modeled and provide strong support for a Fermi-level-controlled point-defect model of phosphorus diffusion with impurity-point-defect pair dissociation creating the tail region. We have essentially duplicated high-concentration phosphorus diffusion results by replacing the highly doped portion of the profile with arsenic. The fact that arsenic diffusion is also predictably enhanced when a phosphorus tail is formed strongly implies that the point-defects involved are vacancies.

KEY WORDS: vacancies, emitter push effect, ion implantation, diffusion

It has been reported that cross-contamination of dopant impurities can occur within mechanically scanned ion implantation systems that do not have a trap for neutral atoms in the ion beam [*1*].[3,4] Thus atoms of other than the desired dopant ions are sputtered or deposited in some way onto the target wafers. Recently, Ahrens[4] showed that phosphorus contamination of

[1] Currently with Microelectronics Center of North Carolina, Research Triangle Park, N.C. 27709; on leave from Bell Laboratories, Reading, Pa. 19604.

[2] Bell Laboratories, Reading, Pa. 19604.

[3] The italic numbers in brackets refer to the list of references appended to this paper.

[4] Private communication from R. E. Ahrens, Bell Laboratories, Reading, Pa., 1980.

arsenic-implanted silicon occurred at levels of 1 to 2% of the arsenic dose in the Varian/Extrion 200-1000 system. MOSFETs whose source/drain implants were performed in this system showed unusually short channel lengths. It was later verified that the phosphorus introduced at low doping levels exhibited a very large diffusivity in the presence of the high-concentration arsenic.[4] This result was unexpected on the basis of SUPREM II and PROSIM simulations of the codiffusion of arsenic and phosphorus.

As a result of these initial observations, additional data have been obtained on the codiffusion of phosphorus in a high-concentration arsenic background in silicon. These results and analysis of the results are presented here. It will be shown that phosphorus diffuses the same way in a highly doped arsenic background as it does in its own surroundings in a highly doped phosphorus layer (excluding misfit strain effects). And in the presence of a shallow arsenic layer, phosphorus can diffuse through the arsenic-doped layer and exhibit enhanced diffusion beyond the arsenic junction (emitter push effect). These results provide strong support for a Fermi-level-controlled point-defect model of phosphorus diffusion with impurity-point-defect pair dissociation creating the emitter push, as described by Fair and Tsai [*2*]. A predictable enhancement of arsenic diffusion was observed when the codiffusing phosphorus penetrated through the arsenic junction. This result strongly suggests that the point defects involved are vacancies.

Experimental Procedure

The wafers used in this experiment were prepared by epitaxially growing 10 to 12 μm of *p*-type silicon (10 to 20 Ω-cm) on a 0.01 Ω-cm ⟨100⟩ *p*-type substrate. To produce the five initial arsenic profiles and one boron profile used in the codiffusion experiments, 54 wafers were implanted and driven-in under the conditions listed in Table 1 (nine wafers per initial profile). Following the drive-in step, eight wafers per group were implanted with phosphorus at 5×10^{14} cm^{-2}, 30 keV. Each of the eight wafers per group was then subjected to a different drive-in condition. The drive-ins were performed at temperatures of 900, 1000, 1050, and 1200°C for times of 360, 120, 30, and 15 min, respectively. Half of the wafers were driven-in in 100% O_2 and half in

TABLE 1—*Conditions for establishing the initial background profiles for codiffusion experiments.*

	Implantation		Drive-In (N_2 + 0.5% O_2)	
Ion	Dose, cm^{-2}	Energy, keV	T, °C	Time, min
As	4×10^{16}	30	1200	15
As	2×10^{16}	150	1050	30
As	2×10^{16}	30	1200	120
As	1×10^{16}	30	1200	120
As	5×10^{15}	30	1200	360
B	1×10^{16}	30	1200	30

N_2 + 0.5% O_2 ambients, although only the N_2 drive-in results are reported here.

The initial arsenic profiles obtained per Table 1 were measured using the spreading resistance technique. The codiffused arsenic/phosphorus and boron/phosphorus profiles were obtained through secondary ion mass spectrometry (SIMS). Measurements on three different SIMS machines[5] were compared with spreading resistance data to be certain of the results. When discrepancies were found, additional SIMS analyses were performed.

The Boltzmann-Matano analysis was used to obtain phosphorus diffusion coefficients as a function of arsenic or boron background doping density. The samples were prepared so that phosphorus diffusion into a uniformly doped background layer could be studied as well as diffusion through a shallow, highly doped arsenic layer.

Experimental Results

The measured profiles shown in Fig. 1 represent a 2×10^{16} As/cm^2, 150-keV implant that was diffused at 1050°C for 30 min in N_2 + 0.5% O_2, followed by

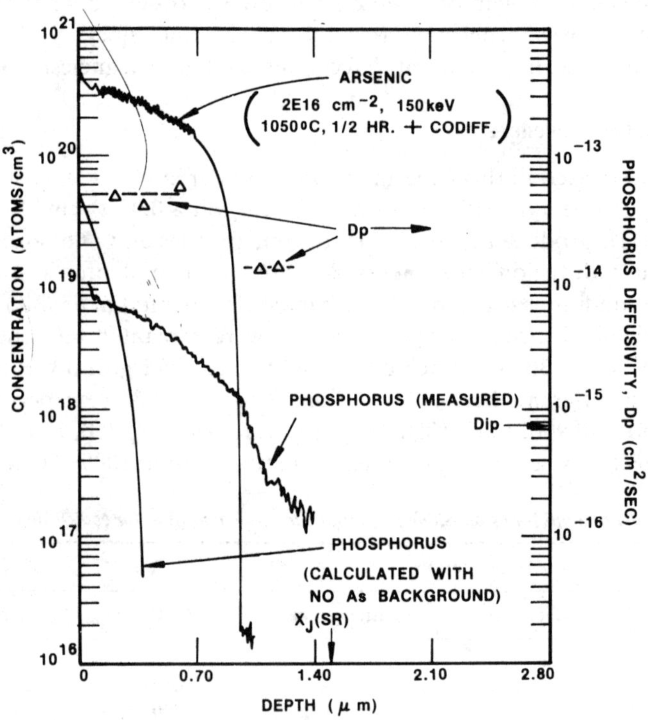

FIG. 1—*Phosphorus codiffusion with arsenic at 900°C for 6 h.*

[5]The SIMS machines used were located at Western Electric Company in Reading, Pa., Bell Laboratories in Allentown, Pa., and Charles Evans Associates in San Mateo, Calif.

a 5×10^{14} P/cm^2, 30-keV implant. The wafer was then annealed at 900°C for 6 h in the same ambient. For comparison, the calculated phosphorus profile that would be obtained without the arsenic background is also shown. It can be seen that the phosphorus diffusivity is greatly enhanced both under the arsenic and beyond the arsenic junction. The results of a Boltzmann-Matano analysis of the phosphorus profile yielded the phosphorus diffusivity data D_P versus depth, also plotted in Fig. 1. It can be seen that under the arsenic profile $D_P \simeq 5 \times 10^{-14}$ cm^2/s, and beyond the arsenic junction $D_P \simeq 1.5 \times 10^{-14}$ cm^2/s. These results can be compared with the intrinsic phosphorus diffusivity D_{iP} at 900°C, which is 7.6×10^{-16} cm^2/s.

A similar result is shown in Fig. 2 for the case where the arsenic/phosphorus codiffusion was performed at 1000°C for 2 h in N_2 + 0.5% O_2. The measured diffusivities of phosphorus are shown to be 10 to 50 times greater than D_{iP} at 1000°C.

Phosphorus diffusions in both shallow and deep initial arsenic-doped silicon layers at 1050°C are compared in Fig. 3. The enhancement of phosphorus diffusion is similar under both arsenic layers, since the background doping levels are similar. However, diffusion beyond the shallow arsenic junction is reduced relative to the deep layer results by a factor of four.

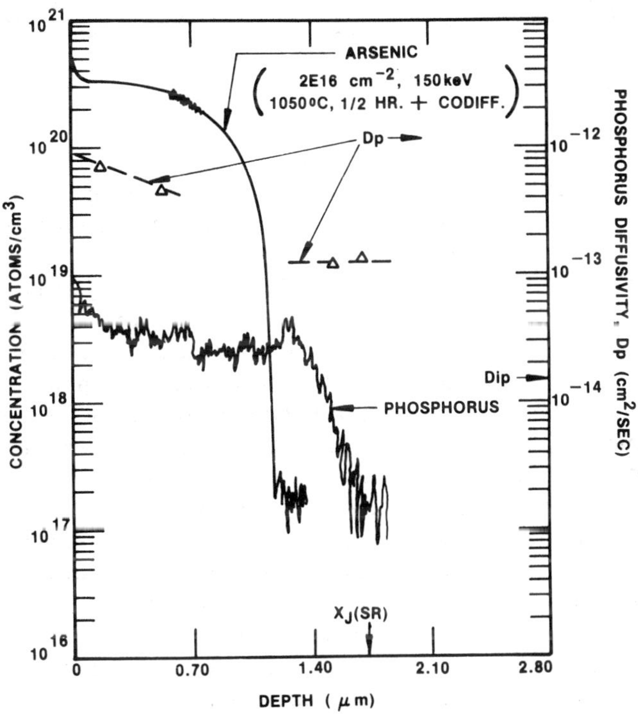

FIG. 2—*Phosphorus codiffusion with arsenic at 1000°C for 2 h*

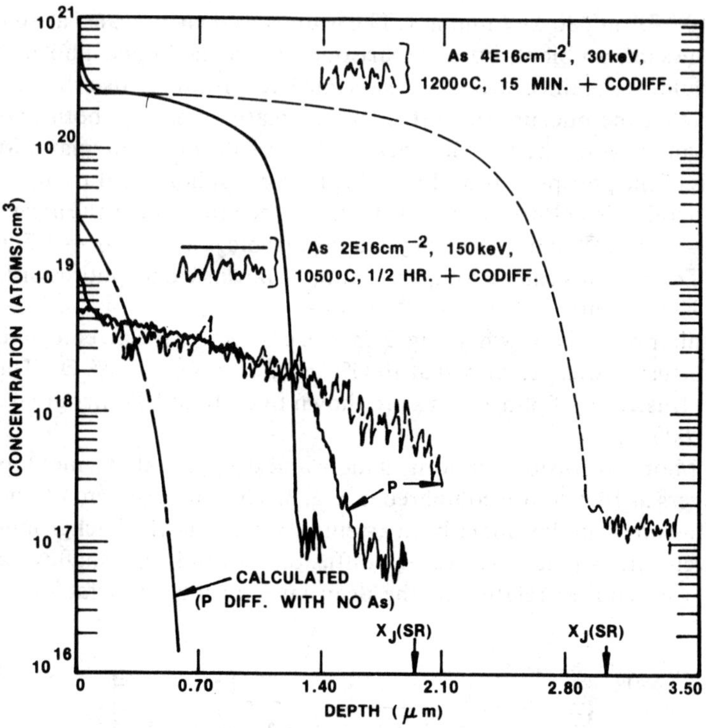

FIG. 3—*Phosphorus codiffusion with shallow and deep arsenic profiles at 1050°C for ½ h.*

Nevertheless, both cases show that phosphorus diffusion at depths greater than 1.2 μm is enhanced by factors of 10 to 40 over the intrinsic value D_{iP}.

The average enhanced phosphorus diffusivity in the region beyond the arsenic junction is shown plotted in Fig. 4 versus $1/T$ where T = temperature. For comparison, D_{iP} is also shown. The data fit reasonably well to calculated points connected by a dashed line that will be described in a later section.

In Fig. 5 D_P is plotted versus the local arsenic background concentration normalized to n_i. These data represent low concentration phosphorus diffusions in silicon doped with arsenic. For data obtained from the shallow junction arsenic samples, the electric field enhancement factor was divided out in order to display isoconcentration results. It can be seen that for $C_{As}/n_i > 10$, $D_P \propto C_{As}^2$. However, the 900°C data can also be fitted by $D_P \propto C_{As}$. For the case of concentration-dependent arsenic diffusion, it is known that $D_{As} \propto C_{As}$ over a wide temperature range [*3*]. This discrepancy is discussed in more detail below. It should also be noted that D_P at 900°C is considerably enhanced for codiffusion in 100% dry O_2 relative to the 0.5% O_2 in N_2 results.

The ambiguity in the D_P versus C_{As}/n_i data is eased when the arsenic doping density is converted to electron concentration at the diffusion temperature. Since only a certain fraction of the total arsenic is "electrically active,"

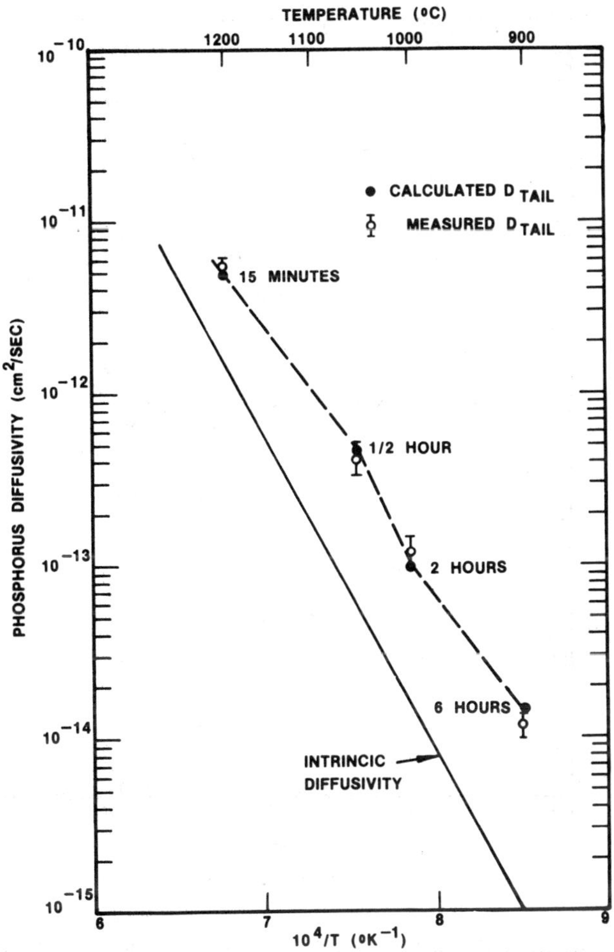

FIG. 4—*Tail region phosphorus diffusion beyond the arsenic junction.*

it is necessary to convert from C_{As} to n (diffusion temperature), since n determines the concentration of ionized vacancies believed to be responsible for phosphorus diffusion in a neutral furnace ambient. Using the data of Murota et al [4], D_P is shown in Fig. 6 versus n/n_i. Some 1000°C data are included. Also included is a 1050°C datum at $n/n_i = 0.4$ that was obtained by diffusion in a highly-doped boron background (see Table 1). Shown for comparison are calculated curves of D_P versus n/n_i based upon the equation

$$D_P = D_i^x + D_i n/n_i + D_i^= (n/n_i)^2 \quad (1)$$

The diffusion coefficients in Eq 1 represent phosphorus diffusion via V^x, V^-, and $V^=$ vacancies according to the Fair-Tsai model [2]. The values of D_i^x, D_i^-

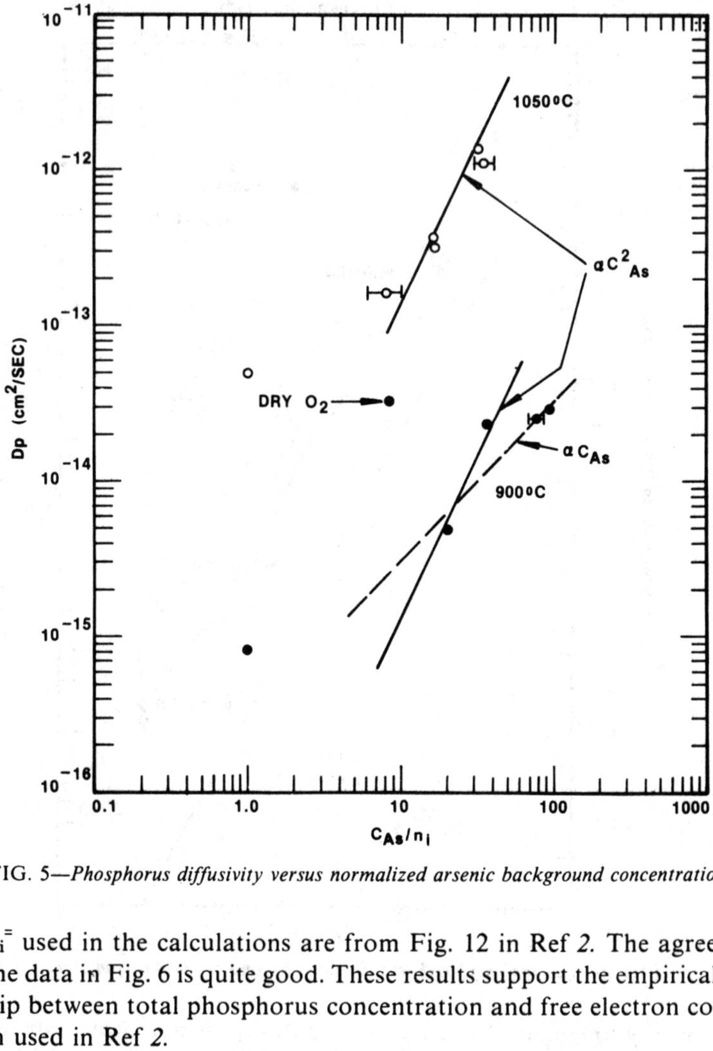

FIG. 5—*Phosphorus diffusivity versus normalized arsenic background concentration.*

and $D_i^{=}$ used in the calculations are from Fig. 12 in Ref *2*. The agreement with the data in Fig. 6 is quite good. These results support the empirical relationship between total phosphorus concentration and free electron concentration used in Ref *2*.

Coincident with the enhanced phosphorus diffusion observed in Figs. 1, 2, and 3 is the enhancement of the arsenic diffusion in the presence of the phosphorus "tail." Examples of 2×10^{16} As/cm^2, 150-keV implants codiffused with and without the 5×10^{14} P/cm^2, 30-keV implant are shown in Figs. 7 and 8. Spreading resistance data and concentration profiles of arsenic are shown to illustrate the significant effect that the emergence of the phosphorus tail has on arsenic diffusion. Since the concentration profiles of the arsenic-only samples were not measured on SIMS, calculated profiles are shown for comparison with the measured codiffusion results. These calculated profiles are consistent with the measured spreading resistance data in showing the enhanced diffusion effect. No enhancement of arsenic was observed when

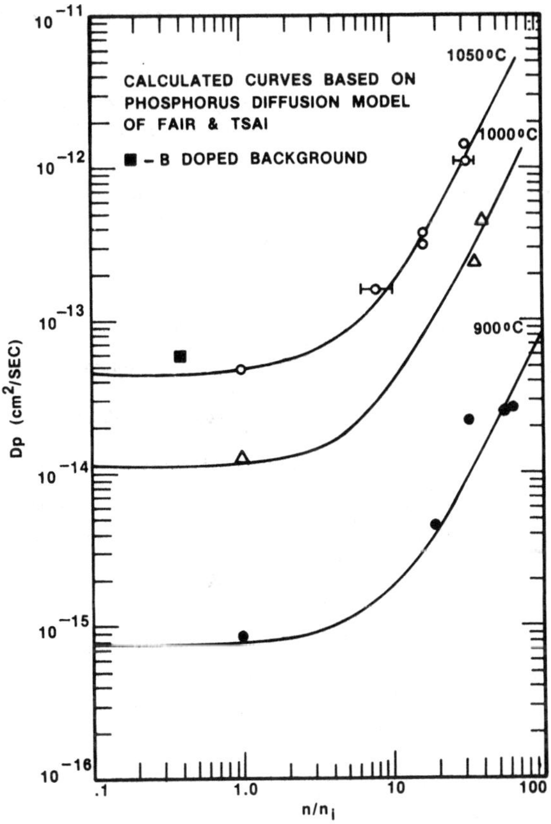

FIG. 6—*Phosphorus diffusivity versus* n/n_i *in an arsenic-doped background.*

the arsenic junction remained deeper than the phosphorus during codiffusion (no phosphorus tail).

Discussion

Phosphorus Diffusion Models

The results obtained in this study have a significant impact upon our understanding of phosphorus diffusion in silicon. By creating "pseudo" high-concentration phosphorus profiles in which the highly doped regions are replaced by a shallow arsenic profile, we have duplicated the well-known enhanced phosphorus diffusivity results. The mechanism that enhances the phosphorus diffusion beyond the arsenic junction (tail region) also appears to enhance the high concentration arsenic diffusion. These results are noteworthy because ion-implanted-diffused arsenic by itself does not cause enhanced diffusion of underlying boron base layers at temperatures above 700°C [*5,6*] (no emitter push), nor does the arsenic have a tail.

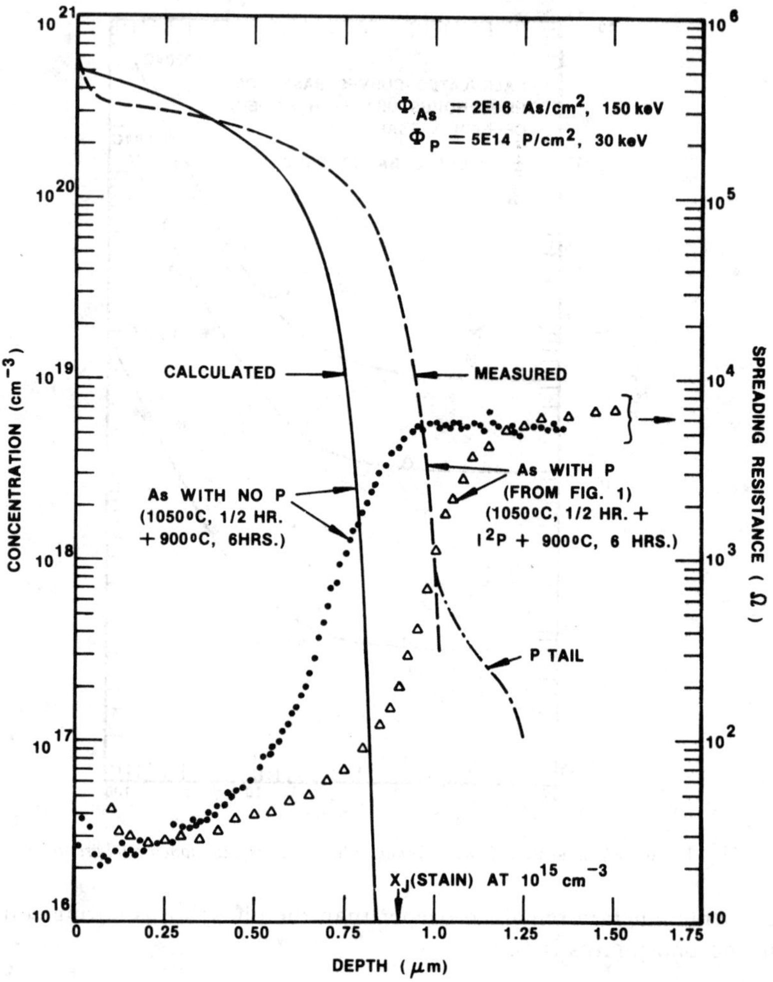

FIG. 7—*Comparisons of arsenic concentration and spreading resistance profiles with and without phosphorus diffusion through the arsenic junction.*

Because the tetrahedral covalent radius of arsenic is almost the same as that of silicon, the lattice contraction coefficient due to arsenic is almost zero. Thus, the "anomalous" enhancement of phosphorus in an arsenic background cannot be related to lattice strain. This result refutes the model of Panteleev [*7*], who proposed that the enhanced diffusion in the tail region of a phosphorus profile is the result of a "two-stream" mechanism involving phosphorus interstitials. Panteleev suggested that the fraction of interstitials would increase with surface doping concentration as a result of the misfit stresses of phosphorus in silicon. The fact that arsenic can enhance phosphorus diffusion beyond its junction also refutes Yoshida's model [*8*] accounting for the concentration dependence of diffusivity in a phosphorus tail region.

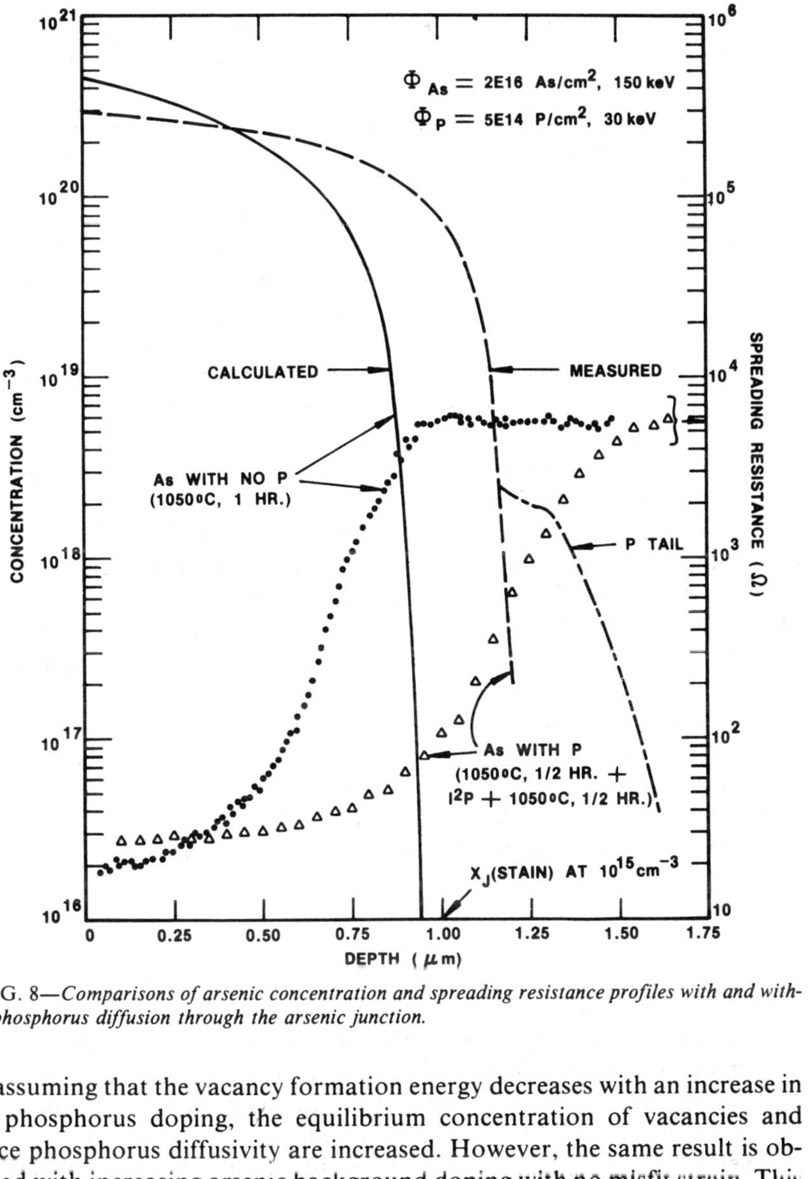

FIG. 8—*Comparisons of arsenic concentration and spreading resistance profiles with and without phosphorus diffusion through the arsenic junction.*

By assuming that the vacancy formation energy decreases with an increase in the phosphorus doping, the equilibrium concentration of vacancies and hence phosphorus diffusivity are increased. However, the same result is obtained with increasing arsenic background doping with no misfit strain. This observation also refutes the proposal of Kendall and Carpio [9], who suggested that misfit stresses at the silicon surface generate excess vacancies that cause the enhanced phosphorus tail diffusion.

Phosphorus Diffusion with High-Concentration Arsenic

The concentration dependence of phosphorus diffusion in a uniform arsenic background is the same as if the background doping were phosphorus, once a correction is made for the "electrically active" doping density. This concentration dependence is, therefore, clearly a Fermi-level related phenomenon. Also, there can be little doubt that the enhanced phosphorus diffusivity beyond the arsenic junction is caused by phosphorus-point defect pairs created in the arsenic region. The 5×10^{14} P/cm^2 implant is too small to create the enhanced diffusion conditions by itself. At 900°C, the phosphorus doping density must exceed $\sim 10^{20}$ cm^{-3} in the surface in order for the tail on the profile to occur [*2*]. Therefore it is believed that the role of the arsenic layer in our experiments is to create a region rich in point defects with which the more lightly doped phosphorus atoms can associate. As these atom-point defect pairs diffuse through the arsenic junction, the equilibrium conditions abruptly change. This may cause the pair to change charge state and become more mobile; or the pair may dissociate, creating a supersaturation of point defects. We believe that pair dissociation does occur since the arsenic diffusion is also enhanced beyond the amount normally encountered in high-concentration arsenic diffusion [*3*].

The point defects involved are believed to be vacancies. It is universally agreed that arsenic diffuses in silicon predominantly by a vacancy mechanism. When silicon self-interstitials are intentionally created in silicon by oxidizing the surface at high temperature, very little oxidation-enhanced diffusion of arsenic is observed [*10*]. This is a result of the large energy required (2.5 eV) to move arsenic off a substitutional lattice site in silicon as part of the interstitial diffusion process [*11*].

It has been proposed by Fair and Tsai [*2*] that phosphorus diffusion can be explained in terms of phosphorus-vacancy pair diffusion and dissociation. The details of the model are shown in Fig. 9. When phosphorus atoms enter the surface, some fraction will become paired with $V^{=}$ vacancies according to the reaction

$$P^{+} + V^{=} \rightleftharpoons P^{+}V^{=} \tag{2}$$

Accordingly, the diffusivity of the $P^{+}V^{=}$ pair will show an n^2-dependence at high doping levels since $[V^{=}] \propto n^2$. The data in Fig. 6 also show that $D_P \propto n^2$ if arsenic is the background dopant. This points out the fact that phosphorus diffusion is controlled by the Fermi-level dependence of the $V^{=}$ vacancy concentration through the free electron concentration in the silicon.

When $P^{+}V^{=}$ pairs leave the highly doped region of a phosphorus profile, it is proposed that they do not dissociate significantly until the Fermi level is ~0.11 eV below the conduction band edge [*2*], corresponding to $n = n_e$. This level corresponds to the second acceptor level of $V^{=}$. When the $V^{=}$ vacancy loses an electron, the binding energy of the phosphorus-vacancy pair decreases by ~0.3 eV. Thus the probability for dissociation is enhanced significantly. This is the "vacancy generation" source of the Fair-Tsai model —$P^{+}V^{=}$ dissociation when $V^{=} \rightarrow V^{-} + e^{-}$ at $n = n_e$ (Fig. 9) and the subsequent

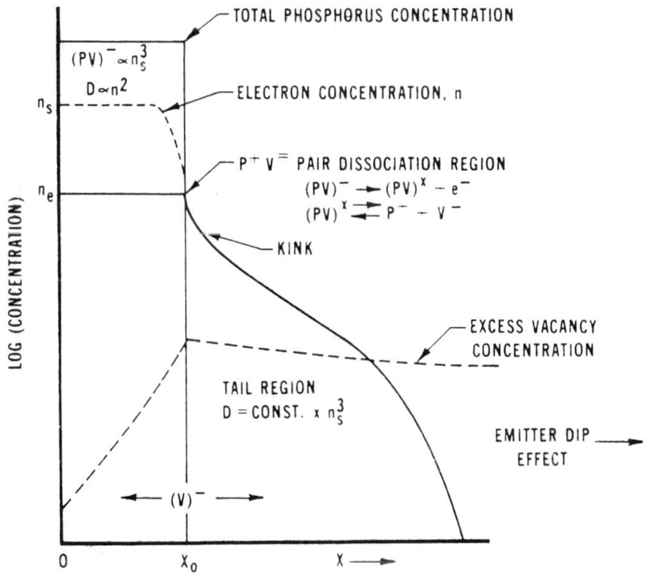

FIG. 9—*Vacancy generation model of phosphorus diffusion* (*after Fair and Tsai*).

creation of the more mobile P^+V^- pair. Based upon this model, the phosphorus diffusivity in the tail region (or beyond the arsenic junction in the present case) is proportional to the concentration of $P^+V^=$ pairs, $C_{PV^=}$ in the surface region.

The mass action that occurs in Eq 2 gives the relation

$$C_{PV^=} = 4\ C_P \exp\left(\frac{E_b^= - \Delta H_v^x}{kT}\right)(n/n_i)^2 \qquad (3)$$

where $E_b^=$ is the $P^+V^=$ binding energy (~1.57 eV [*12*]) and $\Delta H_v^x \simeq 2.57$ eV [*13*] is the neutral vacancy formation enthalpy. For the profiles in Fig. 1, $T = 900°C$ and $n/n_i \simeq 60$, since $C_{As}/n_i \simeq 100$. From Eq 3 $C_{PV^=} \simeq 0.74\ C_P$; most of the underlying phosphorus is complexed with vacancies.

We will assume that $P^+V^=$ pair dissociation occurs at the arsenic junction; then

$$P^+V^= \rightarrow P^+V^- + e^- \rightarrow P^+ + V^- + e^- \qquad (4)$$

will be favorable as the $V^=$ vacancy loses an electron. In its new charge state, the P^+V^- pair is more mobile and more likely to dissociate.

Following the arguments of Ref *2*, the flux of $P^+V^=$ pairs entering the arsenic junction where dissociation can occur equals the flux of V^- vacancies moving into the silicon bulk. Thus

$$\frac{D_V(C_{V^-}^{\,g} - C_{V^-}^{\,i})}{L_V} = \frac{D_P \partial C_{PV^-}}{\partial x} \tag{5}$$

where D_V is the vacancy diffusivity and $C_{V^-}^{\,g}$ and $C_{V^-}^{\,i}$ are the excess generated and equilibrium intrinsic V^- vacancy concentrations, respectively. The vacancy diffusion length is L_V and $\partial C_{PV^-}/\partial x$ is the gradient of phosphorus at the arsenic junction. Assuming a Gaussian distribution of ion-implanted phosphorus under the arsenic background doping, the gradient at $x = x_J$ is

$$\left.\frac{\partial C_{PV^-}}{\partial x}\right|_{x_J} = -\frac{Q_P x_J}{2\sqrt{\pi}(D_P t)^{3/2}} \exp\left(\frac{-x_J^2}{4D_P t}\right) \tag{6}$$

The junction depth is controlled by the ion implanted arsenic and obeys the relation [3]

$$x_J = 2\left(\frac{Q_{As} D_{iAs} t}{n_i}\right)^{1/3} \tag{7}$$

where Q_{As} is the arsenic implant dose and D_{iAs} is the intrinsic diffusivity. Integrating Eq 4, the average excess vacancy concentration during phosphorus/arsenic codiffusion is

$$C_{V^-}^{\,g} - C_{V^-}^{\,i} = \frac{L_V Q_P (Q_{As} D_{iAs}/n_i)^{1/3}}{D_V \sqrt{\pi D_P}\, t_{CO}} \int_0^{t_{CO}} \frac{\exp\left[\dfrac{-(Q_{As} D_{iAs}/n_i)^{2/3}}{D_P t^{1/3}}\right]}{t^{7/6}}\, dt \tag{8}$$

where t_{CO} is the codiffusion time in s. The integral in Eq 8 can be approximated as

$$\int_0^{t_{CO}} \frac{\exp\left(\dfrac{-K}{t^{1/3}}\right)}{t^{7/6}}\, dt \simeq \frac{\beta t^n}{K^2}\ (\mathrm{s}^{-1/6}) \tag{9}$$

where n varies from 0.35 to 0.475 as K goes from 5 to 30. Finally, Eq 8 can be written approximately as

$$\langle C_{V^-}^{\,g} - C_{V^-}^{\,i}\rangle \simeq \beta \frac{L_V n_i}{D_V\sqrt{\pi}} \left(\frac{Q_P}{Q_{As}}\right) \frac{D_P^{3/2}}{D_{iAs}}\, t_{CO}^{\,n-1}\ (\mathrm{cm}^{-3}) \tag{10}$$

where $\beta = 1\ \mathrm{s}^{1/2-n}$ and is a dimensional constant.

The V^- vacancy generation enhances the phosphorus tail region diffusivity which is given by the equation [2]

$$D_{TAIL} = D_i^{\times} + D_i^{-}\,(C_{V^-}^{\,g}/C_{V^-}^{\,i}) \tag{11}$$

where D_i^x and D_i^- were previously used in Eq 1. For example, D_{TAIL} can be calculated for the 900°C codiffusion shown in Fig. 1. Using $L_V = 30$ μm [*14*], $D_V = 1.4 \times 10^{-4}$ exp(−0.33 eV/kT), and $D_P = 5 \times 10^{-14}$ cm²/s, we can solve Eq 10 for $C_{V^-}{}^g - C_{V^-}{}^i$. Using the expression $C_{V^-}{}^i = 5.5 \times 10^{22}$ exp(−2.79 eV/kT) for the equilibrium V^- vacancy concentration, we find $C_{V^-}{}^g/C_{V^-}{}^i = 490$. Since $D_i^x = 7.5 \times 10^{-16}$ cm²/s and $D_i^- = 3 \times 10^{-17}$ cm²/s,

$$D_{TAIL} = 1.56 \times 10^{-14} \text{ cm}^2/\text{s} \tag{12}$$

This calculated value agrees well with the measured result in Fig. 1. Calculated results for 1000, 1050, and 1200°C are shown in Fig. 4 for comparison with the measured data. Since D_{TAIL} is a function of time, it is not possible to extract an activation energy for D_{TAIL}. However, the good agreement between theory and measurement suggests that the temperature dependences built into Eqs 10 and 11 are approximately correct.

Arsenic Diffusion During Enhanced Vacancy Generation

From the previous theory, it is expected that the enhanced arsenic diffusivity depicted in Fig. 7 and 8 is a result of $P^+V^=$ pair dissociation at the arsenic/phosphorus junction. At high concentrations, the arsenic diffusion coefficient depends upon the free electron concentration through the V^- vacancy concentration. Thus, from mass action effects,

$$\begin{aligned} D_{As} &\simeq D_i^- \frac{C_{V^-}}{C_{V^-}{}^i}, \\ &\simeq D_i^- \frac{n}{n_i}. \end{aligned} \tag{13}$$

If the concentration of generated vacancies from Eq 10 is additive with the Fermi-level related V^- concentration, then the enhanced arsenic diffusivity is

$$D_{As}(\text{enh}) = D_i^- \left(\frac{C_{V^-} + C_{V^-}{}^g}{C_{V^-}{}^i} \right) \tag{14}$$

The ratio of Eqs 14 and 13 is the enhancement factor f_{As}:

$$f_{As} = \frac{D_{As}(\text{enh})}{D_{As}} = 1 + \frac{C_{V^-}{}^g}{C_{V^-}{}^i} \frac{n_i}{n} \tag{15}$$

where n is the free electron concentration in the arsenic layer.

Using the surface electron concentrations corresponding to C_{As} ($x = 0$) in our measured profiles, we calculated f_{As} and compared it with the estimated f_{As}-values in Table 2. The estimated values were obtained by calculating an effective D_{As} required to account for the deeper codiffused arsenic profiles

TABLE 2—*Calculated and measured arsenic diffusion coefficient enhancement factors during phosphorus codiffusion.*

Temperature, °C	f_{As} (Calculated)	f_{As} (Measured)
900	10.8	13.0
1000	5.0	5.8
1050	8.0	8.0
1200	1.3	1.3

(Figs. 7 and 8, for example). The agreement between theoretical and measured values of f_{As} is quite good. This shows that the arsenic diffusion was enhanced by the same amount of point defect supersaturation as was the phosphorus tail diffusion. This result strongly supports the contention that the point defects are vacancies.

Summary and Conclusions

We have shown that phosphorus contamination of high-dose arsenic-implanted silicon can result in significantly deeper junction depths after diffusion than when no contamination is present. The highly doped arsenic layer has the same effect on phosphorus diffusion as would a similar phosphorus layer doped to the same level. In the presence of a shallow arsenic-doped layer, phosphorus can diffuse through the junction and form a tail with enhanced diffusion properties. This is particularly interesting since arsenic diffusion alone does not produce a tail region, nor is there any significant emitter push effect.

These results provide strong support for a Fermi-level-controlled point-defect model of phosphorus diffusion with impurity-point-defect pair dissociation. We find that under a highly doped arsenic background, phosphorus diffusivity depends on $(n/n_i)^2$, just as if the background doping were phosphorus. Also, the enhanced phosphorus tail diffusivity beyond the arsenic junction can be calculated from the Fair-Tsai phosphorus model. The coincident enhanced arsenic diffusion when there is a phosphorus tail can be calculated by estimating the point-defect supersaturation created by dissociating *E*-centers. Since arsenic diffusion is little affected by supersaturations of silicon self-interstitials (little oxidation-enhanced diffusion), the point defects involved are most likely vacancies.

References

[1] Haas, E. W., Glawischnig, H., Lichti, G., and Bleier, A., *Journal of Electronic Materials*, Vol. 7, 1978, p. 525.

[2] Fair, R. B. and Tsai, J. C. C., *Journal of the Electrochemical Society*, Vol. 124, 1977, p. 1107.

[3] Fair, R. B. and Tsai, J. C. C., *Journal of the Electrochemical Society*, Vol. 122, 1975, p. 1689.

[4] Murota, J., Arai, E., Kobayashi, K., and Kudo, K., *Journal of Applied Physics*, Vol. 50, 1979, p. 804.

[*5*] Fair, R. B., *Solid State Electronics,* Vol. 17, 1974, p. 17.
[*6*] Shibayama, H., Masaki, H., Ishikawa, H., and Hashimoto, H., *Journal of the Electrochemical Society,* Vol. 123, 1976, p. 742.
[*7*] Panteleev, V. A. *Soviet Physics Solid State,* Vol. 21, 1979, p. 1956.
[*8*] Yoshida, M., *Journal of Applied Physics,* Vol. 48, 1977, p. 2169.
[*9*] Kendall, D. L. and Carpio, R. RNP 396, presented to Fall Meeting, The Electrochemical Society, Pittsburgh, 1978.
[*10*] Antoniadis, D. A., Lin, A. M., and Dutton., R. W., *Applied Physics Letter,* Vol. 33, 1978, p. 1030.
[*11*] Fair, R. B., *Journal of Applied Physics,* Vol. 51, 1980, p. 5828.
[*12*] Fair, R. B. in *Impurity Doping Processes in Silicon,* F. F. Y. Wang, Ed., North-Holland, Amsterdam, The Netherlands, 1981, pp. 315–442.
[*13*] Messmer, R. P. in *Lattice Defects in Semiconductors—1974,* Institute of Physics, London, 1975, p. 44.
[*14*] Lecrosnier, D., Gauneau, N., Paugan, J., Pelous, G., Richou, F., and Henoc, P., *Applied Physics Letters,* Vol. 34, 1979, p. 224.

Defects, Oxygen Precipitation, and Carrier Lifetime

Lubek Jastrzebski[1]

Silicon Crystals for VLSI Technologies

REFERENCE: Jastrzebski, L., "**Silicon Crystals for VLSI Technologies,**" *Silicon Processing, ASTM STP 804,* D. C. Gupta, Ed., American Society for Testing and Materials, 1983, pp. 309–331.

ABSTRACT: The need for a new generation of electronic circuits with a high degree of integration and a large area of silicon imposes a new demand on the quality of silicon used for their fabrication. This paper reviews the generation process of crystallographic defects (in the context of material characteristics), their effect on device/circuit performance, and their relationship to crystal growth conditions.

The defect formation during processing, effectiveness of internal gettering, and mechanical strength of silicon (resistance to warpage) are mainly controlled by characteristics of oxygen in silicon. For a given process, oxygen precipitation kinetics depends on the crystal growth conditions, which control oxygen incorporation into silicon, formation of active sides for oxygen nucleation, and initial stage of oxygen precipitation. It is believed that proper understanding of required silicon characteristics and proper control of the crystal growth process will lead to higher yield of integrated circuits.

KEY WORDS: defects in silicon, silicon crystals, silicon wafers, defect generation in wafers, effects of defects on circuit performance, oxygen precipitates, thermal donors

The relationship between circuit performance/yield and the crystal growth process is not always obvious and straightforward. Present very-large-scale integration (VLSI) circuits contain 10^5 active devices, and a failure of any of them to perform or deviation of some of them from required characteristics can cause the failure of circuits to perform within the limits of specifications. Crystallographic defects present in silicon wafers, as well as defects introduced to the superstructure during processing (such as problems with polysilicon, metal, or dielectric definition), can be responsible for circuit failure; in addition, it is not always obvious which defect causes yield degradation. Different types of defects present in the substrate can affect circuit performance differently [*1*].[2] Even for the same type of defect, however, the effect on circuit performance can vary owing to the difference in defect decoration and circuit design [*1–3*]. The crystal growth process determines the characteristics of as-grown material which, in turn, influence the defect generation process; their interdependence, however, is complex and usually not obvious. Therefore analysis of the relationship between circuit performance

[1] Member Technical Staff, RCA Laboratories, Princeton, N.J. 08540.
[2] The italic numbers in brackets refer to the list of references appended to this paper.

and crystal growth conditions is usually laborious, and the results are often proprietary and relevant only to particular types of circuits and processing. As a result, the straightforward relationship between crystal growth conditions and circuit performance/yield is difficult to establish and only some general comments on this subject can be made.

We will first address the issue of how and which material defects affect circuit performance and why their influence increases during the transition from LSI to VLSI. Next, we shall discuss the defect generation mechanism, particularly the role of oxygen in silicon. In order to understand the relationship between crystal growth conditions and material parameters, phenomena taking place during crystal growth which affect the incorporation of oxygen into crystals and its precipitation kinetics will be described. Internal gettering procedures used to reduce the effect of defects on device performance and mechanical strength of silicon wafers (resistance to warpage) will be reviewed in relationship to required compromises between different material characteristics.

Effect of Material Defects on Circuit Performance

The ways in which material can influence circuit performance can be divided into four categories:

1. Warpage of silicon wafers, which will generate catastrophic defects in the superstructure during photolithographic processes (for example, shorts between metal lines, closed contact holes).
2. Resistivity changes generated by oxygen thermal donors, which will shift threshold voltages of MOS transistors.
3. Functional failure of individual devices.
4. Excessive leakage current caused by degradation of the characteristics of a number of devices, even if the leakage of each one is still within the limit of tolerance.

Yield of VLSI circuits will be affected by warpage of silicon wafers; thus control of flatness during processing is a critical issue for VLSI. To print 2.5-μm lines or contact holes with projection printing, an 8-μm flatness of wafers is required.[3,4] To print 1.5 μm, flatness should be better than 4 μm.[4] Warpage-causing forces [*4*] can be minimized by optimizing processing, although for 12.7-cm (5-in.)-diameter wafers process optimization becomes more and more difficult and mechanical strength of silicon starts to be an important issue.

Thermal donors introduced by oxygen [*5*] can change the resistivity of silicon during processing; this effect will shift threshold voltages of metal-oxide-

[3] Specifications for Perkin-Elmer Model 500 Projection Mask Aligner.
[4] L. White, RCA Laboratories, Princeton, N.J., unpublished data.

semiconductor (MOS) transistors and affect drive currents. The effect becomes significant when the dopant concentration in the virgin wafers or the concentration of threshold adjusting implants starts to be comparable with thermal donors concentration. If the concentration of thermal donors varies across the wafer, its effect cannot be easily compensated by adjustment of processing (for example, additional annealing steps or ion implants). To improve the VLSI circuit-operating speed, there is a trend to reduce parasitic capacitances by increasing the substrate resistivity, which emphasis the issue of thermal donors control.

Crystallographic defects present in the active area of MOS transistors or capacitors can cause deterioration in device characteristics by (*1*) introducing localized generation current, when present in the depletion region; and (*2*) by increasing *p-n* junction leakage due to the presence of diffusion spikes generated by enhanced impurities diffusion along crystallographic defects [*5*]. If the dark generation current introduced by a defect is large enough to collapse the depletion region of MOS transistors that are part of the dynamic circuit during a storage time, the defect will destroy a written information and cause functionality failure of dynamic memory. If the defect gives a large enough current to fill the potential wells of MOS capacitors that are part of a charge-couple-device (CCD) imager, it will wipe out any optical information and introduce white lines or spots in the displayed image. Even if defects do not give dark current strong enough to cause functional failure of individual devices in the circuit, they can affect circuit performance by giving high circuit leakage, an accumulative effect of individual device leakage. Ten percent (10^4) of slightly leaky (10^{-9} A) transistors at 16K CMOS memory would give total leakage current around 10^{-5} A, which for the same application is unacceptably high.

With transition from LSI to VLSI circuits the dimensions of devices shrink [*6*]. With the decreasing dimensions, storage capacitance of MOS devices decreases; this increases the probability that the dark current generated by defects will become large enough to cause functionality failure in VLSI circuits even if it did not cause it in LSI circuits. The transition from LSI to VLSI also increases the amount of devices in the circuit area, which will increase the volume of the space-charge region of transistor junctions. Assuming constant size and density of material defects in the surface region, one can expect that the leakage generated by surface defects will be higher in the VLSI circuit than in the LSI circuit (Fig. 1) [*7*]. Defects marked by arrows intersect the space-charge region of MOS transistors and contribute to leakage. Double arrows mark defects that, depending on the depth of the junction and the defect distance from the surface, may also contribute to leakage (Fig. 1*c*). With a decreasing junction depth (required by VLSI), the probability increases that the double-arrow defects will contribute to leakage.

Material defects known to affect circuit performance can be divided into

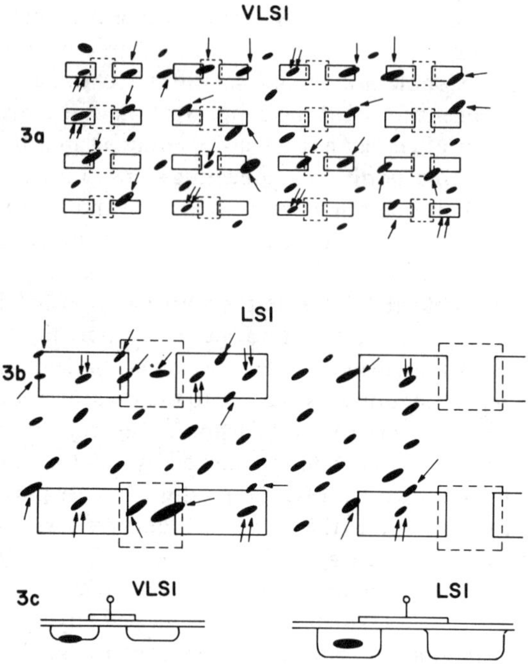

FIG. 1—*Effect from subsurface defects on VLSI* (3a) *and LSI* (3b) *circuits. Defects marked by arrows intersect space-charge regions of transistors and contribute to leakage. Defects marked by double arrows may contribute to leakage, depending on junction depth.* (3c) *An example of the double-arrow marked defect contributing to leakage in VLSI and not in LSI.*

three categories: dislocations, precipitates, and deep levels. The electric activity of the different types of defects is different [*2*], but even in the same category electric activity can be significantly enhanced by the decoration of crystallographic defects by contaminates (for example, heavy metals introduced to wafers during processing) [*1*]. The decoration process seems to be statistical in nature, as shown in Fig. 2, where only two out of four crystallographic defects (revealed by etching) seem to be electrically active (as revealed by EBIC) [*8*]. The probability of decoration is a function of the total amount of recombination centers in the material, as shown in Fig. 3, where the percentage of electrically active (decorated) dislocations observed as white spots in CCD imagers is plotted as a function of the minority carrier lifetime [*9*], which is inversely proportional to the concentration of deep centers ($\tau \sim 1/N_{Recom}$). Different forms of gettering [*10*] are commonly used during processing to decrease concentrations of contaminates in the wafer and to prevent decoration of crystallographic defects.

So far it has been shown that the presence of decorated dislocations, stacking faults, or other forms of oxygen precipitates contributes to leakage currents or functionality failures of CMOS circuits [*11,12*], CCD imagers [*13*],

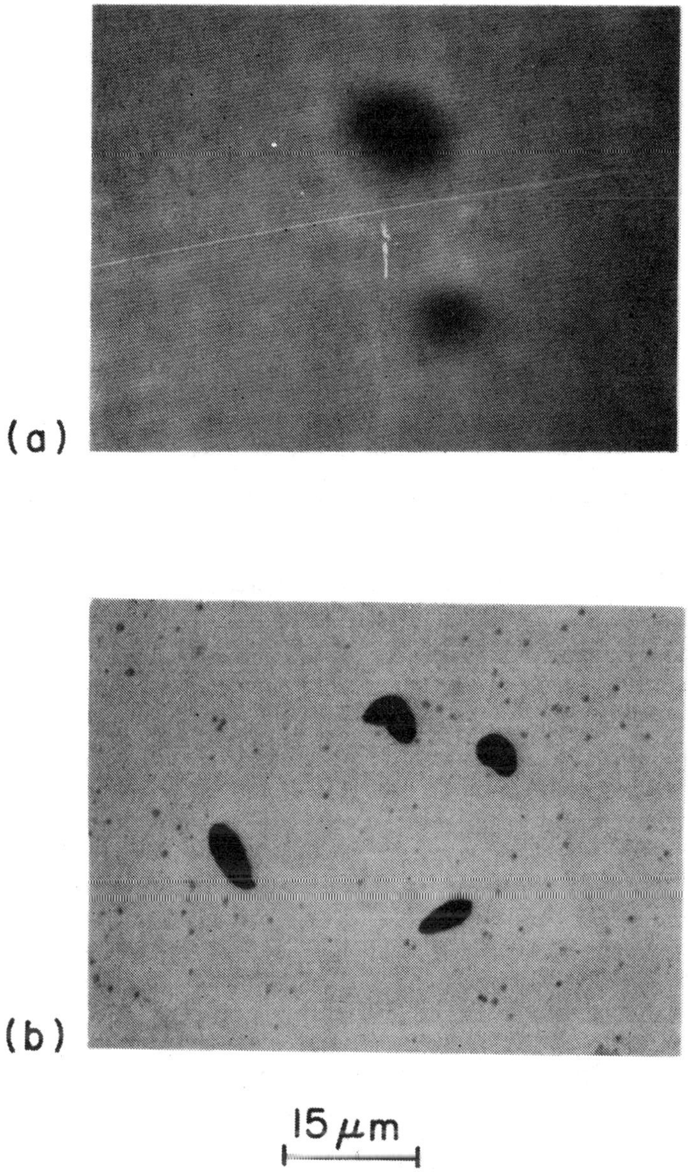

FIG. 2—*Comparison between* (a) *electrically active dislocations observed by EBIC and* (b) *dislocations in the same area revealed by a Dash etch.*

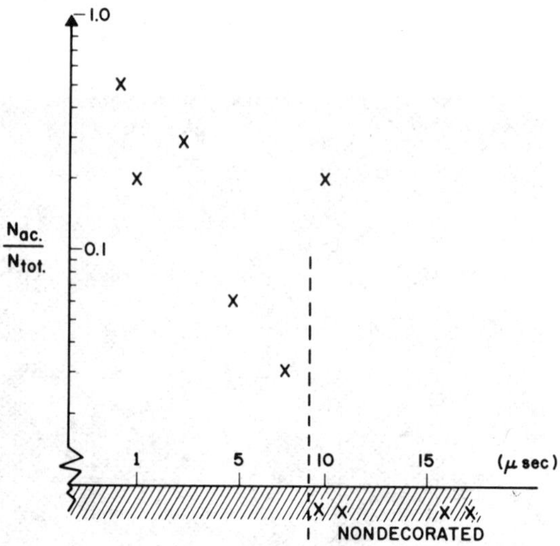

FIG. 3—*Ratio of electrically active dislocations observed as white spots in the CCD imagers to total amount of dislocations revealed by etching in the imaging register as a function of minority carrier lifetime. When dislocations did not generate any white spots, the data point was placed in the "nondecorated" region.*

or NMOS memories [*3*]; in addition, foreign precipitates such as iron needles introduce collector-emitter shorts in bipolar transistors with submicron bases [*14*].

Defect Generation

Defect generation during processing depends on the conditions of high-temperature steps [*15*], superstructure geometry [*16,17*], method of doping [*18*], and concentration of impurities in silicon wafers [*19*]. The present work will not describe the forces responsible for defect generation; defect generation will be discussed only in relation to material parameters (carbon and oxygen). The role of carbon is not so clear as oxygen. The carbon concentration required to influence defect formation still remains a controversial issue.[5] It appears that the effect of carbon is indirect, through the effect on oxygen precipitation kinetics. Therefore the role of carbon in defect generation process will be discussed only in this context.

Oxygen in silicon controls to a large extent the process of defect formation and can play positive as well as negative roles (Table 1). Oxygen can provide a dislocation pinning effect that will improve mechanical strength (resistance to warpage) of silicon wafers. Oxygen can also generate thermal donors [*5*] that will cause undesirable changes of resistivity during processing. Oxygen precipitates and stacking faults present in the surface region will adversely affect circuit performance, but their presence in the bulk of the wafer will

[5] Compare, for example, results from Refs *20* and *21*.

TABLE 1—*Role of oxygen in defect generation.*

Defect Type	Effect on Device/Circuit	
	Negative	Positive
Surface oxide precipitates	functionality failure, leakage	. . .
Oxygen bulk precipitates	dislocation generation/ warpage	internal gettering/decontaminations-reduction of functionality failure and leakage
Oxygen donors	resistivity change/threshold voltage shift	. . .
Oxygen clusters	. . .	dislocation pinning/ resistance to warpage

improve circuit performance due to internal gettering, which removes contaminations from device active areas [*20,21*]. On the other hand, defects providing internal gettering may also pur ch dislocations during processing; this will increase the probability of warpage [*22*]. It seems that effective internal gettering and good mechanical strength are opposite requirements, but it is still not clear to what extent they can be compromised. To understand the complex relationship between oxygen and defect generation, process of thermal oxygen donors, dislocations, and oxygen precipitates formation will be discussed in the context of oxygen precipitation kinetics.

Thermal Donors

Carrier concentrations in silicon can change as much as 10^{15} to 10^{16} cm^{-3} due to the generation of thermal oxygen donors [*5*]. In high-resistivity material, changes caused by oxygen donors can be large enough to significantly shift threshold voltages of MOS transistors. The concentration of thermal donors can vary significantly across the wafer [*23*] or from wafer to wafer [*24*]; thus oxygen donors can seriously hamper the circuit fabrication process.

Two types of oxygen donors have been reported: old donors generated by 450°C annealing [*5*] and new donors generated by 550 to 800°C annealing [*25*]. The process of the oxygen donors formation depends on the oxygen concentration [*5,24*], the thermal history of silicon [*23*], and the concentration and type of shallow donors and acceptors [*23*]. Formation of thermal donors is usually a long process (~100 h), which can be at least partially reversed by subsequent high-temperature annealing procedures [*5*]. Large variations in donors formation kinetics has been observed among different crystals [*24*], across the silicon wafers (peripheral region behaves differently than center) [*23*], and on the microscale in the same wafer [*23*]. Old thermal donors have been usually explained by the formation (during 450°C annealing) of complexes between substitutional oxygen, vacancies, and other impurities

[*5*]. It seems that new donors are formed due to clustering of oxygen (precipitating from interstitial positions) [*25*] during 550 to 800°C annealing. Generation of new donors is strongly affected by carbon [*25*].

It appears as if none of the present models [*5,25*] can provide a detailed and comprehensive explanation of donors formation that accounts for the observed inhomogeneities in donors generation kinetics across the wafer or the effect from shallow impurities on the thermal donors formation. The interaction of oxygen atoms with impurities is poorly understood even for concentrations which are too low to expect any near neighbor effects ($\sim 10^{15}$ cm^{-3}) [*26,27*]. It seems that thermal donors formation is one of the stages of oxygen precipitation in silicon and is controlled by oxygen precipitation kinetics.

Oxygen Precipitates

Stacking faults or oxygen precipitates in the surface region of processed wafers may cause serious problems owing to their adverse effect on circuit performance. Much work has been done in recent years on the formation of surface-stacking faults and oxygen precipitates in silicon wafers during processing [*28–30*]. The proposed model [*28*] describes well the dependence of stacking faults density and morphology on temperature [*29*], oxidizing ambient [*30*], and oxygen content in silicon wafers [*31*]. It has been shown that the formation of bulk stacking faults and oxygen precipitates in the wafer interior usually prevents formation of surface stacking faults or oxygen precipitates [*20,21*]. This observation is the basis for the internal gettering procedure, the major development in defect engineering, which uses defects in the wafer to improve device performance.

In the two-step internal gettering procedure small oxygen precipitates are induced in the wafer by long, low-temperature annealing (600 to 800°C at 16 to 64 h). During subsequent high-temperature annealing (for example, 1050°C at 4 h) oxygen is outdiffused from the surface region and precipitates grow in the bulk of the wafer. A defect-free zone is formed at the wafer surface region when the bulk of the wafer contains a high density of oxygen precipitates; these precipitates act as gettering sides for point defects and metallic contaminates [*32–34*]. There are some indications [*21*] that in order to have an effective internal gettering action a plastically deformed region in the bulk of the wafer is required. An example of gettering action by precipitates present in the bulk of the wafer is shown in Fig. 4.[6] The removal of point defects and metallic contaminates from the surface region by bulk precipitates prevents formation of surface stacking faults that are present in nongettered wafers. It has been shown that the efficiency of internal gettering depends on the time of low-temperature annealing, carbon, and especially interstitial oxygen concentration in silicon wafers [*35,36*]. Even for

[6] Lubek Jastrzebski, RCA Laboratories, Princeton, N.J., unpublished data.

wafers with similar carbon and oxygen concentrations, however, there are some problems with reproducibility and uniformity of internal gettering in commercially available wafers.[6]

An example of nonuniform distribution of precipitates in the wafers after two-step annealing is shown in Figs. 5 and 6.[6] Figure 5 shows the precipitation gradient across the wafer, high-precipitation density in the center of the wafer and low precipitation density at the wafer edge. Interstitial oxygen concentration prior to annealing was similar in both parts of the wafer (Fig. 5) and carbon concentration was below the detection limit (4×10^{16} cm^{-3}).[7] Figure 6 shows an example of other types of non-uniform oxygen precipitates distribution across the wafer after two-step annealing. The middle of the wafer has a high density of precipitates when the center has lower density and the edge is practically precipitation free. The measured interstitial oxygen concentration before and after annealing, as well as the concentration of oxygen present in noninterstitial position in as-grown wafers, are also given in Fig. 6. Carbon concentration in this wafer was below the detection limit (estimated to be around 4×10^{16} cm^{-3}). The amount of interstitial oxygen prior to annealing was similar in all three positions when any significant difference was only in the oxygen present in noninterstitial positions in the as-grown wafers. The problem with reproducibility of internal gettering in wafers from different suppliers is demonstrated in Fig. 7, where precipitation density as a function of distance from the surface is plotted for two groups of wafers having similar initial interstitial oxygen concentration (Fig. 7*b*) and carbon concentration below detection limit.[6] Large differences in precipitation density have been observed in wafers belonging to the same group; in some cases the surface region does not contain any precipitates (below 10^4 cm^{-3}), whereas in other instances a high density of precipitates exists in the denuded zone ($\sim 10^7$ cm^{-3}). Large differences in precipitates density, especially in the region of the denuded zone, exist between the two groups of wafers. It seems that the only noticeable difference between these two groups of wafers is the concentration of noninterstitial oxygen present in as-grown wafers (Fig. 7*c*).

Dislocation Generation

Although presently used commercial silicon crystals are initially dislocation free, dislocations are generated during processing as a result of stress forces such as heavily diffused regions, dielectric films deposited over silicon, and temperature gradients across the wafer during cooling or heating cycles [*4*]. When a dislocation network is introduced in the wafer, a plastic deformation/warpage of the wafer takes place. For a given stress force the probability of dislocation generation is a function of material resistance to warpage (value of upper yield point), which decreases with increasing temperature,

[7] Oxygen concentration is always expressed using old ASTM calibration.

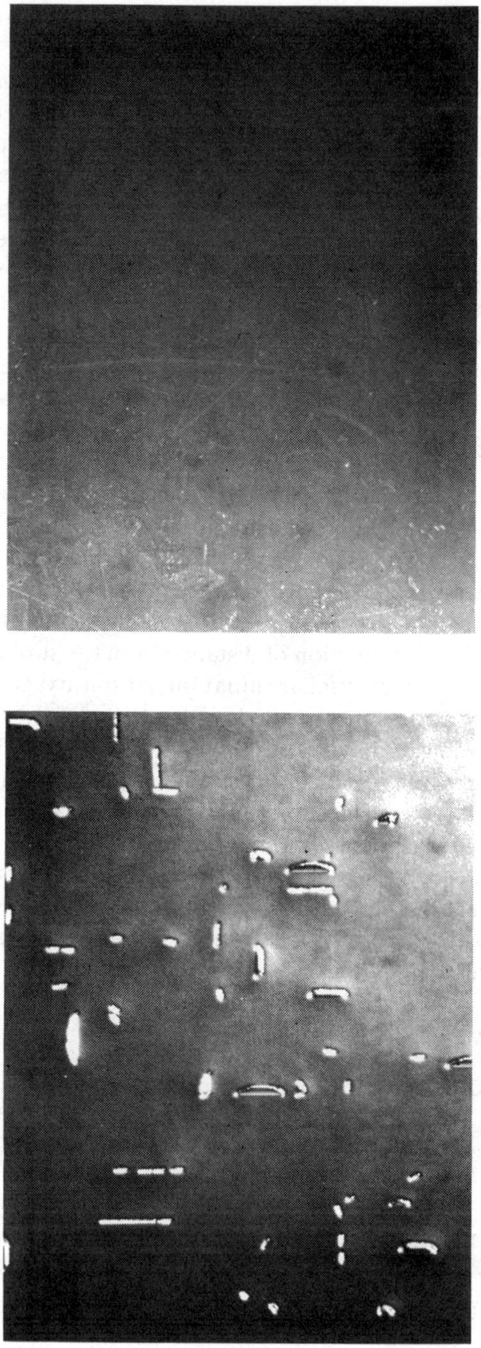

b

a

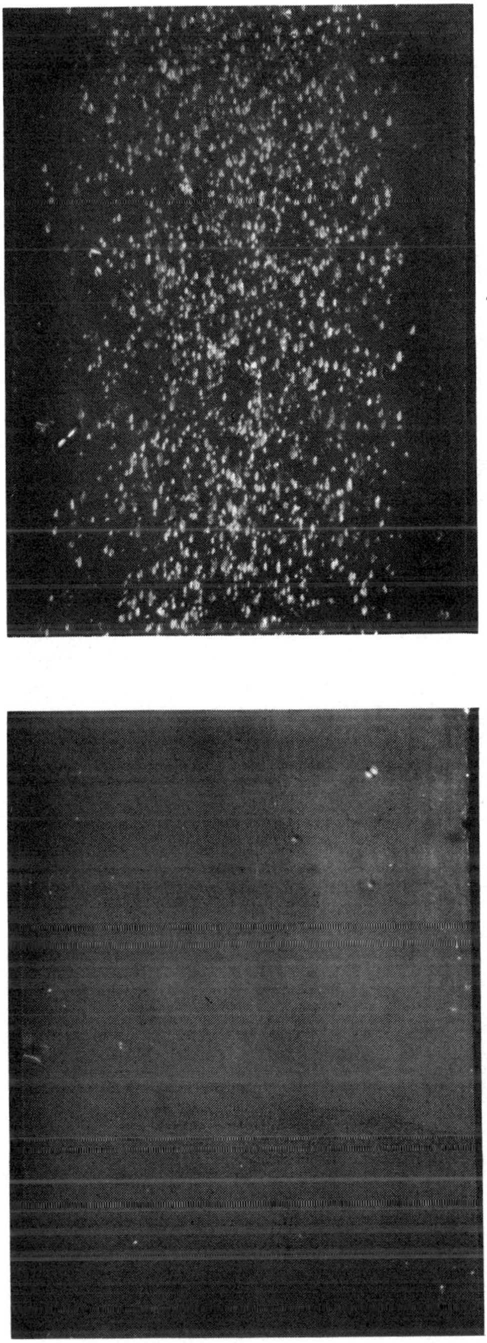

FIG. 4—*Nomarski micrographs of etched (100) surface* (a,b) *and* (110) *cleaved and etched plane* (c,d) *of the wafer after 1100°C, 90 min, wet oxidation. One half of the wafer* (b,d) *was subjected to two-step annealing; the other half* (a,c) *did not receive any treatment prior to oxidation (reduced 15%).*

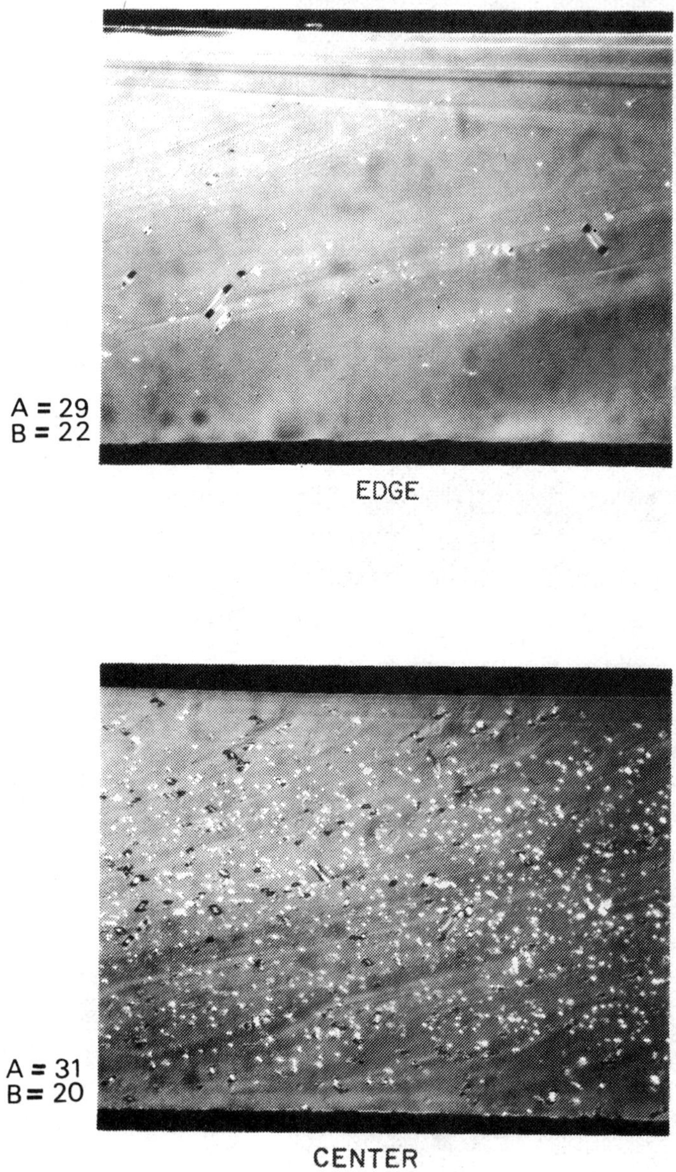

FIG. 5—*Bulk oxygen precipitates observed on the cleaved and etched edge of (100) wafer after two-step annealing at wafer center* (top) *and at wafer edge* (bottom). *Nomarski photomicrograph.* A = *interstitial oxygen concentration as grown;* B = *interstitial oxygen concentration after two-step annealing.*

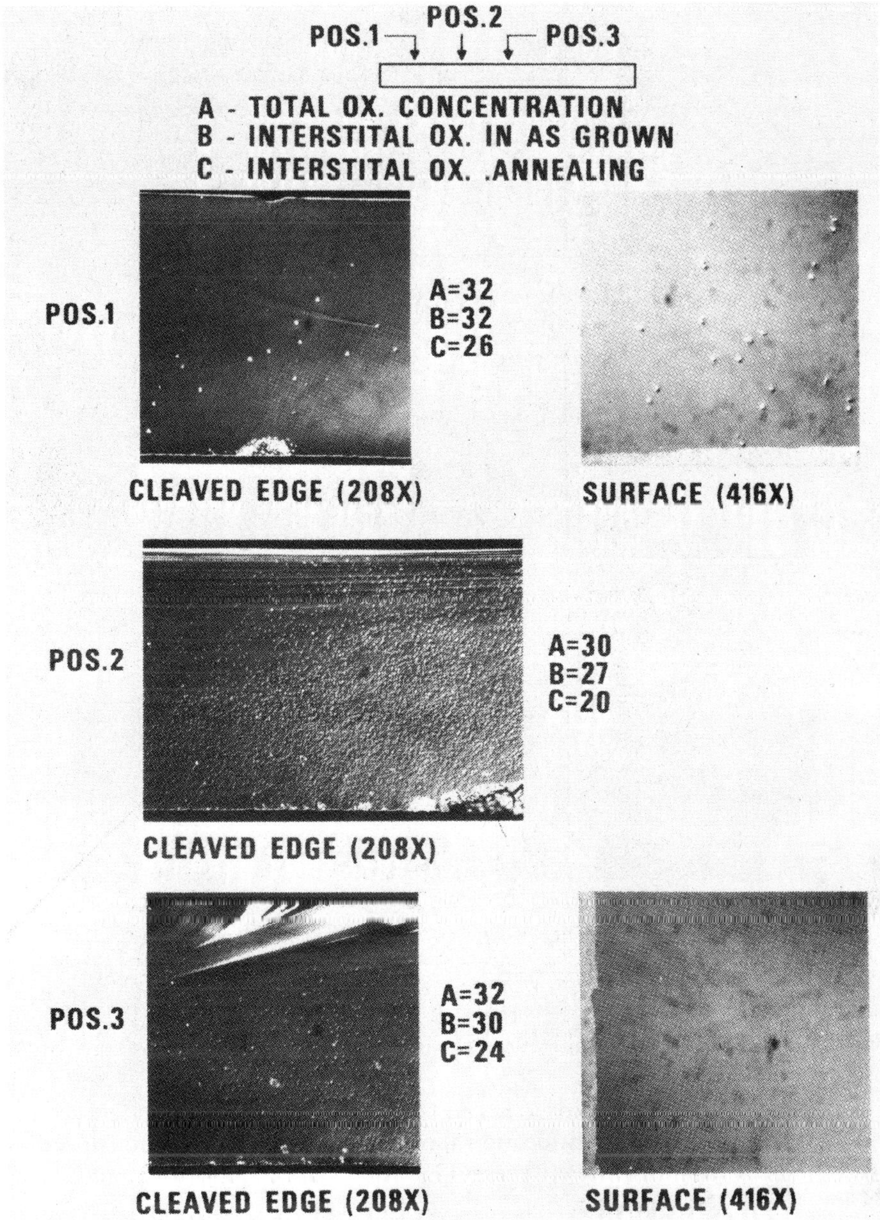

FIG. 6—*Bulk oxygen precipitates in the wafer after two-step annealing. Nomarski photomicrograph of the cleaved and etched (100) wafer.* A = *total oxide concentration,* B = *interstitial oxide concentration as grown;* C — *interstitial oxide concentration after two-step annealing.*

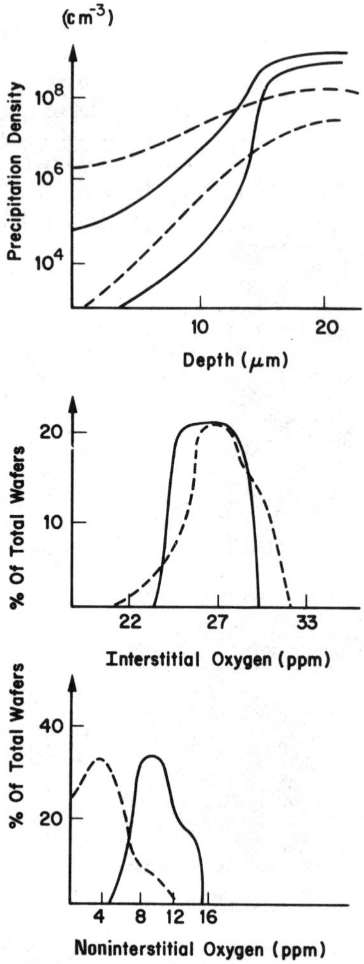

FIG. 7—(a) *Precipitation density as a function of distance from the surface in the two groups of wafers after two-step annealing.* (b) *Distribution of interstitial oxygen in as-grown wafers.* (c) *Distribution of noninterstitial oxygen in as-grown wafers.*

and is a function of material parameters such as concentration and state of oxygen and carbon [*4*].

Dislocation generation processes can be divided into two steps: (*1*) dislocation nucleation and (*2*) dislocation movement and multiplication. In the stress field, dislocations already present in the crystal can move and multiply by the Frank-Read mechanism [*37*]. The dislocation velocity and critical stress field required to start dislocation movement are functions of oxygen concentration [*38*]. It has been demonstrated that oxygen in silicon causes pinning of dislocations; this increases the value of stress threshold required to start dislocation movement and multiplication process [*39*]. This explains the better mechanical strength of CZ wafers (0 ≈ 10^{18} cm^{-3}) compared with float zone wafers (0 ≈ 10^{15} cm^{-3}) [*38,39*]. It is not completely clear in which state oxygen contributes to the pinning effect. Work based on the rosette propagation technique [*40,41*] indicates that small oxygen clusters, not interstitial or precipitated oxygen, provide dislocation pinning. Other published results relate wafer resistance to warpage/mechanical strength to interstitial oxygen concentration [*42*]. A histogram of warpage observed in silicon wafers subjected to the same stress during CMOS processing as a function of interstitial and clustered oxygen concentration (no SiO_2 precipitates were present in analyzed wafers) is presented in Fig. 8.[6] The method used to determine quantitatively the concentration of oxygen in clusters has been previously described [*43*]. The data do not show a clear correlation to interstitial oxygen concentration, but confirm the pinning effect by clustered oxygen (oxygen in noninterstitial state).

It must be pointed out that all results have been obtained when SiO_2 precipitates, giving 1235 cm^{-1} absorption band [*44*], have been absent from the silicon. SiO_2 precipitates provide regions in the wafer with high internal

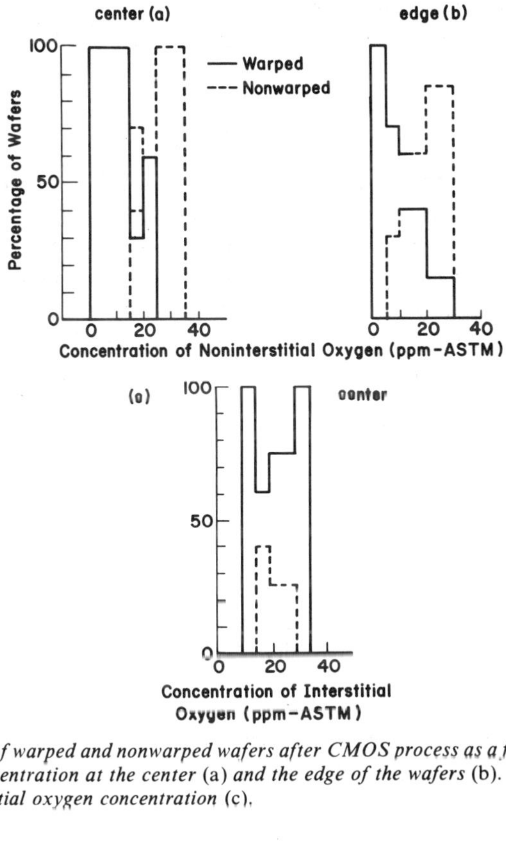

FIG. 8—*Ratio of warped and nonwarped wafers after CMOS process as a function of noninterstitial oxygen concentration at the center* (a) *and the edge of the wafers* (b). *The same ratio as a function of interstitial oxygen concentration* (c).

stress from which dislocations can be punched out (nucleated) at elevated temperatures in the external stress field. If oxygen is present in the form of SiO_2 precipitates, wafer resistance to warpage decreases; this effect is shown in Fig. 9, where warpage is plotted as a function of the amount of oxygen forming SiO_2 precipitates.[6] The amount of oxygen participating in SiO_2 precipitates has been measured by the intensity of the 1235 cm^{-1} absorption band [*44*]. The relationship shown in Fig. 9 is similar to that reported previously [*42*]. The high density of SiO_2 precipitates (required for effective internal gettering) contradicts the requirement for good mechanical strength of silicon wafers.

The amount of oxygen in states which improve or deteriorate the mechanical properties of silicon may change during heat treatments and depends on oxygen precipitation kinetics. As will be discussed later, heat-treatment temperatures, time, atmosphere, and sequence, as well as oxygen and carbon concentration and state in as-grown crystals, determine oxygen precipitation kinetics. A possible future goal of oxygen precipitation kinetics engineering will be to devise a procedure that will provide oxygen in the clusters (increasing mechanical strength of silicon) during high-temperature processing steps having the highest potential for warpage generation, and subsequently to induce precipitates, required for internal gettering, during the following processing steps with lower stress forces.

Oxygen Precipitation Kinetics Model

Only general features of oxygen precipitation kinetic models relevant to parameters of as-grown crystals will be discussed here.

Most modeling of oxygen precipitation kinetics has been done under the assumption of homogeneous nucleation [*45–47*]; oxygen precipitation is spontaneous and caused by oxygen supersaturation in silicon during high-temperature annealing. Oxygen precipitation takes place when oxygen supersaturation exceeds certain critical threshold values [*48*]. The oxygen

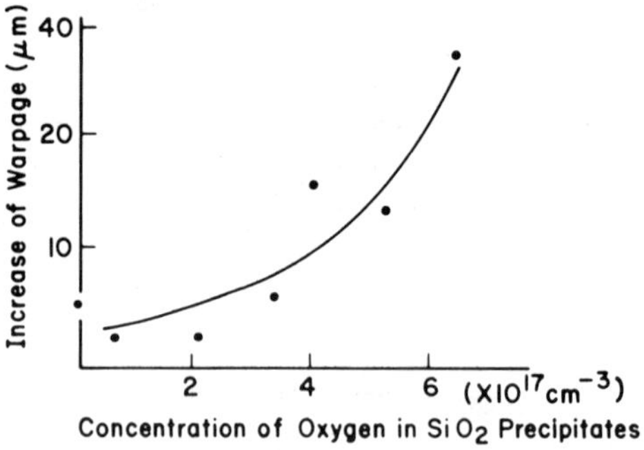

FIG. 9—*Changes of warpage as a function of concentration of oxygen forming SiO_2 precipitates. Wafers after CMOS process.*

atoms will condensate spontaneously, and only oxygen clusters exceeding a certain critical size will become stable and grow further. Supersaturation of oxygen, which is a function of the interstitial oxygen concentration and temperature (oxygen solubility), is the driving force for the precipitation processes, whereas oxygen diffusion is a step which controls the precipitation rate [*46*]. This model explains well the time and temperature dependence of oxygen precipitation kinetics [*47*]. To explain the observed effect of carbon for concentrations above 5×10^{16} cm^{-3} [*20,21*] or shallow donors and acceptors[8] on oxygen precipitation kinetics, it has been proposed that these impurity clusters influence the precipitation process by providing heterogeneous sides for nucleation of oxygen precipitates (heterogeneous model). To explain the dependence of oxygen precipitation kinetics on annealing sequence, it has been pointed out that active centers for oxygen nucleation can also be provided by clusters of point defects, small vacancy complexes [*49*] introduced to silicon during the growth process.

It seems as if the variations in the oxygen precipitates density in material with similar oxygen concentration (Figs. 5 to 7) can be explained only by inhomogeneous distribution of active centers for nucleation in silicon. Phenomena taking place during crystal growth control formation and distribution of active centers for nucleation, as well as oxygen concentration in silicon crystals.

Relationship to Crystal Growth Conditions

The relationship between crystal growth conditions and concentration and initial state of oxygen in as-grown silicon wafers is discussed in this section. In order to understand this relationship, the issues of (*1*) incorporation of oxygen into silicon crystal, (*2*) formation of active centers for oxygen precipitation, and (*3*) evolution of oxygen states in silicon crystals during cooling after the growth process must be addressed.

Incorporation of Oxygen into Silicon

The concentration of oxygen in silicon crystals is controlled by the concentration of oxygen in the silicon melt and the interfacial segregation coefficient of oxygen. The driving force that controls the oxygen concentration is the dissolution of the silica crucible by the silicon melt and the oxygen concentration in the melt [*50,51*] depends on the amount of oxygen dissolved from the crucible wall and evaporated through the free surface of the melt. The amount of evaporated oxygen is a function of the gas pressure (P) over the melt and area of the free surface (S). The amount of dissolved oxygen is a function of the wall area in contact with the melt (A), the wall temperature (T), and the thickness of solute boundary layer in the melt at the crucible

[8] D. Huber, Wacker-Chemitronic, Burghausen, West Germany, private communication.

wall (δ), which depends on the magnitude of convective flow in the melt. Table 2 shows the relationship between these parameters and oxygen concentration in the melt. By adjusting these parameters, different oxygen concentrations in the melt can be obtained. During crystal growth the amount of the melt decreases (A↘); therefore, in order to maintain constant oxygen concentration in the melt in front of the growing interface, double-crucible techniques [*52*] should be used or the dissolution rate should be increased by increasing the convective flow (δ↘); the latter is achieved by adjusting the crucible and crystal rotation rates [*51*]. Control of the oxygen concentration in the melt has received much attention during the last two years [*50–52*] and, as a result, silicon crystals with controlled oxygen concentration are presently commercially available.[9]

Incorporation of oxygen from the melt into the silicon crystal cannot be simply described by BSP theory [*53*], as has been recently pointed out [*54*]. It is likely that silicon-oxygen clusters are formed in the melt in front of a growing interface and that these clusters are incorporated into growing crystals [*55*]. Therefore the effect of growth kinetics on the interfacial segregation coefficient of oxygen is through the effect which growth velocity or temperature fluctuations have on the distribution of these clusters [*54*].

Formation of Active Centers for Nucleation—Interfacial Phenomena

The presence of temperature fluctuations during crystal growth increases the amount of active centers for oxygen nucleation during subsequent annealing [*54*]. It seems that formation of active centers for nucleation is controlled by two phenomena taking place on the growing interface: formation of heterogeneous nucleation centers and point defects. Even in lightly doped crystals, during recrystallization of droplets trapped by the growth interface [*56*], due to segregation phenomena, impurity clusters can form which can act as heterogeneous nucleations centers for oxygen [*57*]. In addition, impurity clusters may also act as heterogeneous condensation centers for point de-

TABLE 2—*Relationship between oxygen concentration in the melt and growth parameters.*

Oxygen in Melt	Gas Pressure (P)	Area of Free Surface (S)	Wall Area in Contact with Melt (A)	Wall Temperature (T)	Thickness of Solute Boundary Layer in the Melt at Crucible Wall (δ)
↗	↙	↙	↗	↗	↙

[9] Silicon wafers with control oxygen concentration have been available since 1980 from different commercial silicon wafer suppliers.

fects; point defect clusters "active" centers for oxygen precipitation [*49*] will grow around them during crystal cooling (see the section of Defect Generation).

Concentration of point defects on the growing interface is a strong function of crystal growth conditions, as was shown in the case of crystals containing swirls [*58*]. Different models have been proposed to explain the effect of growth kinetics on point defect formation at the growth interface: droplet recrystallization [*56*], interfacial reconstruction [*59*], nonequilibrium concentration of point defects during remelting [*60*], and trapping caused by growth velocity acceleration [*61*]. Although interfacial phenomena during crystal growth are still poorly understood, one can conclude that growth velocity and temperature fluctuations during crystal growth affect formation at the growing interface of the centers controlling oxygen precipitation kinetics. Changes of growth kinetics during the crystal growth process are responsible for changes in the concentration of these centers in silicon crystals.

Effect of Cooling

After solidification, a silicon crystal is cooled from the solidification temperature of 1420°C to room temperature. The solubility of point defects and of oxygen in silicon decreases with decreasing temperature; therefore when their concentration during cooling exceeds the solubility limit they will start to precipitate. The precipitating point defect clusters provide the active centers for oxygen precipitation.

In the presence of supersaturation, the condensation process of point defects can be homogeneous (spontaneous) or heterogeneous (on localized centers). In the first case, the initial stage is random and controlled by the value of supersaturation; in the second case, it is controlled by the distribution of condensation centers (for example, carbon atoms) and by diffusion of defects to these centers [*62*]. It seems that heterogeneous and homogeneous condensation can happen in parallel. Point defects close enough to diffuse to heterogeneous centers will condense there, whereas others will condense homogeneously. Cooling rates of the crystal may significantly affect the condensation process. Slower cooling rates allow more time for defects diffusion to heterogeneous condensation centers, which favors the heterogeneous condensation mechanism. An increase in cooling rate increases local supersaturation, which favors the homogeneous mechanism, spontaneous condensation. As the result of condensation, point defect clusters acting as active sites for oxygen precipitation [*49*] will grow. During cooling, point defects from the peripheral region of silicon boule diffuse to the surface of the crystal, where they will annihilate [*63*]. The outdiffusion will introduce point defect gradients across the boule, which can significantly affect the condensation process across the crystal. The outdiffusion of point defects in small diameter silicon crystals grown with slow pulling rates can completely suppress the condensation process and prevent swirls formation [*63*]. The difference in concentration of the point defect clusters (acting as the active centers for oxygen precipitation) in the center and at the edge of the crystal caused by outdiffusion may account in some instances for observed differences in oxygen precipitation kinetics across the wafer during device processing (Fig. 5). Inhomogeneous distribution of defect clusters on the microscale can be caused by (*1*) inhomogeneous distribution of heterogeneous condensation centers such as carbon [*62*], whose segregation coefficient is controlled by growth kinetics; and (*2*) variations in the concentration of point defects on the growing interface caused by the previously discussed instability present during crystal growth. This may explain the observed variations in precipitates density in Fig. 6.

Oxygen precipitation kinetics during processing depends critically on the time which a crystal spends between 850 and 650°C during the cooling cycle [*64*]. By adjusting this time, high or low density of oxygen precipitates during subsequent heat treatments can be obtained. This observation clearly demonstrates that this temperature range is crucial for the initial stage of oxygen precipitation kinetics and probably controls to a large extent the concentration of noninterstitial (clustered) oxygen in as-grown crystals.[6] The seed and tail part of the crystal usually have different thermal histories. The seed end is cooled during the pulling process prior to growth termination, because the temperature in the top part of the crystal pullers is lower than in the bottom. Therefore one can expect that wafers cut from different parts of the crystal will show large differences in the amount of point defect clusters and the amount of oxygen in noninterstitial states, even if the total oxygen concentration is well controlled in these wafers. Different pullers have various thermal configurations and various cooling schedules can be used after termination of growth; this can result in large variations in concentration and state of the initially precipitated oxygen and the point defect clusters. As a result, large differences in the oxygen precipitation kinetics during circuit processing (Fig. 7) can be expected.

Summary and Conclusions

The most important problem with silicon used for VLSI circuits is control of the oxygen precipitation kinetics during processing. Oxygen present in silicon affects formation of oxygen precipitates, stacking faults, dislocations, and thermal donors. It has been pointed out that during the defect generation process, oxygen can play a positive (dislocation pinning and internal gettering) or negative role (thermal donors, decrease of mechanical strength, and precipitates in the device area). Even in the same state at the same time, oxygen can be an asset or liability; for example, SiO_2 precipitates introduce sites for internal gettering but also decrease the mechanical strength of silicon. It is uncertain to what extent internal gettering can be compromised with good mechanical strength.

The material problem of VLSI technologies is not how to increase or decrease oxygen concentration, but how to establish an optimum concentration range for a given process and the means of inducing oxygen into a desirable state. To accomplish this, the initial stage of oxygen precipitation kinetics has to be understood in more detail. It seems that the key issue is the effect which point defect clusters and other nucleation sites have on the initial stage of oxygen precipitation kinetics. Consequently, the mechanism of their formation during the crystal growth process should be the main subject of future investigations. Phenomena taking place at the growing interface control their formation, but unfortunately our knowledge about the relationship between growth kinetics and material properties is still very inadequate. Only after understanding this relationship in more detail can a comprehensive quantitative model of oxygen precipitation kinetics be developed. For known crystal growth conditions, the oxygen state in the wafers can then be predicted for each stage of the wafer process schedule. It is hoped that this will lead to the optimization of material parameters by inducing oxygen in the desired state during certain critical process steps; for example, inducing oxygen clusters during steps with a high probability of warpage and subsequently changing them to SiO_2 precipitates during steps requiring efficient gettering.

Whether the foregoing suggestions will fulfill all requirements for future VLSI technology is uncertain, because of possible conflicting demands that may be placed on materials, processing, and circuit design.

Acknowledgments

I would like to express my gratitude to Dr. L. Goodman, Dr. P. Zanzucchi, and Dr. A. Ipri for helpful suggestions during the preparation of this paper and to Dr. A. Kokkas for interesting discussions and critical proofreading of the manuscript.

References

[*1*] *Proceedings,* Third Microelectronics Measurement Technology Seminar, Session on Defects in Silicon Substrates, San Jose, Calif., 17–18 March 1981.

[*2*] Pearce, C. W. in *Proceedings,* Third Microelectronics Measurement Technology Seminar, San Jose, Calif., 17–18 March 1981, Benwill Publishing Corp., p. V-27.

[*3*] Kishino, S. in *Proceedings,* Third Microelectronics Measurement Technology Seminar, San Jose, Calif., 17–18 March 1981, Benwill Publishing Corporation, p. V-17.

[*4*] For review see Thebault, D. and Jastrzebski, L., *RCA Review,* Vol. 41, No. 4, 1980, p. 592.

[*5*] See, for instance, Helmveich, D. and Sirtl, E. in *Semiconductor Silicon 1977, Electrochemical Society Proceedings,* Vol. 77-2, 1977, p. 626.

[*6*] Block, E. in *Semiconductor Silicon 1981, Electrochemical Society Proceedings,* Vol. 81-5, 1981, p. 20.
[*7*] Jastrzebski, L. in *Proceedings,* Third Microelectronic Measurement Technology Seminar, San Jose, Calif., 17–18 March 1981, Benwill Publishing Corporation, p. V-1.
[*8*] Jastrzebski, L., Levine, P. A., Fisher, W. A., Cope, A. D., Savoye, E. D., and Henry, W. N., *Journal of the Electrochemical Society,* Vol. 128, 1981, p. 885.
[*9*] Jastrzebski, L. and Lagowski, J., *Journal of the Electrochemical Society,* Vol. 128, 1981, p. 1957.
[*10*] Pearce, C. W. in *Semiconductor Silicon 1981, Electrochemical Society Proceedings,* Vol. 81-5, 1981, p. 705.
[*11*] Dishman, J. M., Hasska, S. E., Marcus, R. B., Murarka, D. P., and Sheng, T. T., *Journal of Applied Physics,* Vol. 50, 1979, p. 2689.
[*12*] Murarka, S. P., Seidel, T., Dalton, J., Dishman, J., and Read, M., *Journal of the Electrochemical Society,* Vol. 127, 1980, p. 716.
[*13*] Hayaguji, Y., Shimada, T., and Kawado, S. in *Semiconductor Silicon 1977, Electrochemical Society Proceedings,* Vol. 77-2, 1977, p. 750.
[*14*] Mader, S. in *Proceedings,* Third Microelectronics Measurement Technology Seminar, San Jose, Calif., 17–18 March 1981, Benwill Publishing Corp., p. V-24.
[*15*] Suezawa, M., Sumino, K., and Yonenaga, I., *Physics Status Solidi* (*A*), Vol. 51, 1979, p. 217.
[*16*] Bohg, A. and Gaind, A. K., *Applied Physics Letters,* Vol. 33, 1978, p. 895.
[*17*] Tammura, M. and Sunami, H., *Japanese Journal of Applied Physics,* Vol. 11, 1972, p. 1097.
[*18*] Blackstone, S., Henry, W. N., Jastrzebski, L., and Fisher, W., *Journal of the Electrochemical Society,* Vol. 129, No. 3, 1982, p. 667.
[*19*] Yakano, Y., Namba, N., and Maki, M. in *Semiconductor Silicon 1977, Electrochemical Society Proceedings,* Vol. 77-2, 1977, p. 695.
[*20*] Kishino, S., Matsushita, Y., and Kanamon, M., *Applied Physics Letters,* Vol. 35, 1979, p. 213.
[*21*] Yasutake, K., Vmemo, M., and Kawabe, H., *Applied Physics Letters,* Vol. 37, 1980, p. 789.
[*22*] Shimura, F., Tsuya, H., and Kawamura, T., *Journal of the Electrochemical Society,* Vol. 128, 1981, p. 1579.
[*23*] Rava, P., Gatos, H. C., and Lagowski, J. in *Semiconductor Silicon 1981, Electrochemical Society Proceedings,* Vol. 81-5, 1981, p. 232.
[*24*] Komamori, A., *Applied Physics Letters,* Vol. 34, 1979, p. 287.
[*25*] For example, Kanamori, A. and Kanamori, M., *Journal of Applied Physics,* Vol. 50, 1979, p. 8059.
[*26*] Jastrzebski, L. and Zanzucchi, P. in *Semiconductor Silicon 1981, Electrochemical Society Proceedings,* Vol. 81-5, 1981, p. 138.
[*27*] Zanzucchi, P. and Jastrzebski, J. submitted to *Journal of Applied Physics.*
[*28*] Wada, K., Takaoka, H., Inone, N., and Kohra, K., *Japanese Journal of Applied Physics,* Vol. 18, 1979, p. 1629.
[*29*] Hu, S. M. and Patrick, W. J., *Journal of Applied Physics,* Vol. 46, 1975, p. 1969.
[*30*] Hu, S. M., *Applied Physics Letters,* Vol. 36, 1980, p. 561; Vol. 31, 1977, p. 53.
[*31*] Pearce, C. W. and Rozgornyi, G. A. in *Semiconductor Silicon 1977, Electrochemical Society Proceedings,* Vol. 77-2, 1977, p. 606.
[*32*] Nagasawa, K., Matsushita, Y., and Kishino, S., *Applied Physics Letters,* Vol. 37, 1980, p. 622.
[*33*] Yamamoto, K., Kishino, S., Matsushita, Y., and Lizuka, T., *Applied Physics Letters,* Vol. 30, 1977, p. 31.
[*34*] Kishino, S., Matsushita, Y., and Kamamon, M., *Applied Physics Letters,* Vol. 35, 1979, p. 213.
[*35*] Kishino, S., Nagaswaa, K., and Lizuka, T., *Japanese Journal of Applied Physics,* Vol. 19, 1980, p. 466.
[*36*] Ogino, M., Vsami, T., Watanabe, M., Sekine, H., and Kawaguchi, T., presented at 158th Electrochemical Society Meeting, Hollywood, Fla., Oct. 1980, Abstract 435.
[*37*] Frank, F. C. and Read, W. T., *Physical Review,* Vol. 79, 1980, p. 722.
[*38*] *Semiconductor Silicon 1981, Electrochemical Society Proceedings,* Vol. 81-5, 1981, K. Sumino, p. 208 and Y. Kondo, p. 220.
[*39*] Sumino, K., Harada, H., and Yonenage, I., *Japanese Journal of Applied Physics,* Vol. 19, 1980, p. 49.
[*40*] Hu, S. M., *Applied Physics Letters,* Vol. 31, 1977, p. 53.
[*41*] Hu, S. M. and Patrick, W. J., *Journal of Applied Physics,* Vol. 46, 1975, p. 1869.

[*42*] Takasu, S. in *Proceedings,* Third Microelectronic Measurement Technology Seminar, San Jose, Calif., 17–18 March 1981, Benwill Publishing Corp., p. V-53.
[*43*] Jastrzebski, L., Zanzucchi, P., Thebault, D., and Lagowski, J., *Journal of the Electrochemical Society,* Vol. 129, No. 7, 1982, p. 638.
[*44*] Hu, S. M., *Journal of Applied Physics,* Vol. 51, 1980, p. 5945.
[*45*] Craven, R. A. in *Semiconductor Silicon 1981, Electrochemical Society Proceedings,* Vol. 81-5, 1981, p. 254.
[*46*] Osaka, J., Inone, N., and Wada, K., *Applied Physics Letters,* Vol. 36, 1980, p. 288.
[*47*] Patel, J. in *Semiconductor Silicon 1981, Electrochemical Society Proceedings,* Vol. 81-5, 1981, p. 189.
[*48*] Wada, K., Inone, N., and Kohra, K., *Journal of Crystal Growth,* Vol. 49, 1980, p. 749.
[*49*] Hu, S. M., *Journal of Applied Physics,* Vol. 52, 1981, p. 3974.
[*50*] Benson, K. E., Lin, W., and Martin, E. P. in *Semiconductor Silicon 1981, Electrochemical Society Proceedings,* Vol. 81-5, 1981, p. 33.
[*51*] Murgai, A. in *Semiconductor Silicon 1981, Electrochemical Society Proceedings,* Vol. 81-5, 1981, p. 113.
[*52*] Lin, W. and Pearce, C. W., *Journal of Applied Physics,* Vol. 51, 1980, p. 5540.
[*53*] Burton, J. A., Prim, R. C., and Slichter, W. P., *Journal of Chemical Physics,* Vol. 21, 1953, p. 154.
[*54*] Abe, T., Kikuchi, K., Shirai, S., and Muraoka, S. in *Semiconductor Silicon 1981, Electrochemical Society Proceedings,* Vol. 81-5, 1981, p. 54.
[*55*] Chikawa, J. and Sata, F. in *Proceedings,* Materials Research Meeting, Boston, 1980.
[*56*] Chikawa, J. and Shirai, S., *Japanese Journal of Applied Physics,* Vol. 18, Supplement 1, 1979, p. 153.
[*57*] Shirai, S., *Applied Physics Letters,* Vol. 36, 1980, p. 15.
[*58*] de Kock, A. J. R. and Van de Wijgert, W. M., *Journal of Crystal Growth,* Vol. 49, 1980, p. 718.
[*59*] Van Vechten, J. A., *Applied Physics Letters,* Vol. 26, 1975, p. 593.
[*60*] de Kock, A. J. R. in *Semiconductor Silicon 1977, Electrochemical Society Proceedings,* Vol. 77-2, 1977, p. 508.
[*61*] Jastrzebski, L. Lagowski, J., and Gatos, H. C., *Journal of the Electrochemical Society,* Vol. 128, 1981, p. 697.
[*62*] Foll, M., Gosele, V., and Kolberisen, B. P., *Journal of Crystal Growth,* Vol. 40, 1977, p. 90.
[*63*] Kim, K. M., *Journal of Applied Physics,* Vol. 50, 1979, p. 1135.
[*64*] Nakanishi, H., Khoda, H., Hirata, H., and Hoshikawa, K., *Japanese Journal of Applied Physics,* Vol. 19, 1980, p. 561.

John H. Matlock[1]

Material Defect Factors Affecting MOS Device Performance

REFERENCE: Matlock, J. H., "**Material Defect Factors Affecting MOS Device Performance,**" *Silicon Processing, ASTM STP 804,* D. C. Gupta, Ed., American Society for Testing and Materials, 1983, pp. 332–361.

ABSTRACT: Substrate-related defects that affect MOS yield center around either bulk impurities and precipitates or surface contamination. Uncontrolled bulk defects can cause significant refresh or leakage problems, but it has been shown that with proper processing the defects can form the basis for internal gettering that improves production performance. Surface-related yield problems reflect a joint responsibility of the wafer producer and the user, since residues, particularly organics, may also cause device performance problems. Control of these problems requires a careful analysis of all chemicals used in processing and cleaning the wafer prior to oxidations.

KEY WORDS: carbon, contamination, crystal, denuding, dynamic RAMs, gettering, internal gettering, laser gettering, leakage, MOS, nucleation, oxide, oxygen, oxygen donors, oxygen gettering, precipitation, refresh time, silicon, stacking faults, surface

A great deal of attention is being focused on the effect of the properties of the starting material on MOS and bipolar device performance. Any specific data referred to here will be in relation to NMOS processing on P (1-0-0) substrates, but the general character of the defects and the possible effects on device performance of these defects can be considered valid for a fairly wide range of processing.

It is not intended to provide all the answers for materials specification for optimum device performance. The goal here is to identify some areas for investigation and provide some guidelines for such work. As has always been the case through the years, when understanding increases on certain aspects of material defects more new questions are generated.

The primary focus will be on bulk and surface defect factors due primarily to impurities and subsequent enhancement by thermal processing. First, however, a brief overview of all important parameters for silicon wafer specification will be presented.

[1] Manager, Materials Technology Group, UTC/Mostek Corporation, Carrollton, Tex. 75006.

Silicon Specification Categories

The parameters of silicon wafers important for device preparation can be separated into those properties associated with the shaping of a crystal into thin wafers and those associated with growing the crystal itself. Some specification areas are well established, while others are just being developed.

The basic diameter, of course, is chosen to be consistent with all processing equipment and productivity needs. The most common diameter chosen today in integrated circuit processing is 100 mm, but some use of 125 mm wafers is beginning. The thickness of the wafer is chosen to be able to sufficiently stand up to mechanical and thermal handling. Parameters such as flatness, warp, taper (thickness variation), diameter tolerance, and flat dimensions are set to comply with the photolithography equipment utilized for a given device geometry size. While much work continues on the improvement of flatness and other mechanical parameters, their relationship to needs is fairly straightforward and well understood [*1*].[2] Polishing imperfections and wafer cleanliness can affect the integrity of the oxides and resultant device properties.

With respect to the bulk crystal the first specification is the establishment of electrical properties by the addition of a dopant. Boron is generally added to create a P-type conductivity, while phosphorous (or sometimes antimony or arsenic) is added to get N-type material. Specifications include type, resistivity level, and doping uniformity on a wafer or wafer to wafer in a lot. Microvariations in resistivity, called "striations", are also sometimes of concern. Structural parameters addressed are dislocations (although crystals are generally grown dislocation free) and defect clusters referred to as "swirl". This swirl is a visual pattern of defects seen on etching that relates to impurity agglomeration and precipitation [*2*]. Also, a bulk lifetime measurement (by photodecay) is used as a general measure of the purity and structural integrity of the crystal [*3*].

The crystal process type can be chosen from the float-zone or the Czochralski methods. The Czochralski silicon is the preferred choice for most state-of-the-art integrated circuit applications. This silicon has about 100 times as much oxygen as the float-zone crystals due to the fact that the crystal is drawn from molten silicon in a dissolving quartz crucible [*4*]. Specification of the oxygen level and precipitation rate is just now coming in vogue.

This paper will concentrate on two areas where it is believed specification development is needed: (*1*) the choice of oxygen levels as it is integrated into a total gettering system, and (*2*) the condition and cleanliness of the wafer surface.

[2]The italic numbers in brackets refer to the list of references appended to this paper.

Oxygen Internal Gettering

The interest in the specification of oxygen levels in silicon has increased tremendously over the last year. As will be detailed later, much has been established as to how to make the oxygen work for the device manufacturer, rather than against him, as a method of internal (or intrinsic) gettering. Designed use of the oxygen for this purpose is still limited, however, owing to incomplete knowledge of the sensitivities of the exact oxygen level in relation to the complete thermal process involved.

Desired Conditions

When no specification is placed on oxygen in silicon the user can expect to receive, over a period of time, material that averages about 14 ppm and can be anywhere between 8 and 22 ppm. All values used here for oxygen content in silicon are consistent with ASTM F 121-80 and DIN 50 438/1. They are essentially one half (½) of the values that would be determined from ASTM F 121-79, which are the most common numbers previously quoted in the literature. A typical distribution for oxygen content that a user could expect from major silicon vendors is shown in Fig. 1. The point to remember is this: If a device process is at all sensitive to silicon defects, then oxygen specification is essential; it cannot be ignored. At about 12 to 15 ppm, depending on the particular thermal cycle, the oxygen will begin to precipitate and form clusters and stacking faults. As will be shown, this can be helpful or harmful depending on the thermal cycling involved. This threshold of precipitation (O_i*) is right in the middle of this typical distribution. It is easy to see that one would need to be either all above or all below this level to optimize performance in

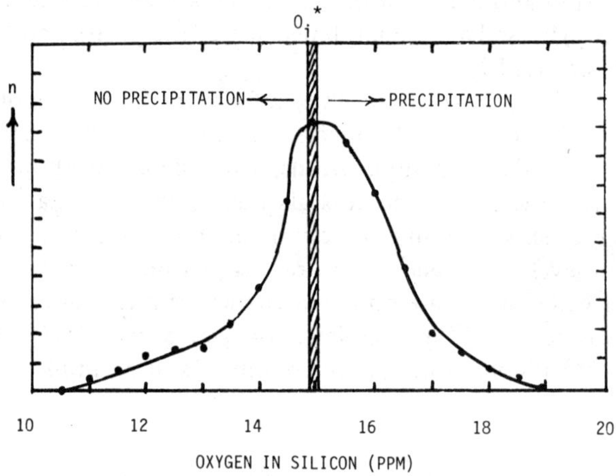

FIG. 1—*Typical distribution for oxygen in Czochralski silicon.* n *is a relative frequency on a linear scale.*

a particular process. Silicon manufacturers have progressed significantly in their ability to control the oxygen content to the reasonably tight ranges which makes such oxygen control feasible [5–8].

The advantages of gettering for device performance have been well established [9,10]. When that gettering is obtained from oxygen, two conditions are essential:

1. A surface region stripped of oxygen to below the precipitation level to eliminate stacking faults at the surface.
2. Sufficient precipitation in the interior of the wafer to provide sites to getter heavy metals and other undesired impurities.

Such a denuded zone is shown diagramatically in Fig. 2. Since the first demonstration and explanation of internal gettering [11,12], a number of reports have been made experimentally establishing conditions for good precipitation and denuded zones [8,13–19]. Generally, these studies were for a particular level of oxygen and did not emphasize the problems of implementing the utilization of a denuded-zone concept in production.

It is not intended here to present phenomenological kinetics of the process, since this has been covered previously [20,21]. Rather, the emphasis will be on conditions necessary for implementation.

To create the right conditions for the wafer a high-temperature step (>1000°C) must be used to deplete the surface of oxygen. The exact temperature depends on the history of the silicon being treated [8]. Higher temperatures may be necessary if precipitation has already occurred in order to redissolve the clusters. Denuding has been shown to occur in both oxygen and inert gas atmospheres.

Also required is some lower temperature processing (600 to 800°C) for the nucleation of oxygen precipitates. This has been shown to maximize at about

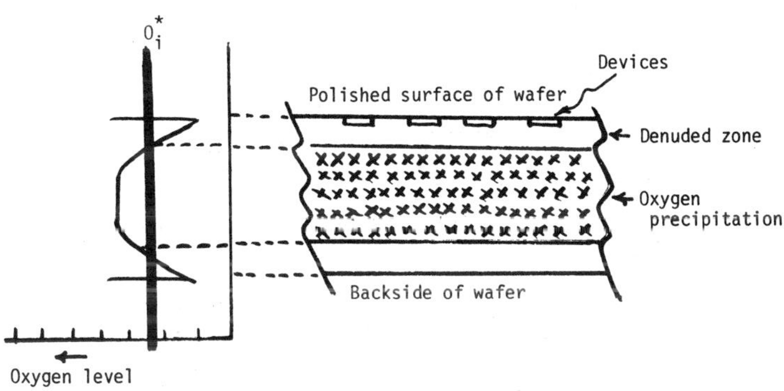

FIG. 2—*Desired denuded-zone conditions.*

700 to 750°C [*21*]. Whether the nucleation or the denuding cycle should come first is confusing; both have been reported. It will be shown here that both have application under certain conditions.

It should be noted that while it is possible to provide the necessary denuding-precipitation cycles previous to actual device processing, sometimes a certain amount of internal gettering occurs without the device manufacturer planning it. Many MOS processes begin with an initial oxidation in the 1000 to 1100°C range. This can serve to do a reasonable amount of denuding. If this is followed by a lower temperature process, such as for polycrystalline silicon deposition or nitride deposition, sufficient nucleation of precipitates should occur, at least for certain oxygen levels. A choice is then available to the user: either add a denuding precycle or tailor the oxygen levels of the silicon to fit the device process.

Denuded-Zone Experiments

In order to better understand the sensitivities involved in optimizing the thermal cycling for device processing, a series of different cycles was tested for denuding the wafer surface and precipitating the internal oxygen. These cycles are shown in Fig. 3 and Table 1. All tests were carried out on P-type (1-0-0) 100-mm wafers. Material was evaluated from eight different silicon vendors at initial oxygen levels between 9 and 22 ppm. All oxygen measure-

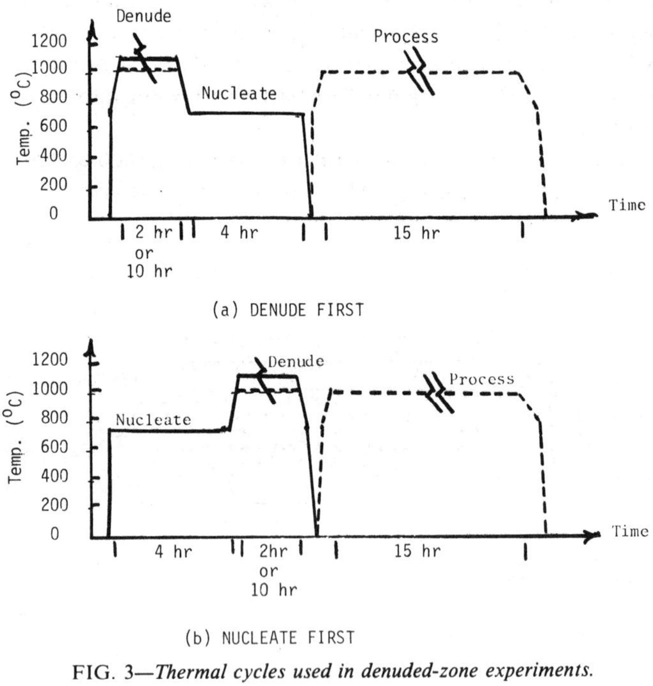

FIG. 3—*Thermal cycles used in denuded-zone experiments.*

TABLE 1—*Thermal cycles used in denuded-zone experiments.*

	Thermal Cycles (1)	(2)	(3)	Ambients	Diagram
(1)	1100°C 2 h	725°C 4 h	1000°C 15 h	O_2, N_2, Ar	(a)
(2)	725°C 4 h	1100°C 2 h	1000°C 15 h	O_2	(b)
(3)	1000°C 2 h	725°C 4 h	1000°C 15 h	O_2	(a)
(4)	725°C 4 h	1000°C 2 h	1000°C 15 h	O_2	(b)
(5)	1100°C 10 h	725°C 4 h	1000°C 15 h	O_2, Ar	(a)
(6)	725°C 4 h	1100°C 10 h	1000°C 15 h	O_2	(b)

ment was done on wafers on a Fourier Transform Infrared Spectrometer, which measures only the dissolved interstitial oxygen.

This group of experimental runs allows examination of the following parameters:

1. Oxygen level.
2. Vendor variations at a given oxygen level.
3. Denuding temperature (1100 versus 1000°C).
4. Denuding time (2 versus 10 h).
5. Denuding atmosphere (O_2, N_2, Ar).
6. The sequence of processing of the denuding cycle and the nucleation cycle.

In all cases, 725°C for 4 h was used for the nucleation cycle. The term "precycle" will be used to refer to the combination of denuding and nucleation cycles previous to device processing. All precycles were followed with a very simplified "process simulation cycle" of 15 h at 1000°C to allow precipitation to grow.

As has been noted, both denuding the surface below the oxygen precipitation threshold and adequately nucleating sufficient precipitation in the interior of the wafer are necessary for device yield improvement from internal gettering. The denuding time most used in these experiments was 2 h.

Figure 4 shows the denuded zones as a function of oxygen level created with all combinations in dry O_2 by using a 2-h denuding cycle. The charts show measurements made after the combination denude/nucleation cycle and the 15-h 1000°C "process simulation cycle". In all cases there was a lower limit where adequate precipitation did not occur to measure the zone

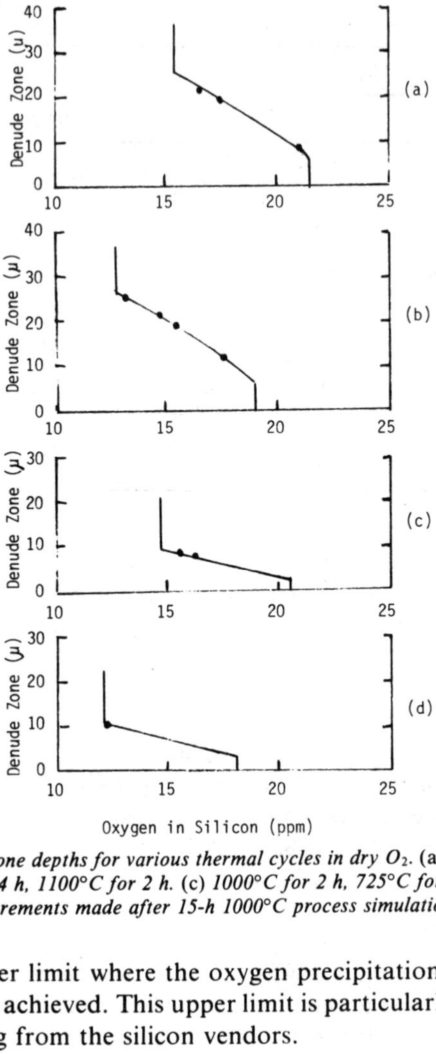

FIG. 4—*Denuded-zone depths for various thermal cycles in dry* O_2. (a) *1100°C for 2 h, 725°C for 4 h.* (b) *725°C for 4 h, 1100°C for 2 h.* (c) *1000°C for 2 h, 725°C for 4 h.* (d) *725°C for 4 h, 1000°C for 2 h. Measurements made after 15-h 1000°C process simulation cycle.*

width and an upper limit where the oxygen precipitation was such that no denuded zone was achieved. This upper limit is particularly sensitive to variables in processing from the silicon vendors.

Figure 5 shows the oxygen precipitation as measured by the change in dissolved oxygen content after thermal processing. Each chart shows the change after the precycles only and after the subsequent 15-h 1000°C "process simulation cycle". Again, only the various combinations for 2-h denuding cycles in dry O_2 are shown. From these charts it can be seen that there is a substantial difference in the precipitation that occurs, depending on the order of the nucleation and precipitation cycles. However, the precipitation that occurs after the process simulation cycle is essentially the same for both precycle combinations doing the denuding first, irrespective of whether it is at 1000 or 1100°C. Likewise, the precipitation was similar for both cycles using the 725°C nucleation cycles first, again independent of denuding temperature.

Figure 6 shows samples from three silicon vendors at various oxygen levels for the two cases of denuding at 1100°C for 2 h before and after the 725°C 4-h nucleation cycle. Each pair of pictures shown is at the same oxygen level from the same crystal lot. The samples were put through the noted thermal cycle, cleaved, and etched for 5 min in Wright Etch [*22*]. The etch removes some of the original surface layer, so the location of the original wafer surface is also defined on the pictures. The effects of the alternate cycles on the extreme oxygen levels were clearly noted. At 13 ppm the 725 to 1100°C cycle showed reasonable precipitation, while the 1100 to 725°C cycle showed nothing. At 20 ppm the 1100 to 725°C cycle showed some denuding, while the 725 to 1100°C cycle showed heavy defect densities all the way to the surface. This is an indication that the right choice of precycle processing for optimum internal gettering will depend on the oxygen range being used.

The necessary depth for a denuded zone is something that must be established for a particular device and a particular thermal process. If the active device extends further into the substrate, the denuded zone must be clearly below that. Figure 7 shows the change in measured denuded-zone depth for 1100°C cycles from 2 to 10 h for silicon at 17 ppm. The depth can be based on the device design.

Another method of generating different depths of denuded zones is to use a different annealing atmosphere. Figure 8 shows denuded zones obtained for 20 ppm oxygen in silicon with a denude cycle of 2 h. No detectable difference was noted in denuded zone depths for 14 to 17 ppm oxygen in silicon, while at 20 ppm the argon and nitrogen both seemed to allow deeper zones. This may be due to the atmosphere being more conducive to the retrogrowth of stacking faults in argon or nitrogen [*23*].

Silicon Vendor Variations

Differences between the methods silicon vendors use to make the crystals can complicate the implementation of oxygen specifications. Figure 9 shows the amount of precipitation obtained from several silicon vendors after subjecting them to a 15-h 1000°C cycle in dry O_2 (*no* precycling was employed). All samples compared were for oxygen contents of 15 ± ½ ppm. This wide variation can be due to differences in the donor annihilation heat treatment which is done by all silicon manufacturers. Some oxygen converts to an N-type electrically active state at 450°C. These donors are made inactive at temperatures above 600°C. Therefore the silicon vendors use a process which involves taking the silicon to a temperature above 600°C for a short period of time (on the order of 1 h) and then cooling the silicon as rapidly as possible through the 500 to 300°C range to keep the oxygen inactive. Since this

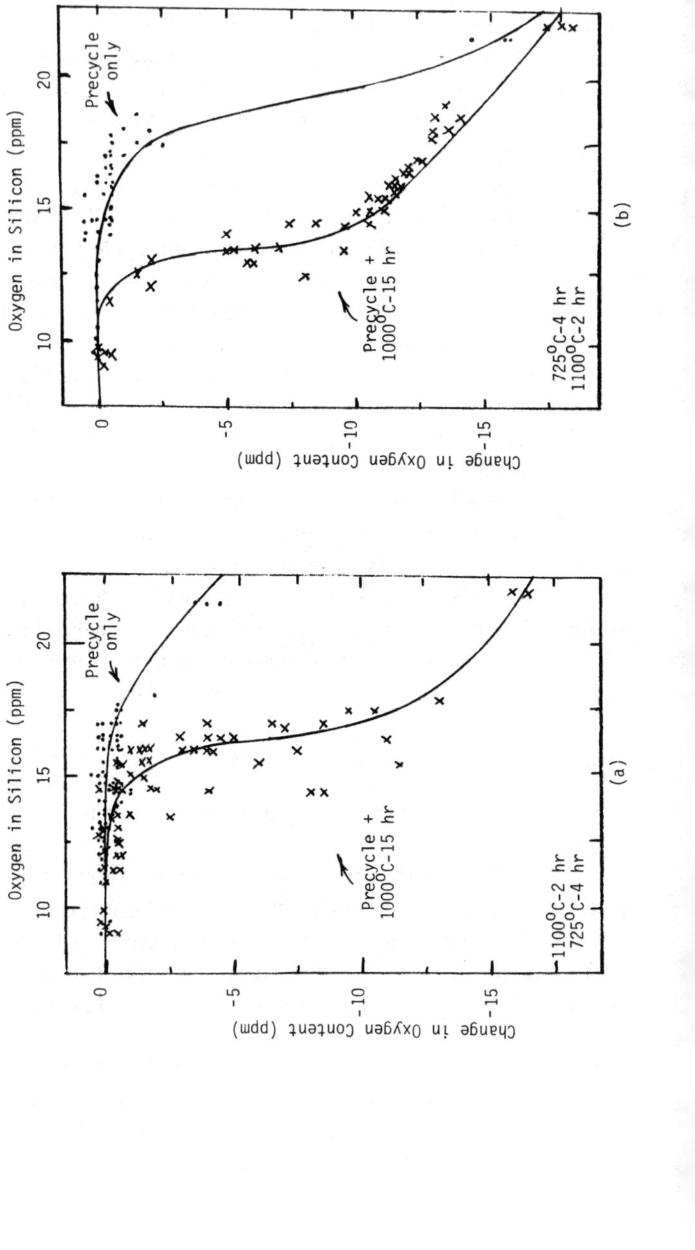
Oxygen in Silicon (ppm)
Change in Oxygen Content (ppm)
Precycle only
Precycle + 1000°C-15 hr
725°C-4 hr
1100°C-2 hr
(b)
Oxygen in Silicon (ppm)
Change in Oxygen Content (ppm)
Precycle only
Precycle + 1000°C-15 hr
1100°C-2 hr
725°C-4 hr
(a)

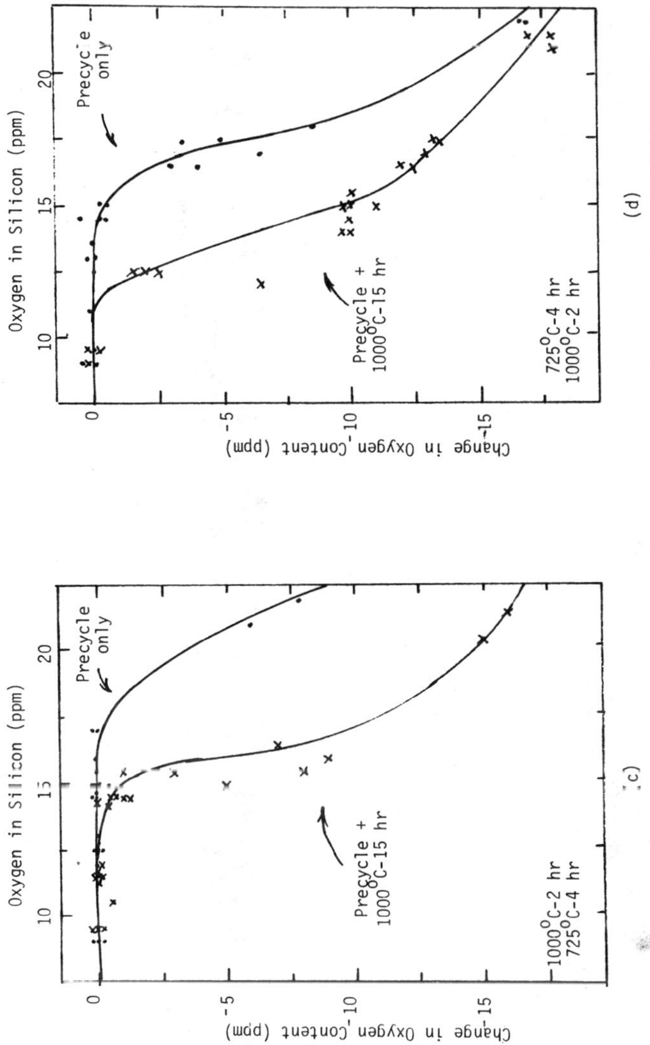

FIG. 5—*Precipitation as a function of oxygen level in dry* O_2. (a) *1100°C for 2 h, 725°C for 4 h.* (b) *725°C for 4 h, 1100°C for 2 h.* (c) *1000°C for 2 h, 725°C for 4 h.* (d) *725°C for 4 h, 1000°C for 2 h.*

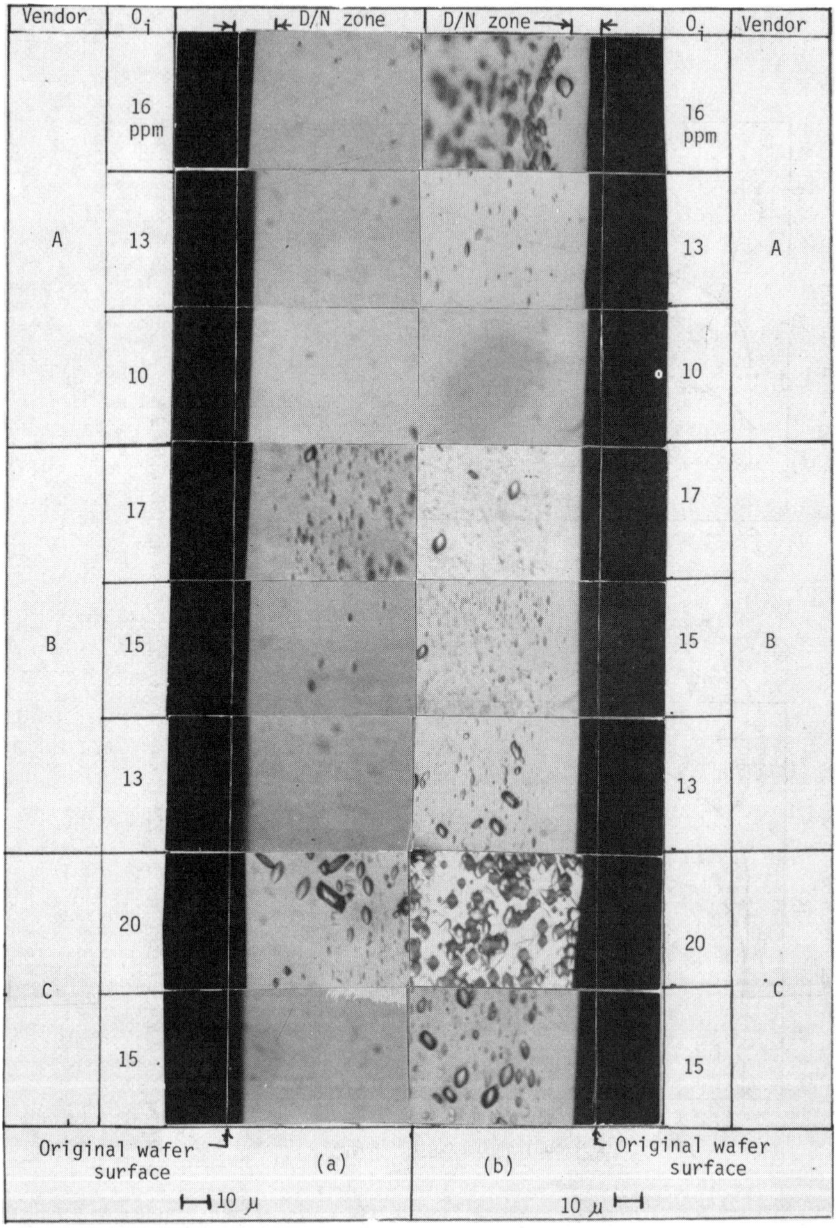

FIG. 6—*Examples of precipitation and denuded zones obtained for thermal cycles in dry* O_2. (a) *1100°C for 2 h, 725°C for 4 h, 1000°C for 15 h.* (b) *725°C for 4 h, 1100°C for 2 h, 1000°C for 15 h.*

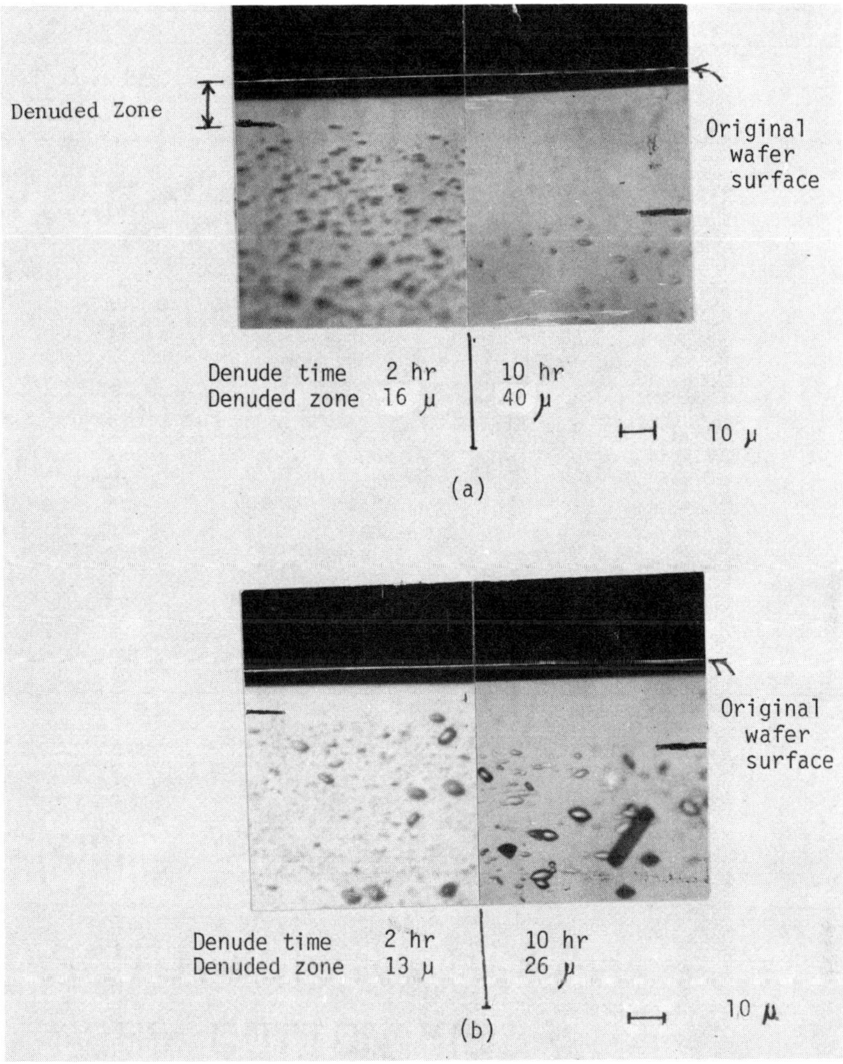

FIG. 7—*Denuded-zone depth as a function of anneal time in dry O_2 (O_i = 17 ppm).* (a) *Denude first.* (b) *Nucleate first.*

cycle is in the same range that nucleates precipitates, significant differences in precipitation rate may result due to different times or temperatures used. The second source of difference in heat treatment comes from the crystal pullers. There are between five and ten significantly different crystal pullers used today by major silicon vendors. These can impart very different *in situ* heat treatment on the crystal immediately after it is grown, owing to the relative proximity to hot-zone parts in the puller.

Another source of difference between vendors can arise from the content

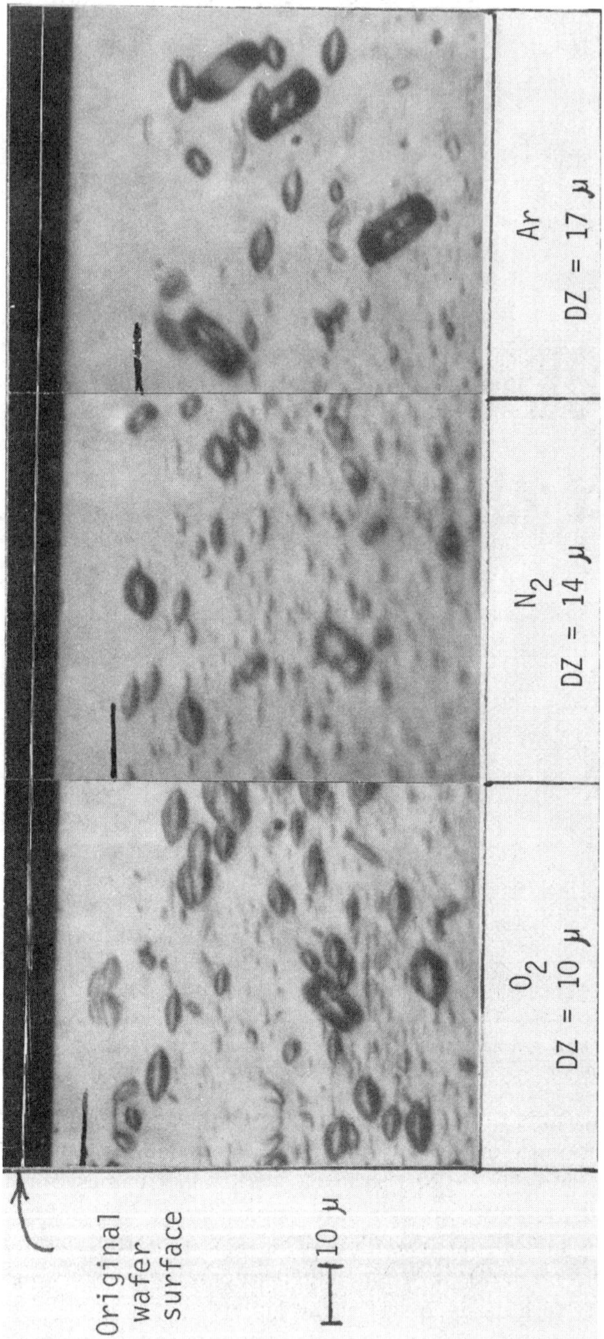

FIG. 8—Denuded zones as a function of anneal atmosphere (2 h) for 20 ppm oxygen in silicon.

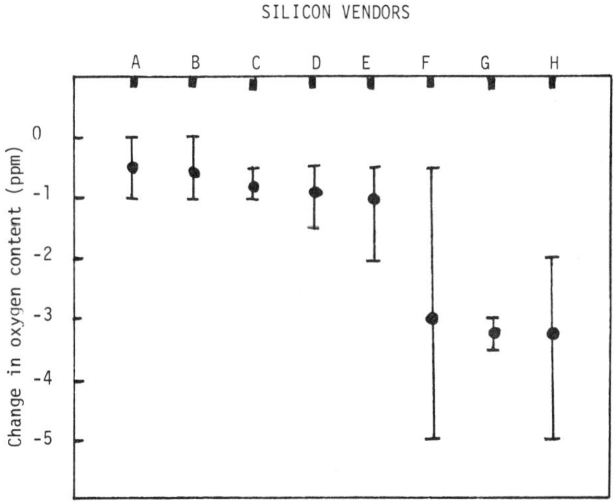

FIG. 9—*Variation in precipitation between silicon vendors for 1000°C, 15 h in dry* O_2 *(oxygen content = 15.0 ± 0.5 ppm).*

of other impurities in the silicon, particularly carbon. It has been shown that the carbon can serve as sites for heterogeneous nucleation of oxygen, thus increasing the effective precipitation rate [*8,24,25*]. Figure 10 shows three samples from one vendor crystal lot. All were treated identically by the silicon manufacturer, and processed together in further thermal cycles. All three measured 14.5 ± 0.2 ppm oxygen. These wafers were processed through cycles of 1100°C for 2 h, 725°C for 4-h, and 1000°C for 15-h. As can be seen, wafers (a) and (c) show significant precipitation while wafer (b) shows very little. The only difference between these wafers noted was that wafers (a) and (c) had about 1 ppm carbon while wafer (b) read below the detection limit of 0.05 ppm. The effect of carbon only seems to show up when the oxygen level is just on the marginal threshold of precipitation for a given silicon and thermal cycle. When higher oxygen material is used, or more nucleation thermal treatment to create conditions that are more ideal, the effect of different carbon levels (at least in the low ppm range) seems to be negligible.

Another source of variation is the position in the crystal from which the wafers are cut. Recent reports have shown nonuniform precipitation that occurs towards the seed end of the crystal [*26*]. This is particularly true in the first 2 to 4 cm of crystal, where the growth rates are undergoing some violent changes before the freezing interface stabilizes in a slightly concave configuration. Figure 11 shows this diagramatically. This extreme seed end material has been shown in certain cases to be deleterious to device performance [*27*]. On the other hand, the seed end of the crystal is the purest, with respect to other impurities. Most impurities, particularly heavy metals and carbon, have segregation coefficients much less than one, so they tend to build up towards the tang end of the crystal. As will be mentioned later, the portion of

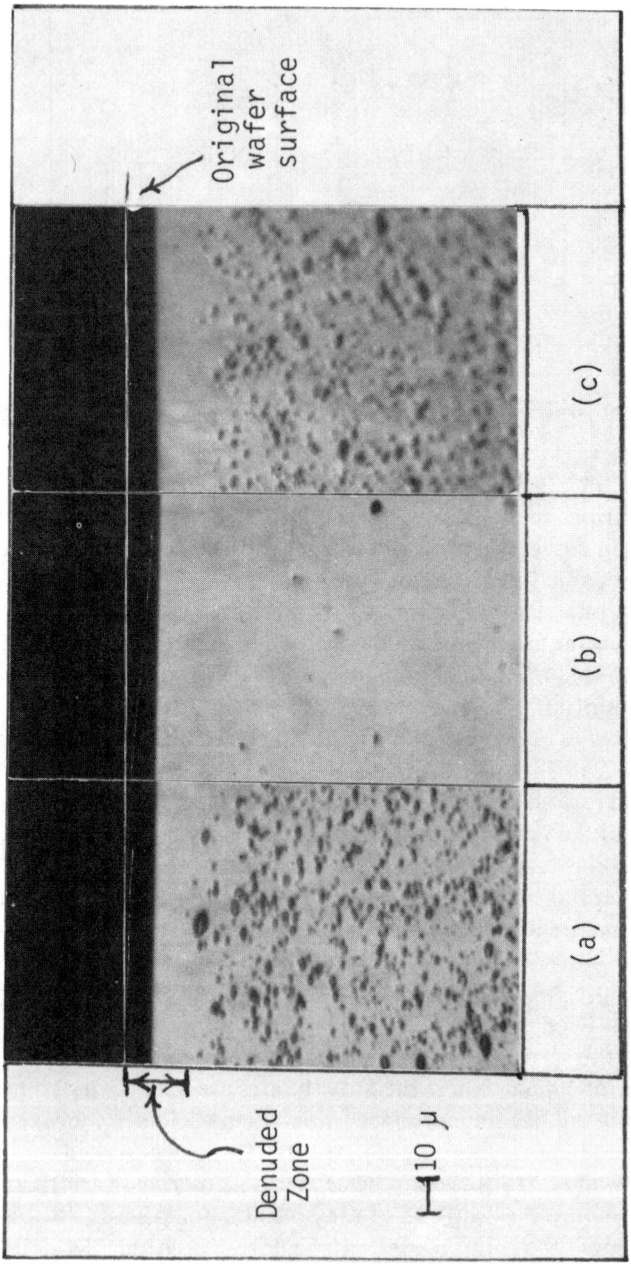

FIG. 10—*Effect of carbon on oxygen precipitation for thermal cycles of* (1) *1100°C for 2 h,* (2) *725°C for 4 h,* (3) *1000°C for 15 h in dry* O_2. (a) *Carbon content* = *1 ppm.* (b) *Carbon content* = *<0.05 ppm.* (c) *Carbon content* = *1 ppm.*

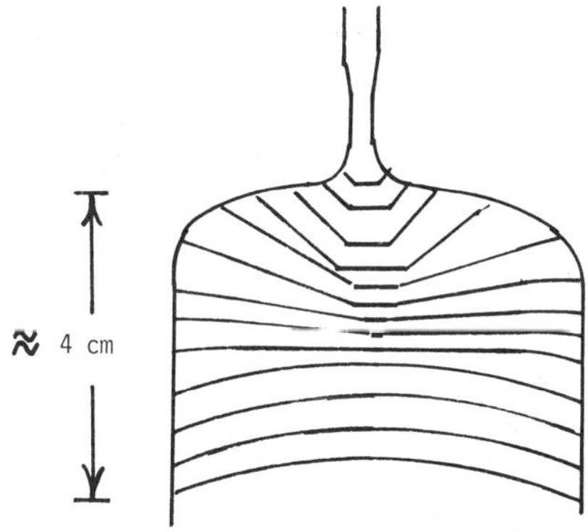

FIG. 11—*Typical crystal growth interfaces at the beginning of the crystal growth process.*

crystal utilized can have an effect on performance in a particular process. All the variables involved are not yet well understood.

Device Results

It is important to consider the actual effects of the proposed internal gettering on device performance. Data are available that relate device problems caused by both insufficiently gettered processes [*28*] and stacking faults precipitated but intersecting the device [*29*]. MOS dynamic RAM memories provide one of the most sensitive mediums for studying these effects.

Only limited amounts of actual device results have been published related to internal gettering. This scarcity of data is due to inadequate methods of sorting out substrate effects from other process effects in analyzing the performance, and the general proprietary nature of the device yields. Reports have been made public that show improvement of leakage/refresh yield losses in dynamic RAMS due to internal gettering; this should translate directly into overall improvement of device yields. Otsuka et al [*30*] have shown the clear advantage of internally gettered silicon over standard processed wafers for the holding time of dynamic RAM cells. Figure 12 shows the holding time versus drain voltage that they obtained. Their work used silicon at the 18 ppm oxygen level. It is not clear from their published results whether the control wafers referred to as "bulk" wafers were of the same oxygen level or of a lower level. In the former case, they would likely be precipitated, but not denuded, and the results would be representative of stacking faults intersecting the wafer surface. In the latter case, the material might well not be precipitated and the results would then represent insufficient gettering of impurities in the process.

Figure 13 gives results of leakage-limited yield of 64K DRAM devices reported by Steinbeck [*31*]. His results show much less refresh yield losses at the seed end of the crystal than at the tail end. Since oxygen is generally higher towards the seed end, this effect is likely caused by internal gettering.

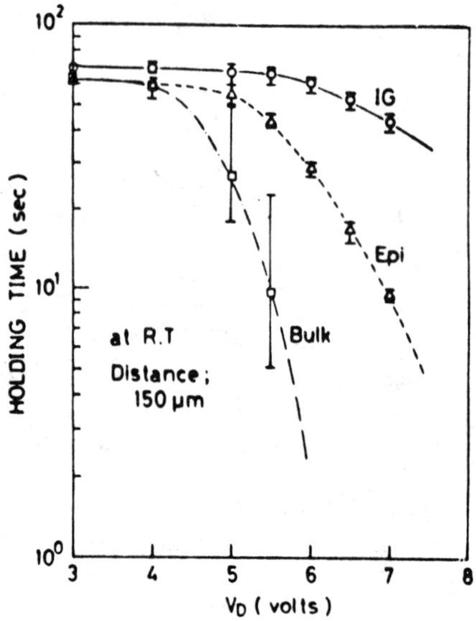

FIG. 12—*Holding time versus drain voltage* (V_D) *for intrinsic gettering* (IG), *epitaxial* (Epi), *and bulk wafers* [30].

The tail end material would then be too low in oxygen for good precipitation. As Steinbeck notes in his paper, these results are very dependent on running a device process which creates a denuded zone.

Figure 14 shows results for this work comparing relative refresh yield losses (as checked at a given drain voltage and time) for 16K dynamic RAM memories with the oxygen level of the wafer. It can be seen that there are

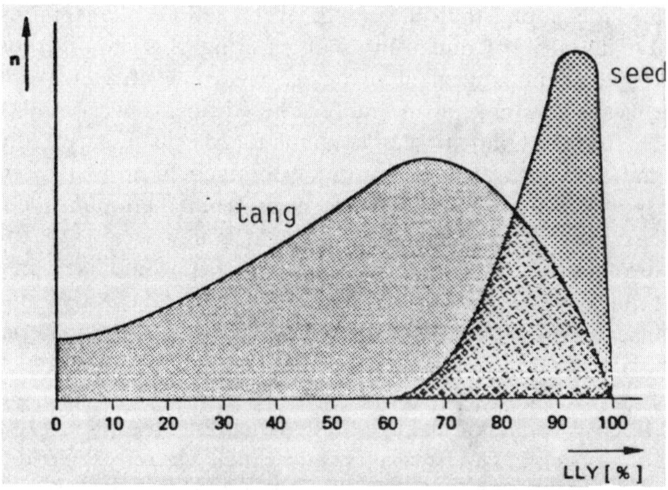

FIG. 13—*Leakage-limited yield distribution separated in Cz-wafers from crystal head and tail* (*50% each of total length*) [31].

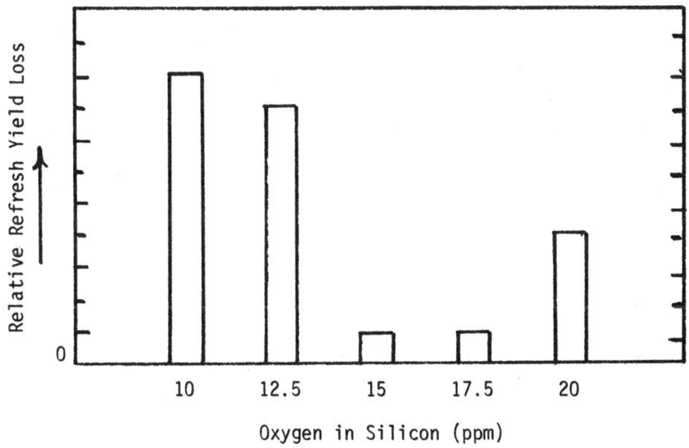

FIG. 14—*Relative average refresh yield loss as a function of oxygen level for 16K DRAMs.*

higher yield losses below the oxygen precipitation threshold and above the oxygen level where the silicon can be cleared of defects which grow into stacking faults. The exact position of these limits is not unique but would be very sensitive to the particular device process being run. Figure 15 shows the distribution of refresh yield losses obtained for a group of wafers at 12 ppm and a group at 17 ppm in this 16K DRAM test. The distributions look very similar to the seed end and tang end experiments reported by Steinbeck.

Similar effects have recently been shown for 64K DRAMS. Figure 16

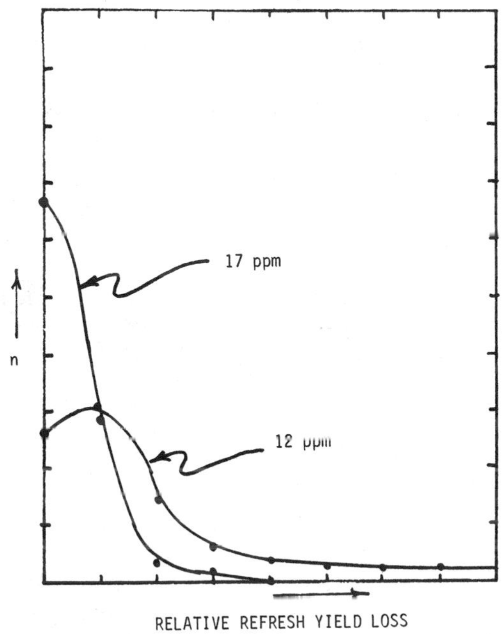

FIG. 15—*Comparison of refresh yield loss distribution for oxygen levels of 12 and 17 ppm.* n *is a relative frequency on a linear scale.*

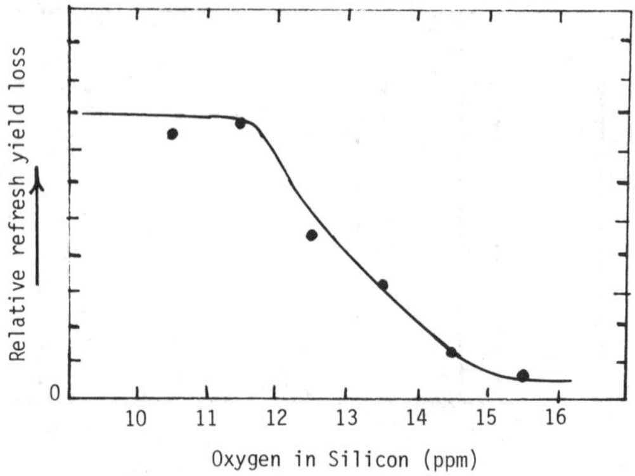

FIG. 16—*Relative average refresh yield loss as a function of oxygen level for 64K DRAMs.*

shows in a general way results obtained for a number of lots for which oxygen level and refresh yield losses were carefully characterized. It can be seen that, for this particular process, the leakage limited yield is greatly improved for the region above the threshold of precipitation (O_i*). The threshold for a particular thermal process could be between 13 and 16 ppm, depending on the temperatures involved.

Alternative Gettering Mechanisms

The use of oxygen for internal gettering is only one of several mechanisms that have been used to draw unwanted impurities away from the device surface. Again, which is best for a particular case can only be determined by evaluating the complete thermal process. Several reviews of gettering mechanisms have recently been issued [*9,10*].

The use of heavy phosphorous diffusions has been shown to be an effective getter [*32*]. The phosphorous generates dislocation in the lattice which then becomes gettering sites. Pre-process treatment would be done to the backside of a wafer, but during processes phosphorous gettering can be used under certain circumstances on the front or back side of the wafer.

Controlled mechanical damage on the backside of the wafer has been done in several ways. Damage on the order of 10 to 20 μm deep has generally been achieved by using a grinding mechanism on a spinning wafer, while lighter damage can be controlled by a slurry lap process or by blasting the backside of the wafer with fine particles such as glass beads or silicon carbide.

More recently, a process of laser damage gettering has been introduced [*33–35*]. Laser irradiation damage on the backside of the wafer is used to generate gettering sites. High-power laser pulses are claimed to give some defect regions which are thermally stable and which do not have the opportunity for contamination as do some of the other damaging methods.

Backside ion implantation is yet another way to create strained regions to

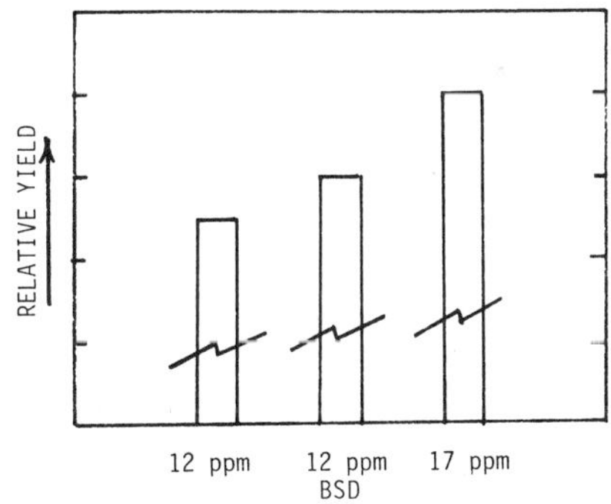

FIG. 17—*Comparison of backside damage versus internal gettering effectiveness for 16K DRAMs.*

draw the impurities [*36,37*]. The use of deposited films, such as boron silicide or silicon nitride, on the backside can also cause strain on the backside and therefore some gettering, but this method may not be as effective as the other treatments [*38*]. On the other hand, such film deposition has less opportunity to affect the fundamental strength of the wafer in processing.

Results obtained by the author for 16K dynamic RAMS comparing lower oxygen material with and without a mechanical backside treatment to higher oxygen intrinsically gettered material are shown in Fig. 17. The mechanical treatment helped, but in this particular process did not do as well as the material achieving internal gettering.

The question arises as to whether there is a need or advantage to have some form of backside treatment combined with higher oxygen in the substrate. The oxygen can provide significant intrinsic gettering. If the intrinsic gettering is created in the device process, the backside damage could provide gettering during initial processing while precipitation is being established. On the other hand, the possibility of the backside treatment being effective enough to getter the oxygen itself could cause a problem with competing mechanisms which might not be as effective as either are alone [*39*].

Another possible concern is the effect of process annealing on the stability of a given backside treatment. Certain damages may tend to anneal out and lose their ability to getter, while others may do a controlled amount of regeneration of defects through the process. A danger with very high temperature (~1200°C) heavy backside damage is the possibility of damage propagation through the wafer to the front side.

A Gettering Scenario

There are no truly uniformly applicable gettering systems. Tailoring must fit the particular process. However, it is intended here to present some general trends that may be applicable to different ranges of temperature process-

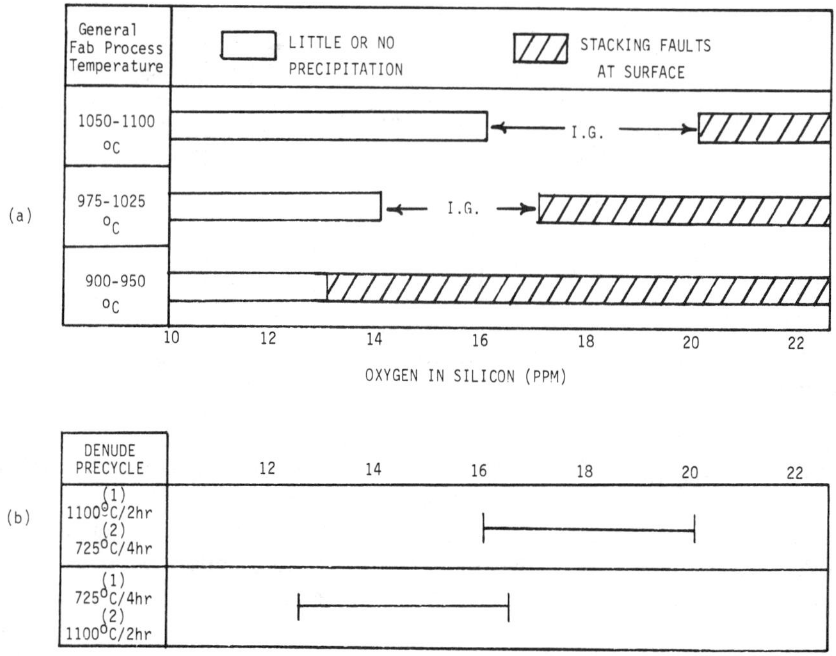

FIG. 18—(a) *Range of in-process internal gettering expected as a function of oxygen level for various fabrication processing temperatures.* (b) *Range of general applicability of precycles studied.*

ing. Figure 18 shows a possible range of applicability of in-process intrinsic gettering.

At higher MOS processing temperatures (1050 to 1100°C) it is thought that silicon in the 16 to 20 ppm range will give acceptable internal gettering. Below 16 ppm there may be difficulty in achieving good precipitation; above 20 ppm stacking faults will tend to be at the device surface. At medium temperatures (975 to 1025°C) the range of in-process denuding/precipitation is narrow, while at low temperatures (900 to 950°C) there will be essentially no denuding of the wafer and precipitation will proceed down to a somewhat lower level (~13 ppm).

Therefore, for 1050 to 1100°C processing, one will likely choose the 16 to 20 ppm range of oxygen for optimum performance and then depend on it for intrinsic gettering. If a short time at lower temperature (600 to 800°C) does not occur early in the process this may have to be added. If backside damage material is chosen for this high-temperature processing, the strain on the backside may either anneal out and loose its effectiveness or, if heavy enough, contribute to propagating damage through the wafer. An additional concern with high-oxygen material is the potential of warpage if care is not taken during cycling.

For medium temperatures (975 to 1025°C) several alternatives exist. One can use processed-induced intrinsic gettering in a fairly narrow oxygen range. This, of course, requires a very careful correlation with the silicon vendors. One can also use lower oxygen (<13 ppm) silicon and add a stable backside getter, possibly like the laser damage. Alternatively a denude/nucleation precycle such as those presented earlier could be applied to the silicon previous to device processing using an oxygen range of 16 to 20 ppm. Which of these choices is best cannot be predicted in advance; they must be evaluated carefully for each particular case.

For lower processing temperatures (900 to 950°) there is no chance to obtain good dependable internal gettering from the process alone. One must either choose the lower oxygen (<13 ppm) with a backside damage or the higher oxygen (16 to 20 ppm) with a higher temperature denuding precycle before processing. It is believed here that the lower oxygen with backside damage may be the best choice for these operating conditions. At these temperatures the damage should be more stable with less chance of annealing out or causing a catastrophe.

Also shown in Fig. 18 are the expected applicable ranges of processing for the primary denuding/nucleation cycles studied: 1100°C for 2 h, 725°C for 4 h, and the reverse order. There is nothing unique about the exact choice of times used here for these precycles. Moreover, it should be remembered that the experiments mentioned earlier used an extremely simplified process simulation cycle of 1000°C for 15 h. For effective application one must use the actual cycles of the device of concern.

Details of the device being run must be considered in the design of the optimum thermal process for internal gettering. The denuded-zone depth must be set deeper than where any active interaction with the device occurs, at room temperature and at the device operation temperature. This zone depth can be adjusted by changing the time or temperature of the denuding cycle. The threshold of precipitation and the precipitation rate can be adjusted with the time and temperature of the nucleation cycle and the sequence of the two cycles.

The question of precipitation rate is an interesting one. How long after complete precipitation of the oxygen can effective gettering be expected to continue? Does one want the precipitation to be fully established at the start of a device process, or on more of a "time-released" basis throughout the process? The latter would provide fresh areas of local strain as new precipitation grows. The answers to these questions are unclear and more experimentation is needed on the life cycle of intrinsic gettering effectiveness.

Oxygen Donors

It has long been established that oxygen can complex with silicon to behave as a donor under certain conditions. The primary case is the one noted previously where donors are created at 450°C and annihilated at above 600°C [*10,11*]. The rate of formation of these complexes goes up rapidly with oxygen level, particularly above 13 to 15 ppm. Silicon vendors all use some type of heat cycle to minimize these donors as received by the user, but there are variations between them. Moreover, device processes which have steps in the 400 to 500°C range at the end can be subject to a resistivity shift from these effects. Use of such temperatures are common with current metalization and glassification procedures.

Recently another oxygen donor complex has also been identified [*42, 43*]. This "new" oxygen donor is generated above 600°C and is stable up to 950°C. To examine the formation of this donor an experiment was conducted. Samples from eight lots of material from four silicon vendors were run for 10 h at 700°C. Wafers used were again 100 mm, P-type (1-0-0), with resistivities of about 40 ohm-cm. Figure 19 shows the net change in resistivity after heat cycling for this material. As can be seen, only two of the eight lots showed any measurable generation of these new donors. One showed a moderate increase, the other was much larger. In one case one wafer in the lot actually had enough of the donors to convert to N-type. The initial oxygen levels measured on all lots are noted in the figure. It can be seen that these two lots are not particularly different from the others with respect to oxygen levels. Therefore these data indicate that there is another variable involved that is not identified. Since this experiment involved only a 700°C cycle and not a higher temperature to allow precipitation to grow, no measur-

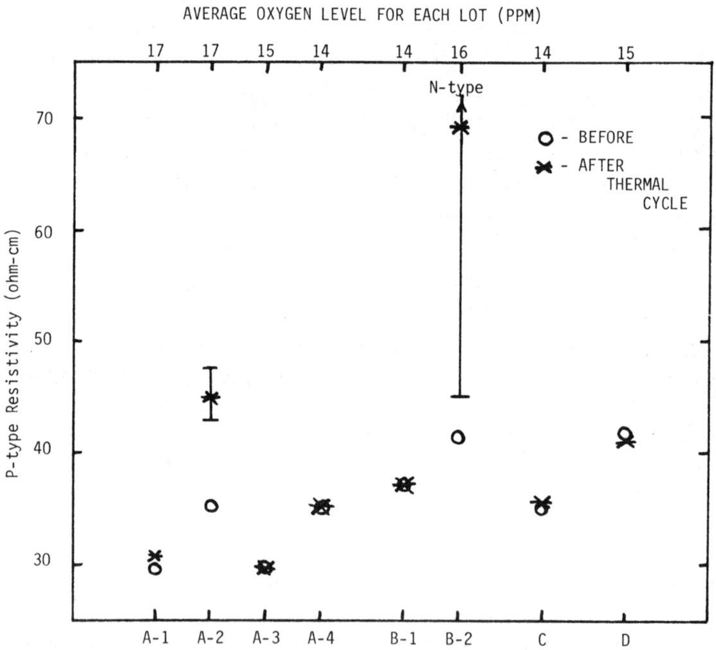

FIG. 19—*Resistivity before and after 10-h 700°C thermal cycle, demonstrating creation of "new" donors.*

able change in interstitial oxygen content was detected. It is possible that this propensity of forming "new" donors can be correlated with position in the crystal and therefore possible thermal anomalies in the growing process. More controlled experimentation is needed to determine the cause of differences in behavior between lots. Just as the other donors are of concern in relation to processing in the 400 to 500°C range late in the device process, concern about "new" donors is strongest where the process includes a step between 600 and 850°C which is not followed by a step at 1000°C or greater.

In the case of both oxygen donor states, precipitated oxygen cannot be converted to an electrically active condition. Therefore, use of an oxygen level and thermal cycle designed for significant precipitation will tend to leave the residual dissolved oxygen at about 5 ppm at the end of the process. At this oxygen level the resultant donor creation is insignificant.

Wafer Cleanliness

A large amount of attention has been given in this paper to oxygen specifications. The condition of the wafer surface is also a very important specification area for which significant development is needed. With the trend towards thinner oxides in MOS device processing this becomes even more critical. Much less has been established as to what truly are the parameters of concern. It is believed that understanding the "25 Å of mystery" between the actual silicon wafer and the oxide grown at the beginning of the device process may well be the next frontier in wafer specification. Results have been obtained that relate some device leakage to particular surface conditions which are typical of certain wafers available today [*44*].

Current evaluation of surfaces is generally limited to bright collimated light visual inspection. Cause for rejection may be stains, spots, particles, scratches, polishing haze, and other polishing imperfections. The point of concern here are those items related to post-polishing cleanup and not the polishing process itself. Generally, all major companies supply silicon having a high degree of reliability of passing such inspections. Specifications are generally written in relation to the absence of stains, the limit on the number of particles, or the total length of scratches. A recent improvement on visual inspection is the advent of laser scanning devices that finally appear to be capable of replacing and improving on human visual inspection. This has been an industry goal for years, but only recently has equipment capability been demonstrated.

A large percentage of device users still reclean wafers before inserting them into the process, though it is the goal of every silicon manufacturer to provide a wafer clean enough to be used directly. Therefore the whole topic of what cleaning is the ideal is a joint responsibility of the silicon manufacturer and the device user. (The possible application of a precipitation precycle as discussed in previous sections certainly affects this. If the precycle is done at the device house, then the main concern is the cleanliness after this cycle.)

Clearly, new techniques are needed to better describe how the condition of the surface affects the device. The use of oxide growth rates is one indirect measure that may shed light on the situation. Previous reports have shown some variation of oxide growth rates, depending on the cleaning system used [*45*]. We conducted an experiment to evaluate the difference in oxide growth rate on wafer surfaces as received from the silicon vendors. Wafers from six silicon companies were subjected to a 900°C dry oxidation for 2 h. Other wafers from each vendor were first recleaned by a proprietary process and then tested. The results are shown in Fig. 20. As can be seen, there was about a 40% difference in growth rate between the high and low vendors. Even if the oxides were of equivalent quality this variation of growth rate is unacceptable for device processing. After the wafers were recleaned, all exhibited the same growth rate, which was equivalent to the slowest growth rate from the vendors. The wafers of the two companies where the highest growth rate was seen had a hydrophobic surface (no native oxide, water beads up) whereas the others were hydrophillic (thin native oxide from cleaning, water sheets off). The recleaning that was used also left the wafers hydrophillic. The question of whether a surface should be hydrophobic or hydrophillic is a controversial one. When a cleaning process had left the surface hydrophobic then

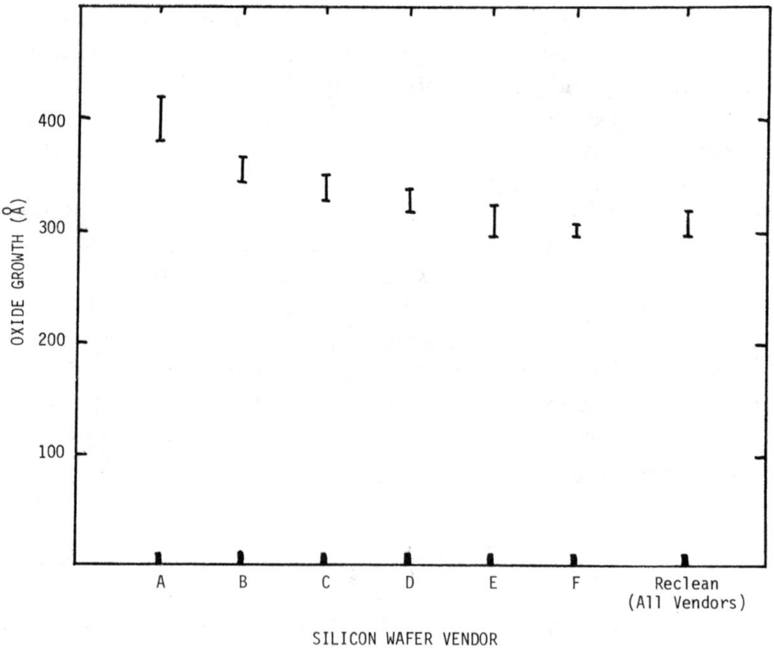

FIG. 20—*Oxide growth versus silicon vendor after 2-h 900°C dry oxidation.*

all native oxide has been stripped. Theoretically this surface should be the cleanest, but it does not show this in production. The reason is that this surface tends to be very attractive to particles and other contamination; thus it does not stay clean. On the other hand, a surface cleaned in a process that reforms a thin native oxide appears to be much less attractive to subsequent contamination.

As noted, the oxide growth rate variation itself is unacceptable even if the quality is the same. To evaluate the oxides grown, capacitor caps were placed on them and oxide breakdowns evaluated. Results were somewhat unclear, but the higher growth rate oxides had much more variation in the measurements and more capacitors exhibiting very low breakdowns. Overall, it is felt that the enhancement of an oxide growth rate may be a measure of thin organic films of contamination that are invisible to inspection.

Another technique that has shed new light on surface cleanliness is the use of a scanning electron microscope on polished surfaces. Spots and thin films not visible in bright light examination show up readily. Figure 21 shows several examples of contamination. Figure 21*a* is particularly interesting. Under bright light only the particles were visible. In the scanning electron microscope all particles were shown to be adhering to patchy areas of organic films. In this case the film was later identified to be wax used by the vendor in mounting wafers in the polishing process. A point should be made here about the common practice of using particle counts in surface specifications. The fundamental cause of the problem is really the areas of the wafer that are particularly attractive to particles rather than the particles themselves. The other examples shown in Fig. 21 may be polishing compound or residue from packaging. It should be noted that nearly every wafer examined directly out of sealed packages showed some areas of removable contamination when examined in the SEM.

This paper does not intend to provide a comprehensive study of possible cleaning processes; these have been recently reviewed [*46*]. The purpose is rather to emphasize the need for further study. Probably one of the most widely used cleaning systems is based on variations of ammonium hydroxide-hydrogen peroxide mixtures published many years ago [*47*]. Recently, cleaning solutions based on chlorine have been proposed [*48*]. Whether any of these systems or one totally different is really the optimum will require much future work—work that provides a real study of the surface chemistry involved. Any ultimate cleaning system must address the removal of a variety of organic films as well as the removal of ions (such as sodium) and heavy metal atoms (such as iron and copper).

Conclusions

Silicon specifications on oxygen and on surface chemistry were identified as key areas for development. The benefit of good internal gettering from the oxygen was demonstrated.

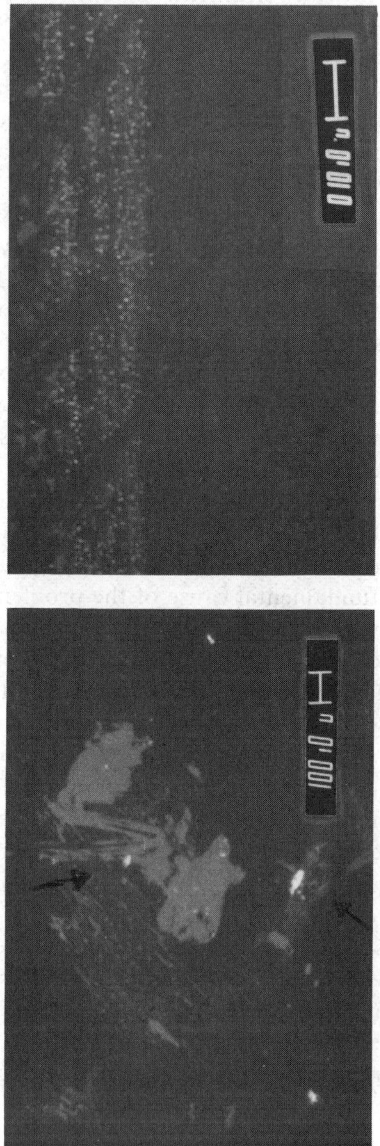

(a) (b)

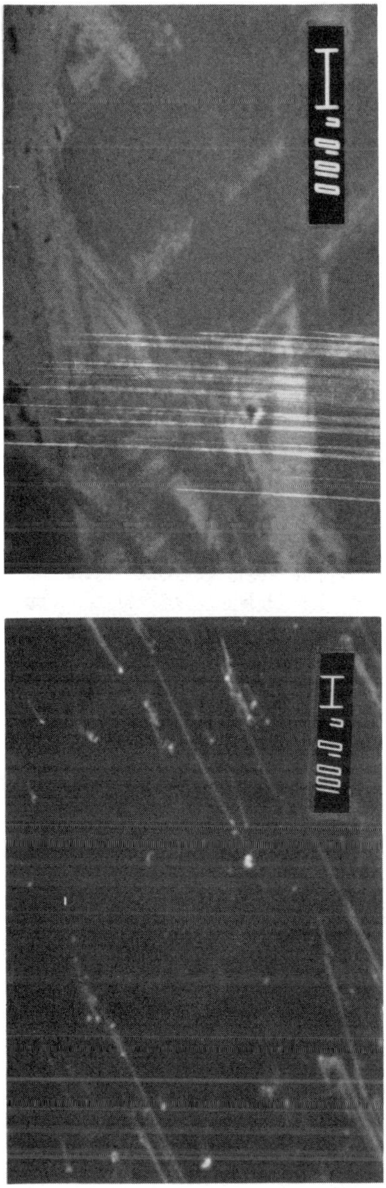

(c) (d)

FIG. 21—*Examples of contamination of polished surface as seen by scanning electron microscope.*

Items noted where silicon product can vary between vendors included oxygen precipitation rate, "new" donor formation, and oxide growth rates. These highlight the need for more control on the silicon wafers.

Some suggested areas for future work include:

1. Study of the gettering life of oxygen internal gettered material.
2. More work on the precipitation relationship between oxygen and carbon.
3. Methods for tighter control of the oxygen level in the crystal.
4. Study of the *in situ* heat treatment of crystal pullers.
5. Identification of variables involved in "new" donor creation.
6. Understanding and specification of the surface chemistry of a wafer as it affects device performance.

Acknowledgments

The author would like to thank his co-workers at MOSTEK: Shirley Duncan for carrying out the thermal cycle experiments and Milt Gosney for helpful discussions and review of the manuscript.

References

[*1*] Baylis, W. A., *Solid-State Technology,* Vol. 24, No. 2, Feb. 1981, pp. 132–135.

[*2*] Hu, S. M., *Journal of Vacuum Science and Technology,* Vol. 14, No. 1, Jan./Feb. 1977, pp. 17–31.

[*3*] Vieweg-Gutberlet, F. G. in *Lifetime Factors in Silicon, ASTM STP 712,* American Society for Testing and Materials, 1979, pp. 7–13.

[*4*] Matlock, J. H. in *Semiconductor Silicon 1977,* H. R. Huff and E. Sirtl, Eds., The Electrochemical Society, 1977, pp. 32–51.

[*5*] Benson, K. E., Lin, W., and Martin, E. P. in *Semiconductor Silicon 1981,* H. R. Huff, R. J. Kriegler, and Y. Takeishi, Eds., The Electrochemical Society, 1981, pp. 33–48.

[*6*] Hoshikawa, K., Hirata, H., Nakanishi, N., and Ikuta, K. in *Semiconductor Silicon 1981,* H. R. Huff, R. J. Kriegler, and Y. Takeishi, Eds., The Electrochemical Society, 1981, pp. 101–112.

[*7*] Murgai, A. in *Semiconductor Silicon 1981,* H. R. Huff, R. J. Kriegler, and Y. Takeishi, Eds., The Electrochemical Society, 1981, pp. 113–125.

[*8*] Craven, R. A. in *Semiconductor Silicon 1981,* H. R. Huff, R. J. Kriegler, and Y. Takeishi, Eds., The Electrochemical Society, 1981, pp. 254–271; and Korb, H., *Solid-State Technology,* Vol. 24, No. 7, July 1981, pp. 55–61.

[*9*] Pearce, C. W., Katz, L. E., and Seidel, T. E. in *Semiconductor Silicon 1981,* H. R. Huff, R. J. Kriegler, and Y. Takeishi, Eds., The Electrochemical Society, 1981, pp. 705–723.

[*10*] Monkowski, J. R., *Solid-State Technology,* Vol. 24, No. 7, July 1981, pp. 44–51.

[*11*] Tan, T. Y., Gardner, E. E., and Tice, W. K., *Applied Physics Letters,* Vol. 30, No. 4, 1977, pp. 175–176.

[*12*] Rozgonyi, G. A. and Pearce, C. W., *Applied Physics Letters,* Vol. 32, No. 11, 1978, pp. 747–749.

[*13*] Shimizu, H., Yoshinaka, A., and Sugita, Y., *Japanese Journal of Applied Physics,* Vol. 17, No. 5, May 1978, pp. 767–771.

[*14*] Nagasawa, K., Matsushita, Y., and Kishino, S., *Applied Physics Letters,* Vol. 37, No. 7, Oct. 1980, pp. 622–624.

[*15*] Kugimiya, K., Akiyama, S., and Nakamura, S. in *Semiconductor Silicon 1981.* H. R. Huff, R. J. Kriegler, and Y. Takeishi, Eds., The Electrochemical Society, 1981, pp. 294–303.

[*16*] Series, R. W., Barraclough, K. G., and Bardsley, W. in *Semiconductor Silicon 1981,* H. R. Huff, R. J. Kriegler, and Y. Takeishi, Eds., The Electrochemical Society, 1981, pp. 304–312.
[*17*] Ogino, M., Usami, T., and Watanabe, M., *Extended Abstracts,* Vol. 80-2, Abstract 435, The Electrochemical Society, 1980, pp. 1097–1099.
[*18*] Shimura, F., Tsuya, H., and Kawamura, T., *Journal of the Electrochemical Society,* Vol. 128, No. 7, 1981, pp. 1579–1583.
[*19*] Sasaki, M., Nishi, K., and Ushio, S., *Extended Abstracts,* Vol. 81-2, Abstract 393, The Electrochemical Society, 1981, pp. 954–956.
[*20*] Patel, J. R. in *Semiconductor Silicon 1981,* H. R. Huff, R. J. Kriegler, and Y. Takeishi, Eds., The Electrochemical Society, 1981, pp. 189–207.
[*21*] Inoue, N., Wada, K., and Osaka, J. in *Semiconductor Silicon 1981,* H. R. Huff, R. J. Kriegler, and Y. Takeishi, Eds., The Electrochemical Society, 1981, pp. 282–293.
[*22*] Wright-Jenkins, M., *Journal of the Electrochemical Society,* Vol. 124, 1977, pp. 757–762.
[*23*] Huber, D., Stallhofer, P., and Blatte, M. in *Semiconductor Silicon 1981,* H. R. Huff, R. J. Kriegler, and Y. Takeishi, Eds., The Electrochemical Society, 1981, pp. 756–765.
[*24*] Wang, J. and Kulkarni, M., *Extended Abstracts,* Vol. 80-2, Abstract 532, The Electrochemical Society, 1980, pp. 1330–1332.
[*25*] Abe, T., Kichuchi, K., Shirai, S., and Muraoka, S. in *Semiconductor Silicon 1981,* H. R. Huff, R. J. Kriegler, and Y. Takeishi, Eds., The Electrochemical Society, 1981, pp. 54–71.
[*26*] Hearn, E. W., Markovits, G. A., and Kulkarni, M. V., *Semiconductor International,* Vol. 4, No. 9, Sept. 1981, pp. 101–112.
[*27*] Matlock, J. H., unpublished data.
[*28*] Sun, R. C., and Clemens, J. T., *International Electron Devices Meeting Technical Digest,* IEEE, 1977, pp. 254–257.
[*29*] Strack, H., Mayer, K. R., and Kolbesen, B. O., *Solid-State Electronics,* Vol. 22, 1979, pp. 135–140.
[*30*] Otsuka, H., Watanabe, K., and Nishimura, H., *Extended Abstracts,* Vol. 81-2, Abstract 396, The Electrochemical Society, 1981, pp. 962–964.
[*31*] Steinbeck, H. H., *Extended Abstracts,* Vol. 80-2, Abstract 530, The Electrochemical Society, 1980, pp. 1325–1326.
[*32*] Rozgonyi, G. A., Petroff, P. M., and Read, M. H., *Journal of the Electrochemical Society,* Vol. 122, No. 12, 1975, pp. 1725–1729.
[*33*] Pearce, C. W. and Zaleckas, V. J., *Journal of the Electrochemical Society,* Vol. 126, No. 8, 1979, pp. 1436–1437.
[*34*] Katz, L. E., Schmidt, P. F., and Pearce, C. W., *Journal of the Electrochemical Society,* Vol. 128, No. 3, 1981, pp. 620–624.
[*35*] Hayafuji, Y., Yanada, T., and Aoki, Y., *Journal of the Electrochemical Society,* Vol. 128, No. 9, 1981, pp. 1975–1980.
[*36*] Geipel, H. J. and Tice, W. K., *IBM Journal of Research and Development,* Vol. 24, No. 3, 1980, pp. 310–317.
[*37*] Prussin, S., *Solid-State Technology,* Vol. 24, No. 3, 1980, pp. 310–317.
[*38*] Petroff, P. M., Rozgonyi, G. A., and Sheng, T. T., *Journal of the Electrochemical Society,* Vol. 123, No. 4, 1976, pp. 565–570.
[*39*] [Anonymous], *Semiconductor International,* Vol. 4, No. 5, May 1981, pp. 11–12.
[*40*] Fuller, C. S. and Logan, R. A., *Journal of Applied Physics,* Vol. 28, No. 12, 1957, pp. 1427–1436.
[*41*] Baker, J. A., *Extended Abstracts,* Vol. 73-1, Abstract 82, The Electrochemical Society, 1973, pp. 208–209.
[*42*] Kanamori, A. and Kanamori, M., *Journal of Applied Physics,* Vol. 50, 1979, p. 8095.
[*43*] Craven, R. A., *International Electron Device Meeting Technical Digest,* IEEE, 1981, pp. 228–231.
[*44*] Matlock, J. H., unpublished data.
[*45*] Schwettmann, F. N., Chiang, K. L., and Brown, W. A., *Extended Abstracts,* Vol. 78-1, Abstract 276, The Electrochemical Society, 1978, pp. 688–689.
[*46*] Burkman, D., *Semiconductor International,* Vol. 4, No. 7, July 1981, pp. 103–116.
[*47*] Kern, W. and Puotinen, D. A., *RCA Review,* Vol. 31, No. 2, June 1970, pp. 187–206.
[*48*] Muraoka, H., Kurosawa, K., Hiratsuka, H., and Usami, T., *Extended Abstracts,* Vol. 81-2, Abstract 238, The Electrochemical Society, 1981, pp. 570–572.

Jane C. Allison,[1] *Oral F. Maxwell,*[1] *Roland Schindler,*[1] *and Marcello Domenici*[2]

Oxygen Precipitation Studies at Various Impurity Levels Using Thermal Donor Activation

REFERENCE: Allison, J. C., Maxwell, O. F., Schindler, R., and Domenici, M., "**Oxygen Precipitation Studies at Various Impurity Levels Using Thermal Donor Activation,**" *Silicon Processing, ASTM STP 804,* D. C. Gupta, Ed., American Society for Testing and Materials, 1983, pp. 362–368.

ABSTRACT: Thermal donor activation and related intrinsic *pn*-junctions are used to characterize outdiffusion and precipitation of oxygen in silicon wafers. It is found that the depth of the *p*-layers and the depth of the denuded zones do not coincide. Both are found to depend on the wafer position within the crystal and are not solely related to the oxygen content.

KEY WORDS: thermal donor, carbon, oxygen, denuded zone

Spreading resistance profiling (SRP) has been used [*1–4*] as a means of obtaining detailed information on oxygen distribution via thermal donor (TD) formation at 450°C.[3] A fairly detailed analysis of oxygen profiles is reported in experiments by Hu [*1,2*], who studied the outdiffusion of oxygen in the course of denuded-zone formation. Hu made use of a calibrated interstitial oxygen-thermal donor relationship to calculate local distribution of interstitial oxygen.

This work uses the same methodology to study oxygen precipitation and denuded-zone formation in wafers originating from different positions on the ingots, which involves axial gradients in oxygen and carbon levels.

Experimental Procedure

The specimens tested were 100 mm-diameter P(111) CZ wafers with a resistivity range of 3.1 to 6.6 Ω-cm. The specimens were obtained from two different ingots. One ingot received a typical resistivity anneal prior to slicing (Wafer Groups 1 to 3). From the second ingot every other wafer received a

[1] National Semiconductor Corporation, Santa Clara, Calif. 95051. Dr. Schindler is currently with Wacker Chemitronic, Burghausen, W. Germany.

[2] Dynamit Nobel Silicon, S.P.A., Novara, Italy.

[3] The italic numbers in brackets refer to the list of references appended to this paper.

TABLE 1—*Characteristics of wafers used in experiment.*

Wafer Group	Thermal Treatment	Oi,[a] ppma	C,[b] ppma	% Melt Solidified	Resistivity
1	Xtal	20 to 24	<0.3	0 to 20	6.3 to 6.5
2	Xtal	17 to 18	<0.3	50 to 70	4.9 to 5.0
3	Xtal	16 to 17	1.5 to 2.8	extreme tang end	3.1 to 3.6
4	wafer	18.5 to 22	<0.24	0 to 35	5.9 to 6.4
5	none	18.5 to 22	<0.24	0 to 35	...
6	wafer	17 to 18.5	<0.24	35 to 70	5.0 to 5.4
7	none	17 to 18.5	<0.24	35 to 70	...
8	wafer	16 to 18	0.7 to 1.5	extreme tang end	3.7 to 4.2
9	none	16 to 18	0.7 to 1.5	extreme tang end	...

[a] ASTM Test for Interstitial Atomic Oxygen Content of Silicon by Infrared Absorption.
[b] ASTM Test for Substitutional Atomic Carbon Content of Silicon by Infrared Absorption.

resistivity stabilization, which was similar to the anneal in ingot form. This provided exact knowledge of the resistivity in the nonstabilized wafers. Wafer identity, heat treatment, oxygen and carbon concentrations, and fraction of melt solidified (that is, crystal positions) are listed in Table 1.

The process thermal treatments were performed in a Thermco Brute diffusion system. All material was processed through two typical manufacturing operations and two low-temperature annealing steps. The first of these process steps was an initial oxidation in which the wafers were inserted into the diffusion furnace at 840°C. The system was then ramped up to approximately 1250°C, where the wafers remained for 800-nm silicon dioxide growth. At completion of the high-temperature sequence the furnace was returned to 840°C and the wafers were withdrawn.

The second operation performed was an antimony collector diffusion, standard at our factory, utilizing an oxidizing ambient. It was ramped identically to the other process. Some of the material received a 6-μm collector and others a 1.5 μm collector.

The low-temperature annealing sequence involved a 100-h anneal at 450°C in nitrogen ambient followed by an angle lap of 5 deg. The SRP measurements were taken on all samples and the carrier concentrations were plotted. The samples were then reinserted into the furnace at 450°C for an additional 250 h, and a second set of SRP measurements was taken after another 5-deg angle lap. The SRP system used was a Solid State Measurements Inc. ASR-100 Automatic Spreading Resistance Probe, which measures resistance and converts the data into carrier concentration. After SRP measurements the samples were Wright [5] etched for 5 min.

Results and Discussion

All the wafers exposed to the 450°C anneal showed intrinsic *pn* junctions due to thermal donor formation. An area close to the surface remained *p*-type, whereas the bulk of the wafer turned *n*-type. The SRPs clearly showed sharp junctions (Fig. 1), which could be seen after staining for 5 min in Wright etch (Fig. 2), and which has been identified as *p*- or *n*-type with a hot probe. This formation of a *pn*-junction is a result of oxygen depletion near the surface. During high-temperature processing, oxygen redistributes near the surface (for example by outdiffusion), resulting in an oxygen concentration gradient towards the surface. Therefore we observe a gradient in thermal donor formation which should reflect the actual oxygen concentration. Additionally, the Wright etch revealed defects such as stacking faults or other precipitation-related defects. In most cases a clear denuded area could be seen, that is, an

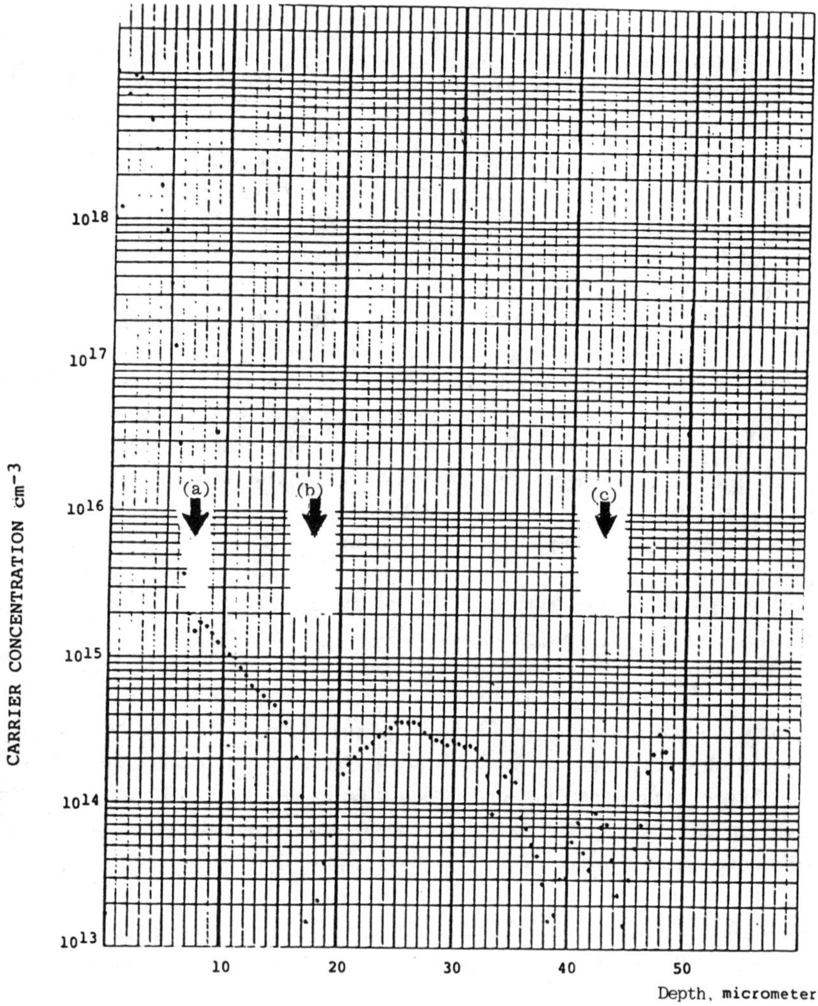

FIG. 1—*Spreading resistance profile of wafer after initial oxidation, collector diffusion, and 100-h anneal at 450°C.* (a) np-*junction between collector and bulk.* (b) *Intrinsic* pn-*junction due to thermal donor formation.* (c) *Noise due to oxygen precipitation.*

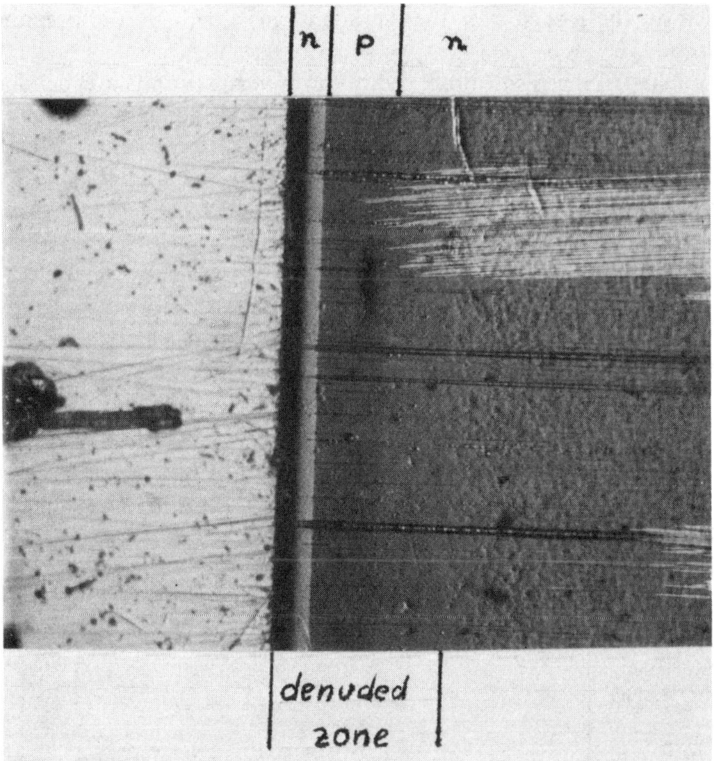

FIG. 2—*Angle-lapped, etched wafer showing intrinsic* pn-*junction and denuded zone* (*same SRP as in Fig. 1*).

area essentially without etch defects. This denuded zone extended into the bulk 10 to 20 μm deeper than *p*-type layers.

This can be easily understood when one bears in mind that the *pn*-junction occurs when the thermal donor concentration is equal to the acceptor concentration and does not give an actual measure of crystal lattice defects related to oxygen. For high-resistivity starting material, the difference between the depth of the *p*-layer and the depth of the denuded zone should be even larger than in the present study. In some cases the oxygen precipitates can be seen as fluctuations of the thermal donor concentration in the SRP (Fig. 1).

The defects revealed by the Wright etch were more pronounced in wafers originating from the seed and center part of the ingot than in wafers from the tang end. Nor is there any indication of precipitates in the SRP from tang end wafers. At present this observation is not totally understood.

Prior to processing, some of the wafers used in this study underwent a resistivity stabilization anneal in crystal form (30 min at 650°C) and some in wafer form (30 min at 650°C). We could therefore compare the influence of preannealing procedures. In our case there was no, or only a very marginal, influence on the *p*-region or the denuded-zone depth by the preannealing procedure.

The most striking result observed in our investigation is a dependence of the *pn*-junction depth on the wafer position within the ingot. There is a very dramatic increase of the depth of the *p*-layers with the amount of silicon melt solidified in the crystal growth process (Fig. 3). This coincides with two noticeable changes in impurity concentration, namely a decrease in oxygen concentration and an increase in carbon concentration.

Examining a typical wafer with a large radial gradient in oxygen concentration (less than 6 ppm from center to 6-mm distance from the edge), we did not find any radial change in the depth of the *p*-layer which would coincide with the change in oxygen concentration. We therefore ruled out the axial change of oxygen concentration in our ingots as the sole major factor in the increase of depth of the *p*-layer. An explanation of this observation has to take into account the increase in carbon concentration, the absence of visible defects after the Wright etch, and the absence of evidence of precipitates in the SRP profiles.

It has been suggested [6,7] that the thermal donor formation is related not only to the oxygen content in silicon, but also to intrinsic point defects, spe-

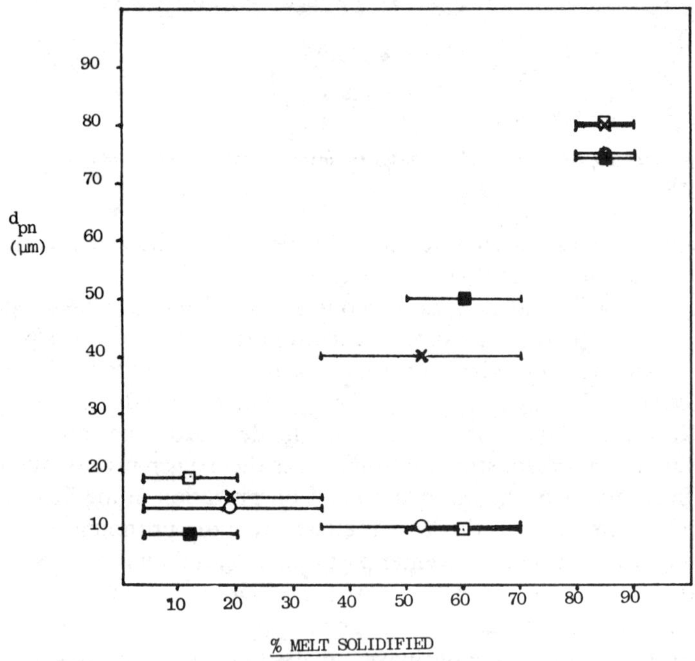

FIG. 3—*Depth of intrinsic* pn-*junction as a function of fraction of melt solidified after 100-h anneal at 450°C.* (□) *Samples with 6-μm collector.* (■) *Samples with 4.5-μm collector, preannealed in crystal form.* (○) *Samples with 4.5-μm collector, preannealed in wafer form.* (×) *Samples with 4.5-μm collector, not preannealed.*

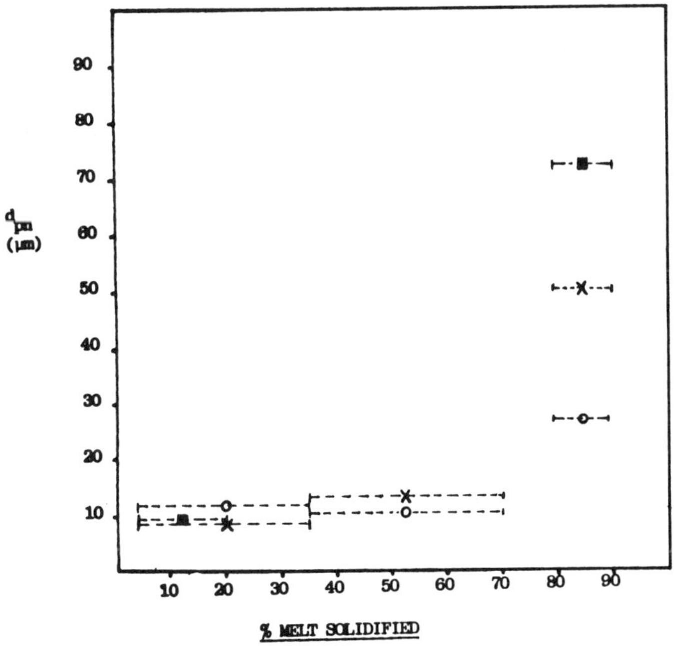

FIG. 4—*Depth of intrinsic* pn*-junction as a function of fraction of melt solidified after 350-h anneal at 450°C. See Fig. 3 for legend.*

cifically vacancies. In the light of findings by Rava and co-workers [*7*], we may explain our results by assuming that the thermal donor formation is controlled not only by the oxygen concentration, but also by the vacancy concentration. One possible explanation may be that vacancy concentration is strongly influenced by the presence of carbon. According to Sirtl and Helmreich [*6*] the small covalent radius of substitutional carbon is responsible for the generation of silicon self-interstitials which in turn recombine with vacancies. Thus with increasing carbon content the vacancy concentration is reduced, and hence the thermal donor formation rate becomes lower. This results in less thermal donors formed for a given time; therefore a wider *p*-layer remains. For very long annealing times, however, thermal donor formation will reach saturation [*8*]. This can result in more donors formed, which will be seen in a smaller *p*-layer. This shift is actually observed in the results of the additional 250-h anneal following the 100-h anneal (Fig. 4).

The observations presented indicate that, although oxygen, carbon, and intrinsic point defects are incorporated into the wafers during crystal growth, their interaction during device processing still reflects the way they have been incorporated into the crystal. Thermal donor formation as a tool to obtain fine profiles of interstitial oxygen concentration must therefore take this interaction into account.

Acknowledgments

The authors appreciate the technical assistance of Treva Johnson, K. Lowe, and A. Williams. They are indebted to their colleagues L. Beckwith and M. Christ for their critical support in the preparation of this paper.

References

[*1*] Hu, S. M., *Applied Physics Letters,* Vol. 36, 1980, p. 561.
[*2*] Hu, S. M., *Applied Physics Letters,* Vol. 31, 1977, p. 53.
[*3*] Domenici, M. and Molinari, E., *Materials Chemistry,* Vol. 4, 1980, p. 291.
[*4*] Gaworcewki, P. and Riemann, H., *Kristall und Technik,* Vol. 112, 1977, p. 191.
[*5*] Wright-Jenkins, M., *Journal of the Electrochemical Society,* Vol. 124, 1977, p. 757.
[*6*] Sirtl, E. and Helmreich, D. in *Semiconductor Silicon 1977,* H. R. Huff and E. Sirtl, Eds., The Electrochemical Society, Pennington, N.J., 1977, p. 626.
[*7*] Rava, P., Gatos, H. C., and Lagowski, J. in *Semiconductor Silicon 1981,* H. R. Huff, R. J. Kriegler, and Y. Takeishi, Eds., The Electrochemical Society, Pennington, N.J., 1981, p. 232.
[*8*] Cazcarra, V. and Zunino, P., *Journal of Applied Physics,* Vol. 51, 1980, p. 4206.

Charles J. Varker,[1] *James D. Whitfield,*[1] *and Peter L. Fejes*[1]

Effects of Oxygen Precipitation on Minority Carrier Lifetime in Silicon Crystals

REFERENCE: Varker, C. J., Whitfield, J. D., and Fejes, P. L., **"Effects of Oxygen Precipitation on Minority Carrier Lifetime in Silicon Crystals,"** *Silicon Processing, ASTM STP 804,* D. C. Gupta, Ed., American Society for Testing and Materials, 1983, pp. 369–388.

ABSTRACT: Infrared (IR) absorption measurements of the interstitial oxygen concentration [O] for two-step annealed wafers, with reverse recovery time T_{rr} and capacitance relaxation time T_s measurements on processed devices, have been used to investigate the influence of oxygen on the minority carrier lifetime (τ) in a Czochralski-grown silicon crystal. Measurements of T_{rr} and T_s show a strong dependence on the axial distribution of the initial interstitial oxygen concentration and on the concentration of precipitated oxygen. For a two-step anneal at 800°C and 1050°C for 16 h, the maximum average lifetimes of 4 μs and 180 μs are obtained from T_{rr} and T_s respectively in wafers with an initial oxygen concentration $[O]_i \simeq 1.3 \times 10^{18}$ cm^{-3}, where oxygen precipitation is minimal. Minimum lifetime values of 0.31 μs and 10 μs for the same respective measurements are obtained for wafers with $[O]_i \geq 1.8 \times 10^{18}$ cm^{-3}, where oxygen precipitation dominates. The experimental results indicate that oxygen precipitation is the dominant mechanism contributing to the degradation of minority carrier lifetime. The detection of an SiO_2 phase in the infrared spectrum and the detection of precipitate platelets in these wafers subsequent to the two-step anneal provide additional supporting evidence of a direct relation between oxygen precipitation and minority carrier lifetime degradation.

KEY WORDS: Czochralski silicon, carrier lifetime, reverse recovery, oxygen precipitation, microdefects, infrared absorption

In recent years, the emphasis in large scale integration has been placed on high-quality manufacturing processes and dislocation-free silicon crystals. It is anticipated that these considerations will produce very high yields on large area chips where a large fraction of the chip area is devoted to active device structures. In the manufacture of high-density metal-oxide semiconductor (MOS) memories, improved processing technology has reduced the incidence of "process-induced" defects in dislocation-free silicon crystals to a point where crystal inhomogeneities are potentially the major yield-limiting

[1] Senior Member of Technical Staff, Engineer, and Senior Staff Scientist, respectively, Semiconductor Research and Development Laboratories, Semiconductor Sector, Motorola Inc., Phoenix, Ariz. 85008.

defects. Axial and radial impurity inhomogeneities in dislocation free crystals result from "normal" segregation during freezing and from microsegregation effects that are characteristic of the crystal growth technique.

For silicon crystals grown with the Czochralski (CZ) method, oxygen and carbon are the most dominant unintentionally added impurities. The incorporation of oxygen and carbon in silicon by the Czochralski growth method has been the subject of many recent investigations [*1–3*].[2] Problems associated with CZ silicon crystals that arise from the fundamental properties of oxygen in silicon such as its solubility, diffusion coefficient, and precipitation kinetics have recently been reviewed by Patel [*4*]. The concentration of oxygen in typical CZ crystals is 1 to 2×10^{18} cm^{-3} and is normally well above its solubility limit during processing. The axial and radial distribution of oxygen in particular, and carbon to a much lesser extent, have been characterized [*3,5,6*] for silicon crystals grown with the Czochralski method.

The most dominant electrical effect arising from interstitial oxygen in silicon is the generation of thermal donors [*7,8*]. The concentration of thermal donors is maximized when silicon is annealed for extended periods of time at temperatures near 450°C. Electron traps at 0.13 eV below the conduction band edge that are attributed to oxygen donors in silicon have been characterized by using transient capacitance spectroscopy [*9*]. More recently, "new donors" in silicon have been detected in the temperature range of 650 to 800°C [*10*]. The concentration of oxygen donors often introduces an upper limit on the resistivity of CZ silicon crystals and creates unpredictable inhomogeneities of the free carrier concentration, particularly at high resistivities (>50 Ω-cm).

The major problem arising from the high concentration of interstitial oxygen in dislocation-free silicon crystals is its tendency to precipitate during thermal processing. Microdefects consisting of oxygen precipitates and their associated secondary defects, such as dislocation loops and stacking faults, have been the subject of extensive investigations [*11*]. Recent studies of oxygen precipitation in silicon using two-step thermal anneals have produced evidence in support of both homogeneous [*12*] and heterogeneous nucleation [*13*], as well as a modified homogeneous nucleation model [*14*]. Although the detrimental effects of oxygen precipitation in silicon are readily apparent from investigations of defects in silicon, recent claims regarding the beneficial effects of oxygen precipitation are not as obvious and are often difficult to reproduce.

In a recent study [*15*] using two-step thermal anneals, the authors present results that indicate that oxygen precipitation and oxygen out-diffusion can be controlled to provide lifetime improvements in CZ silicon wafers. The objective of this investigation is to characterize the effects of oxygen precipitation on minority carrier lifetime in a typical dislocation-free silicon crystal

[2] The italic numbers in brackets refer to the list of references appended to this paper.

grown with a Czochralski method. Computerized test methods are used on diode and capacitor arrays to investigate the axial and radial distribution of minority carrier lifetime. The relationship between the minority carrier lifetime, the initial concentration of oxygen, and the concentration of precipitated oxygen is investigated for two-step thermal anneals, which are designed to *force oxygen precipitation* over an extended region in the ingot.

Experimental Procedures

Polished silicon wafers were prepared from a boron-doped (acceptor concentration $N_A \approx 1 \times 10^{15}$ cm^{-3}) dislocation free crystal with a [100] orientation grown by the Czochralski method. A cylindrical ingot was prepared in a conventional manner with a diameter of ~75 mm and length of ~45 cm and sawed into five sections. The ingot spans the total useful length of the crystal between the tapered regions at the seed and tang-ends. The individual sections were prepared with a standard [110] orientation flat and a minor flat that was tapered along the ingot axis to provide a position reference for the wafers. Wafers were sliced, chemically thinned, and slurry-polished on one side to a final thickness of 500 μm. The wafers were then sorted and labeled with a laser marking tool for permanent identification.

Wafers used for this investigation were selected from three sections, each having a length of 10 cm, which represent the seed (S), center (C), and tang (T) regions of the crystal. The two intermediate sections (~7 cm each) were reserved for carbon measurements. Neither the initial ingot nor the finished wafers were subjected to a conventional thermal donor anneal at 650°C. This procedure minimized the precipitation of oxygen prior to the desired two-step thermal anneal and it permitted resistivity measurements of the residual thermal donor concentration following crystal growth. Initial resistivity measurements were made in accordance with ASTM Test for Measuring Radial Resistivity Variation on Silicon Slices (F 81). The measurements were obtained on polished wafers subjected to the donor anneal at 650°C for 1 h and were in the range from 10 to 20 Ω-cm. The oxygen concentration [O] was obtained from infrared absorption (IR) measurements at 1106 cm^{-1} with a Perkin Elmer Model 180 spectrometer, on single-side polished wafers before and after the two-step anneal. The absorption peak at 1106 cm^{-1} is attributed to interstitial atomic oxygen in silicon [*16–18*]. The IR method used in this study requires optical thickness matching between the sample and a reference wafer of known thickness [*19*]. The calibration factor of 4.81×10^{17} cm^{-3} given in ASTM Test for Interstitial Atomic Oxygen Content of Silicon by Infrared Absorption (F 121) was used to calculate the oxygen concentration. The substitutional carbon concentration [*20*] was calculated by using the calibration factor of 1.0×10^{17} cm^{-3} given in ASTM Test for Substitutional Atomic Carbon Content of Silicon by Infrared Absorption (F 123).

A two-step thermal anneal was designed, consisting of 16 h at 800°C in nitrogen and 16 h at 1050°C in dry O_2 and argon, to force oxygen precipitation over an extended region of the crystal. This procedure is based partly on the results of other investigators [*15*] and on oxygen precipitation studies at our laboratory. When reduced oxygen precipitation was desired a shorter time period was chosen at 800°C.

Optical microscopy was used to examine the distribution of microdefects resulting from oxygen precipitation. Microdefects were delineated by chemical decoration etching with a chromic/hydrofluoric acid etchant on selected polished wafers after the two-step precipitation anneal. The wafers were photographed with an MP3 photographic unit and oblique optical illumination. The micrographs were used for comparison with the infrared spectra to monitor the distribution of microdefects associated with changes in the spectrum resulting from oxygen precipitation.

The effects of oxygen precipitation on the axial and radial distribution of minority carrier lifetime was determined from measurements on gated n^+p diode arrays fabricated on selected annealed wafers from the three ingot sections. The diodes were fabricated using the field oxide grown during the precipitation anneal at 1050°C for the diffusion mask. The array fabrication process and the measurement technique using computerized testing and data graphics are described in detail elsewhere [*21*]. The electrical test system consisted of a Tektronix S-3260 equipped with a Pacific Western wafer prober. Parametric testing, data reduction, and graphics programs were developed by using system software (Tektest III). For the radial distribution studies, the lifetime test data are presented as a three-character range map or frequency distribution histogram for minority carrier lifetime τ. The characters in the test map approximate the locations of each test diode. The dots (●) represent diodes with measured values within the specified range limits. Diodes having values above or below this specified range are indicated with positive (+) and negative (−) range characters respectively (see, for example, Fig. 9).

Average "bulk" lifetime values (τ_B) for individual wafers were calculated from reverse recovery measurements on 1250 diodes. This average represents a statistical sample of ~25%, covering 17% of the wafer surface, and corresponds to a sampling depth that is represented by the average diffusion length for minority carriers, typically 100 μm. The diode arrays provide test data with sufficient spatial resolution to display the radial lifetime inhomogeneities resulting from oxygen precipitation [*21*].

An array of NMOS test structures containing both capacitors and diodes were also fabricated on annealed wafers by using NMOS processing technology. These wafers were measured to obtain capacitance relaxation time (T_s) and reverse recovery time (T_{rr}) for comparisons between the near surface and "bulk" lifetimes, respectively, for varying degrees of [O] precipitation. The areas of the capacitor and diode in the NMOS test arrays are 5×10^{-4} cm^2 and 1.3×10^{-3} cm^2 respectively.

Near surface lifetime (τ_s) was calculated from measurements of the capacitance and current transients from a biased deep depletion state on selected NMOS capacitors. The calculated values of τ_s were obtained by using the Zerbst method [*22,23*]. Since the capacitors are surrounded by a thick local oxide and a channel stop, perimeter effects are minimized. Under these conditions, the bulk generation current density is assumed to be proportional to the depletion width (W) below the capacitor field plate multiplied by the thermal generation rate $n_i(2\tau_s)^{-1}$ for minority carriers; the surface generation current is assumed to be proportional to the surface generation velocity (s). This total current density J_g is given by the equation

$$J_g = qn_i s + \frac{qn_i}{2\tau_s}(W - W_f) \tag{1}$$

Values for τ_s are calculated from the reciprocal slope obtained from a linear regression fit between the measured values of J_g and the normalized depletion width ($W/W_f - 1$).

The measurement system for the near-surface lifetime measurements consisted of an HP 85 desk-top computer, a PAR 410 C-V plotter, and a Keithley 619 electrometer/multimeter. Relaxation measurements were obtained with a wafer probe at 40 to 50°C in a dark ambient. Although the elevated temperature decreases the measurement time significantly, it was not feasible to map the entire wafer. Consequently, measurements were limited to twelve capacitors per wafer, representing a 3 to 4% sample. The small sample size was offset by selecting four capacitors from each of three population groups on a given wafer. Each group of four represents a sample taken from ~33% of the total distribution of reverse recovery lifetime measurements. By sampling the total range of bulk lifetime values represented by the three population groups, a good "spatial" average for τ_s was obtained. The average values for τ_s and τ_B from both test arrays were used to plot axial lifetime profiles for correlation with the oxygen distribution and the degree of oxygen precipitation in the crystal resulting from the two-step anneal.

In addition to the lifetime measurements, the reverse dark current (I_R) was measured on all diodes by using a test voltage (V_R) in the range 15 V $< V_R <$ 25 V. This value represents approximately 50% of the avalanche voltage of the diode. The radial and axial distributions of I_R were investigated for correlation with the oxygen distribution in the same manner as that discussed previously for the lifetime measurements. Since the execution time for these measurements is relatively short (<20 ms per device), a sample containing ~1250 measurements was generally achieved.

Experimental Results

IR Absorption Measurements of the Oxygen Concentration

Figure 1 plots the initial ($[O]_i$) and final ($[O]_f$) values of the oxygen concentration for the two-step anneals against the axial position of the corresponding wafers in the ingot. The resulting $[O]_i$ distribution shows that the initial oxygen

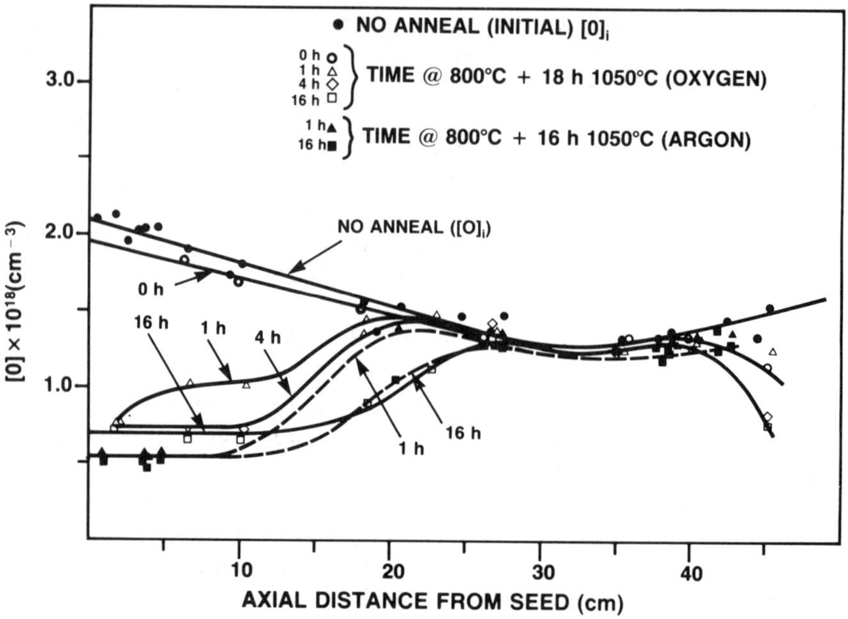

FIG. 1—*Axial distribution of oxygen concentration in the crystal before and after a two-step anneal for 0 h, 1 h, and 4 h at 800°C in argon and 16 h at 1050°C in dry oxygen (solid curves); and for 1 h and 16 h at 800°C in argon and 16 h at 1050°C in argon (dashed curves).*

concentration for the unannealed ingot (solid circles) decreases linearly with distance from the seed-end of the ingot. The linear region extends from a maximum value of $[O]_i = 2.1 \times 10^{18}$ cm^{-3} to a broad minimum where $[O]_i = 1.3 \times 10^{18}$ cm^{-3}. In the tang-region (35 cm $< X <$ 45 cm), $[O]_i$ increases slightly to a value of 1.5×10^{18} cm^{-3}. The degree of oxygen precipitation is evident by the change in the oxygen concentration $\Delta[O]$ between the $[O]_i$ curve and the curves representing the final oxygen concentration $[O]_f$ subsequent to the two-step anneals. In all cases, the temperature and time for the second annealing step was held constant.

The curves representing the 4-h and 16-h anneals at 800°C indicate that the oxygen concentration has decreased to a minimum value for $[O]_f = 7 \times 10^{17}$ cm^{-3} for the initial 10-cm section of the ingot. This value represents the solubility limit for dissolved oxygen at 1050°C for the anneals in an oxygen ambient. For the 1-h anneal at 800°C only partial precipitation is indicated. In the region corresponding to the broad minimum (28 cm $< X <$ 38 cm) where $[O]_i \approx 1.3 \times 10^{18}$ cm^{-3} the [O] curves merge, which indicates that no significant precipitation has occurred for all anneals. In the tang-region, however, where $X >$ 40 cm and $[O]_i > 1.3 \times 10^{18}$ cm^{-3}, significant oxygen precipitation is evident for the 4-h and 16-h anneals, with slight precipitation indicated for the 0-h and 1-h anneals.

In Fig. 1 a comparison is drawn between the anneals in an oxygen ambient (solid curves) discussed previously and a 1-h and 16-h anneal at 800°C followed by a 16-h anneal in argon at 1050°C (dashed curves). For the argon anneal it is evident that the precipitation rate has increased and the oxygen solubility limit has decreased. The precipitated oxygen $\Delta[O]$ is plotted against $[O]_i$ in Fig. 2 for a group of wafers annealed in argon. The equation describing the linear relationship shown in Fig. 2 is given by

$$\Delta[O] = [O]_i - [O]_{sol} \tag{2}$$

where the precipitated oxygen concentration $\Delta[O]$ is equal to the difference between the initial oxygen concentration $[O]_i$ and its solubility $[O]_{sol}$ at 1050°C. A linear regression fit of this data yields a value of 5.3×10^{17} cm^{-3} for the solubility limit at 1050°C. This value is in agreement with the value shown in Fig. 1 for the argon anneals.

Optical Microscopy and IR Spectra

Optical micrographs of chemically etched wafers that have previously been subjected to the two-step precipitation anneals are shown along with their infrared spectra in Fig. 3. The oxygen precipitation pattern is evident

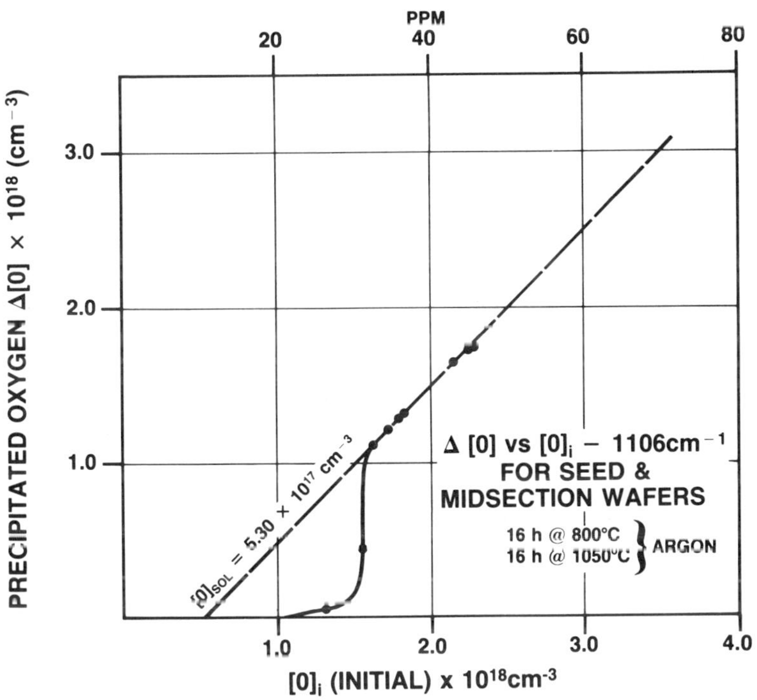

FIG. 2—*Concentration of precipitated oxygen $\Delta[O]$ versus the initial oxygen concentration $[O]_i$ for a two-step anneal in argon for 16 h at 800°C and 16 h at 1050°C.*

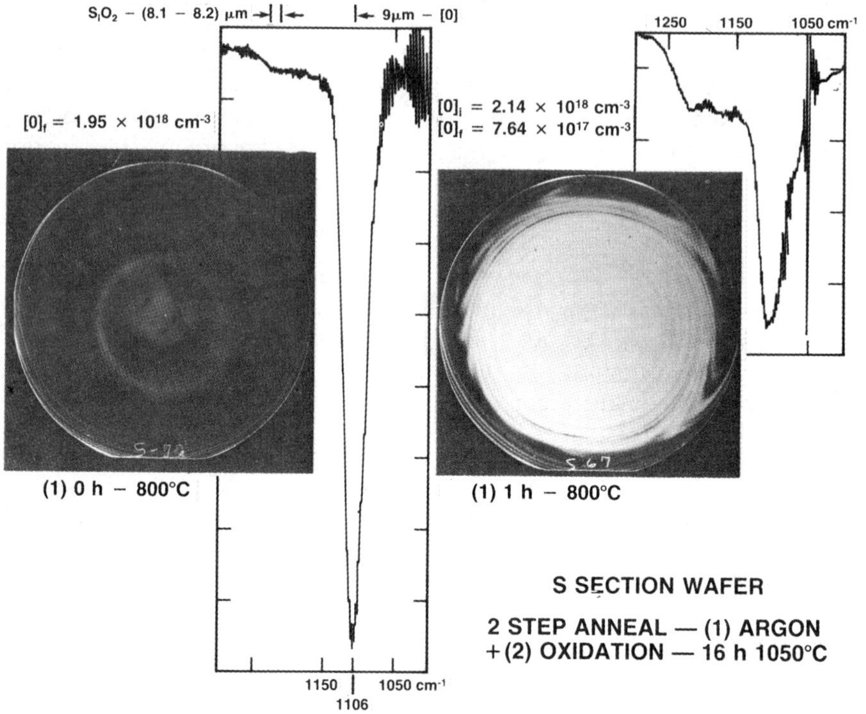

FIG. 3—*Optical micrograph showing the distribution of oxygen precipitates and microdefects and their associated IR spectra for seed-section wafers corresponding to a two-step anneal (0 h and 1 h at 800°C followed by 16 h at 1050°C in dry oxygen).*

by the formation of etch pits formed during chemical decoration etching, which results from the selective etching of oxygen precipitates and their secondary defects. The interstitial oxygen concentration [O] that is calculated from the IR absorption peak at 1106 cm^{-1} is roughly proportional to the peak height. The micrograph and absorption spectra in Fig. 3 correspond to a pair of wafers from the seed-end of the ingot where $[O]_i \approx 2.1 \times 10^{18}$ cm^{-3}. The wafer that was unannealed shows a light distribution of etch pits, which suggests that some oxygen precipitation has occurred at 1050°C. The absorption peak, which is approximately 10% smaller than its initial value, $[O]_i = 2.14 \times 10^{18}$ cm^{-3}, indicates that a decrease of the soluble oxygen concentration has also occurred. The wafer representing the 1-h anneal shows extensive etch pit formation with the exception of a 1-cm-wide band along the perimeter. The absorption peak has decreased significantly and the value corresponding to $[O]_f = 7.64 \times 10^{17}$ cm^{-3} is near the solubility limit for this group. The wafers for the 4-h and 16-h anneals at 800°C shown in Fig. 4 reveal a more extensive distribution of etch pits. Their spectra contain well-developed absorption peaks at 1215 to 1230 cm^{-1}, which indicates the pres-

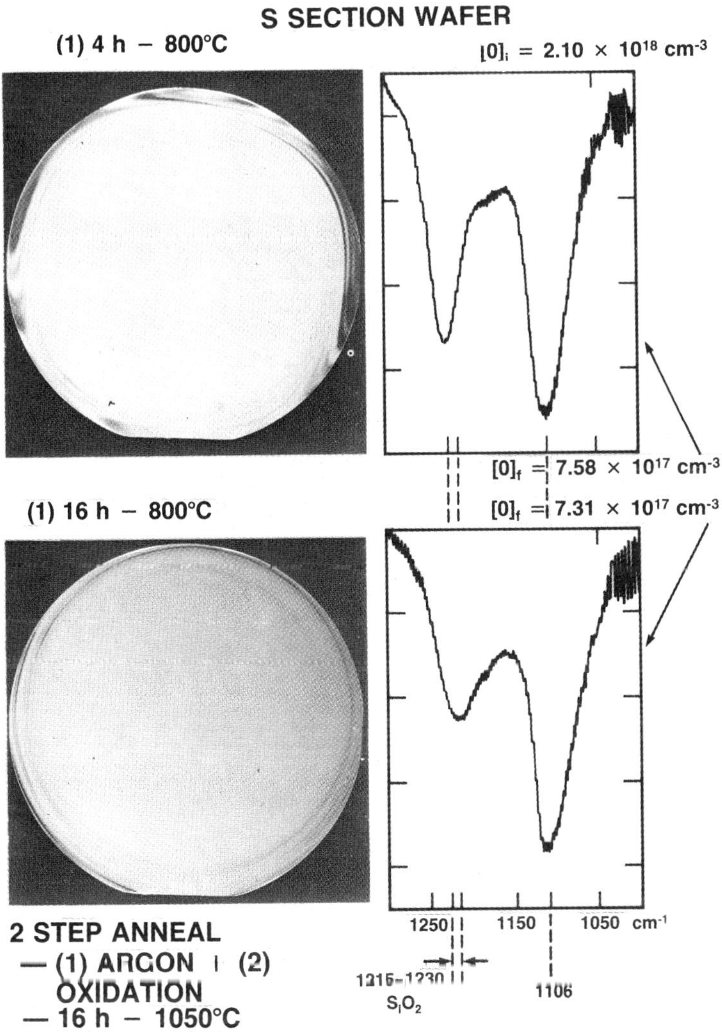

FIG. 4—*Optical micrograph showing the distribution of oxygen precipitates and microdefects and their associated IR spectra for seed-section wafers corresponding to a two-step anneal (4 h and 16 h at 800°C followed by 16 h at 1050°C in dry oxygen).*

ence of an SiO_2 phase [*24*], and the oxygen concentration is approximately at the solubility limit.

In contrast with these results, Figs. 5 and 6 reveal the effects of reduced oxygen concentration $[O]_i$ on the degree of precipitation. This group of wafers was selected closer to the oxygen minimum in the ingot, at $X = 18$ cm and where $[O]_i \approx 1.5 \times 10^{18}$ cm^{-3}. The micrographs representing the 0-h and 1-h anneals show little evidence for precipitation; their absorption peaks at 1106

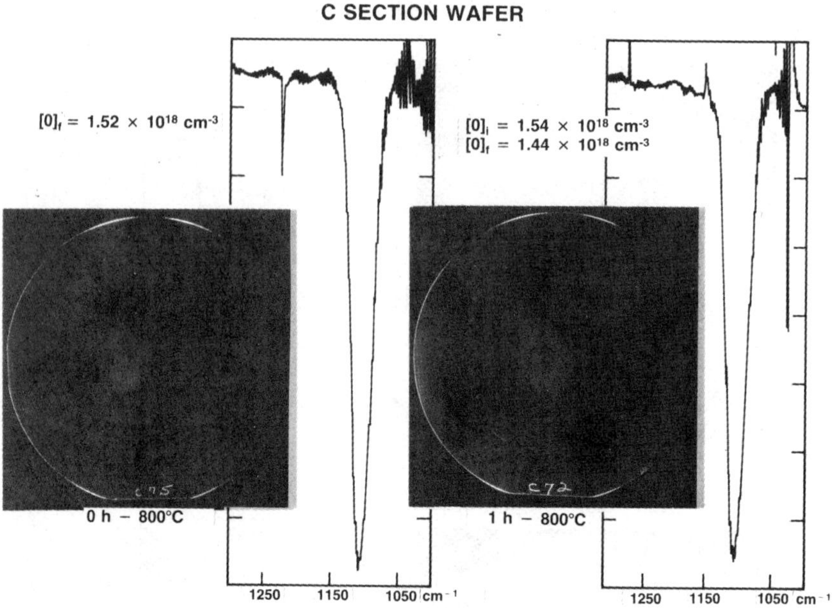

FIG. 5—*Optical micrograph showing the distribution of oxygen precipitates and microdefects and their associated IR spectra for center-section wafers corresponding to a two-step anneal (0 h and 1 h at 800°C followed by 16 h at 1050°C in dry oxygen).*

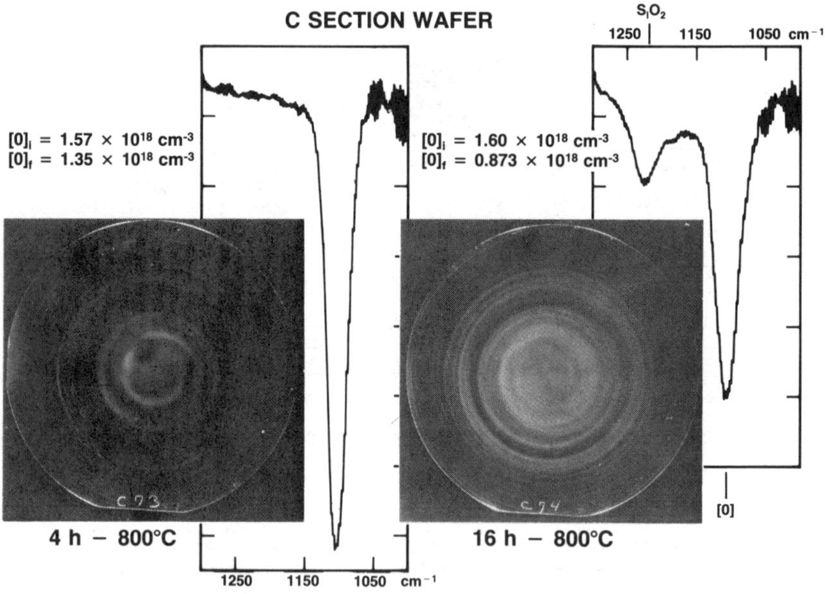

FIG. 6—*Optical micrograph showing the distribution of oxygen precipitates and microdefects and their associated IR spectra for center-section wafers corresponding to a two-step anneal (4 h and 16 h at 800°C followed by 16 h at 1050°C in dry oxygen).*

cm^{-1} are within 10% of their initial values. In Fig. 6 the micrograph for the 4-h anneal shows some evidence for precipitation; its absorption peak has decreased more than 10%, corresponding to a value of $[O]_f = 1.35 \times 10^{18}$ cm^{-3}. At 16 h, the typical inhomogeneities characteristic of nonuniform oxygen precipitation are observed in the optical micrograph and the absorption peak at 1215 to 1230 cm^{-1} is well defined. The oxygen concentration is below its initial value, although it remains slightly above the solubility limit of 7.3×10^{17} cm^{-3}.

Axial Distribution of Minority Carrier Lifetime and Reverse Current

The axial distribution curves for the τ_B and I_R measurements obtained on the diode arrays are shown in Fig. 7. The τ_B curves show a characteristic distribution, with maximum lifetime values where the initial oxygen concentration $[O]_i$ and the precipitated oxygen concentration $\Delta[O]$ are minimum. The 1-h and 16-h anneals at 800°C show a significant reduction for τ_B toward the seed-end of the ingot where oxygen precipitation has occurred during the two-step anneal. The reference anneal for 0 h at 800°C shows a similar distribution trend but has consistently higher values for τ_B in the seed region of the ingot. The corresponding I_R distribution reveals a sharp increase in the values for I_R toward the seed-end of the ingot for the 16-h anneal at 800°C but a relatively uniform I_R distribution for the 0-h and 4-h curves. This ap-

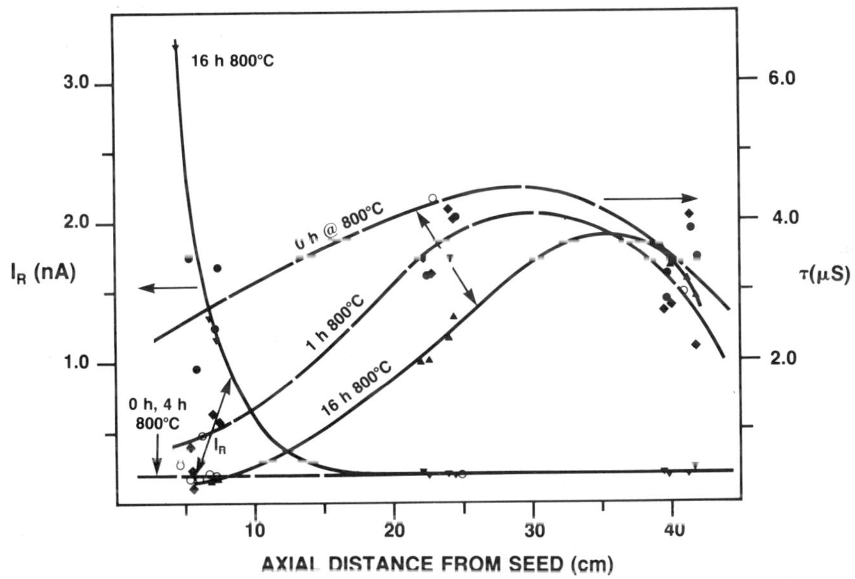

FIG. 7—*Axial distribution plot of the "bulk" lifetime* τ_B *and the reverse current* I_R *for* n^+p *diode array corresponding to a two-step anneal for 0 h, 1 h, and 16 h at 800°C in argon and 16 h at 1050°C in dry* O_2 *(*τ *curves) and 0 h, 4 h, and 16 h at 800°C in argon and 16 h at 1050°C in dry* O_2 *(*I_R *curves).*

parently uniform distribution of I_R for the 0-h and 4-h anneals at 800°C results from a surface generation current limit at $I_R \approx 200$ pA for the gated diode structures. For $I_R < 200$ pA the diode current is dominated by a surface component of current at the perimeter of the diode rather than the "normal" depletion region current that is generated in the depletion volume bounded by the area of the diode. Consequently, no valid correlations between $[O]_i$ or $\Delta[O]$ and I_R were obtained for $I_R < 200$ pA.

Figure 8 compares the axial distributions for the "near-surface" τ_s and "bulk" τ_B lifetimes obtained from T_s and T_{rr} measurements on the NMOS capacitor/diode arrays. The τ-curves correspond to a two-step precipitation anneal for 16 h at 800°C in argon followed by 16 h at 1050°C in oxygen. The dashed curves represent the initial and final oxygen distribution corresponding to this two-step anneal. Both τ-curves show a lifetime maximum at the 30 to 40 cm location in the ingot, where $[O]_i$ and $\Delta[O]$ are minimum. The maximum average values for τ_s and τ_B are approximately 180 μs and 4μs respectively. Both curves show a decrease in lifetime at the seed and tang-ends of the ingot, where oxygen precipitation is dominant.

At the seed-end of the ingot, the minimum average values for τ_s and τ_B are 10 μs and 0.31 μs respectively. The values represented by the τ_s profiles are more than a factor of ten greater than the corresponding values for the τ_B profile over the entire length of the ingot. In the regions where oxygen precipitation dominates, for $[O]_i > 1.3 \times 10^{18}$ cm^{-3} the oxygen reduction ratio ($[O]_i/[O]_f$) is greater than 2, and the corresponding lifetime reduction ratio (τ_{max}/τ_{min}) is greater than 10. This result compares favorably with Fig. 7 for the diode arrays which were fabricated by a relatively simple process. The comparison provides strong evidence for the influence of oxygen precipitation independent of the subsequent process thermal history.

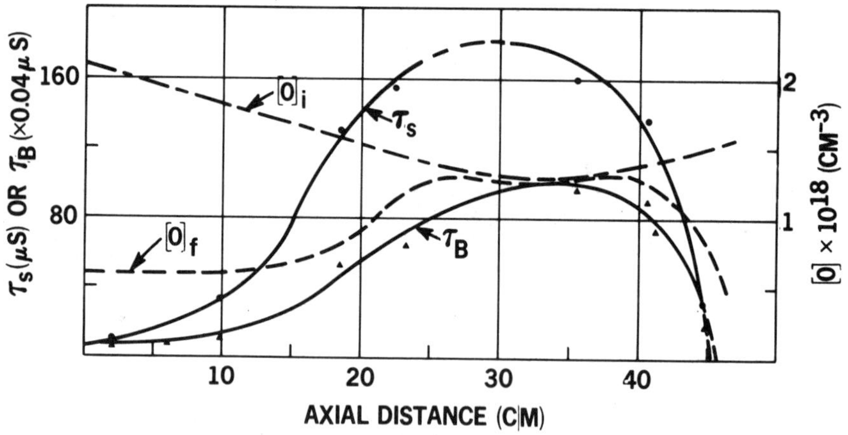

FIG. 8—*Axial distribution plot of the near surface τ_s and "bulk" τ_B lifetimes for NMOS capacitor and diode arrays and the initial* $[O]_i$ *and final* $[O]_f$ *oxygen concentrations for a two-step anneal for 16 h at 800°C in argon and 16 h at 1050°C in dry oxygen.*

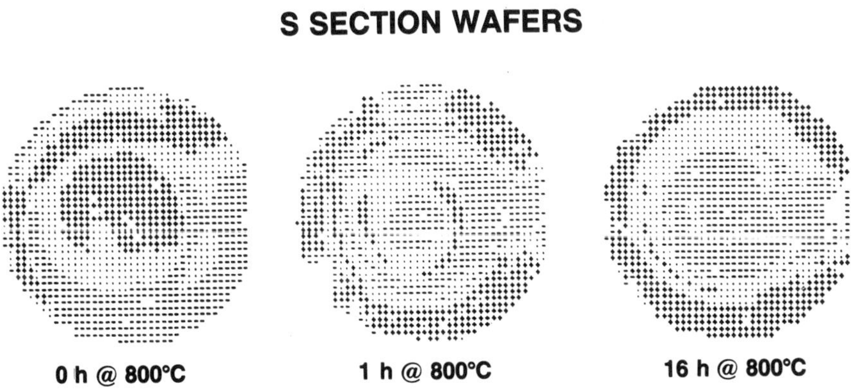

FIG. 9—*Minority carrier lifetime range maps for seed-end wafers subjected to a two-step anneal for 0 h, 1 h, and 16 h at 800°C followed by 16 h at 1050°C in dry oxygen.*

Radial Distribution of Minority Carrier Lifetime

The radial distribution of minority carrier lifetime is shown in Fig. 9 for three adjacent seed-end wafers selected from a region of the crystal having an oxygen concentration $[O]_i = 1.9 \times 10^{18}$ cm^{-3}. The range characters in the τ maps are partitioned at 33% each. The range character limits for Fig. 9, which represents the 0-h anneal at 800°C, are defined by the following range inequality: 2.96 μs (−) $< \tau$ (·) $<$ 3.93 μs (+). The (+) characters represent the upper range values of τ. For the 1-h anneal at 800°C the map which is now defined by the range inequality 900 ns (−) $< \tau$ (·) $<$ 1.36 μs (+) shows that a spatial redistribution has occurred with a tendency for the (+) characters to form an annular distribution at the perimeter of the wafer. Subsequent to the 16-h anneal at 800°C, the central region of the wafer is populated by the (−) characters representing the lower range values, whereas the annular region now represents predominantly the higher range values of τ. In this case, the range limits are defined by 257 ns (−) $< \tau$ (·) $<$ 359 ns (+).

The redistribution of the upper and lower range values of τ represented by the (+) and (−) characters respectively in Fig. 9 reveals the effects of the inhomogeneous distribution of oxygen in the crystal and the nonuniform nature of the precipitation process resulting from crystal growth conditions and the two-step anneals. Wafers selected from the region of the ingot where no precipitation is evident from their infrared measurements show no similar redistribution of τ-values in their respective range maps.

Discussion

The axial distribution of oxygen represented by the $[O]_i$ curve in Fig. 1 is typical of silicon crystals with a diameter of ~75 mm grown from a medium size melt (~7 kg). The linear decrease of $[O]_i$ with distance along the ingot (melt fraction solidified) indicates that the effective segregation coefficient for oxygen $k_e > 1$. Published values for the equilibrium segregation coefficient k_0 are in the range $1.0 < k_0 < 1.5$ [*2*]. The oxygen concentration increase in the tang region of the ingot is associated with a change of the surface area–to-volume ratio of the melt at the base of the crucible, which has a hemispherical shape. Geometry effects and thermal conditions other than "normal" segregation during freezing, which determine the oxygen concentration and its axial distribution in the ingot, are generally associated with the oxygen transport and flow characteristics of the melt. The radial distribution of oxygen is also influenced by factors associated with melt dynamics and is dependent on the stability and thickness of the boundary layer at the solid/liquid interface during the growth process [*2*].

During crystal growth, it is postulated that nuclei are formed in the crystal either by the agglomeration of oxygen atoms during cooling or at existing microprecipitates such as carbon, SiC, or SiO_2 [*2,13,25*]. The axial and radial distribution of oxygen is inhomogeneous [*5*] and therefore one can anticipate an inhomogeneous spatial distribution for the oxygen precipitates and microdefects [*21*]. At 800°C small precipitates are formed that grow very slowly and are essentially rate-limited by the diffusion of oxygen [*26*]. During subsequent anneals at 1050°C rapid growth of the precipitates occurs due to the increased diffusion rate for oxygen. Secondary defects in the form of prismatic dislocation loops are often generated [*27*] by the precipitation-induced stresses. The oxygen precipitates, their local strain field, and their attendent secondary defects are factors that can directly influence the local and distributed minority carrier lifetime.

Figure 1 shows that the initial oxygen concentration throughout the ingot is above the solubility limits of 7.3×10^{17} cm^{-3} and 5.3×10^{17} cm^{-3} as determined from the two-step precipitation anneals in oxygen and argon, respectively. The [O] curve for the wafer that is not annealed at 800°C shows a small decrease for the oxygen concentration after the 18-h anneal at 1050°C. The optical micrograph and absorption spectrum shown in Fig. 3 for an unannealed S-wafer show slight evidence for oxygen precipitation in the center of the wafer. A careful examination of its IR spectrum reveals the development of a broad band at 1150 to 1250 cm^{-1} adjacent to the [O] absorption peak at 1106 cm^{-1}. The τ-curve shown in Fig. 7 for the 0-h anneal shows a slight decrease from its peak value of 4.3 μs at ~28 cm to 2.8 μs in the S-region of the ingot. The corresponding C wafer shown in Fig. 5, which is selected near the peak of the τ-curve, shows no evidence for oxygen precipitation and its IR spectrum reveals no broad spectral band at 1150 to 1250 cm^{-1}.

For the 1-h anneal, oxygen precipitation is dominant in the initial 20 cm of the ingot, as shown in Fig. 1, although the curve remains above the solubility limit. The optical micrograph in Fig. 3 for an S-wafer shows strong evidence for oxygen precipitation and the absorption spectrum shows further development of a broad band at 1150 to 1250 cm^{-1}.

The annular band observed in Fig. 3 for the 1-h anneal corresponds to a region of reduced oxygen precipitation. This results from the typical radial profile of oxygen in silicon in which the oxygen concentration decreases near the perimeter of the ingot [3]. This effect results from oxygen transport near the surface of the melt during growth, and is characteristic of wafers from the seed-end of the ingot, where growth conditions favor this type of distribution. A reexamination of Fig. 9 for the 16-h anneal indicates that the annular region containing diodes having the greater relative values of τ corresponds to the location of the annular band in Fig. 3 where oxygen precipitation is reduced.

The τ curve for the 1-h anneal in Fig. 7 shows a sharp decrease for τ in the S-region of the ingot. The optical micrograph for a C-wafer corresponding to the 1-h anneal shown in Fig. 5 reveals no similar evidence for precipitation and its IR spectrum is relatively unchanged. The apparent reduction of the peak value for the τ-curves with increased anneal time relative to the τ-curve for the unannealed group shown in Fig. 7 is probably not significant, since the data distribution did not permit an accurate determination of τ-values in the peak region of the distribution.

The [O] distribution for the 4-h anneal (Fig. 1) shows that oxygen has precipitated to the solubility limit over a 10-cm section of the ingot. A comparison of the optical micrographs in Figs. 4 and 6 shows that whereas extensive precipitation has occurred in the S-wafers shown in Fig. 4, only slight precipitation is evident for the C-wafer in Fig. 6. The broad spectral band at 1150 to 1250 cm^{-1} has developed into a well-defined absorption peak at 1215 to 1230 cm^{-1}, which indicates the presence of an SiO_2 phase during precipitation.

The 16-h anneal at 800°C in Fig. 1 shows an extended range of precipitation (~15 cm) in which the oxygen concentration is at the solubility limit. The micrographs for the S and C wafers shown in Figs. 4 and 6 respectively both reveal oxygen precipitation, and their absorption spectra show strong SiO_2 peaks. The degree of oxygen precipitation, however, differs significantly. For the S-wafer the precipitated oxygen concentration $\Delta[O] = 1.4 \times 10^{18}$ cm^{-3}, whereas for the C-wafer a value for $\Delta[O]$ of 7.3×10^{17} cm^{-3} is obtained. The corresponding τ-curve for the 16-h anneal shown in Fig. 7 clearly shows this distinction. In the S-region of the ingot, where $\Delta[O] = 1.4 \times 10^{18}$ cm^{-3}, the lifetime reduction ratio due to oxygen precipitation $\tau_{max}/\tau_{min} = 7$, whereas for the C-region of the ingot, where $\Delta[O] = 7.3 \times 10^{17}$ cm^{-3}, $\tau_{max}/\tau_{min} = 1.5$. The abrupt increase for the I_R curve in Fig. 7 for the 16-h anneal can be correlated with the decreasing lifetime in the region where $\tau < 2$ μs. In this region, the thermal generation current in the depletion region of the diodes exceeds the surface generation current at ~200 pA.

The relationship between τ and I_R and the concentration of precipitated oxygen $\Delta[O]$ is shown in Fig. 10. This graph reveals the strong relationship between the measured lifetime values and the concentration of precipitated oxygen resulting from the anneals at 800°C. The maximum lifetimes are observed in the region where $\Delta[O] < 1 \times 10^{17}$ cm^{-3} for this ingot. The apparent plateau for the τ-curve in Fig. 10 suggests that the carrier lifetime may be limited by the soluble oxygen concentration in the crystal, which is 1.3×10^{17} cm^{-3} for this region of the ingot. This value is above the solubility limit at 1050°C; consequently, an early stage of precipitation cannot be ruled out, although it is not detectable with infrared absorption methods.

It is apparent in Fig. 8 that the near-surface lifetime represented by the τ_s curve is significantly greater than the "bulk" lifetime represented by the τ_B curve; however, the effects of oxygen precipitation are common to both the τ_s and τ_B distributions. If we assume that oxygen precipitation has not occurred to any appreciable degree in the region corresponding to the broad minima 28 cm $< X <$ 38 cm, then it is interesting to speculate that enhanced lifetime in the near-surface layer is due primarily to a reduction of the oxygen concentration. The concentration of oxygen, above the solubility limit, at 1050°C, which would be reduced at the surface due to outdiffusion, may be the limiting factor controlling the minority carrier lifetime at the surface. This interpretation suggests that improved lifetime would result primarily from a reduction of the excess soluble oxygen concentration.

The progressive decrease of minority carrier lifetime with increasing anneal time at 800°C (Fig. 7) is associated with an increasing concentration of precipitated oxygen (Fig. 10) and the progressive development of an SiO_2 absorption peak at 1215 to 1230 cm^{-1} (Figs. 4 and 6). This provides strong experimental evidence for a direct relationship between oxygen precipitation

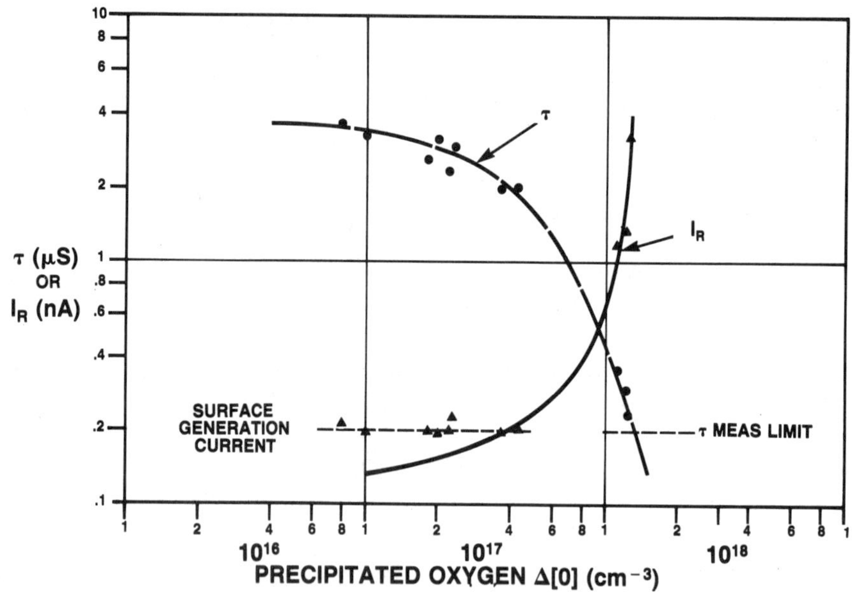

FIG. 10—*Plot showing the relationship of the "bulk" lifetime* τ_B, *reverse current* I_R, *and the concentration of precipitated oxygen* Δ[O] *for a two-step anneal for 16 h at 800°C in argon and 16 h at 1050°C in dry* O_2.

and the resultant degradation of minority carrier lifetime in both the near-surface and "bulk" regions of the wafer (Fig. 8).

The consistently larger values obtained for τ_s relative to τ_B suggest that a lifetime gradient may exist between the near-surface and "bulk" region of the wafer. This effect, which can be attributed to oxygen outdiffusion, is currently under investigation at our laboratory with transient capacitance spectroscopy [*28*]. The wafers used in this investigation do not contain any well-defined denuded zones on wafers which have been cleaved and chemically etched, though this was reported by other investigators [*29–31*] for similar two-step thermal anneals. Published results on the existence of denuded zones, however, generally apply to wafers subjected to two-step thermal anneals without subsequent device processing, as reported in this investigation. The precipitation of oxygen is clearly responsible for the reduced carrier lifetime on S-wafers, as indicated by the τ_s and τ_B data. The C-wafers show little evidence for precipitation and exhibit much greater near-surface and "bulk" lifetimes, although an early stage of precipitation consisting of very small precipitates cannot be entirely discounted.

Transmission electron microscopy (TEM) analysis has been conducted at the University of Arizona[3] on annealed seed-end wafers. The wafers, annealed for 16 h at 800°C, contain precipitates measuring 100 to 200 nm in diameter that are surrounded by prismatic dislocation loops punched out in [110] directions in the crystal. The precipitates appear to be platelets lying on (100) planes normal to the wafer surface, as reported by Wada et al [*26*]. The volume density of the precipitates as calculated from a typical TEM micrograph is $\sim 2 \times 10^{10}$ cm^{-3}. The total defect density including the dislocation loops is estimated to be $\sim 5 \times 10^{14}$ cm^{-3}. Detailed results of this TEM study will be published elsewhere.

Minority carrier lifetime reduction can be attributed to at least three factors: (*1*) the concentration of oxygen above the solubility limit during processing, which represents a metastable condition; (*2*) defects that result directly from oxygen precipitation; and (*3*) segregation of metallic impurities due to solubility enhancement or the precipitation of metallic impurities at defect sites during cooling. On the basis of the experimental results presented here, those conditions that appear to minimize the precipitation of oxygen maximize the minority carrier lifetime. Reduction of the oxygen concentration in the near-surface region of the wafer or a general reduction of the oxygen concentration in the crystal minimizes oxygen precipitation and secondary defect generation during subsequent heat treatment. The use of processing technology referred to by the generic term *gettering* [*32*] to prevent undesired segregation of metallic impurities in or near the active region of the device should minimize the effects of metallic impurities.

[3] Private communication, L. J. Demer and L. H. Chang, University of Arizona, Phoenix, 1981.

Summary and Conclusion

The effects of oxygen precipitation on the axial and radial distribution of minority carrier lifetime has been investigated for a dislocation-free, boron-doped CZ crystal with a (100) growth orientation. Axial profiles for the oxygen concentration and the minority carrier lifetime were obtained from measurements on selected unannealed, annealed, and annealed-and-processed wafers which were marked to identify their original position in the ingot. A two-step anneal was used consisting of a 16-h anneal at 800°C in argon followed by a 16-h anneal at 1050°C in argon or dry oxygen to force oxygen precipitation over an extended range of oxygen concentration. Axial profiles of the oxygen concentration were obtained by using infrared absorption spectroscopy on selected wafers before and after annealing. The concentration of interstitial oxygen was calculated from measurements of the absorption coefficient at 1106 cm^{-1}. The presence of an SiO_2 phase resulting from precipitation was monitored from changes in the absorption band at 1215 to 1230 cm^{-1}. The average lifetime values for the wafers were calculated from the reverse recovery time T_{rr} and capacitance relaxation time T_s measurements on n^+p diode and NMOS capacitor arrays.

At the seed-end of the ingot the initial oxygen concentration is near its maximum, $\sim 2 \times 10^{18}$ cm^{-3}, and oxygen precipitates to its solubility limit of 5 to 7×10^{17} cm^{-3} during the 16-h anneal at 1050°C. In this region, the average lifetimes are minimal, corresponding to 10 μs and 0.31 μs for the near surface and "bulk" respectively. In the region where the initial oxygen concentration is minimum, 1.3×10^{18} cm^{-3}, oxygen precipitation is minimal and the average lifetime is at its maximum, 180 μs and 4 μs for the near surface and "bulk" respectively. The experimental results indicate that oxygen precipitation is the dominant mechanism contributing to the degradation of minority carrier lifetime. The excess oxygen concentration in solution above its solid solubility apparently introduces an upper limit to the minority carrier lifetime. This effect may result from an early stage of precipitation and requires further investigation. The radial distribution of the minority carrier lifetime is inhomogeneous and suggests that radial oxygen inhomogeneities which are known to exist are the cause of these nonuniformities. The detection of an absorption peak in the infrared spectrum at 1215 to 1230 cm^{-1}, which indicates the presence of an SiO_2 phase during precipitation, and the detection of precipitate platelets in these wafers provide strong additional evidence in support of a direct relation between oxygen precipitation and the minority carrier lifetime degradation.

Acknowledgments

The authors thank Ross Boyle, Jerry Maki, and Pat Huegle for their assistance with the experimental work, Marlene Scott for her help in preparing the manuscript, Ed Hall for discussions and review of this manuscript—all with SRDL, Motorola Inc., Phoenix, Arizona—, and to J. B. Price for his help with NMOS processing in APRDL, Motorola Inc., Mesa, Arizona. Special thanks are due Lou Demer and Li Chang at the University of Arizona for their TEM analysis.

This work was supported in part by the National Science Foundation under its Industry/University Cooperative Program (IUC) Grant DMR-79-18023.

References

[*1*] Benson, K. E., Lin, W., and Martin, E. P. in *Semiconductor Silicon 1981,* Vol. 81-5, The Electrochemical Society, Pennington, N.J., 1981, p. 33.

[*2*] Abe, T., Kikuchi, K., Shirai, S., and Muraoka, S. in *Semiconductor Silicon 1981,* Vol. 81-5, The Electrochemical Society, Pennington, N.J., 1981, p. 54.

[*3*] Liaw, H. M., *Semiconductor International,* Vol. 2, No. 8, Oct. 1979, p. 71.

[*4*] Patel, J. R. in *Semiconductor Silicon 1981,* Vol. 81-5, The Electrochemical Society, Pennington, N.J., 1981, p. 189.

[*5*] Abe, T., Kikichi, K., and Shirai, S. in *Semiconductor Silicon 1977,* Vol. 71-1, The Electrochemical Society, Pennington, N.J., 1977, p. 95.

[*6*] Murgai, A. in *Semiconductor Silicon 1981,* Vol. 81-5, The Electrochemical Society, Pennington, N.J., 1981, p. 113.

[*7*] Fuller, C. S., Ditzenberger, N. B., Hannay, N. B., and Buehler, E., *Physical Review,* Vol. 96, No. 3, Nov. 1954, p. 833.

[*8*] Kaiser, W., Frisch, H. L., and Reiss, H., *Physical Review,* Vol. 112, No. 5, Dec. 1958, p. 1546.

[*9*] Kimerling, L. C. and Benton, J. L., *Applied Physics Letters,* Vol. 39, No. 5, Sept. 1981, p. 410.

[*10*] Kanamori, A. and Kanamori, M., *Journal of Applied Physics,* Vol. 50, No. 12, Dec. 1979, p. 8095.

[*11*] "Oxygen in Silicon," Chapter 3 of *Silicon Material Processing (II), Semiconductor Silicon 1981,* The Electrochemical Society, Pennington, N.J., 1981, pp. 187–328.

[*12*] Osaka, J., Inoue, N., and Wada, K., *Applied Physics Letters,* Vol. 36, No. 4, Feb. 1980, p. 288.

[*13*] Kishino, S., *Journal of the Physics Society of Japan,* Vol. 49, Supplement A, 1980, p. 49.

[*14*] Hu, S. M., *Journal of Applied Physics,* Vol. 52, No. 6, June 1981, p. 3974.

[*15*] Yamamoto, K., Kishino, S., Matsushita, Y., and Iizuka, T., *Applied Physics Letters,* Vol. 36, No. 3, Feb. 1980, p. 195.

[*16*] Kaiser, W., Keck, P. H., and Lange, C. F., *Physical Review,* Vol. 101, No. 4, Feb. 1956, p. 1264.

[*17*] Hrostowski, H. J. and Kaiser, R. H., *Physical Review,* Vol. 107, 1957, p. 966.

[*18*] Corbett, J. W., McDonald, R. S., and Watkins, G. C., *Journal of Physics and Chemistry of Solids,* Vol. 25, No. 8, Aug. 1964, p. 873.

[*19*] Boyle, R., "The Measurement of Interstitial Oxygen in Single-Sided Polished Wafers by Infrared (IR) Spectroscopy," Internal Technical Report, Motorola Inc., 1981.

[*20*] Newman, R. C., and Willis, J. B., *Journal of Physics and Chemistry of Solids,* Vol. 26, No. 2, Feb. 1965, p. 373.

[*21*] Varker, C. J., Whitfield, J. D., Rao, K. V., and Demer, L. J. in *Semiconductor Silicon 1981,* Vol. 81-5, The Electrochemical Society, Pennington, N.J., 1981, p. 313.

[*22*] Zerbst, M., *Zeitschrift für Angewandte Physik,* Vol. 22, No. 1, Jan. 1966, pp. 30–33.

[*23*] Heiman, F. P., *Transactions of IEEE,* Vol. ED-14, No. 11, Nov. 1967, p. 781.

[*24*] Hu, S. M., *Journal of Applied Physics,* Vol. 51, No. 11, Nov. 1980, p. 5945.

[*25*] Shimura, F., Tsuya, H., and Kawamura, T., *Applied Physics Letters,* Vol. 37, No. 5, Sept. 1980, p. 483.

[*26*] Wada, K., Inoue, N., and Kohra, K., *Journal of Crystal Growth,* Vol. 49, No. 4, Aug. 1980, p. 749.

[27] Tan, T. Y., and Tice, W. K., *Philosophical Magazine,* Vol. 34, 1976, p. 614.
[28] Miller, G. L., Lang, D. V., and Kimerling, L. C., *Annual Review of Material Science,* Vol. 7, 1977, p. 377.
[29] Inoue, N., Wada, K., and Osaka, J. in *Semiconductor Silicon 1981,* The Electrochemical Society, Pennington, N.J., 1981, p. 282.
[30] Kugimiya, K., Akiyama, S., and Nakamura, S. in *Semiconductor Silicon 1981,* The Electrochemical Society, Pennington, N.J., 1981, p. 294.
[31] Series, R. W., Barraclough, K. G., and Bardsley, W. in *Semiconductor Silicon 1981,* The Electrochemical Society, Pennington, N.J., 1981, p. 304.
[32] Pearce, C. W., "International Symposium on Gettering in Semiconductors," *Extended Abstracts,* Abstract 483, The Electrochemical Society, Vol. 79-2, Oct. 1979, p. 1224.

A. Rohatgi[1] *and P. Rai-Choudhury*[1]

Defects and Carrier Lifetime in Silicon

REFERENCE: Rohatgi, A. and Rai-Choudhury, P., **"Defects and Carrier Lifetime in Silicon,"** *Silicon Processing, ASTM STP 804,* D. C. Gupta, Ed., American Society for Testing and Materials, 1983, pp. 389–404.

ABSTRACT: The effects of substrate cleaning, high-temperature processing, epitaxial growth, and impurities on carrier lifetime and defects in silicon are investigated. Because of the possible accumulation of particulate contamination from deionized water, there appears to be an optimum rinse time for the final substrate-cleaning step. $POCl_3$, HCl, and back-surface damage gettering techniques improve carrier lifetime and reduce defect density, provided the samples are slow cooled. Generation lifetimes of several milliseconds are obtained from the combined use of back-surface damage and HCl or $POCl_3$ gettering. Carrier lifetimes in epitaxial silicon can be improved by treating the substrates with a chelating solution that removes heavy metals from the surface. Epitaxial layers grown on heavily phosphorus-doped substrates have higher lifetimes than those grown on lightly phosphorus-doped or heavily antimony-doped substrates. Metallic impurities in the concentration range of 10^{11} to 10^{16} cm^{-3} can be detrimental to the carrier lifetime, and their influence can vary with the mode of incorporation and thermal history of the wafer. High-temperature gettering can mitigate the adverse effect of a fast-diffusing impurity in silicon like chromium, but it has negligible influence on a very slow diffuser like molybdenum.

KEY WORDS: defects in silicon, deep-level transient spectroscopy, carrier lifetime, deep levels in silicon, effects of defects on device performance, gettering

Minority-carrier lifetime in silicon is a very important material parameter for device designers. It determines speed, leakage current, and conduction losses of most devices. High carrier lifetime is desirable for solar cells and charge-coupled devices, while devices like fast-switching thyristors demand low carrier lifetime in selected regions. Therefore it is important to be able to control the carrier lifetime, which can be affected directly or indirectly by stacking faults, dislocations, carbon, oxygen, and metallic impurities [*1–7*].[2] Some of these defects may be present in the starting material or can be generated or incorporated during device fabrication. Crystal-growth technology has advanced sufficiently to provide starting wafers with very high carrier lifetimes. The major problem is to preserve or control the carrier lifetime

[1] Westinghouse R&D Center, Pittsburgh, Pa. 15235.
[2] The italic numbers in brackets refer to the list of references appended to this paper.

during subsequent processing such as cleaning, oxidation, heat treatments, epitaxial growth, etc. In this paper, we will discuss the influence of some important process variables on carrier lifetime and defects in silicon. In particular, we will address the following areas: (*1*) optimum deionized-water rinse after chemical treatments; (*2*) effects of cooling rate, oxidation in HCl/O_2 ambient back-surface damage, and $POCl_3$ gettering; (*3*) effects of chelating, epitaxial deposition temperature, and substrate doping; and (*4*) effects of metallic impurities before and after $POCl_3$ diffusion.

Experimental Work

Wafers were cleaned by various solvent and acid treatments. Specific chemical mixtures used prior to the final deionized-water rinse were of secondary importance. Chelating and final water rinse were found to be important and are investigated in more detail. Substrate-cleaning studies were conducted on 3 to 4 Ω-cm *p*-type silicon, which gave a photoconductive decay (PCD) lifetime of ~40 μs prior to any heat treatment. The wafers were cleaned by using standard solvents and acids followed by hydrogen peroxide/ammonium hydroxide (H_2O_2/NH_4OH) and hydrogen peroxide/hydrogen chloride (H_2O_2/HCl) treatments. In order to see the effect of final water-rinse time on post-heat-treatment lifetime, the specimens were oxidized in an atmosphere of 95% O_2 + 5% HCl at 1000°C for 15 min to distribute the fast-diffusing residual surface contamination, if any, throughout the bulk. The samples were slow-cooled at a rate of 1°C/min. The oxide layer was removed and the measured PCD lifetime was used as a relative indication of surface contamination from the deionized-water rinse.

High-temperature process-induced effects were investigated by fabricating metal-oxide-semiconductor (MOS) capacitors on 4 to 8 Ω-cm *n*-type (100) float zone silicon [*8*]. Approximately 1500-Å oxides were grown at various temperatures (900 to 1200°C) in different ambients (0 to 6% HCl/O_2 mixtures). Samples were annealed for 20 min in N_2 at the oxidation temperature prior to cooling. The cooling rate in the N_2 ambient was varied in the range of 1 to 10°C/min, and a 1°C/min rate was selected for all the slow-cool experiments. Back-surface damage was provided by lapping the back side of the wafer with 60-μm Al_2O_3 powder prior to oxidation. In order to obtain enhanced process sensitivity, 100 Ω-cm *n*-type silicon was used for $POCl_3$ gettering performed at 900°C for 45 min. Back-surface damage for the $POCl_3$ gettering experiment was provided by a 180-keV phosphorous implant with a dose of 10^{16} cm^{-2}.

High-temperature process effects were evaluated by measuring generation lifetime (τ_g) in the bulk silicon by the pulse MOS-C technique [*9,10*]. The MOS capacitors were sintered at 425°C for 1 h in H_2 to keep the surface generation small, and the following simple approximation [10] was used to calculate the lifetime:

$$\tau_g = \frac{n_i}{8N_D} \cdot \frac{C_F}{C_o}\left(1 + \frac{C_i}{C_F}\right) \cdot t_F \qquad (1)$$

where C_o is the oxide capacitance, C_i is the initial capacitance in deep depletion, C_F is the inversion capacitance, n_i is the intrinsic carrier concentration, N_D is the donor concentration, and t_F is the time it takes to recover from deep depletion to inversion. In selected instances, oxidation-induced defects in silicon were delineated by removing the metal and the oxide, and then preferentially etching ~3-μm silicon with Seeco etch.

Epitaxial deposition was carried out in a horizontal rf-heated reactor by hydrogen reduction of $SiCl_4$. The substrate-cleaning procedure consisted of degreasing in trichloroethylene and acetone followed by boiling in $3H_2SO_4$:HNO_3. This was followed by two methods of removing heavy metals from the substrate surface. The first method uses the ammonium salt of EDTA (ammonium ethylenediaminetetracetate) as a chelating agent; the second method uses H_2O-H_2O_2-NH_4OH to remove organic contaminants and H_2O-H_2O_2-HCl to remove metallic impurities. All wafers were given an *in situ* HCl etch prior to epitaxial growth. After epitaxial growth, MOS capacitors were formed by chemical-vapor deposition of SiO_2 at 750°C, using a silane-oxygen process followed by evaporation of aluminum dots. Minority-carrier generation lifetime was determined by the pulse MOS-C technique [9, 10] described earlier. The capacitors were annealed in hydrogen to minimize the effect of the Si-SiO_2 interface on the pulsed MOS-C response.

Metallic impurities can be incorporated in the starting material during subsequent processing. It is often difficult to identify the impurity and its true concentration in a processed wafer. Moreover, it may not be the only factor controlling carrier lifetime. Therefore, in order to evaluate the impurity effects, we intentionally incorporated controlled amounts of various impurities in 4 Ω-cm, (111) *p*- and *n*-type Czochralski crystals. These impurities were grown into the crystals by contaminating the silicon melt prior to crystal growth [11]. The impurity concentration range was 10^{11} to 10^{17} cm^{-3}. Impurity-induced defects or deep levels were delineated by deep-level transient spectroscopy (DLTS) [12, 13]. DLTS measurements were performed on *p*-type Si by fabricating 0.76-mm (30-mil) diameter Ti-Si Schottky barrier diodes with Au on top of Ti. Aluminum metal evaporation followed by a 450°C/30 min N_2 anneal was used for *n*-type silicon. In order to see the effect of impurities on carrier lifetime, 1 by 1 cm n^+-*p* diodes or solar cells were fabricated by an 825°C/50 min $POCl_3$ diffusion. Solar cell efficiency and open-circuit voltage-decay measurements were used to evaluate the impurity effects [11].

In selected instances, the phosphorus-rich n^+ region was removed by etching followed by a series of steps etched into the p-silicon. DLTS measurements were performed on each step via Schottky barrier diodes to study the influence of $POCl_3$ diffusion on grown-in impurities in the bulk silicon.

Results and Discussion

Substrate Cleaning

As mentioned earlier, final deionized-water rinse time was found to be a crucial step in the substrate cleaning. Therefore this is studied in more detail.

The effect of final water rinse time on the recombination lifetime is shown in Fig. 1. The presence of a maximum in the curve indicates two competing mechanisms. The first mechanism is likely to involve dissolution into the water of metallic impurity salts left on the wafer surface as a result of prior cleaning steps. This phenomenon will cause an increase in the bulk lifetime. The opposing effect of reduction in lifetime could come from the accumulation of particulate contamination on the wafer coming from the deionized water. A mathematical model using both mechanisms has been fitted to the data points, and the agreement is quite good [*14*]. The fact that these contaminants originate at the wafer surface and redistribute during subsequent heat treatment was also demonstrated by examining the effect of wafer

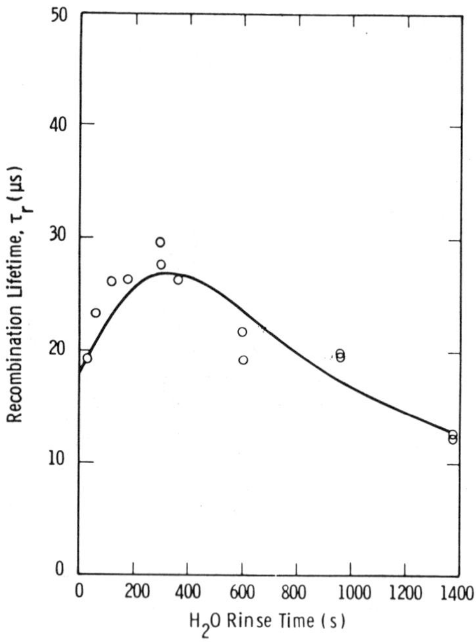

FIG. 1—*Experimental data on postanneal recombination lifetime compared to theoretical curve.*

thickness on postanneal lifetime. Lifetime was found to increase with wafer thickness, as may be expected from the redistribution of a fixed amount of surface contaminants.

High-Temperature Processing

Float-zone silicon used in this study did not show oxidation-induced stacking faults, and dislocation density was also quite small (~25 cm^{-2}). The dominant defects were saucers, which are small, circular, shallow, and flat-bottomed pits (Fig. 2).

Table 1 shows the effects of cooling rate, damage gettering, and HCl oxidation on lifetime and defects for an 1100°C oxidation. The data show that when a sample is quenched after oxidation, lifetime is very small (~20 μs), regardless of damage gettering or HCl oxidation. Figure 2 and Table 1 show that damage gettering reduces oxidation-induced saucer density by two orders of magnitude $10^6 \rightarrow 10^4$ cm^{-2}, that HCl oxidation reduces saucer den-

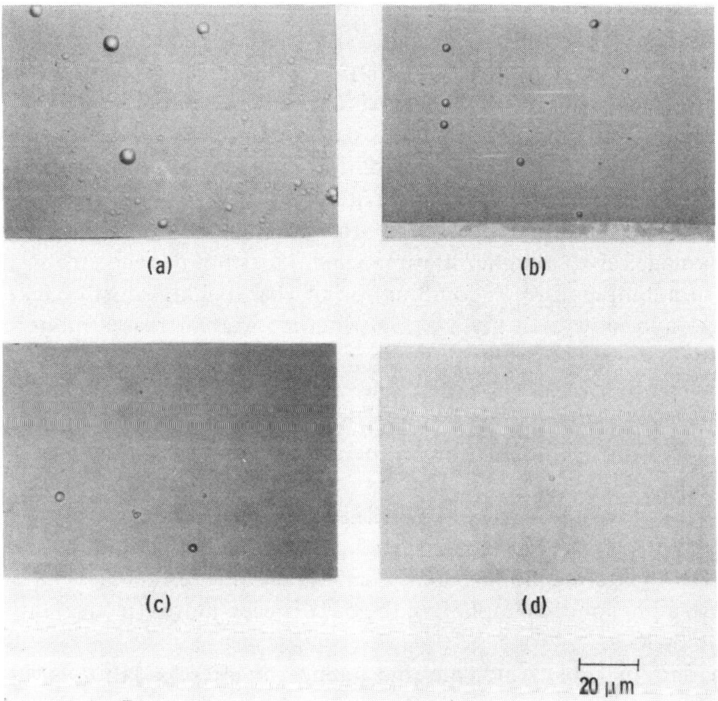

FIG. 2—*Effect of back-surface damage gettering and HCl oxidation on saucer defects:* (a) *0% HCl oxidation without back-surface damage, ~1 × 10^6 saucers/cm^2;* (b) *1% HCl oxidation without back-surface damage, ~1 × 10^5 saucers/cm^2;* (c) *0% HCl oxidation with back-surface damage, 1 × 10^4 saucers/cm^2;* (d) *1% HCl oxidation with back-surface damage, ~1 × 10^3 saucers/cm^2.*

TABLE 1—*Effect of cooling rate, back-surface damage gettering, and HCl oxidation on lifetime and defects in FZ silicon.*[a]

Process	Lifetime, μs	Saucer Density, cm^{-2}
No BSD[b] 0% HCl quench	20	10^6 to 5×10^6
BSD 0% HCl quench	20	10^4 to 5×10^4
BSD 1% HCl quench	20	10^3 to 5×10^3
No BSD 0% HCl slow-cool	100	10^6 to 5×10^6
BSD 0% HCl slow-cool	500	10^4 to 5×10^4
BSD 1% HCl slow-cool	2000	10^3 to 5×10^3
BSD 6% HCl slow-cool	2100	10^3 to 5×10^3

[a] Oxidation temperature 1100°C.
[b] BSD = back-surface damage.

sity by an order of magnitude $10^6 \rightarrow 10^5$ cm^{-2}, and that a combination of the two brings down saucer density to $\sim 10^3$ cm^{-2}. The saucer density on the starting wafer was in the range of 500 to 1000 cm^{-2}. Slow cooling, without damage and HCl gettering, gives somewhat higher lifetime ($\sim$100 μs), but saucer density remains in the range 1 to 5×10^6 cm^{-2}. Introducing back-surface damage gives a substantial increase in lifetime ($\sim$500 μs), if the sample is slow-cooled. One percent HCl oxidation with damage and slow cool gives lifetime on the order of 2 ms. Increasing the HCl concentration from 1 to 6% has very little effect on lifetime and saucer density.

Figure 3 shows the effect of oxidation temperature on lifetime. When a sample is quenched after oxidation, lifetime is very low and relatively insensitive to the oxidation temperature. Slow cooling; with or without back-surface damage, gives a higher lifetime which does not depend strongly on the oxidation temperature. A combination of HCl oxidation with back-surface damage and slow-cool gives a high lifetime that increases monotonically with the increase in oxidation temperature. Increasing the HCl concentration from 1 to 6% has no appreciable effect on lifetime. This is in agreement with Wang et al [*15*]. Table 2 shows that saucer density remains unchanged with increasing HCl oxidation temperature, despite the systematic increase in lifetime.

Table 3 shows the results of $POCl_3$ gettering after an 1100°C thermal oxidation. Without the $POCl_3$ treatment, generation lifetime is quite small ($\sim$25 μs). $POCl_3$ gettering alone increases carrier lifetime to 1400 μs. $POCl_3$ gettering with a phosphorus implant in the back is even more effective, raising lifetime to 3600 μs.

The data in Table 1 indicate that quenching after oxidation is very detrimental to device lifetime, regardless of back damage and HCl gettering. It has been shown by Wang [*16*] and Yau [*17*] that quenching gives rise to two deep levels, at approximately $E_c - 0.26$ eV and $E_c - 0.49$ eV, which may result from the dissolved impurities quenched in the interstitial sites. Our observation of no improvement in lifetime with the HCl or back-surface dam-

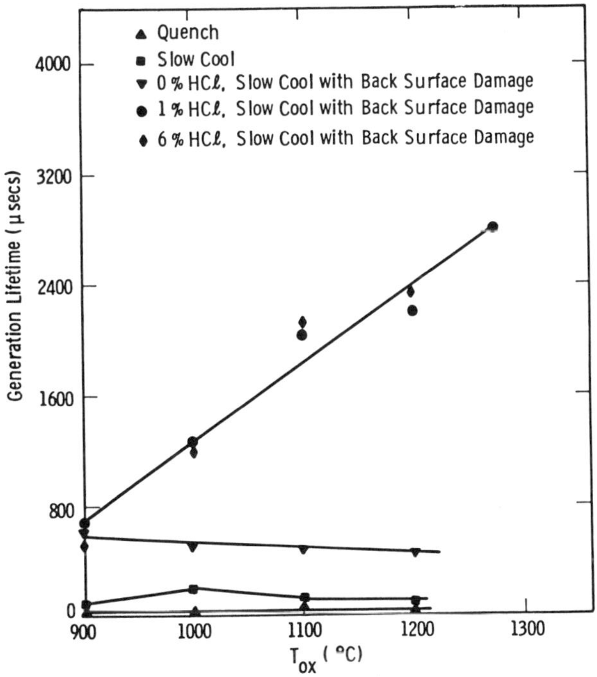

FIG. 3—*Minority-carrier lifetime as a function of oxidation temperature for different oxidation conditions.*

age gettering of the quenched samples is consistent with Yau and Sah [*17*], who found that gettering was ineffective in reducing the concentration of the quenched-in levels. In quenched samples, lifetime is probably controlled by these levels and has no relationship with saucer density (Table 1). On the other hand, in the slow-cooled samples, lifetime is inversely related to saucer density. During slow cooling, the dissolved impurities have enough time to migrate, precipitate, and become electrically inactive; therefore they no longer limit lifetime as in the case of quenched samples. This is supported by two separate experiments which showed a significant loss in carrier lifetime with the increase in quench temperature (Table 4) or the cooling rate (Fig. 4). Yau [*17*] showed that increasing the cooling rate indeed results in an increase

TABLE 2—*Effect of HCl oxidation temperature on lifetime and defects in FZ silicon.*[a]

Oxidation Temperature, °C	Lifetime, μs	Saucer Density, cm^{-2}
900	660	10^3 to 5×10^3
1000	1275	10^3 to 5×10^3
1100	2000	10^3 to 5×10^3
1200	2200	10^3 to 5×10^3

[a] 1% HCl oxidation with back surface damage and slow-cool.

TABLE 3—*Effect of $POCl_3$ gettering on carrier lifetime.*

- 100 Ω-cm, (111), *n*-type wafers
- 1100°C, 90 min, 1% HCl oxidation
- 900°C, 45 min, $POCl_3$ gettering
- 180 keV phosphorus implant, 10^{16} cm^{-2}

Gettering	Generation Lifetime, μs
None	25
$POCl_3$ only	1400
Phosphorus implant + $POCl_3$ (as a back-surface damage)	3600

in the quenched-in deep-level concentration. Moreover, according to Pearce et al [*18*], the oxidation-induced saucers can be generated from heavy metals incorporated during handling, cleaning, and processing of the wafer. They found that saucers are the result of a dislocation loop formed by heavy metal precipitates, although no satisfactory model has yet been proposed to account for their behavior. These saucers, being indicative of metallic contaminants, can reduce lifetime. Therefore, in the absence of other lifetime-limiting sources, lifetime is expected to vary inversely with saucer density (Table 1).

Table 2 shows that HCl oxidation with damage and slow-cool is a very effective process for achieving high lifetime. The lifetime increases monotonically with the increase in HCl oxidation temperature; however, saucer density remains unchanged, which indicates that this systematic increase in lifetime cannot be attributed to saucer density. Several investigators [*19,20*] have pointed out that the defects should be decorated with impurities in order to be detrimental to device performance. It appears from the data that higher HCl oxidation temperatures clean the bulk silicon more effectively by extracting the impurities or by preventing them from getting into the bulk during the oxidation. Therefore, in spite of a constant saucer density, the number of decorated saucers diminishes at higher temperature, which may account for the monotonic increase in lifetime.

The model in Fig. 5 illustrates these process-induced effects on lifetime and defects. It shows that when a sample is quenched after oxidation, lifetime is very low, primarily due to the dissolved impurities in the silicon lat-

TABLE 4—*Effect of quench temperature of lifetime and defects in FZ silicon.*[a]

Quench Temperature, °C	Lifetime, μs	Saucer Density, cm^{-2}
500	2012	10^3 to 5×10^3
600	2000	10^3 to 5×10^3
700	850	10^3 to 5×10^3
800	73	10^3 to 5×10^3
900	40	10^3 to 5×10^3

[a] Oxidation was performed at 1100°C with 1% HCl by using back-surface damaged wafers and a cooling rate of 1°C/min.

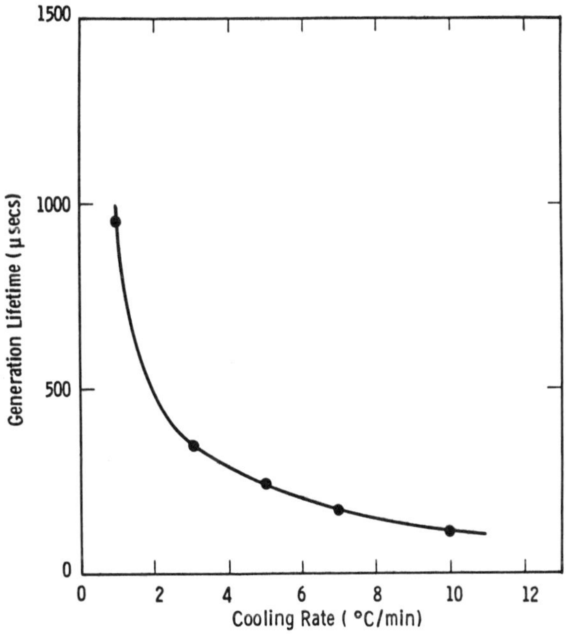

FIG. 4—*Effect of postoxidation cooling rate on generation lifetime.*

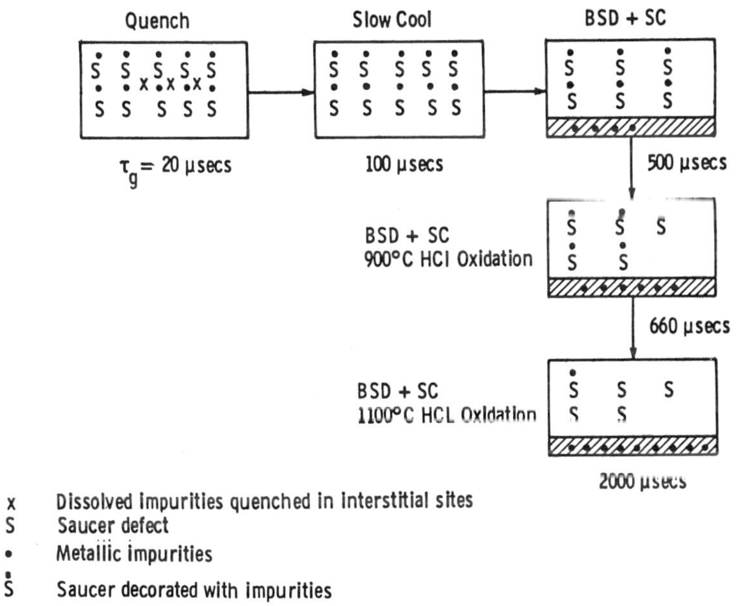

FIG. 5—*Schematic representation of process-induced effects on lifetime and defects.*

tice. When a sample is slow cooled, some of the dissolved impurities are able to precipitate and become electrically inactive, resulting in an increase in lifetime. Slow cooling by itself has no effect on saucer density. However, slow-cooling with back-surface damage reduces saucer density by two orders of magnitude because the damage provides a sink for the point defects and fast-diffusing impurities. This results in a substantial increase in lifetime. HCl oxidation reduces saucer density by an order of magnitude, resulting in some additional increase in lifetime. Increasing the HCl oxidation temperature has no appreciable effect on the net saucer density but decreases concentration of the decorated saucers to improve lifetime further. Thus, by selecting the right process, lifetime can be increased from 20 μs to 2 ms.

Epitaxial Growth

Lifetime data for only the *n*-epitaxial layers (arsenic-doped) are presented here. These layers were grown on three types of substrates: (*1*) 0.0009 Ω-cm phosphorus-doped, and (*2*) 10 Ω-cm phosphorus-doped, and (*3*) 0.008 Ω-cm antimony-doped, and at three temperatures: 1100, 1150, and 1250°C.

The effect of the cleaning procedure is seen in Fig. 6 by comparing the lifetime values of epitaxial films grown at 1150°C for the three substrates. It is quite obvious that the chelated wafers had significantly higher lifetimes than those cleaned in the same manner but not chelated. This points out the heavy-metal complexing property of the chelating agent. Although it is not obvious from the data (each point is the average of ten measurements), the scatter in the data for wafers treated with the EDTA chelating agent is considerably

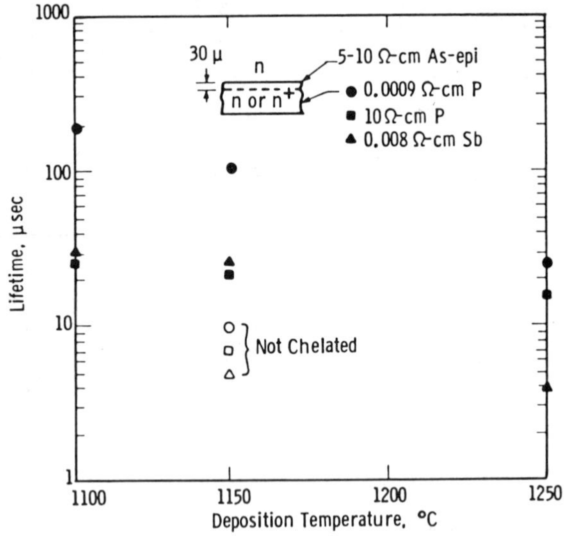

FIG. 6—*Lifetimes in* n*-epitaxial layers as a function of deposition temperature. For 1150°C, lifetimes with and without chelating are shown; all others are chelated.*

less than that for wafers not chelated. For example, when the substrate was not chelated, the lifetime in the layer grown on the heavily phosphorus-doped substrate at 1150°C varied from 1.6 to 28 μs; using chelated substrates, lifetimes varied from 91 to 120 μs. A comparison of these two sets of data points at 1150°C, which were the first to be done in these experiments, prompted us to use a chelating agent for all other runs.

The second important observation to be made from Fig. 6 is that lifetime increases with lower growth temperatures. Although lifetime values do not all change by the same amount, the trend of higher lifetimes for lower temperatures is certainly obvious.

The third observation is that consistently higher lifetimes were obtained on layers grown on degenerately phosphorus-doped substrates. This was found to be the case for all layers (*n*- and *p*-type) grown during this investigation.

In an attempt to improve the lifetime of those layers grown on lightly phosphorus-doped or degenerately antimony-doped substrates, they were gettered by using phosphorus at 950°C. This treatment had essentially no effect on lifetime, which indicated that the impurities responsible for lifetime degradation cannot be easily removed by this gettering procedure.

It is obvious from Fig. 6 that the highest lifetimes were obtained on layers grown on heavily phosphorus-doped substrates. These substrates were characterized by the presence of considerable inhomogeneities, such as impurity striations, phosphorus-induced precipitates, and dislocations. An examination by transmission electron microscopy, however, indicated that epitaxial layers grown on these substrates were of good quality and comparable to layers grown on dislocation-free, high-perfection, high-resistivity substrates. Sirtl etching [*21*] of the layers grown on these substrates revealed only a few dislocations, the number of which decreased with decreasing growth temperature. However, X-ray topographs of epitaxial layers with and without substrates indicated the apparent presence of strain, which Sirtl etch failed to reveal. Although the origin of these apparent strain effects is not known, their presence in layers grown on degenerate phosphorus-doped substrates at all temperatures reduces their importance as lifetime-controlling factors, since the lifetime in these layers was quite high. Thus, with the exception of a slight variation in the dislocation density at a very low level, the layers on all substrates were free from any precipitates and stacking faults. Hence an explanation of lifetime results should primarily come from point-defect considerations. Point defects, such as vacancies, carbon, oxygen, heavy metals, and various complexes, can form numerous deep energy levels [*22,23*] that reduce the carrier lifetime.

The reason for the relatively high lifetime in the epitaxial layers grown on heavily phosphorus-doped substrates is not clearly understood. The following is suggested as a possible explanation. The type and location of the impurity-pinning centers in epitaxial silicon (for instance, thermally induced dislocations extending the entire thickness of the wafer) are such that the impurities cannot be gettered as effectively as in diffused junctions in bulk material. The diffusion-induced dislocations, on the other hand, are formed in the heavily doped side of the junction, with a maximum dislocation density located away from the junction plane [*24–26*], and are decorated with the diffusing dopant. Thus epitaxial silicon is generally more susceptible to contamination, and, once contaminated, affects the device behavior adversely. The diode characteristics (not discussed here) and lifetime results, together with the observed effects of chelating, indicate that lifetime-killing impurities originate primarily from the substrate surface rather than from the epitaxial growth system. High phosphorus concentration contracts the silicon lattice and increases the affinity and solubility of heavy metals such as copper, silver, and gold. Any heavy-metal contamination introduced during substrate preparation may be held more effectively by degenerate phosphorus-doped substrates than by antimony or lightly phosphorus-doped substrates during subsequent epitaxial growth. If the growth temperature is high enough to render these adsorbed impurities mobile, then the grown layer becomes contaminated, causing a decrease in the measured lifetime. Metallic impurities probably constitute the primary mode of layer contamination, either as individual point defects or as complexes with other point defects such as vacancies. Decoration of dislocations with metallic impurities is most likely only a secondary mode of contamination.

Impurity Effects

Figure 7 shows the deep levels induced by various grown-in impurities in silicon. We were unable to detect deep levels due to manganese, silver, tin, nickel, and copper, despite the fact that the impurity concentration in the crystals was several orders of magnitude higher than the DLTS detection limit for these samples. These impurities are generally regarded as fast dif-

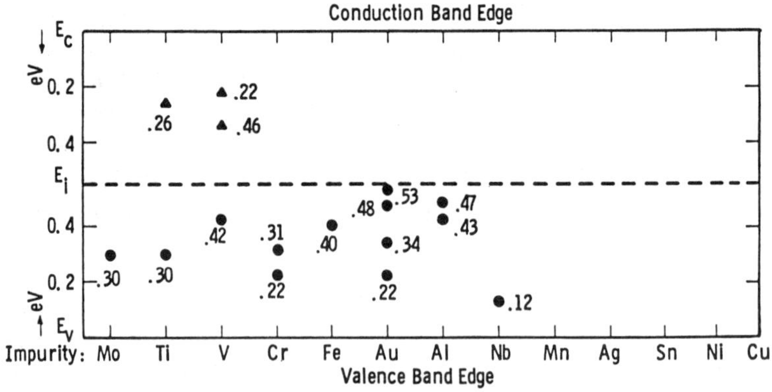

FIG. 7—*Grown-in impurity-induced deep levels in silicon.*

fusers in silicon. Deep levels have been reported [*7,27,28*] due to the above impurities when they were incorporated by diffusion or other methods after crystal growth. We also found cases, such as chromium and aluminum, for which the observed deep levels were not in agreement with values in the literature. For example, reported [*7*] levels for chromium are $E_v + 0.11$ eV, $E_c - 0.23$ eV, and $E_c - 0.41$ eV, but we observed levels at $E_v + 0.22$ eV and $E_v + 0.31$ eV, where E_c is conduction band edge and E_v is valance band edge. Deep levels due to molybdenum, titanium, and vanadium, which diffuse relatively slowly in Si, were found to be in general agreement with the reported values [*7,28*].

The above data indicate that the impurity effect on carrier lifetime may depend upon how it gets into the material. This is expected because deep levels depend on the site or configuration the impurity acquires and the complexes it is able to form. Therefore, the thermal history of the material, solid solubility, and diffusion constant of the impurity in silicon may play an important role in determining the deep levels and their concentrations. Figure 8 indeed shows that a 825°C/50 min heat treatment in a $POCl_3$ ambient can in-

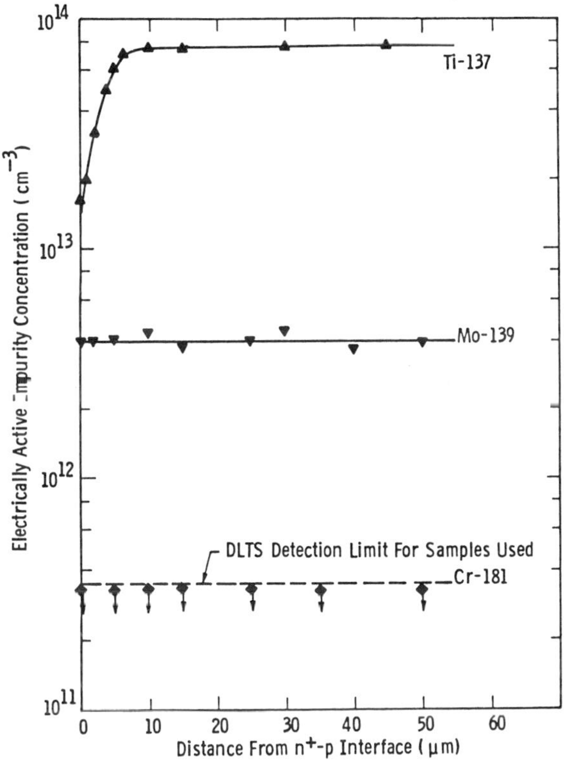

FIG. 8—*Electrically active impurity profiles for several species after an 832°C, 50-min $POCl_3$ treatment.*

fluence the deep-level concentrations of molybdenum, titanium, and chromium impurities. Prior to $POCl_3$ diffusion, molybdenum, titanium, and chromium deep-level concentrations were uniform and equal to 4×10^{12} cm^{-3}, 7.6×10^{13} cm^{-3}, and 1.8×10^{14} cm^{-3}, respectively. Concentration of molybdenum, which diffuses very slowly in silicon, remains unaffected by $POCl_3$ diffusion. Titanium-induced deep-level concentration is lowered at and near the surface, and the deep-level concentration of chromium, which diffuses much more rapidly in silicon compared to molybdenum and titanium, is suppressed by several orders of magnitude throughout the bulk and actually falls below the DLTS detection limit. These data indicate that the adverse influence of fast-diffusing impurities on carrier lifetime can be mitigated by high-temperature gettering techniques.

In order to see the effect of impurities on carrier lifetime, diffused n^+-p solar cells were used as test vehicles. Figure 9 shows the experimental data for the effect of various impurities on solar cell performance and carrier lifetime in the bulk of the cell. The plotted cell efficiencies are normalized with respect to the uncontaminated solar cell efficiency. The normalized cell efficiency η/η_0) and carrier lifetime τ are related according to the empirical equation [*11*]

$$\eta/\eta_0 = 0.872\ I_n{}^{1.128} + 0.128\ I_n{}^{12} \tag{2}$$

where $I_n = C_1(1/C_2/\sqrt{(D\tau)} + 1)$, $C_1 = 1.11$, $C_2 = 19.2\ \mu m$, and D is the diffusivity of carriers.

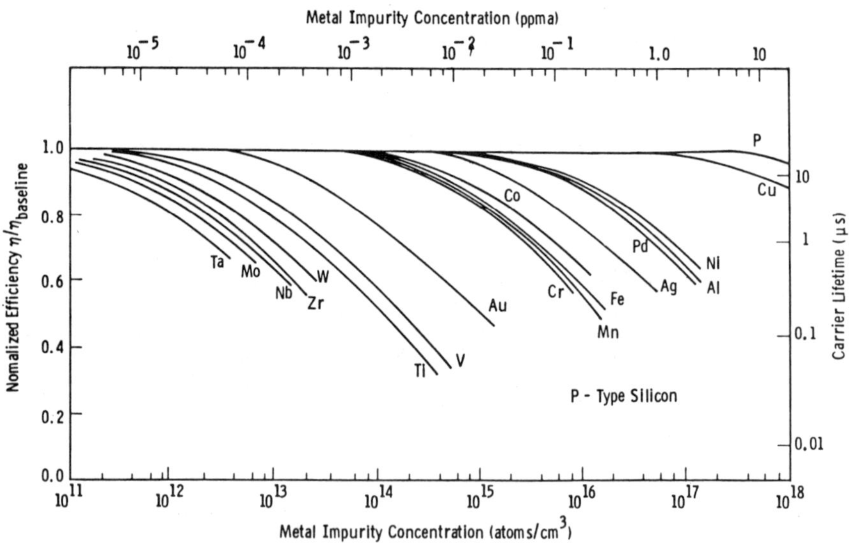

FIG. 9—*Minority-carrier recombination lifetime and normalized efficiency of* n^+-p *solar cells as a function of metallurgical impurity concentration. For clarity, data points are not shown and the curves represent a model fit to the data* [11].

It is important to recognize that these samples have undergone an 825°C/50 min $POCl_3$ diffusion; therefore, the results shown in Fig. 9 do not depict the true impurity effect, but a combination of an impurity and process-induced effect.

Conclusions

Both impurities and defects play an important role in determining the minority carrier lifetime in silicon. This study shows that incorporation or generation of defects and impurities in a processed device can be reduced significantly by appropriate substrate cleaning, high-temperature processing, and controlled cooling. For example, chelating and adequate deionized-water rinse are found to be crucial in keeping the lifetime-killing centers out of the bulk during subsequent high-temperature processing. Rapid cooling results in low carrier lifetime because of the quenched-in deep levels. These levels can be removed by a cooling rate of 1°C/min or less. Process-induced impurities and defects can be minimized by a combination of HCl or $POCl_3$ gettering with back-surface damage and slow-cooling. Such treatments give generation lifetimes of several milliseconds. Gettering reduces the metallic impurity-induced deep level concentration; however, gettering effectiveness diminishes with the decrease in diffusion constant of the impurity in silicon.

Acknowledgments

The authors would like to thank P. D. Blais (deionized-water rinse studies); J. R. Davis and R. H. Hopkins (impurity effects on solar cell); D. K. Schroder ($POCl_3$ gettering and lifetime measurements); S. Karako (DLTS measurements); S. G. Hitchins and R. R. Adams (MOS fabrication and defect-density measurements); D. N. Schmidt, J. B. McNally, and R. Hauser (solar cell fabrication and testing); and G. S. Law (critical reading and preparation of the manuscript).

References

[*1*] Rozgonyi, G. A. in *Semiconductor Silicon 1977*, H. R. Huff and E. Sirth, Eds., The Electrochemical Society, Princeton, N.J. 1977, p. 504.
[*2*] Shiraki, H., *Japanese Journal of Applied Physics*, Vol. 15, No. 1, Jan. 1976, p. 1.
[*3*] Ravi, K. V., Varker, C. J., and Volk, C. E., *Journal of the Electrochemical Society*, Vol. 120, 1973, p. 533.
[*4*] Tan, T. Y. and Tice, W. K., *Philosophy*, Vol. 34, 1976, p. 615.
[*5*] Tice, W. K. and Tan, T. Y., *Applied Physics Letters*, Vol. 28, 1976, p. 564.
[*6*] Monkowski, J. R., *Solid State Technology*, Vol. 44, July 1981.
[*7*] Milnes, A. G., *Deep Impurities in Semiconductors, John Wiley & Sons*, New York, 1973.
[*8*] Rohatgi, A. and Rai-Choudhury, P., *Journal of the Electrochemical Society*, Vol. 127, No. 5, May 1980, p. 1136.
[*9*] Zerbst, M., *Zeitschrift für Angewandte Physik*, Vol. 22, May 1966, p. 30.
[*10*] Schroder, D. K. and Guldberg, J., *Solid State Electronics*, Vol. 14, Dec. 1971, p. 1285.
[*11*] Davis, J. R. et al, *IEEE Transactions on Electron Devices*, Vol. ED-27, No. 4, April 1980, p. 677.

[*12*] Miller, G. L., Lang, D. V., and Kimerling, L. C., *Annual Review of Materials Science*, Vol. 7, 1977, p. 377.
[*13*] Lang, D. V., *Journal of Applied Physics*, Vol. 45, No. 7, July 1974, p. 3023.
[*14*] Blais, P. D. and Seiler, C. F. in *Lifetime Factors in Silicon, ASTM STP 712*, American Society for Testing and Materials, 1979, p. 148.
[*15*] Wang, K. L. and Connery, R. J. in *Semiconductor Characterization Techniques*, P. A. Barnes and G. A. Rozgonyi, Eds., The Electrochemical Society, Pennington, N.J., 1978, p. 386.
[*16*] Wang, K. L., *Journal of the Electrochemical Society*, Vol. 125, Oct. 1978, p. 1664.
[*17*] Yau, L. D. and Sah, C. T., *Journal of Solid State Electronics*, Vol. 17, 1974, p. 193.
[*18*] Pearce, C. W. and McMohan, R., *Journal of Vacuum Science and Technology*, Vol. 14, 1977, p. 40.
[*19*] Claeys, C. L. et al in *Semiconductor Characterization Techniques*, P. A. Barnes and G. A. Rozgonyi, Eds., The Electrochemical Society, Pennington, N.J., 1978, p. 366.
[*20*] Strack, H. et al, *Journal of Solid State Electronics*, Vol. 22, 1979, p. 135.
[*21*] Sirtl, E. and Adler, A., *Z. Metallk*, Vol. 52, 1961, p. 529.
[*22*] Bean, A. R., Newman, R. C., and Smith, R. S., *Journal of Physics and Chemistry of Solids*, Vol. 31, 1970, p. 739.
[*23*] Sze, S. M. and Irvin, J. C., *Solid State Electonics*, Vol. 11, 1968, p. 599.
[*24*] Prussin, S., *Journal of Applied Physics*, Vol. 32, 1961, p. 1876.
[*25*] Cohen, B. G., *Solid State Electronics*, Vol. 10, 1967, p. 33.
[*26*] Mann, J. E. and Sandler, N. P., *Physics Failure Electron.*, Vol. 2, 1964, p. 145.
[*27*] Kimerling, L. C., Benton, J. L., and Rubin, J. J., *Proceedings*, International Conference on Defects in Semiconductors, Oiso, Japan, Sept. 1980.
[*28*] Chen, J. W. and Milnes, A. G., *Annual Review of Material Science*, Vol. 10, 1980, p. 157.

Measurements for Process Control

R. W. Dutton,[1] *P. Fahey,*[1] *K. Doganis,*[2] *L. Mei,*[3] *and H. G. Lee*[4]

Computer-Aided Process Modeling for Design and Process Control

REFERENCE: Dutton, R. W., Fahey, P., Doganis, K., Mei, L., and Lee, H. G., **"Computer-Aided Process Modeling for Design and Process Control,"** *Silicon Processing, ASTM STP 804,* D. C. Gupta, Ed., American Society for Testing and Materials, 1983, pp. 407–421.

ABSTRACT: The use of computer aids for process modeling has grown rapidly over the past five years. Currently, industrial organizations are simulating thousands of process steps each month, both to cut development costs and to create more reproducible results. This paper will review recent progress in the field of process modeling and show typical applications. Both the design and process control aspects of modeling will be discussed.

KEY WORDS: process modeling, device simulation, integrated circuits, N-channel metal-oxide-semiconductor (NMOS), depletion load, phosphorus, polysilicon, diffusion

Process modeling has become an essential part of the design of scaled-down technology for high density metal-oxide-semiconductor (MOS) integrated circuits (ICs). One motivation comes from the reduction in aspect ratio of minimum gate feature size of devices in comparison to source/drain junction depths. In the 1970s this ratio was typically 10:1 or greater and simple threshold voltage models could be applied for MOS devices. During the 1980s it will be typical for the ratio to be reduced, with a junciton depth of 1000 Å becoming common. The classical gradual channel approximation assumes planar junctions of negligible effect on the Poisson solution, which will be violated because of junction depths that are now comparable to thicknesses of the thin dielectrics. Hence the control of electrical parameters of small-geometry MOSFETs will depend critically on both gate dimensions and the bulk junction properties in these devices. The objective of this paper is to summarize the current state of process modeling. The emphasis will be

[1] Stanford University, Stanford, Calif. 94305.
[2] Xerox Corporation, Palo Alto Research Center, Palo Alto, Calif. 94304; formerly at Stanford University, Stanford, Calif. 94305.
[3] Fairchild Research and Development Laboratory, Palo Alto, Calif. 94304; formerly at Stanford University, Stanford, Calif. 94305.
[4] Hewlett-Packard Corporation, Cupertino, Calif. 94304; formerly at Stanford University, Stanford, Calif. 94305.

placed on MOS technology and in particular the modeling of dopant redistribution effects in silicon. Although a major emphasis of this paper will be given to device-related effects, a final section considers the problems of physical parameters related to the prediction of shallow junction properties.

Scaled MOS Devices

The cross-section of a typical N-channel metal-oxide-semiconductor (NMOS) technology with enhancement driver and depletion load is shown in Fig. 1. The typical dimensions include shallow (*n*) junction depths less than 0.5 μm and minimum gate features of about 3.0 μm. Shown in the figure is a gate that traverses some lateral distance to reach the driver gate. From the perspective of current process modeling, there are several labeled cross-sections that reflect design trade-offs and hence process modeling needs. The vertical slices labeled *a* to *c* reflect the channel doping profiles for both driver and load as well as the buried contact profile made by a polysilicon diffusion source. The lateral gate section labeled *d* indicates the need for process modeling in the polysilicon itself, since this resistance can become substantial for large runs of polysilicon for interconnection. Turning to the two-dimensional effects, the sections labeled *1* and *2* indicate two of several critical design effects. In particular, the shallow junction/short channel effects of the driver device require accurate modeling of both process and device effects. To obtain improved short channel threshold control, the junctions are scaled down just in the local region near the channel. However, this requires care in the process design to minimize problems of added series resistance. Similarly, at the sidewall region of the oxide isolation, it is essential that details of the two-dimensional oxidation and diffusion be available as well as the electrical properties of this parasitic be understood. In the next section, Fig.

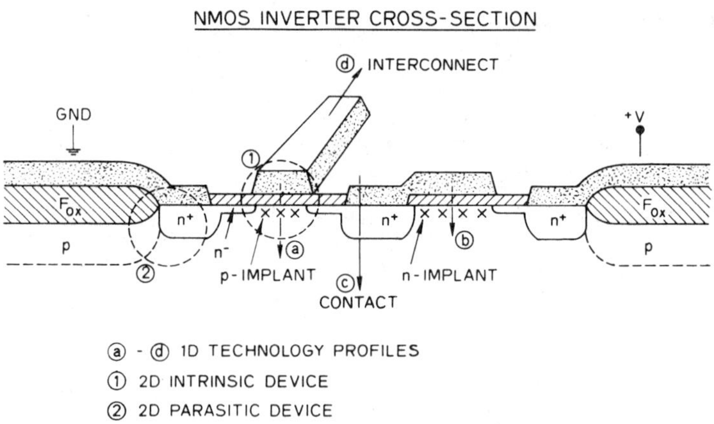

FIG. 1—*Section view of NMOS devices and indications of critical technology and device modeling issues.*

1 will be used further as the motivation in defining process modeling and device analysis.

Tools for Process and Device Analysis

Within the last five years, research in process modeling has increased dramatically and computer tools are now used frequently to simulate IC fabrication. The term *process modeling* is used to describe the numerical simulation of physical and chemical changes to silicon during the IC fabrication process. For example, a typical fabrication sequence involves oxide growth, impurity deposition in the silicon—usually by ion implantation—and thermal redistribution of the dopants. Figure 2 shows schematically three examples. Figure 2*a* shows the gate region for the depletion load of an NMOS inverter. In this case, the implantation, oxidation, and redistribution occur prior to the deposition of the polysilicon region. For Fig. 2*b*, the so-called buried contact region is represented where the polysilicon is used as a dopant source and its upper surface is simultaneously oxidized. Details for both of these cases are given in the next section.

A third example of process modeling is shown in Fig. 2*c*, where the interconnect technology of polysilicon is depicted. Here the polysilicon layer is deposited on an SiO_2 layer; subsequent thermal processing, including dopant redistribution and oxide growth, alter the physical and electrical properties of the layer. While the details of process modeling are given elsewhere [*1,2*],[5] it is useful to briefly show the electrical results as pertinent for circuit applications.

Figure 2 shows typical one-dimensional cross-sections used for process modeling. Third-generation simulators now allow such multilayer systems to be simulated [*3*]. As shown in Fig. 1, these technology cross-sections represent a substantial fraction of the constraints involved in the design of a local oxidation technology. However, from the electrical viewpoint there are several cross-sections shown in Fig. 1 that are inherently two-dimensional and thus require the appropriate analysis—in terms of both process and device effects. The short-channel effects of the enhancement driver device have been discussed extensively in the literature [*4–8*]. Generally, the analysis is based on the assumption of a planar surface, which is not a major limitation for the intrinsic device. With the continued push to scale surface dimensions, there

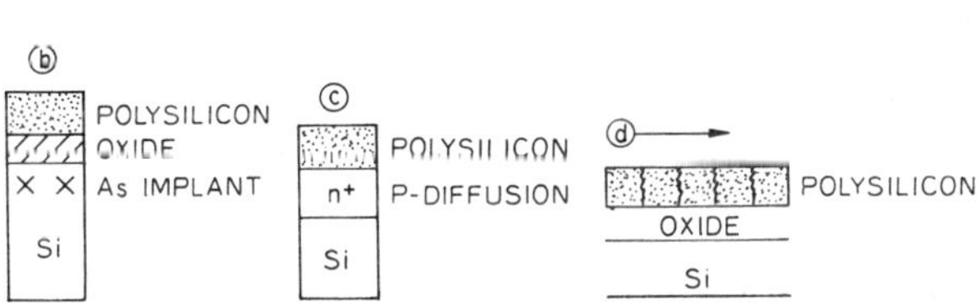

FIG. 2—*One-dimensional technology cross-sections with labels corresponding to those of Fig. 1.*

[5] The italic numbers in brackets refer to the list of references appended to this paper.

is a growing interest in three-dimensional analysis; a few results in this area have been reported recently [*9, 10*]. From the perspective of the intrinsic devices, the length dimension of depletion loads also falls into the category discussed above. The example discussed below illustrates different concerns for narrow-width device effects.

Turning to the parasitic labeled *2* near the locally oxidized region in Fig. 1, we see that the topology is nonplanar and both technology and electrical device effects must be considered. For the NMOS technology the parasitic concerns tend to be with passive devices (sidewall capacitance, for example) [*11*]. With the growing interest in complementary metal-oxide-semiconductors (CMOS), however, it is inevitable that the analysis must consider nonplanar as well as bipolar effects such as those involved in latch-up. There has been limited treatment of nonplanar structures [*12*]. In the domain of general two-carrier analysis, the efficiency and convergence properties of most analysis techniques seems questionable as yet from an engineering viewpoint. The use of matrix inversion techniques, solving for both carriers simultaneously, require excessive amounts of memory. On the other hand, iterative solutions that consider one carrier at a time, while more reasonable in requirements for memory, have convergence properties that are only empirically understood. New hybrid analysis techniques [*13*] using more limited regions of two-carrier solutions coupled with circuit-like approximations to represent the greater spatial extent of the actual device seem attractive as a means to solve the above mentioned two-carrier problems.

Applications of Process Modeling for Device Design

In this section several examples are presented in order to illustrate the points presented in the previous section. The cross-section shown in Fig. 1 is again used for reference and the examples consider both process and electrical device effects.

Figure 3 shows the simulated and measured resistivity of phosphorus-doped polysilicon layers after thermal processing. As indicated in the figure, an 0.25-μm layer of polysilicon was doped to different concentrations as shown and heat-treated for 1 h at 1200°C. The surfaces were covered with 100 nm of low-temperature-deposited oxide prior to heat treatment. The resulting films have different grain structures [*14*] and there are different amounts of dopant segregated at the grain boundaries and oxide interfaces. The results show that for increased dopant concentration, the grain size increases. Moreover, the fraction of dopant segregated to the boundary decreases and hence the resistivity decreases dramatically. Thus, over the dopant range considered the resistivity changes by more than an order of magnitude. The results are in good agreement with the simulated results. The models used for the simulation are discussed in greater detail elsewhere [*3*].

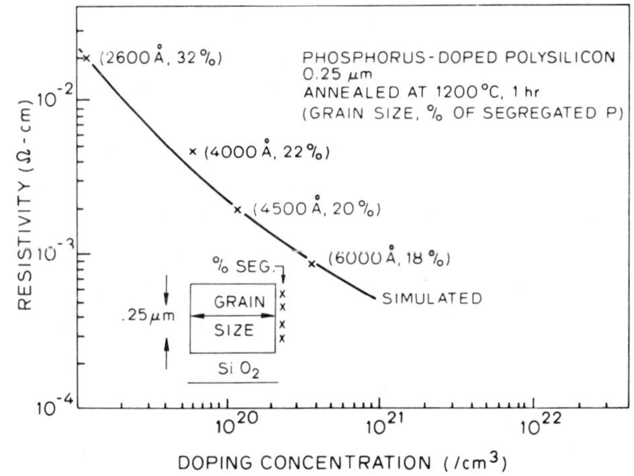

FIG. 3—*Resistivity versus doping for annealed polysilicon films.*

The key point to be emphasized is the fact that grain growth as well as segregation at boundaries must be included in the model.

As a second example, Fig. 4 shows simulated and measured junction depth for phosphorus diffused into bulk silicon from a polysilicon diffusion source. The experiment uses an 0.3-μm layer doped to an initial concentration of 3.3×10^{21} per cm^3. A nitride-patterned region as shown in Fig. 4 protected a portion of the sample from the 1000°C wet oxidizing ambient. The results show that the junction depth in the bulk is substantially greater under

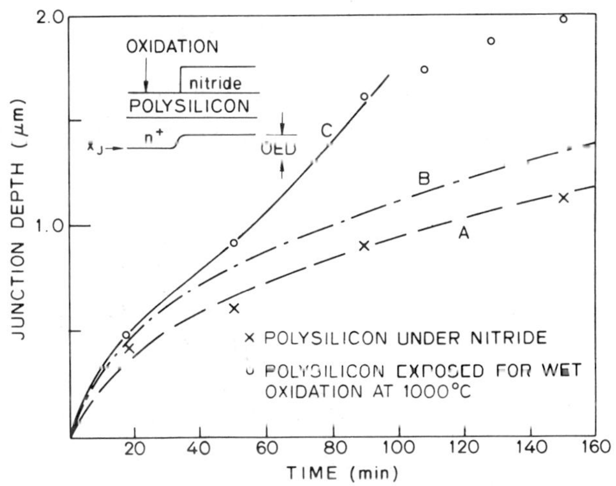

FIG. 4—*Junction depth versus processing time for buried contacts, with and without oxidation enhanced diffusion.*

the region where oxidation occurs. It might appear at first that dopant pile-up could cause this observed junction depth difference. However, the simulation labeled *B* indicates that dopant push is a negligible contribution to the observed increase in junction depth. In fact the oxidation enhancement of diffusion is essentially the same as that observed with bulk silicon. The generation of point defects during the oxidation process is responsible for the observed enhancement of diffusivity. This phenomenon is crucial in process modeling and substantial progress is being made in understanding and modeling the phenomenon in both one and two dimensions, the latter being especially important for locally oxidized isolation (LOCOS) structures [*15–17*].

The final example considered in this section deals with the depletion load shown in Fig. 1. Figure 5 shows the observed drain-source current (I_{DS}) versus gate-source voltage (V_{GS}) curves for several depletion mode devices with different arsenic implant doses. With increased dose, the threshold becomes more negative, which in turn increases the speed of the technology at the expense of increased power dissipation. The control of threshold voltage in the depletion load involves two factors: vertical dopant distribution and dimensional effects related to the LOCOS. Figure 6*a* and *b* shows two aspects of the vertical profile control. The simulated variation of threshold voltage as a function of dose and range are presented. Over a substantial range, the variation of threshold voltage with dose is linear. On the other hand, the dependence of threshold voltage on energy exhibits a nonlinear fall-off, as shown in Fig. 6*b*. This dependence is replotted in terms of sensitivity in Fig. 7. The sensitivity is defined as:

$$\int_{\text{Energy}}^{V_{\text{th}}} = \frac{\partial V_{\text{th}}}{|V_{\text{th}}|}\,\frac{\text{Energy}}{\partial_{\text{Energy}}} \qquad (1)$$

The greatest sensitivity occurs in the region of Fig. 6*b*, where the slope is large and the relative threshold is low. From a process design viewpoint, it is desirable to be sure the implant energy is sufficient to move to a region of

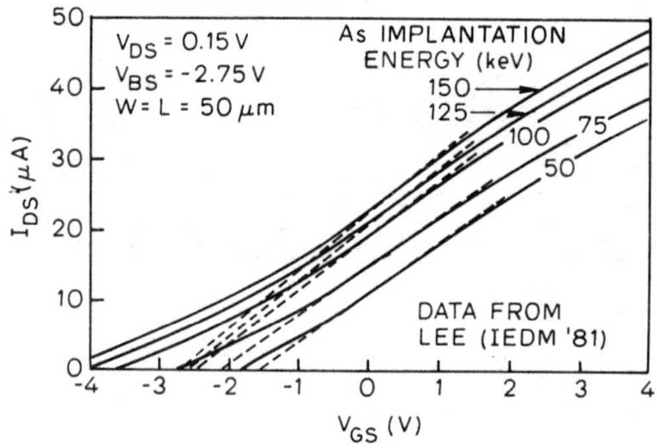

FIG. 5—*Measured* I_{DS} *versus* V_{GS} *for depletion load devices with different implant energies.*

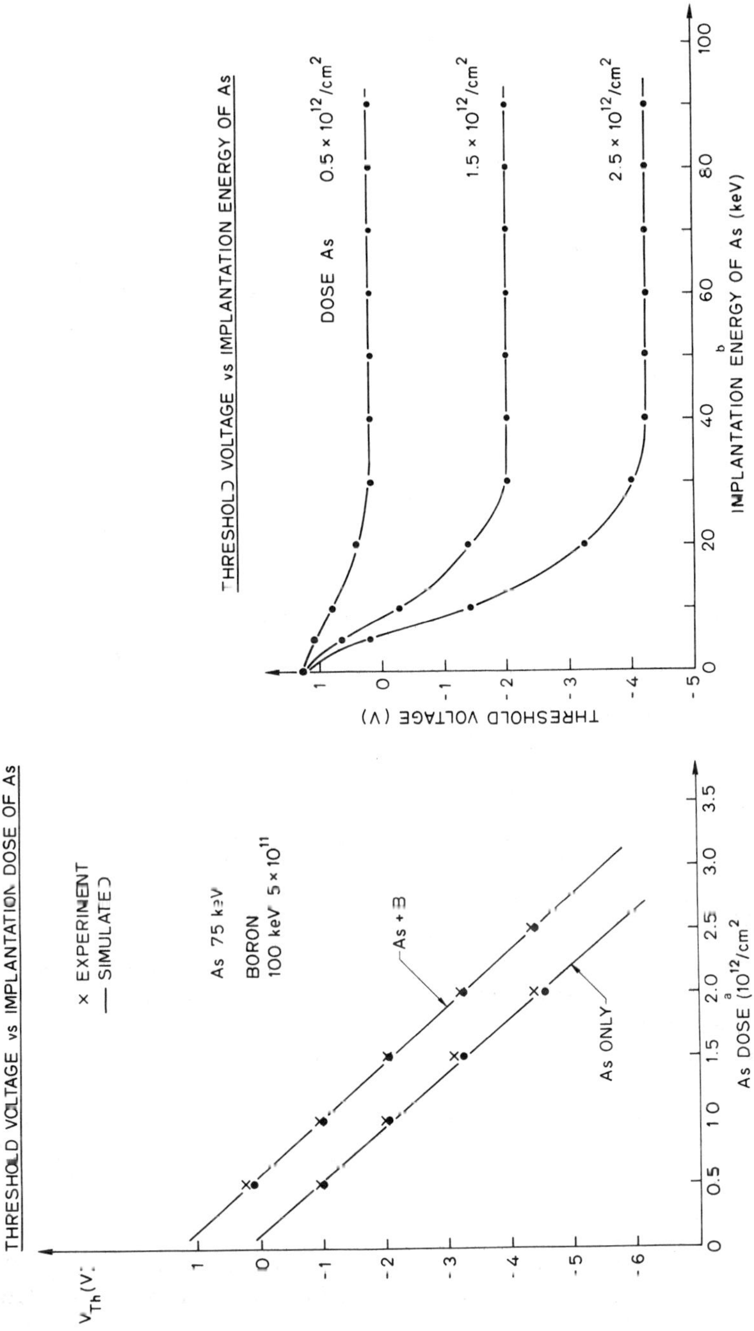

FIG. 5—*Simulated and measured threshold voltage versus* (a) *implant dose and* (b) *implant energy from depletion load devices.*

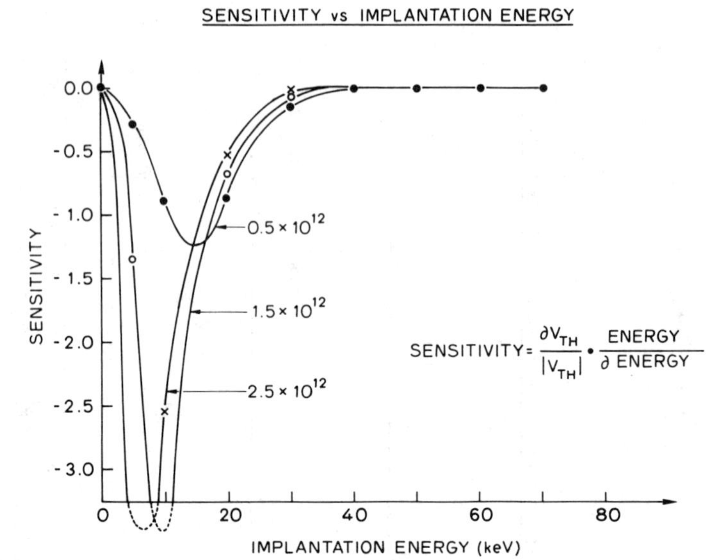

FIG. 7—*Extracted sensitivity of threshold voltage as a function of implant energy for depletion load.*

minimum sensitivity. The extracted dependence of threshold voltage on energy follows the empirical relationship [*18*]:

$$V_{th} = 1.035\,\frac{\text{Dose (As)}}{0.5 \times 10^{12}}\left[e^{-}\left(\frac{\text{Energy}}{12}\right)^2 - 1\right] + 1.2 \qquad (2)$$

where the dominant dependence of implant energy is the "Gaussian" and "linear" terms.

Turning to the cross-sectional view of the device, we note that Fig. 8 shows the width dimension of a depletion device with masked dimension prior to LOCOS of 4 μm. The dashed line shows the depletion edge for a source-bulk bias of $V_{SB} = 2.25$ V and $V_{GS} = -5$ V. It is clear that the p^+ channel-stop regions encroach on the channel width and alter the current-handling capability of the device. To further investigate this effect, the mask width dimension was varied as shown from the nominal. The results are summarized in Fig. 9. It is clear from the simulated data that dimensional control of depletion devices is a critical performance issue for NMOS technology. Moreover, it is clear that accurate process simulation as well as two-dimensional device analysis are necessary in order to correctly understand the design trade-offs.

Physical Parameters for Process Modeling

Implicit in the foregoing discussions is the assumption that accurate data and model coefficients are available and incorporated in the simulation tools. In the early development of SUPREM, models and data were taken

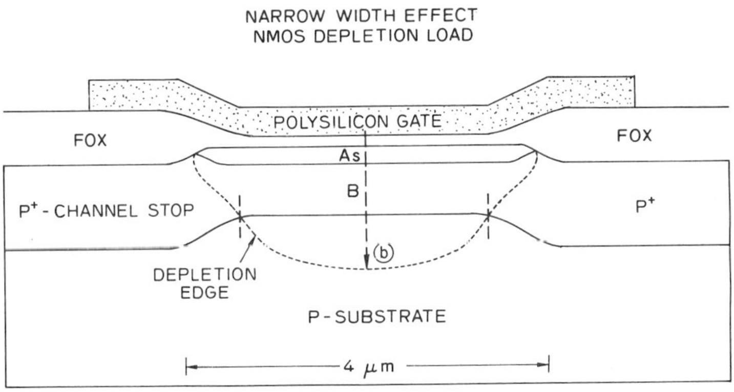

FIG. 8—*Width dimension of depletion load for an NMOS device with local oxide isolation. Two-dimensional process simulation* [11] *and device modeling* [12] *were used to determine depletion edge and surface sheet charge.*

from the literature. With more stringent requirements for scaled-down devices, the physical dimensions and tolerances of shallow junctions are significant factors in process modeling. Moreover, the number of new models and the sophistication of simulation capabilities are growing rapidly. Because of these trends, an assessment of the current process models in SUPREM II is appropriate to determine the potential of each model in meeting future modeling requirements and to ensure consistency between various models. Some factors relating to the phosphorus diffusion model (Fair-Tsai [*19*] model) that have recently come to light will illustrate these points.

Fig. 10 shows a plot of mobility data versus carrier or dopant concentrations obtained from various sources. In Version 03, SUPREM II utilized a mobility curve very similar to the one used by Fair and Tsai [*19*] in developing their *P* diffusion model. This curve was superseded in Version 05 by use of the data shown as open circles, which are based on measurements of mo-

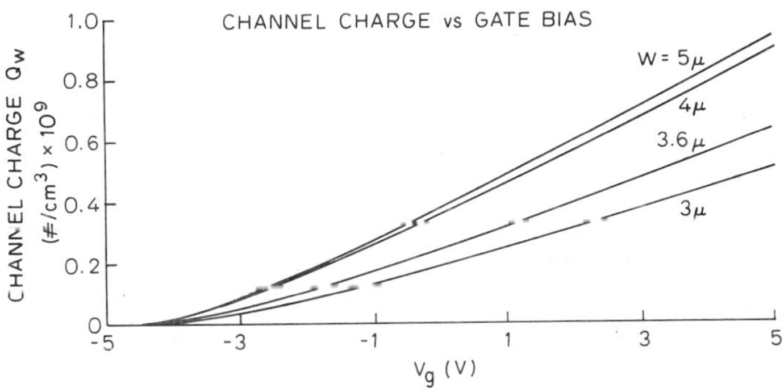

FIG. 9—*Simulated channel sheet charge versus gate bias as a function of masked gate width for depletion load devices.*

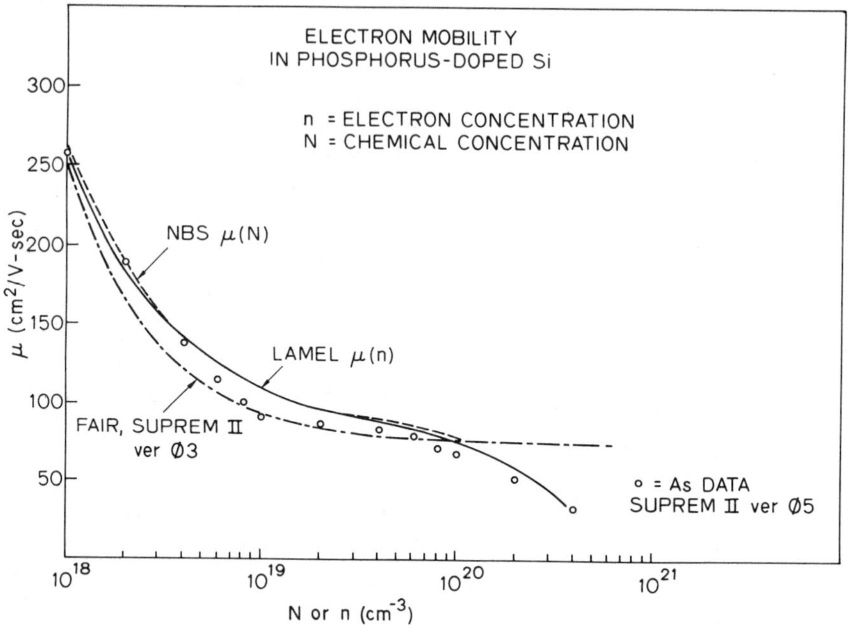

FIG. 10—*Mobility versus doping for electrons in silicon as measured by several groups—data for both phosphorus and arsenic doping.*

bility in As-doped Si samples.[6] For mobility measurements of heavily phosphorus-doped Si, the data from the National Bureau of Standards [*20*] and the LAMEL Laboratory [*21*] are in excellent agreement but definitely indicate higher mobility values than those for the As data at corresponding dopant concentrations. Actually, differences in the carrier concentration dependence of mobility for Sb, P, and As dopants were noted as early as 1960 [*22*]. Recent experiments have confirmed that this effect is also present in heavily doped polysilicon for both thermal and laser annealing treatments [*23*]. It is not clear at this time what effect these differences may have in engineering applications of process modeling. Figure 11 shows a simple example where P implant dose is plotted as a function of sheet resistance; sheet resistances were calculated by using both the As data and the LAMEL results of Fig. 10. These calculations indicate possible errors of 20 to 30% in calculated values of sheet resistance or dose. Since this effect is systematic, it has the virtue of being easily corrected if the differences are important.

Of much more importance is the intimate relationship these results have to the phosphorus model. Fig. 12 shows the dependence of phosphorus diffusivity on carrier concentration at 1000°C as determined by the experimental measurements of Fair and Tsai [*19*] and Makris and Masters [*24*]. The solid line represents Fair's best fit to this data and is also the diffusivity curve cal-

[6] Private communication from D. Antoniadis.

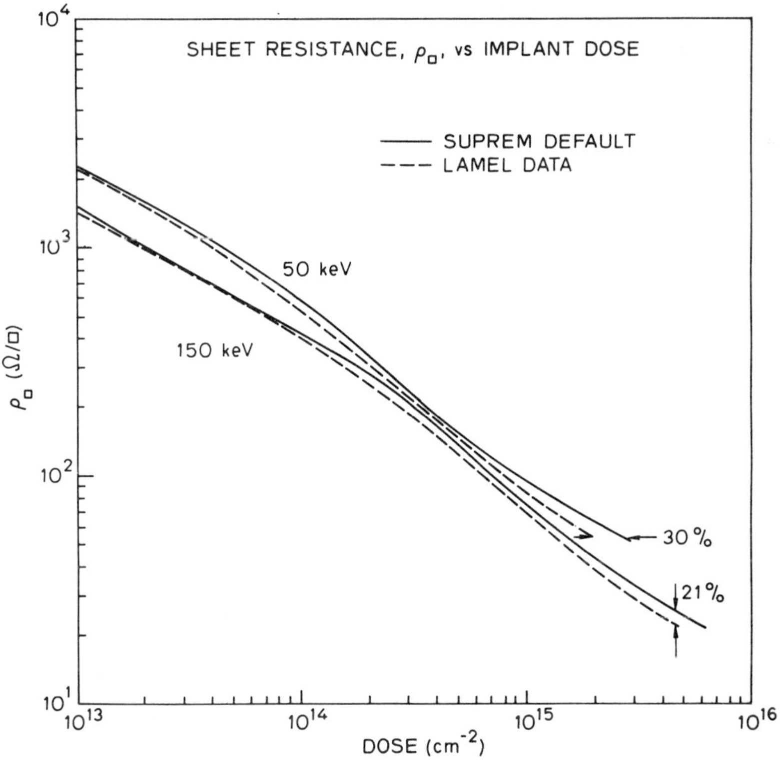

FIG. 11—*Simulated sheet resistance versus dose for different implant energies showing variation resulting from different mobility data.*

culated by SUPREM. When it is realized that Fair and Tsai converted resistivity measurements to carrier concentrations using a relationship that has since proven to be inaccurate at high concentrations, it is obvious that there must be some error associated with any model coefficients calculated from these data. Taking the original data of Fair and Tsai and their resistivity/carrier concentration curve, these data have been converted to values that are consistent with the LAMEL results. These data are plotted as open circles in Fig. 12. This figure should not be interpreted as a correction to the diffusivity (this would necessitate extraction of phosphorus diffusivity (D_P) from the original diffused profiles corrected to the LAMEL or NBS data) but rather as an indication of the errors associated with calculating D_P if the phosphorus concentration is accurately known. As indicated by the dashed line, at high concentrations SUPREM would predict a diffusivity five times too large based on the original experimental data. In Fig. 13 the relationship of total concentration N to electrically active n phosphorus concentration used by SUPREM is plotted as a solid line. Again, this represents a best fit to Fair and Tsai's data. This relationship is crucial to the model as the inactive

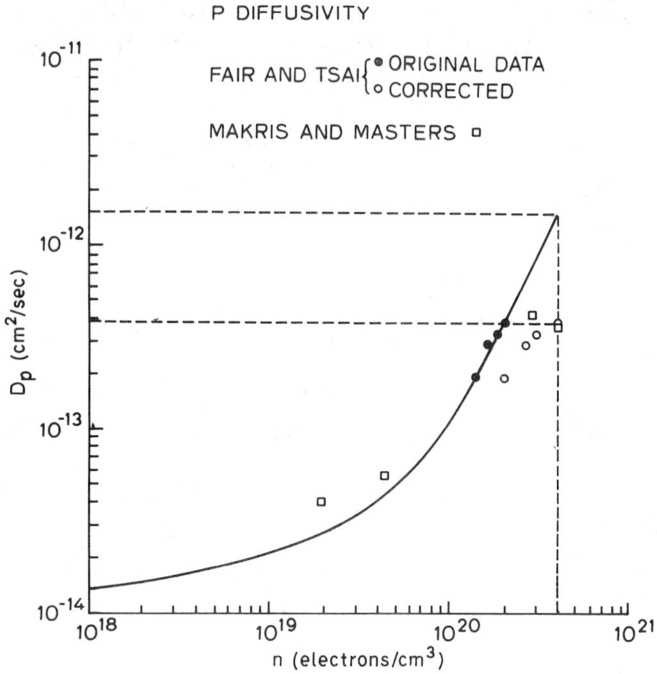

FIG. 12—*Extracted phosphorus diffusivity versus electron concentration. Variations due to mobility data are indicated as well as model results of Fair* [19].

phosphorus is attributed to $P^+V^=$ pair point defects that account for all of the anomalous effects associated with phosphorus diffusion. The open squares in Fig. 13 represent the results of correcting the original data by the procedure outlined above. Evidently a much closer correspondence between electrically active and total concentrations results (up to the apparent solid solubility limit of 4×10^{20} P atoms per cm^3 [*25*]).

It cannot be concluded from these results that the present phosphorus model used in SUPREM II should be abandoned, but clearly the numerical values of the model coefficients are open to question. At present, these results have the most serious repercussions for bipolar process modeling, since junction depths and base-push effect depend sensitively on the cube of the active surface concentration and band-gap narrowing terms (determined from the N versus n relationship). The importance of these types of errors as relates to future modeling can be seen in two ways. Firstly, what appear as junction depth errors in one dimension will become lateral diffusion effects in two-dimensional modeling. This could also affect modeling of polysilicon by introducing errors in the lateral diffusion of phosphorus from the grain into the grain boundaries. Secondly, in our overall scheme of modeling point defects it is becoming clearer that many effects such as base-push effect, enhanced grain growth in poly, impurity enhanced oxidation [*26*], and even

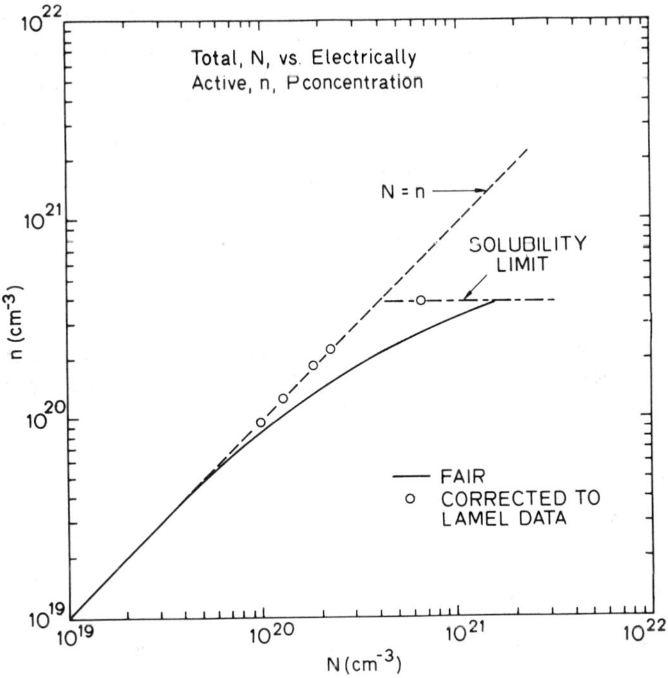

FIG. 13—*Electron concentration versus total dopant concentration for phosphorus. Data are corrected using LAMEL mobility and the curve is the model of Fair* [19].

gettering mechanisms [*27*] can be modeled as Fermi-level effects. That is, at a given temperature, the onset and degree to which the above effects occur is determined by the position of the Fermi level, which in turn depends on the local values of n (or p). Thus not only must the true values of the active concentrations be known but the coefficients of various models must depend consistently on these values.

Phosphorus diffusion has served as a representative model to elucidate some of the problems we will be addressing and our approach to fulfilling future process modeling requirements.

Conclusions

The intent of this paper is to describe computer-aided process modeling from several perspectives. Applications of process modeling for technology design have been demonstrated. Emphasis is given to new multilayer simulation capabilities involving polysilicon. For device design it is clear that process control is a key issue. The examples of depletion load current-voltage characteristics illustrate both one- and two-dimensional process control issues. The final thrust of this paper is intended to emphasize the increasingly important role that accurate and consistent model coefficients will play in future process simulations. The phosphorus diffusion model of Fair and Tsai is chosen to illustrate the types of problems encountered when the physical mechanisms of an existing model are extended to two dimensions or coupled to other models. Results of this investigation have brought to light the previously unsuspected tight coupling between mobility data and extracted diffusion model coefficients. Our conclusion is that the numerical values associated with the phosphorus model need to be reexamined and the compatibility between this and other proposed models assessed.

Acknowledgments

This work was supported under Defense Advanced Research Project Agency (DARPA) Contract MDA 903-79-C-0257 and Army Research Office (ARO) Contract DAAG-29-80-K-0013.

References

[*1*] Dutton, R. W. and Hansen, S. E., *Proceedings of IEEE,* Vol. 69, No. 10, Oct. 1981, pp. 1305–1320.

[*2*] Antoniadis, D. A. and Dutton, R. W., *Journal of Solid-State Circuits,* Vol. SC-14, No. 2, April 1979, pp. 412–422.

[*3*] Mei, L. and Dutton, R. W., "A Process Simulation Model for Multilayer Structures Involving Polycrystalline Silicon," *IEEE Transactions on Electron Devices,* Vol. ED-29, No. 11, Nov. 1982, pp. 1726–1734.

[*4*] Yokoyama, K., Yoshii, A., and Horiguchi, S., *Journal of Solid-State Circuits,* Volume SC-15, No. 4, Aug. 1980, pp. 574–579.

[*5*] Oka, H., Nishiuchi, K., Nakamura, T., and Ishikawa, H., *Journal of Solid-State Circuits,* Vol. SC-15, No. 4, Aug. 1980, pp. 579–585.

[*6*] Dang, L. M. and Konaka, M., *Journal of Solid-State Circuits,* Vol. SC-15, No. 4, Aug. 1980, pp. 598–604.

[*7*] Selberherr, S., Schutz, A., and Potzl, H. Z., *Journal of Solid-State Circuits,* Vol. SC-15, No. 4, Aug. 1980, pp. 605–615.

[*8*] Liu, S., Hoefflinger, B., and Pederson, D. O., *Journal of Solid-State Circuits,* Vol. SC-15, No. 4, Aug. 1980, pp. 615–623.

[*9*] Buturla, E. M., *ISSCC Digest,* International Solid-State Circuit Conference, Feb. 1980, p. 76.

[*10*] Chamberlain, S. G. and Husain, A., *IEDM Digest,* International Electron Devices Meeting, Dec. 1981, p. 592.

[*11*] Chin, D., Kump, M., Lee, H. G., and Dutton, R. W., *IEEE Transactions on Electron Devices,* Vol. ED29, No. 2, Feb. 1982, pp. 336–340.

[*12*] Greenfield, J. A. and Dutton R. W., *Journal of Solid State Circuits,* Vol. SC-15, No. 4, Aug. 1980, pp. 585–597.

[*13*] Engl, W., Laur, R., and Dirks, H., "MEDUSA—A Simulator for Modular Circuits," *IEEE Circuits and Systems,* to be published.

[*14*] Mei, L., Rivier, M., Kwart, Y., and Dutton, R. W., *Journal of the Electrochemical Society,* Vol. 129, No. 8, Aug. 1982, pp. 1791–1795.

[*15*] Lin, A. M., Antoniadis, D. A., and Dutton, R. W., *Journal of the Electrochemical Society,* Vol. 128, No. 5, May 1981, pp. 1131–1137.

[*16*] Lin, A. M., Dutton, R. W., and Antoniadis, D. A., *Applied Physics Letters,* Vol. 35, No. 10, Nov. 1979, pp. 799–801.

[*17*] Hamasaki, M., Suzuki, J., and Shimada, T. F., "Lateral Effect of Oxidation Enhanced Diffusion in Silicon," *IEDM Technical Digest,* International Electron Devices Meeting, Dec. 1981.

[*18*] Doganis, K., Dutton, R. W., and Gonzalez, A. G., "Optimization of IC Processes Using SUPREM," Stanford Electronics Laboratories, Stanford, Calif., June 1981.

[19] Fair, R. B. and Tsai, J. C. C., *Journal of the Electrochemical Society,* Vol. 124, No. 7, July 1977, p. 1107.
[20] Thurber, W. R., Mattis, R. L., Liu, Y. M., and Filliben, J. J., *Journal of the Electrochemical Society,* Vol. 127, No. 8, Aug. 1980, p. 1802.
[21] Masetti, G. and Solmi, S., *Solid-State Electron Devices,* Vol. 3, No. 3, May 1979, p. 65.
[22] Logan, R. A., Gilbert, J. F., and Trumbore, F. A., *Journal of Applied Physics,* Vol. 32, 1961, p. 131.
[23] Solmi, S., Severi, M., and Angelucci, R., "Electrical Properties of Thermally and Laser Annealed Polysilicon Films Heavily Doped with Arsenic and Phosphorus," presented at ECS Conference, Denver, 1981.
[24] Makris, J. S. and Masters, B. J., *Journal of the Electrochemical Society,* Vol. 120, No. 9, Sept. 1973, p. 1253.
[25] Masetti, G., Nobili, D., and Solmi, S., "Profiles of Phosphorus Predeposited in Silicon and Carrier Concentration in Equilibrium with SiP Precipitates," in *Semiconductor Silicon 1977,* The Electrochemical Society, Pennington, N.J., 1977, p. 656.
[26] Ho, C. P. and Plummer, J. D., *Journal of the Electrochemical Society,* Vol. 126, No. 9, Sept. 1979, pp. 1516–1522.
[27] Lecrosnier, D., Paugam, J., Pelous, G., Richou, R., and Salvi, M., *Journal of Applied Physics,* Vol. 52, No. 8, Aug. 1981, p. 5090.

David S. Perloff[1]

Characterizing Process Nonuniformities on Large-Diameter Wafers: An Overview

REFERENCE: Perloff, D. S., **"Characterizing Process Nonuniformities on Large-Diameter Wafers: An Overview,"** *Silicon Processing, ASTM STP 804,* D. C. Gupta, Ed., American Society for Testing and Materials, 1983, pp. 422–444.

ABSTRACT: An overview is presented of the techniques available for characterizing process nonuniformities on large-diameter wafers. Examples are given of the use of two-dimensional and three-dimensional mapping techniques for displaying within-wafer variability, and of relative and cumulative frequency plots for representing process tolerances. Two new display methods are recommended for dealing with small data sets: minicontour maps and folded-axis cumulative frequency plots.

KEY WORDS: resistivity characterization, nonuniformities, silicon wafers, large-diameter wafers, ion implantation, resistivity mapping, two-dimensional maps, three-dimensional maps, contour maps, frequency plots

The continuing trend towards more complex integrated circuits has placed increasing demands on the various technologies employed in semiconductor wafer fabrication. Only by collecting and analyzing data at all stages of manufacturing can one be certain that equipment performance is well-controlled and that operating procedures are being adhered to. Factors such as oxygen content in the starting silicon wafer, dose accuracy of implantation steps, and dimensional control and registration accuracy of masking levels must be continually assessed.

The variations in process parameters that occur from wafer-to-wafer and lot-to-lot are generally monitored by performing measurements at a small number of strategic locations on either product or test wafers. Computer-controlled measurement equipment such as d-c parametric testers, film thickness analyzers, and infrared spectrophotometers are well-suited for this type of process and materials monitoring. Data derived from such measurement tools may be stored, retrieved, analyzed, and presented by using a

[1] Department Manager, Signetics Corporation, Sunnyvale, Calif. 94086; present address: PROMETRIX Corp., Sunnyvale, Calif. 94086.

computerized information management system [*1,2*].[2] In this way, not only can trends in data be detected which would be indicative of poor process control, but more subtle behavior, such as the relationship between process variability and circuit yield, can be investigated [*3*].

Understanding the factors that cause process parameters to vary nonuniformly *within* individual wafers requires a rather different approach than that used for evaluating the variations that occur *between* wafers [*4–9*]. For example, test sites must be distributed so as to provide a "fair" sampling of the wafer surface; measurement errors must be kept well below the variability of the parameter being evaluated; care must be taken to exclude invalid readings; and results must be displayed in a form which makes it possible to perceive and interpret the underlying pattern of nonuniformity. Table 1 summarizes four measurement techniques found to be particularly useful for monitoring within-wafer process nonuniformities: sheet resistance [*10–15*], line width [*16–20*], mask registration accuracy [*17–25*], and film thickness [*26,27*]. Each of the methodologies cited utilizes minimal wafer processing, is compatible with automated data collection, and is capable of isolating the behavior of the parameter of interest.

This paper provides an overview of the techniques available for characterizing process nonuniformity on large-diameter wafers. The first section contrasts the various mapping techniques which may be used for displaying parameter variations. Next, useful methods for presenting the frequency distribution of test data are described. Lastly, an approach to evaluating spatial inhomogeneities using a minimum number of test sites is discussed. Results obtained from a variety of measurement tools and processes are used for the purpose of discussion and illustration.

Mapping Techniques

Numeric maps, consisting of an array of test values, can be generated quite easily using standard line printers. It is rather difficult, however, to interpret or compare maps consisting of a large number of multidigit values. Some method must be employed to simplify the display so that regions of similar magnitude may be recognized with a minimum of mental "decoding".

The following subsections describe the use of two-dimensional and three-dimensional mapping techniques for visualizing spatial nonuniformities associated with various integrated-circuit process parameters. As will be seen, these techniques differ in the complexity of the analysis, the sophistication in the plotting or printing equipment used to generate the display, and the relative ability of the maps to convey rapidly and unambiguously the underlying

[2] The italic numbers in brackets refer to the list of references appended to this paper.

TABLE 1—*Techniques for monitoring process nonuniformities on large-diameter wafers.*

Parameter	Measurement Vehicle	Applications	Concerns	References
Sheet Resistance	• Four-point probe or van der Pauw resistor	• Ion implantation, diffusion processes, expitaxial growth and metal deposition	• Uniformity of deposition processes	[*9–15*]
Line Width	• Bridge-type and van der Pauw resistor	• Advanced optical lithography • Resist development • Dry etching	• Maintaining dimensional control through photo-masking and etching	[*16–20*]
Mask Registration accuracy	• Various potentiometric resistor structures	• Advanced optical lithography	• Performance of automatic alignment methodologies • Mix-match capabilities • Process-induced wafer deformation	[*17–25*]
Film Thickness	• Various optical measurement systems	• Oxidation • Dry etching • Dielectric deposition	• Film uniformity • Etching uniformity	[*26,27*]

pattern of nonuniformity. Specific advantages and disadvantages of the various techniques are outlined in Table 2.

Symbolic Maps

Figure 1*a* shows a very basic symbolic mapping technique used in a commercially available system for measuring wafer flatness.[3] The symbols "+" and "−" correspond to measurement sites exceeding user-defined limits for focal plane deviation, whereas blank portions of the wafer are within limits. This example shows that the wafer under examination is composed of hills ("+") and valleys ("−"), but does not provide information on the magnitude of the focal plane distortion at individual measurement sites.

Another example of a symbolic representation is shown in Fig. 1*b*, which depicts reverse-current inhomogeneities in a gated diode array fabricated on dislocation-free silicon wafers [*28*]. The display reveals spatial inhomogeneities which are dependent on the microdefect distribution in the original unprocessed wafer. Reasonable spatial resolution has been obtained by employing the three symbols "+", "·" and "−" to represent user-defined regions of "high", "medium" and "low" reverse current.

Halftone Maps

A representation capable of providing greater spatial resolution than can be achieved with symbolic plots is illustrated in Fig. 2, which is a map of metal-to-n^+ contact resistance obtained by using a four-terminal resistor test structure at 87 test sites [*7*]. Here eight tones are used to represent the data, which are distributed between two limits that are approximately three standard deviations on either side of the mean. Linear interpolations have been made between test sites to provide a more continuous distribution.

If a sufficient number of test sites is employed, an interpolation scheme will not be necessary. Figure 3 shows the nonuniformity of a phosphorus-implanted polycrystalline silicon layer, for which the sheet resistance varies over a range from about 2 to 11 $\times$ 10^6 Ω/square [*7*]. Each of the 531 test sites has been identified by one of 14 tones, as shown below the map.

Improvements in the resolution of halftone maps require an increase in the number of intervals over which the data are distributed. At some point, however, it will become difficult for the eye to discriminate between adjacent intervals, owing to the small differences in tone. In fact, it is customary to employ a fixed number of intervals, whatever the degree of variability of the test data [*5,7*]. As a result, halftone maps are somewhat inconvenient for making wafer-to-wafer comparisons of process nonuniformity, or for detecting low-level nonuniformities within a given wafer.

[3] Tencor Flatgage, Tencor Instruments, Mountain View, Calif.

TABLE 2—*Techniques for mapping within-wafer process nonuniformities.*

Technique	Typical Representation	Advantages	Disadvantages
Numeric	• Measured ($P_1, P_2, \ldots, P_n$), normalized (e.g. $P_1/\bar{P}$, $P_2/\bar{P}, \ldots, P_n/\bar{P}$) or relative values (e.g. $\Delta P_1/\sigma$, $\Delta P_2/\sigma, \ldots, \Delta P_n/\sigma$) are printed as a numeric array corresponding to the distribution of test sites.	• Fully quantitative • Compatible with standard line printers • Provides a starting point for more advanced representations	• Extensive mental decoding required to identify patterns of nonuniformity • Outliers not readily detected
Symbolic (Fig. 1)	• Test data are distributed among N cells of width Δ/N where, for example, $\Delta = P_{max} - P_{min}$. Each cell is assigned a distinctive alphanumeric character or symbol (e.g. . . . −2, −1, 0, 1, 2 . . . or +, ·, −).	• Enhances ability to detect regions of similar value • Can be implemented using standard line printers • Computational requirements are minimal	• May require a key identifying each symbol • Difficult to derive quantitative information • Cell width may have to be adjusted to optimize appearance
Halftone (Figs. 2 and 3)	• As with symbolic maps, test data are distributed among N cells. Utilizes graphics terminals to create a distinctive tone, shade, or color for each cell.	• Greater spatial resolution than symbolic maps • High-quality graphics can enhance recognition of spatial patterns • Computational requirements are minimal	• A key identifying each tone, shade, or color must be provided • Difficult to derive quantitative value for specific test sites

Equivalue Contour (Figs. 4, 5, and 13)	• Numerical techniques are used to construct a series of contours having preselected equally spaced parameter values. Contours may be individually labeled or identified in relation to a reference contour such as $\bar{P}$.	• No limitation on the number of cells that can be defined by contour values • Variable contour density readily distinguishes between uniform and nonuniform spatial distributions • Especially useful for displaying highly uniform distributions	• May require linear interpolation between test sites to achieve smoothly varing contours • Appearance degraded by outliers and measurement-related errors comparable to contour interval • Requires substantial computational facilities and sophisticated graphics equipment
Vector Maps (Fig. 6*a*)	• An array of vectors is used to represent the x- and y- axis components of level-to-level mask registration error at various locations on the wafer.	• Global sources of registration error (translation, rotation, expansion) are readily observed in appearance of map • Random sources of error (e.g. those associated with step-and-repeat operations) may be detected after removal of global errors	• For each wafer, a large number of maps may be required to fully represent the various source of registration error
Three-Dimensional (Figs. 7 to 9)	• Numerical techniques are used to construct a three-dimensional rendering of the parametric surface $P(x,y)$. Values of particular test sites are determined in relation to a reference plane such as $P(x,y) = P_{min}$.	• Emphasizes shape of the spatial distribution • Particularly effective for qualitative comparisons among a group of wafers • Especially useful for displaying highly nonuniform distributions	• Interpretation may be influenced by amount of rotation and tilt of the reference plane • Appearance may be totally dominated by outlier values, making it impossible to detect intrinsic pattern of nonuniformity • Difficult to identify the location and value of specific test sites

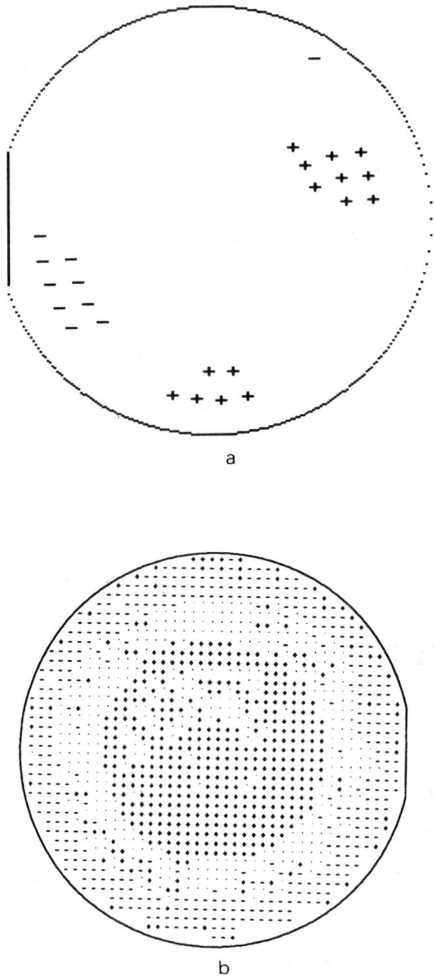

FIG. 1—*Symbolic maps showing variations in* (a) *silicon wafer flatness and* (b) *gated diode reverse leakage current.*

Equivalue Contour Maps

Equivalue contour maps allow one to employ as many intervals as are needed to distinguish inhomogeneities in the data [*9*]. Contour maps may be generated by analytically fitting a surface to the parameter data. One then employs numerical techniques for determining the locus of the intersection of the surface with a series of planes of preselected parameter value. Unfortunately, this procedure requires rather extensive computing capabilities and may result in maps whose appearance depends on the particular method employed for surface fitting.

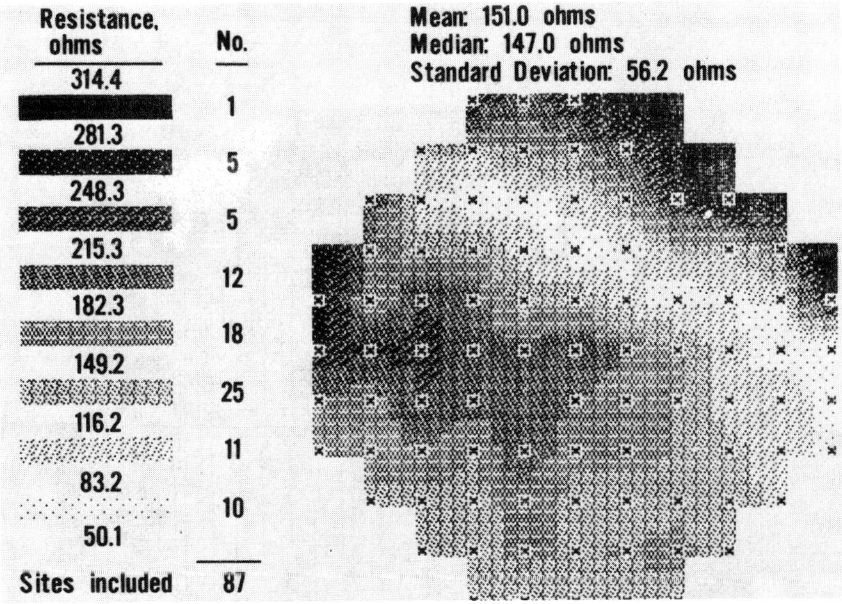

FIG. 2—*Halftone map showing variations in metal-to-n*$^{+}$ *contact resistance.*

An easier approach to deriving contours involves creating a grid of contiguous cells. One determines, by linear interpolation along cell edges, the points of intersection of a preselected parameter value. Line segments are then drawn connecting these points and the procedure repeated for all cells to create continuous equivalue contours [*10*].

Figure 4 is a contour map representing the same sheet resistance data as those depicted in Fig. 3 [*7*]. Much higher spatial resolution has been achieved by employing contours to delineate adjacent intervals. The value of a given contour line is readily determined by reference to the contours marked "A", which represent the mean of 531 data points, and the symbols "+" and "−", which correspond to locations above or below the wafer mean. The contour interval for this particular example is 10% of the wafer mean.

Contour maps are particularly useful for displaying highly uniform distributions, such as those associated with ion beam doping [*11–15*]. Figure 5 represents the variation in sheet resistance of a boron-implanted layer, obtained using a mechanical four-point probe array [*4*]. Linear interpolation and extrapolation of contours to the wafer boundary have been used to improve the appearance and readability of the map [*10*]. The heavy contour is equal to the mean of 118 data points. Specific test site values are readily determined by counting contours, which for this example are in increments of 1% of the wafer mean.

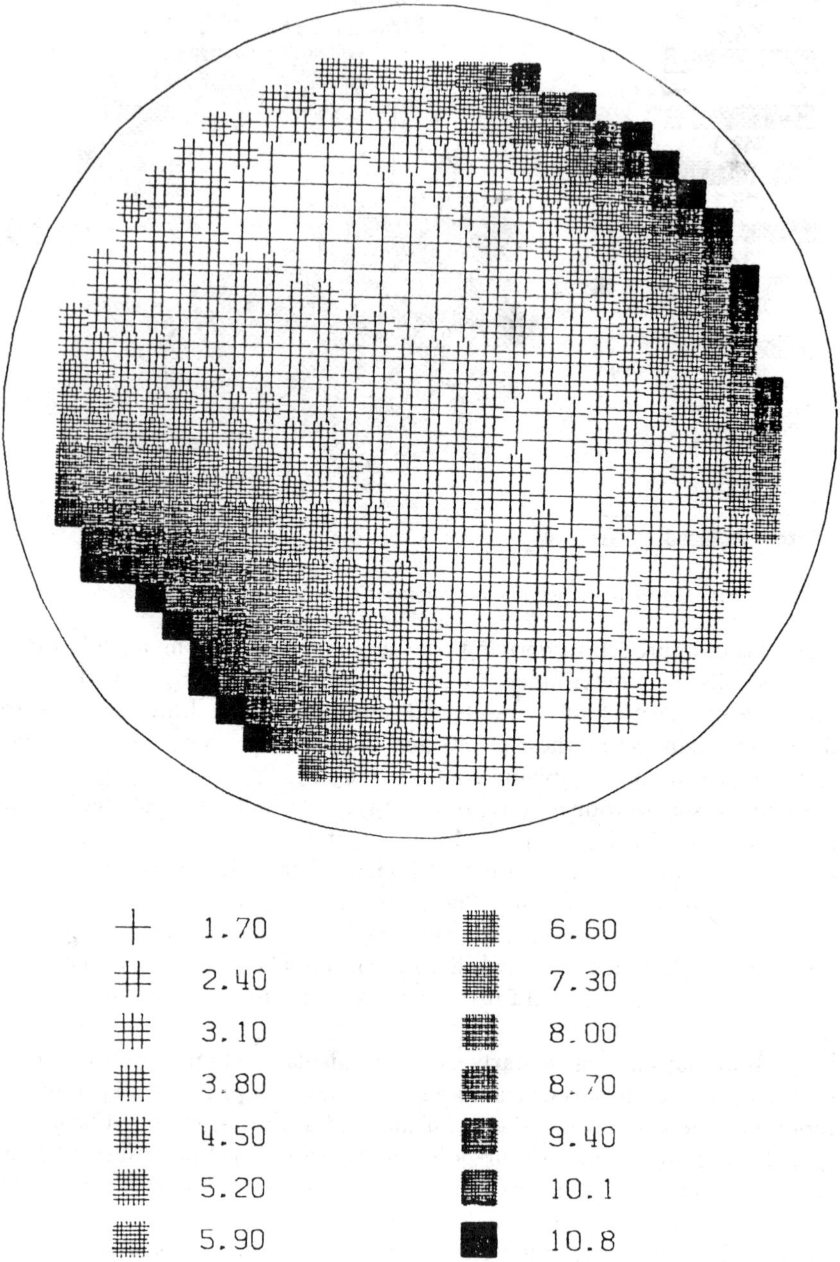

FIG. 3—*Variations in sheet resistance of a phosphorus-implanted polycrystalline silicon layer: halftone map.*

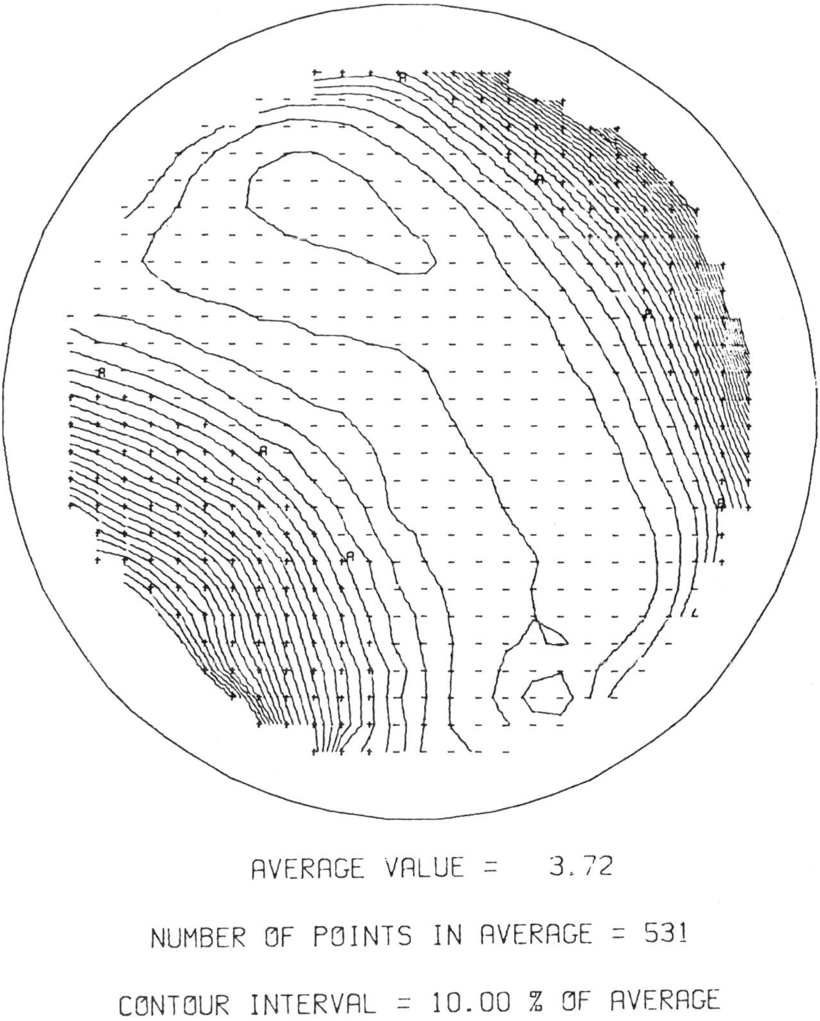

FIG. 4—*Variations in sheet resistance of a phosphorus-implanted polycrystalline silicon layer: equivalue contour map.*

Vector Maps

Figure 6*a* is a useful representation for depicting variations in registration errors occurring between two masking levels [*19*]. The vectors clearly illustrate the direction of misregistration at each test site, the magnitude of the error is estimated by comparing the vector length with the scale bar shown above the map. An analysis of the *x*- and *y*-axis components of registration error is used to determine systematic versus random effects [*22*]. The exam-

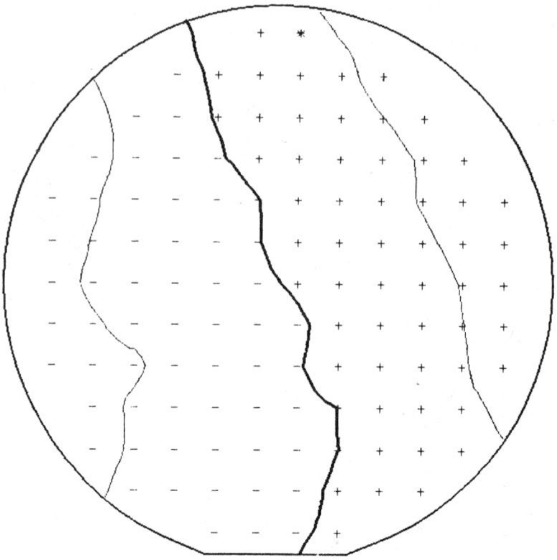

FIG. 5—*Sheet resistance contour map obtained from a boron-implanted silicon wafer (contour interval = 1% of mean sheet resistance).*

ple shown in Fig. 6*a* illustrates the results of linearly scaling the *x*- and *y*-axis step-and-repeat distances in a 10:1 wafer stepper. (The scaling is performed about an axis near the left-hand edge of the wafer.)

An alternative use of vector maps is shown in Fig. 6*b*, which depicts the pattern of linewidth variations associated with exposure and development processes [*20*]. An upward-pointing arrow stands for a measured linewidth larger than the mean, whereas a downward-pointing arrow represents a linewidth smaller than the mean.

Three-Dimensional Maps

Two-dimensional contour maps provide a representation of a three-dimensional surface which facilitates rapid subjective assessments and comparisons between data sets. There are cases, however, in which it is useful to view the surface directly. Figure 7 depicts the relative variation in thickness of a metal film deposited on a group of seven wafers in a planar-geometry sputtering system.[4] The wafers were coated in the same run, and occupied positions on the platen as indicated in the figure. One derives from such a representation a sense of the overall performance of the deposition system which might be more difficult to achieve using the two-dimensional plots described previously.

Three-dimensional maps are most effective when used to display large parameter variations, in which distinctive shapes may be readily discerned.

[4] Manwaring, A. D., Signetics Corporation, Sunnyvale, Calif., private communication.

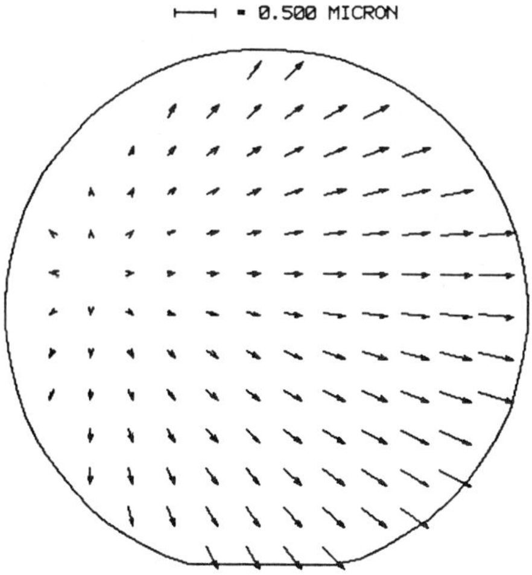

a

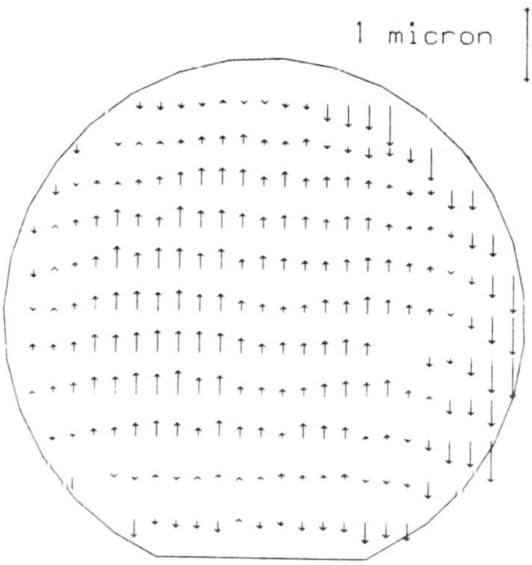

b

FIG. 6—*Vector maps used to represent* (a) *mask registration error and* (b) *variations in linewidth.*

Figure 8*a* shows a three-dimensional plot of sheet resistance for a wafer implanted in an electrostatically scanned system in which the dose delivered to the left-hand side of the wafer is about 10% greater than that delivered to the right-hand side. (The reader is referred to Figs. 15 and 18 in Ref *11*.) The data were collected and displayed by a commercially available measurement system.[5] Clearly, one can only estimate the magnitude of the variability from

[5] Semiconductor Equipment Corporation, Moorpark, Calif. 93021.

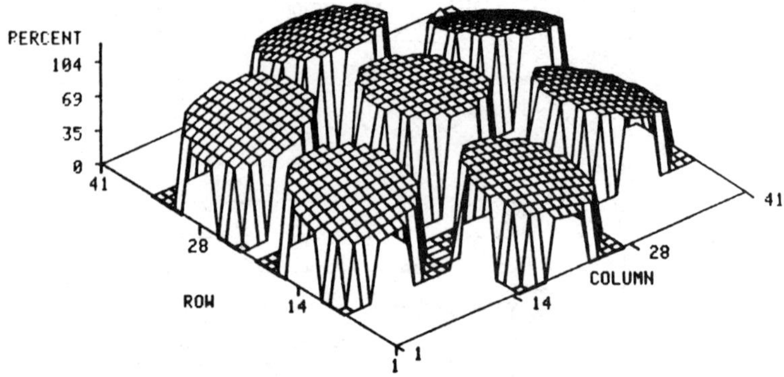

FIG. 7—*Three-dimensional representation of variations in metallization thickness over a group of seven wafers.*

such a plot. In addition, the appearance may be altered significantly depending on the amount of rotation and tilt of the reference plane. The effective use of such a display may therefore require direct interaction with the computer and the high-level graphics device used for generating the output.

A potential problem when employing three-dimensional plots is illustrated in Fig. 8*b*. Here the measured sheet resistance data contain a significant number of erratic readings. The appearance of the map is clearly dominated by these values, so that it is impossible to detect the intrinsic nonuniformity. In contrast, Fig. 9 illustrates the advantageous use of three-dimensional maps [*3*]. Here regions of low *npn* transistor collector-isolation breakdown voltage are prominently displayed, although the representation makes it more difficult to pinpoint exact wafer locations than would a two-dimensional format.

Frequency Distributions

Two-dimensional and three-dimensional maps provide information that can contribute to an understanding of the sources of parameter nonuniformity on large-diameter wafers. Ultimately, however, one is concerned with specifying process *tolerances*, which requires summarizing the data obtained from a number of wafers. For this purpose, several different types of *frequency distributions* [*29*] may be employed, with advantages and disadvantages as outlined in Table 3.

Relative Frequency Plots (Histograms)

Histograms are the most commonly used type of frequency distributions. A binning procedure is employed in which the parameter values $P_1, P_2, \ldots, P_n$ are assigned to bins of width ΔP and the relative frequency of occurrence, or bin count, is displayed. Figure 10 shows two histograms of the same line-

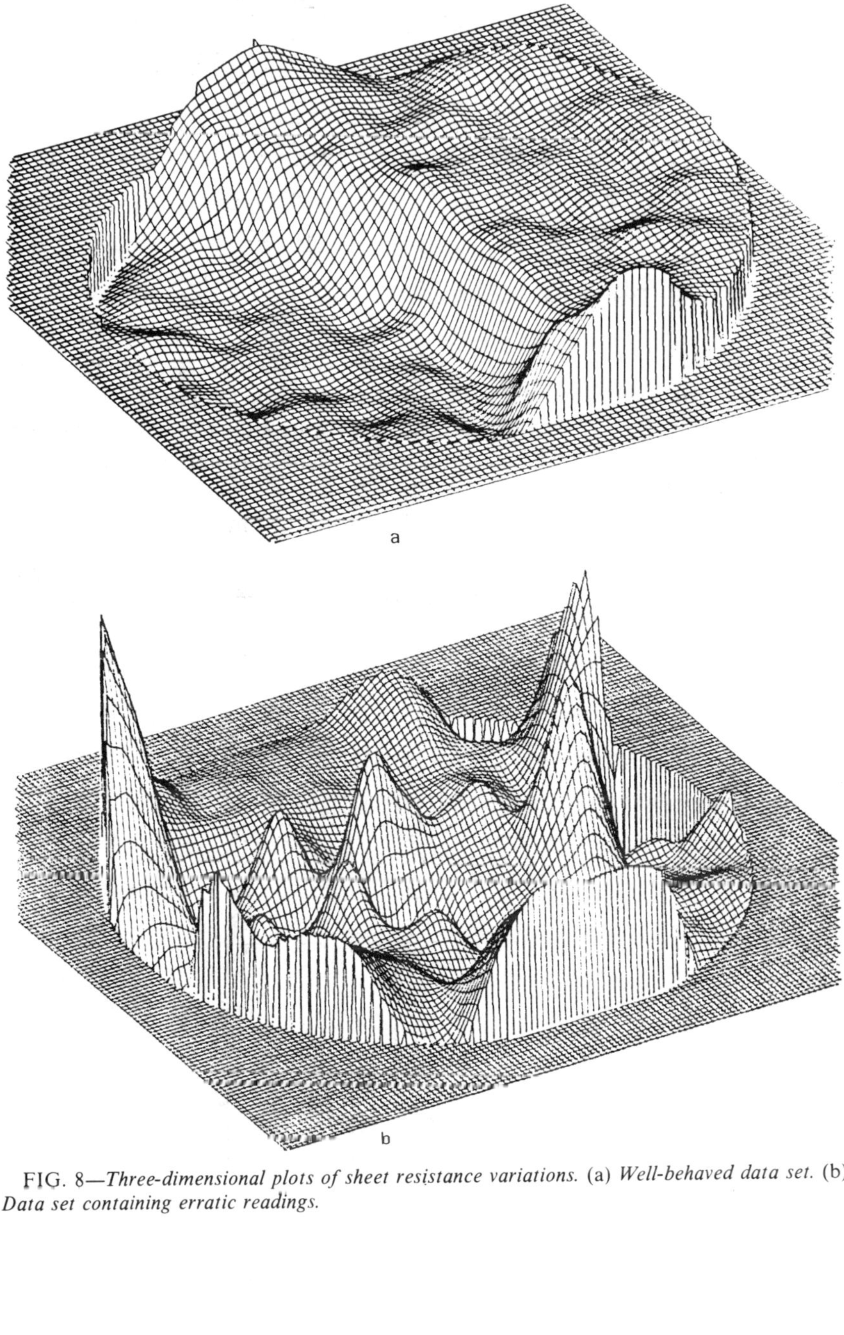

FIG. 8—*Three-dimensional plots of sheet resistance variations.* (a) *Well-behaved data set.* (b) *Data set containing erratic readings.*

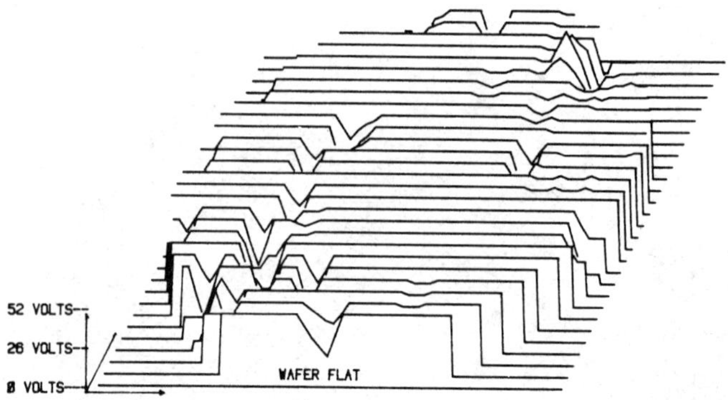

FIG. 9—*Three-dimensional plot of collector-isolation breakdown voltage.*

width data, obtained from a group of six test wafers used to evaluate the dimensional control associated with a 1:1 projection alignment system. As shown in Table 4, the linewidth at a total of nine sites per wafer was measured, with a mean for the 54 readings of $\bar{w} = 3.142$ μm and a standard deviation of $\sigma = 0.072$ μm.

Figures 10*a* and 10*b* differ only in the choice of bin width, 0.02 μm and

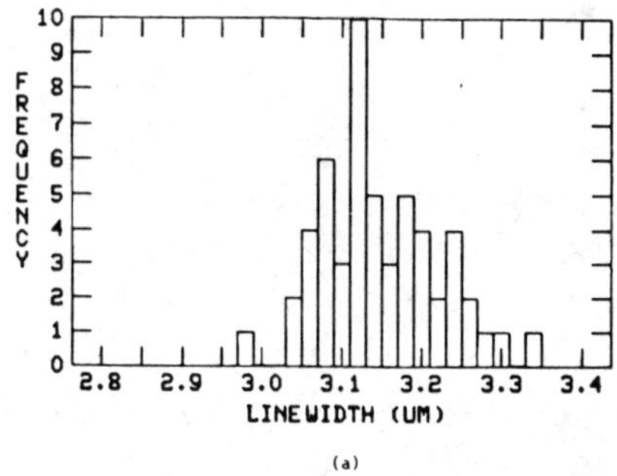

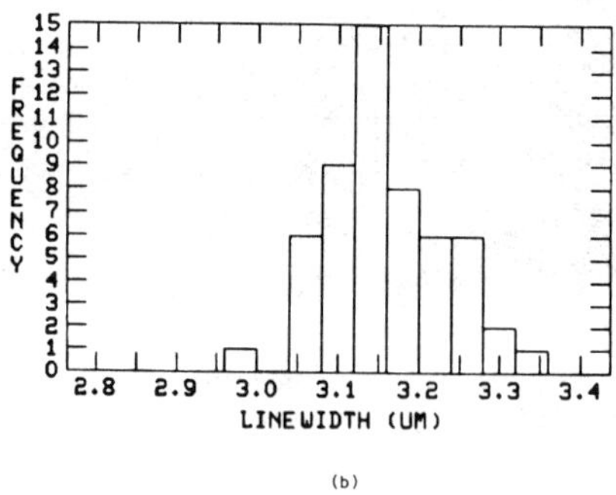

FIG. 10—*Histograms of linewidth data* (*Table 4*) *for two different choices of bin size.* (a) *0.02 μm.* (b) *0.04 μm.*

TABLE 3—*Techniques for displaying the frequency distribution of test data.*

Method	Advantages	Disadvantages
Relative Frequency Distribution (Histogram) (Fig. 10)	• Shows clearly the number or percentage of occurrence in bins of width ΔP • Multimodal behavior apparent by inspection • May be generated manually, without use of a computer	• Appearance varies with choice of bin width, particularly for small number of observations • Difficult to compare two or more frequency distributions by overlaying them
Cumulative Frequency Distribution (Fig. 11)	• Representation utilizes a single monotonically increasing curve, facilitating the overlay of two or more distributions on the same graph • Use of a log-normal scale results in straight-line dependence for a normal distribution, with the slope of the curve inversely proportional to the standard deviation • Multimodal behavior exhibited as change in slope for log-normal scale • Number or percentage of test values falling between particular parameter limits readily determined	• Comparison of several distributions requires careful consideration of shape and location of curves • Number of observations must be specified for each distribution • Generally requires the use of a computer for deriving percentile values, particularly for the log-normal scale
Folded-Axis Cumulative Frequency Distribution (Figs. 12 and 14)	• Highly effective for overlaying several distributions, particularly those containing a small number of observations • Use of a log-normal scale results in an isosceles triangle for a normal distribution • Number or percentage of test values falling between particular parameter limits readily determined	• Number of observations must be specified for each distribution • Generally requires the use of a computer for deriving percentile values, particularly for case of log-normal scale • Multimodal behavior more difficult to detect than with the conventional cumulative frequency distribution

TABLE 4—*Line width data in micrometres for six test wafers.*

NOTE: $\bar{w}$ and σ are the mean and standard deviation, respectively, for the entire data set; $\bar{w}_k$ and σ_k are the mean and standard deviation, respectively, for the kth wafer; $\bar{w}_{ij}$ and σ_{ij} are the mean and standard deviation, respectively, for the ijth test site.

Site	Wafer No. (k)							
(i,j)	1	2	3	4	5	6	$\bar{w}_{ij}$	σ_{ij}
(1,1)	3.25	3.23	3.20	3.20	3.15	3.17	3.200	0.037
(1,2)	3.22	3.14	3.07	3.13	3.10	3.03	3.115	0.049
(1,3)	3.16	3.08	3.07	3.12	3.07	3.05	3.092	0.041
(2,1)	3.17	3.23	3.13	3.23	3.12	3.18	3.178	0.047
(2,2)	3.17	3.11	3.09	3.11	3.22	3.12	3.137	0.065
(2,3)	3.15	3.04	3.07	3.06	2.98	3.06	3.060	0.055
(3,1)	3.33	3.25	3.29	3.27	3.23	3.19	3.260	0.049
(3,2)	3.19	3.14	3.12	3.13	3.07	3.06	3.118	0.048
(3,3)	3.17	3.10	3.11	3.12	3.12	3.11	3.122	0.025
$\bar{w}_k$	3.201	3.147	3.128	3.152	3.118	3.108	$\bar{w} = 3.142$	...
σ_k	0.058	0.074	0.074	0.067	0.077	0.061	...	$\sigma = 0.072$

0.04 μm, respectively. It is clear from these plots that the appearance of a histogram will be influenced by the binning procedure employed. This behavior is particularly troublesome with small data sets, in which the bin count is frequently 0 or 1.

Cumulative Frequency Plots

An alternative method of representing the distribution of test data is to plot the cumulative frequency as a function of parameter value. One can then easily determine the number or percentage of test values falling between particular limits (Fig. 11*a*). By choosing a log-normal scale for the cumulative frequency axis (Fig. 11*b*), one can readily identify normally distributed data by the linearity of the curve [*21*].

The interpretation of cumulative frequency plots is facilitated by "folding" the cumulative frequency axis about the 50th percentile (Fig. 12*a*). Here one reads from 0 to 50% moving *up* the left-hand ordinate, and from 50 to 100% moving *down* the right-hand ordinate. The cumulative frequency axis may also be folded about the 50th percentile for the case of the log-normal scale (Fig. 12*b*), so that normally distributed data are represented as an isosceles triangle.

As noted previously, conventional histograms may exhibit substantial variations in their appearance, depending on the choice of bin width. This tendency is particularly distracting when the number of data points is relatively small. In contrast, since a folded-axis cumulative frequency plot does not require binning of the data, it can be applied to relatively small numbers of data points with much less risk of misinterpretation. In addition, several folded-axis cumulative frequency plots may be overlayed on the same graph,

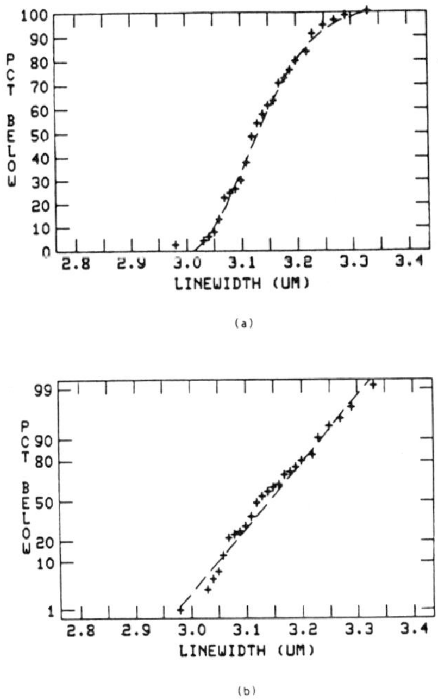

FIG. 11—*Cumulative frequency plots for the data set used in Fig. 10.* (a) *Linear scale.* (b) *Log-normal scale.*

which makes this a particularly economical method of data presentation. In the next section these advantages are applied to the case of a "minimum" array of test sites.

Small Data Sets

General Considerations

The examples employed in previous two sections to illustrate techniques for characterizing process nonuniformities involved the analysis of relatively large data sets. The availability of computer-controlled measurement tools and *x-y* positioning tables now makes it practical to perform such intensive evaluations for a large number of commonly measured device and process parameters. One may, however, be able to collect only limited test data when many wafers must be evaluated within a short period of time. Similarly, when only manual measurement equipment is available, collecting data at a large number of test sites may become impractical, particularly for processes that are ordinarily well-controlled.

The number of test sites required to adequately describe spatial nonuniformities depends on the complexity of the three-dimensional parameter surface. In particular, surfaces with many local fluctuations will require far more test sites for proper identification than will those that are smoothly

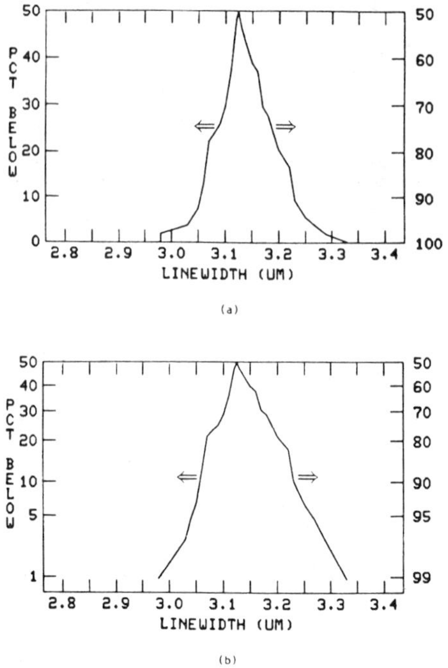

FIG. 12—*Folded-axis cumulative frequency plots corresponding to Fig. 11.* (a) *Linear scale.* (b) *Log-normal scale.*

varying. In many cases, however, the shape of the surface is relatively simple. For example, a linear gradient (*planar* surface) or radial distribution (*concave* or *convex* surface) may be readily distinguished by using data collected from a "loose" grid of test sites. The following example shows that even a minimum 3 by 3 grid of test sites can be extremely useful for detecting spatial inhomogeneities in parameter data.

Example

Figure 13 shows three equivalue linewidth contour maps obtained from three of the six 100-mm-diameter test wafers for which data are given in Table 4. The nine test sites on each wafer are separated by distances of 30 mm in x and y. Smoothly varying contours have been obtained by using linear interpolation to create an artificial 9 by 9 grid. The heavy contour corresponds to the mean of the nine data points. The contour interval for this particular example is 0.025 μm.

The three "minimaps" in Fig. 13 were obtained by measuring nominal 4-μm-wide photoresist features using a commercially available optical linewidth measurement system. The appearance of the maps is indicative of a smoothly varying parameter and well-controlled measurement technique. Of

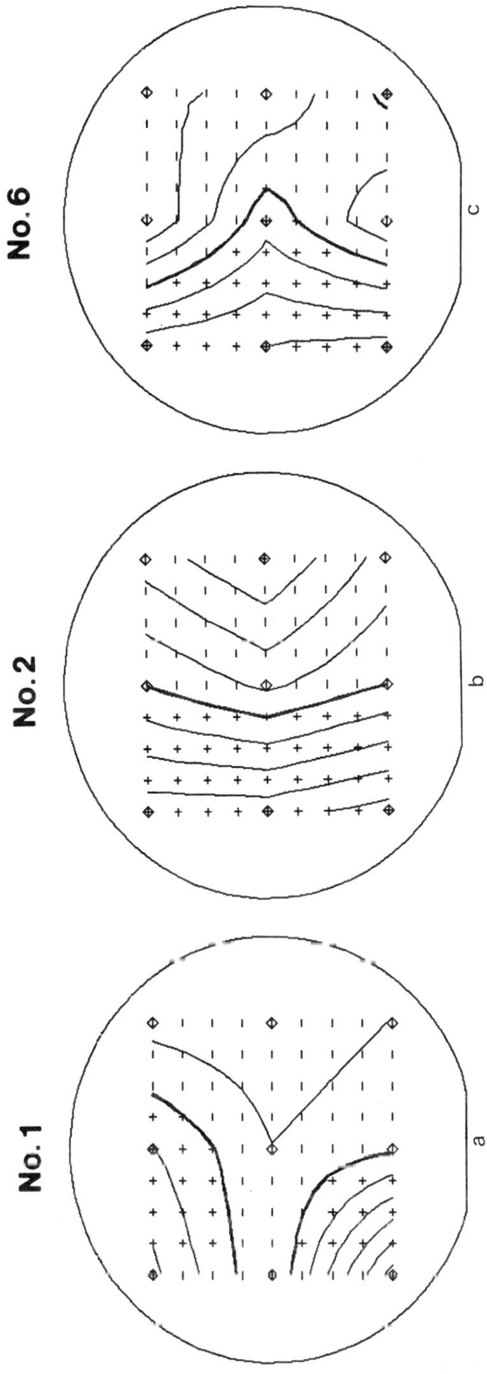

FIG. 13—*Minimaps obtained from a 3 by 3 array of test sites for three of the wafers in Table 4.*

course, using a 4 by 4 or 5 by 5 array of test sites would probably modify somewhat the appearance of the maps in Fig. 13, as well as the values for the mean linewidth $\bar{w}_k$ and standard deviation σ_k given in Table 4. Nevertheless, one can use the 9-point maps as a convenient basis for identifying those wafers for which process parameters vary in an unusual manner, or for which the test data may be inaccurate.

The folded-axis cumulative frequency plots discussed in the section on Frequency Distributions make it possible to perform a rapid visual comparison of wafer-to-wafer and site-to-site variability. Figure 14*a*, which illustrates the within-wafer frequency distributions corresponding to the three data sets used in the maps of Fig. 13, shows that this type of representation is extremely effective despite the small number of readings. Figure 14*b* compares distributions over all six wafers for sites (2,1), (2,2) and (2,3).

Conclusion

This paper has provided an overview of the techniques now in use for characterizing and displaying process nonuniformities on large-diameter wafers. In addition, several new methods of displaying test data for the case of small data sets have been described. The examples used clearly demonstrate the importance of this methodology for understanding and controlling

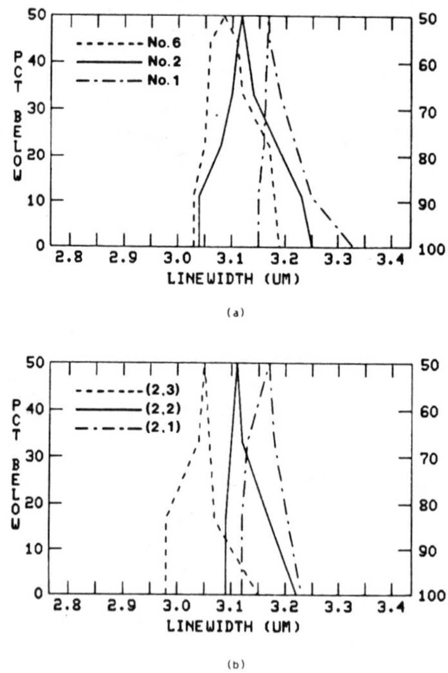

FIG. 14—*Comparison of* (a) *wafer-to-wafer and* (b) *site-to-site variabilities.*

the materials and equipment factors responsible for within-wafer variability. Wider use of these techniques in semiconductor manufacturing should occur now that additional computer-controlled measurement tools, with the capability of collecting and displaying data automatically, are becoming commercially available.

Acknowledgments

The author wishes to thank the following individuals for allowing use of various maps employed in this paper: J. M. Charles (Monolithic Memories), G. G. Evans (Keithley Instruments), J. Hays (Semiconductor Equipment Corp.), L. W. Linholm (National Bureau of Standards), A. D. Manwaring (Signetics), M. H. M. Vervoordeldonk (N. V. Philips), and C. J. Varker (Motorola). The data in Table 4 were provided by I.-W. Huang of Signetics Corporation. Special thanks are due T. Hosseini, for assistance in preparing the manuscript, and T. F. Hasan, C. L. Mallory, and J. Johnson, for carefully reading and commenting on its contents.

References

[*1*] Scott, D. B. H., "PROMIS: A Process Management and Information Management System for Wafer Processing," Microelectronic Measurement Technology Seminar, San Jose, Calif. March 1980.

[*2*] Murakami, W., "The Technical Database and Analysis System: Missing Link in the Semiconductor Information System," Microelectronic Measurement Technology Seminar, San Jose, Calif., March 1981.

[*3*] Charles, J. M. and Lantz, M. W., *IEEE Transactions on Electron Devices,* Vol. ED-27, Aug. 1980, pp. 2299–2303.

[*4*] Perloff, D. S., Hasan, T. F., and Blome, E. R., *Solid-State Technology*, Vol. 23, Feb. 1980, pp. 81–86.

[*5*] Linholm, L. W., Mattis, R. L., Frisch, R. C., and Reeve, C. P., "Characterizing and Analyzing Critical Integrated Circuit Process Parameters," Fourth International Symposium on Silicon Materials Science and Technology, Minneapolis, Minn., May 1981, The Electrochemical Society.

[*6*] Wahl, F. E. and Perloff, D. S., "Techniques for Measurement and Display of VLSI Process Uniformity," Microelectronics Measurement Technology Seminar, San Jose, Calif., Feb. 1979.

[*7*] Evans, G. G., "Semiconductor Parametric Testing: Yesterday, Today and Tomorrow," *Microelectronics Measurement Technology Seminar*, San Jose, Calif., March 1981.

[*8*] Perloff, D. S., Wahl, F. E., Mylroie, S. W., and Mallory, C. L., *Solid-State Technology*, Vol. 24, Sept. 1981, pp. 75–80.

[*9*] Perloff, D. S., Wahl, F. E., and Reimer, J. D., *Solid-State Technology*, Vol. 20, Feb. 1977, pp. 31–36.

[*10*] Perloff, D. S., Wahl, F. E., and Conragan, J., *Journal of the Electrochemical Society*, Vol. 124, April 1977, pp. 582–590.

[*11*] Perloff, D. S., Gan, J. N., and Wahl, F. E., *Solid-State Technology*, Vol. 24, Feb. 1981, pp. 112–120.

[*12*] Perloff, D. S., Wahl, F. E., and Kerr, J. T., "Measurements of the Doping Uniformity of Ion-Implanted Silicon Wafers," in *Proceedings*, 7th International Conference on Electron and Ion Beam Science and Technology, Washington, D.C., 1976.

[*13*] Wahl, F. E. and Perloff, D. S., "An Automated Test System for Measuring and Displaying the Doping Uniformity of Ion-Implanted Silicon Wafers," in *Proceedings*, 8th International Conference on Electron and Ion Beam Science and Technology, Seattle, Wash., May 1978.

[14] Gan, J. N. and Perloff, D. S., "Post-Implant Methods for Characterizing the Doping Uniformity and Dose Accuracy of Ion Implantation Equipment," to be published in *Nuclear Instruments and Methods.*

[15] Perloff, D. S. and Markert, M. J., "The Use and Limitations of Sheet Resistance Measurements for Monitoring Ion Implantation Equipment," Microelectronics Measurement Technology Seminar, San Jose, Calif., March 1981.

[16] Buehler, M. G., Grant, S. D., and Thurber, W. R., *Journal of the Electrochemical Society*, Vol. 125, April 1978, pp. 650–654.

[17] Hasan, T. F., Perloff, D. S., and Mallory, C. L., "Test Vehicles for the Measurement and Analysis of VLSI Lithographic and Etching Parameters," in *Proceedings*, Fourth International Symposium on Silicon Materials Science and Technology, Minneapolis, Minn., May 1981, The Electrochemical Society.

[18] Hasan, T., Shapiro, A., Perloff, D., and Mallory, C., "Characterization of Registration Distortion in Optical Alignment Equipment Using Sub-Micron Measurements," Microelectronic Measurement Technology Seminar, San Jose, Calif., March 1981.

[19] Perloff, D. S., Hasan, T. F., Hwang, D. H., and Frey, J., *Solid-State Technology*, Vol. 24, May 1981, pp. 126–130.

[20] Vervoordeldonk, M. H. M. and van der Pol, A. G. M., "Towards a More Detailed Knowledge of the Quality of Pattern Definition," in *Proceedings*, Conference on Microcircuit Engineering '79, Aachen, West Germany, 1979.

[21] Drum, C. M. and Panousis, P. T., "A Test Pattern for the Evaluation of Bipolar Integrated Circuit Processes," Fourth International Symposium on Silicon Materials Science and Technology, Minneapolis, Minn., May 1981, The Electrochemical Society.

[22] Perloff, D. S., *IEEE Journal of Solid-State Circuits,* Vol. SC-13, Aug. 1978, pp. 436–444.

[23] Hasan, T. F., Katzman, S. U., and Perloff, D. S., *IEEE Transactions on Electron Devices*, Vol. ED-27, Dec. 1980, pp. 2304–2312.

[24] Stover, H., Perloff, D., Hasan, T., Hwang, D., and Blome, G., "Characterizations of Three Wafer Steppers and User Implications," in *Proceedings*, Conference on Microcircuit Engineering '80, Amsterdam, The Netherlands, 1980.

[25] Stemp, I. J., Nicholas, K. H., and Brockman, H. E., *IEEE Transactions on Electron Devices*, Vol. ED-26, April 1979, pp. 729–732.

[26] Perloff, D. S., "Film Thickness Measurements in Silicon Process Technology," Microelectronics Measurement Technology Seminar, San Jose, Calif., March 1980.

[27] Mallory, C. L., "An Automated Small-Spot Film Thickness Mapping System," Microelectronics Measurement Technology Seminar, San Jose, Calif., March 1980.

[28] Varker, C. J., *IEEE Transactions on Electron Devices*, Vol. ED-27, Dec. 1980, pp. 2205–2212.

[29] Dixon, W. J. and Massey, F. J., *Introduction to Statistical Analysis*, McGraw-Hill, New York, 1969.

David G. Mead,[1] *Richard M. Gummer*[1] *and Charles R. Anderson*[1]

Temperature-Dependent Infrared Characterization of Silicon Wafers

REFERENCE: Mead, D. G., Gummer, R. M., and Anderson, C. R., **"Temperature-Dependent Infrared Characterization of Silicon Wafers,"** *Silicon Processing, ASTM STP 804,* D. C. Gupta, Ed., American Society for Testing and Materials, 1983, pp. 445–458.

ABSTRACT: The use of infrared measurements to characterize impurities in doped semiconducting silicon wafers at room and low temperatures is reviewed. Examples include the detection of antimony, arsenic, boron, phosphorus, carbon, and oxygen impurities; the spectral features of these elements are qualitatively discussed. In particular, the design and implementation of a cryostat suitable as an accessory to an infrared interferometer for nondestructive characterization of large-diameter silicon wafers at liquid nitrogen and liquid helium temperatures are also given.

KEY WORDS: cryostat, infrared characterization, silicon dopants, large-diameter silicon wafers, infrared interferometer

Silicon often contains impurities such as carbon and oxygen formed in the growing process. For crystalline silicon the carbon atom actually substitutes for a silicon atom, while the oxygen atom may be situated between the array of silicon atoms. Identification of such substitutional carbon [*1–3*] and interstitial oxygen [*4–7*] bands in the infrared have been reported.[2] The dependence of their respective absorption coefficients on concentration and the necessary procedures for such quantitative analysis are now fairly well established for 2-mm-thick silicon slices. Standards for these analyses include ASTM Test for Interstitial Atomic Oxygen Content of Silicon by Infrared Absorption (F 121) and Parts 1 and 2 of the West German standard DIN 50 438. Also, spectra associated with impurities such as arsenic, antimony, and phosphorus (Group V) donors that make silicon *n*-type, and boron, aluminum, gallium, and indium (Group III) acceptors that make Si *p*-type, have been investigated [*8–16*].

Infrared excitation spectra associated with such Group III or Group V impurities have been thoroughly studied over the last two decades by using dispersive instruments. The first such observations were made by Burstein [*8*] who examined the spectrum of boron in silicon. Colbow [*9*] made a thor-

[1] Nicolet Instrument Corporation, Madison, Wis. 53711.

[2] The italic numbers in brackets refer to the list of references appended to this paper.

ough study at temperatures ranging from 90 to 4.2 K of the far infrared absorption of silicon lightly doped with boron. By allowing for line distortion of the finite spectrometer slit width, the temperature dependence of the boron halfwidths could be accurately determined [*9*]. Onton et al [*10*] also made a comprehensive study of Group III acceptor states, including boron, aluminum, gallium, and indium at liquid helium temperatures and under stress.

There are several other papers [*11–16*] that discuss the absorption spectra of neutral donors or acceptors (for singly charged centers). More recently, White [*11*] and Kolbesen [*12*] reported the simultaneous determination of the total concentrations of B and P in silicon (with resistivity $\rho \simeq 50\ \Omega - \text{cm}$) by a simple infrared spectroscopic technique with the silicon samples at low temperature. The technique involves illumination of the silicon with enough light of energy greater than the band-gap energy ($\tilde{\nu} \geq 8950\ \text{cm}^{-1}$) [*11,12*].

Interferometric spectroscopy should yield improved performance compared with grating spectrometers. Baber [*15*] has described such a method utilizing a Fourier transform infrared (FT-IR) interferometer and low-temperature techniques; he has made extensive studies to determine the calibration factors for the quantitative analysis of boron, phosphorus, and arsenic, including estimates for the other shallow impurities. Detection sensitivities of better than 5×10^{11} atoms$\cdot$cm^{-3} were demonstrated. Pajot and Debarre [*16*] also have made a quantitative determination of B and P in silicon. For phosphorus they considered the accuracy of a neutron-transmuted doped sample to be high enough to use it as a primary standard for calibration purposes.

All the above measurements suffer from the disadvantage that the low temperature techniques allow the use only of small samples that have to be cut to size. Typical cryostat specimen holders hold only specimens with a maximum diameter of about 25.4 mm (1 in.). To overcome this disadvantage a cryostat was developed that houses large-diameter silicon wafers or slices. This paper briefly describes both the spectrometer and cryostat (see section on Experimental Method) and also the results obtained in the 200 to 7000 cm^{-1} region on crystalline silicon containing some impurities (C, O, B, P, As).

The need to control the quality of these silicon wafers and slices by using a nondestructive material characterization technique such as infrared spectroscopy has also brought renewed interest in this technique.

Theoretical Background

Before considering the experimental details of the cryostat and results, we outline a very elementary discussion on the infrared absorption lines observed from Group III and Group V shallow impurities.

Addition of a Group V substitutional donor impurity such as phosphorus into a silicon lattice causes an electron to be bound to the impurity ion. This situation, in which the electrons are bound to specific neutral impurity sites, occurs at helium temperatures when the thermal excitations (phonons) cannot participate. The electron energy levels of the impurity atoms exist between the valence and conduction band of the silicon lattice. Similarly, discrete electron energy levels of a bound hole of an acceptor atom (such as boron) exist in the forbidden gap of Si at very low temperatures.

Figure 1 shows the schematic energy band diagram and the various transitions due to shallow donors (electron transitions) and acceptor (hole transitions) centers. Figure 1 is simplified, since it is well known that the valence band of silicon exhibits spin-orbit coupling. The $P_{1/2}$ valence band lies below the $P_{3/2}$ valence band. Optical transitions from the $P_{1/2}$ impurity ground state to excited states are not depicted in Fig. 1. By observing such transitions, it may be possible to determine the impurity concentrations present, providing the measurement is calibrated with a known standard, as described in the references given in the introduction.

In a perfectly compensated semiconductor with N_D donors $\cdot$ cm^{-3} and N_A acceptors $\cdot$ cm^{-3}, there will ideally be at low temperatures N_A negatively charged acceptors $\cdot$ cm^{-3} and N_D positively charged donors $\cdot$ cm^{-3}. Accordingly, in *n*- or *p*-type silicon samples, there will be $|N_A - N_D|$ neutral atoms $\cdot$ cm^{-3}. When $N_A \simeq N_D$ we have a highly compensated sample. At low temperatures this situation may be realized. IIowcvcr, only the neutral centers can absorb radiation, and the compensated impurities remain ionized, so the spectrum would be characteristic of net impurities only. By generating

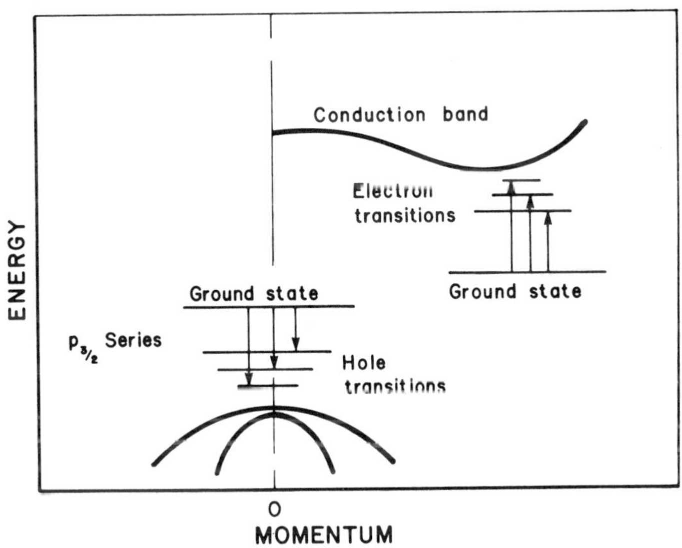

FIG. 1—*Schematic energy-momentum diagram for doped silicon, indicating acceptor and donor states.*

excess electron-holes pairs across the forbidden gap with white light [*11,12*], all the impurities may be considered neutral and the total shallow impurities can be detected.

Experimental Method

The Spectrometer

A Fourier transform infrared spectrometer basically consists of a Michelson interferometer linked to a dedicated computer. The computer controls:

1. The data collection (the interference pattern signal emanating from the detector output).
2. The infrared beam path, so that the infrared signal may be brought out of the main interferometer housing (FT-IR system) to pass through the windows housed in the cryostat and onto a suitable detector housed on the accessory bench.
3. The radial profiling of each large-diameter silicon wafer (or specimen) placed in the Teflon cassette in the cryostat.

To calibrate the transmittance (or absorbance) scale the background infrared interferogram (interference pattern arriving at the detector) was recorded. In this mode, the optical cryostat has no specimens in the infrared beam. The particular wafer or sample whose spectrum is to be recorded is then automatically raised (under computer control) into the beam path and the sample (foreground) interferogram is recorded. The resulting raw interferograms are subsequently Fourier transformed and the ratio of the foreground/background amplitude spectra gives the transmission spectrum through the specimen at the particular temperature of the inner housing of the cryostat.

The Cryogenic Accessory

The cryostat is used as an accessory to the infrared spectrometer. The requirements and operating features are:

1. To build a cryostat to efficiently cool down silicon wafers to liquid nitrogen (77 K) and liquid helium temperatures (1.2 to 20 K).
2. To use a cassette to house the specimens.
3. To build the cryostat large enough to accommodate standard cassettes that would house large-diameter wafers with 51-, 76-, 102-, and 127-mm (2-, 3-, 4-, and 5-in.) diameters.
4. To allow each wafer in the cassette to be radially profiled for impurities at room, liquid nitrogen, or liquid helium temperatures.
5. To incorporate a white light lamp to be able to irradiate the samples if

necessary (Kolbesen method [12]). A fiber optic cable serves to guide the radiation into the cryogenic cavity and onto the sample being studied.

A special operating mechanism (not described) to be able to raise and lower each large-diameter wafer at low temperatures was designed. Figure 2 shows the principle of operation for such a mechanism. Figure 3 shows a plan of the optics of the accessory bench. A deuterated triglycine sulphate detector operating at room temperature was used.

Typical resolution varied from 2 cm^{-1} to 0.5 cm^{-1} for transmission measurements (see also Ref *15*). The temperature of the silicon wafers was measured by a calibrated silicon photodiode.

Specimens

Czochralski-grown crystals were used. For the room temperature and low-temperature studies, silicon wafers of 76-mm (3-in.) diameter and approximately 0.4 mm (0.015 in.) thick were mounted in the cassette. In some cases the specimens studied were believed to have been deliberately doped with phosphorus and arsenic impurities. In some cases, sizes of 25.4-mm (1-in.) diameter from different boules (resistivities of 5 and 700 Ω-cm) were

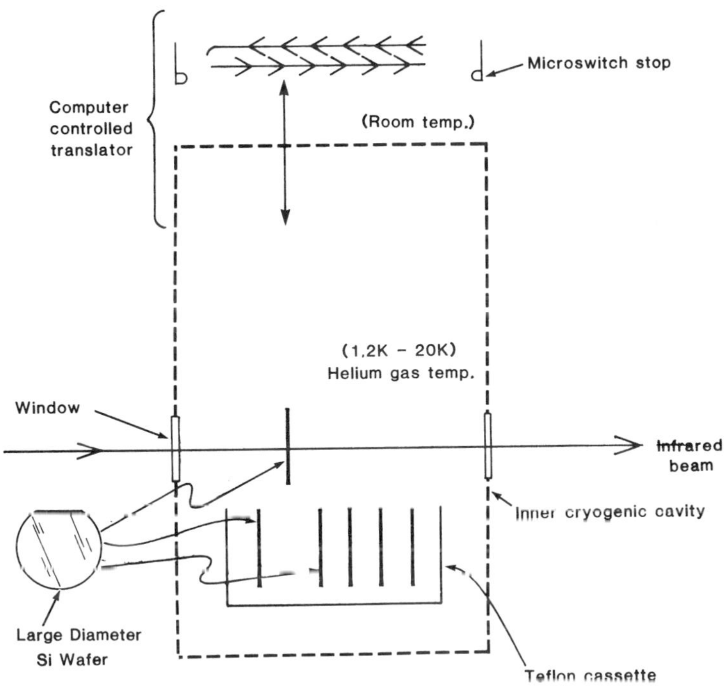

FIG. 2—*Principles of operation of the automated cryostat.*

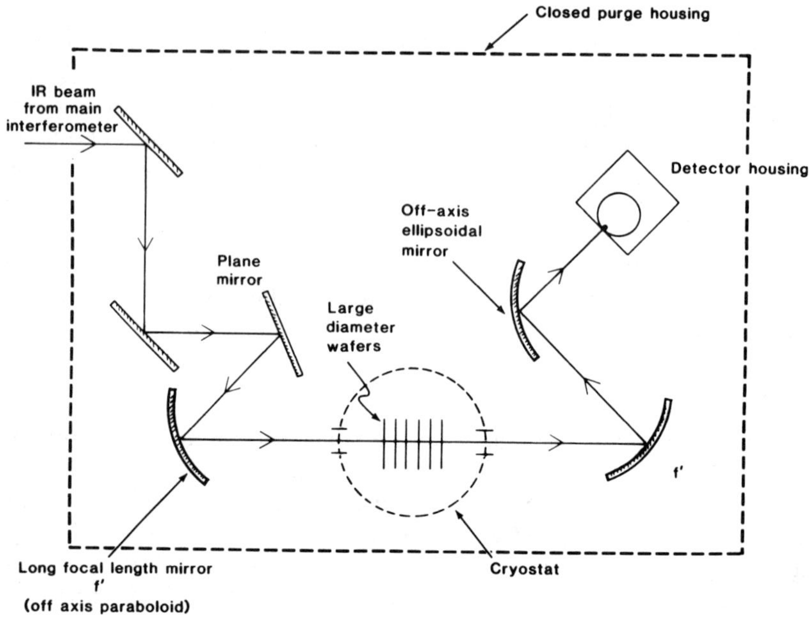

FIG. 3—*Plan of optics of the cryostat accessory bench.*

available, and were mounted concentrically on a dummy aluminum wafer (with a 25.4-mm [1-in.] radial hole). In one case a 4-mm (0.16-in.) thick slice ($\rho \simeq 250$ Ω-cm) of Si was used.

Results

Figures 4 and 5 depict the room temperature transmission spectra for *n*-type and *p*-type silicon samples, respectively. Table 1 lists the samples (A to E) used for Figs. 4 and 5 and their respective impurity characteristics. Figure 5 shows the spectra (of Specimens B and D in Table 1) as essentially featureless until about 2000 cm^{-1} (5 μm), where there is onset of absorption by the silicon lattice and some characteristic bands occur, including the oxygen bands and two and three phonon summation bands [*17*]. Towards the very long wavelength region there occurs an onset where the transmittance values decrease in both samples. For *n*-type silicon, Specimens C and E show similar room temperature behavior (Fig. 4).

However, in the spectrum of Specimen A both the oxygen band at about 1100 cm^{-1} and the two phonon lattice band at about 600 cm^{-1} are prominent, compared to the spectrum of Specimen C, doped with antimony. The antimony impurity tends to mask out the infrared lattice modes. This effect is seen in the spectrum of Specimen E, a phosphorus-doped epitaxial layer on an antimony-doped silicon sample. This overall decrease in transmission is associated with the transition of electrons from the valence band to the con-

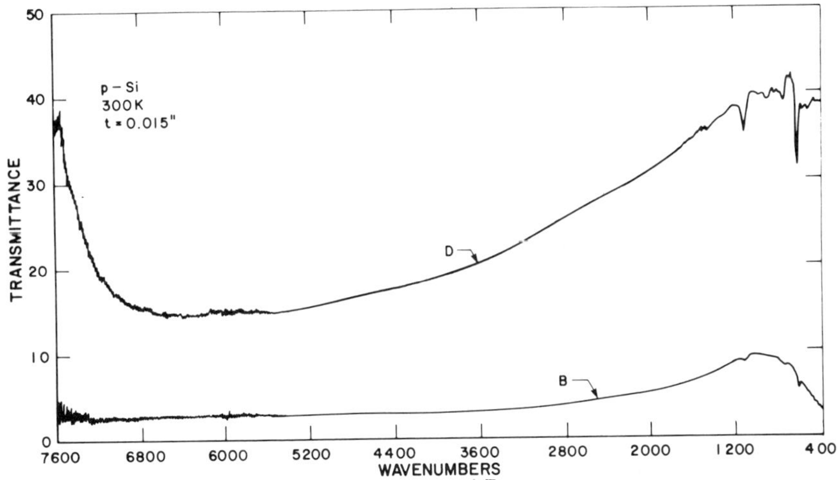

FIG. 4—*Infrared spectra of three* n-*type silicon wafers.*

duction band. Such free electron absorption (from the ionized impurities) might be expected to decrease at low temperatures [*8–16*] and shows the need for cryogenic temperatures. At low temperatures, the infrared-induced impurity lattice-type modes would also be expected to be spectroscopically observed.

The net 300 K and 15 K transmission spectra obtained from a 3-mm (0.12-in.) thick silicon sample are shown in Figs. 6*a* and *b*, respectively. The 300 K oxygen peak at 1104 cm^{-1} is seen to shift to a higher frequency and split upon cooling to 15 K. The prominent two phonon band does not shift frequency

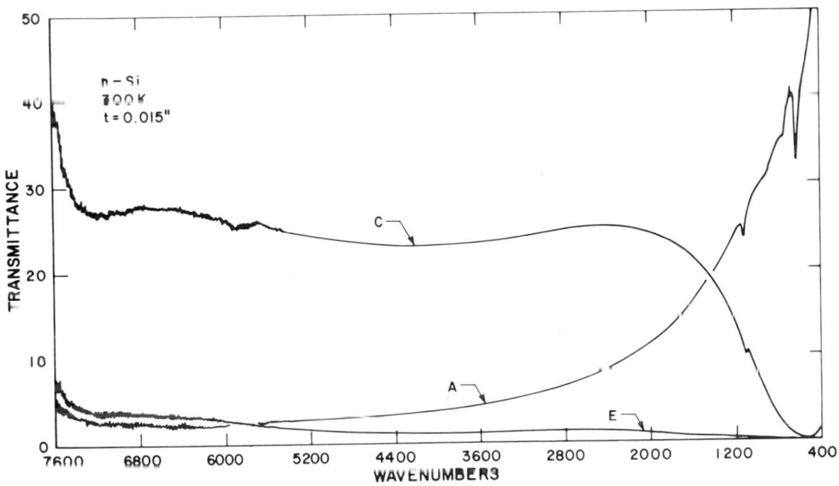

FIG. 5—*Infrared spectra of two* p-*type silicon wafers.*

TABLE 1—*Evaluation of five silicon wafers.*[a]

Specimen	Donor or Acceptor	Dopant	Impurity Density, cm^{-3}	Impurity Density, ppb (a)	Resistivity, Ω-cm
A	*n*	phosphorus	8×10^{14}	0.016	6.0
B	*p*	boron	9×10^{16}	1.8	0.2
C	*n*	antimony	6×10^{17}	12.0	0.06
D	*p*	boron	1.3×10^{16}	0.26	1.05
E	*n*	phosphorus-doped epitaxial layer on antimony-doped substrate	epilayer 1.1×10^{14};	0.0022	42.5
			substrate 3×10^{18}	60	0.014

[a] All 0.4 mm (0.015 in.) thick.

appreciably upon cooling, but decreases in the intensities of the range of lattice bands in the 500 to 1000 cm^{-1} region, reflecting changes in the density-of-states curve for the lattice. The localized impurity mode due to carbon is just visible in Fig. 6*b*, situated on the two phonon peak at 607.5 cm^{-1}. Bands at 519 and 1204 cm^{-1} due to oxygen were also seen in the sample studies (Figs. 6*a* and *b*, respectively). The 15 K spectrum did not show the corresponding room temperature 519 cm^{-1} peak, but rather an antiresonance [*18*] centered at about the same frequency as the room temperature band. Also prominent

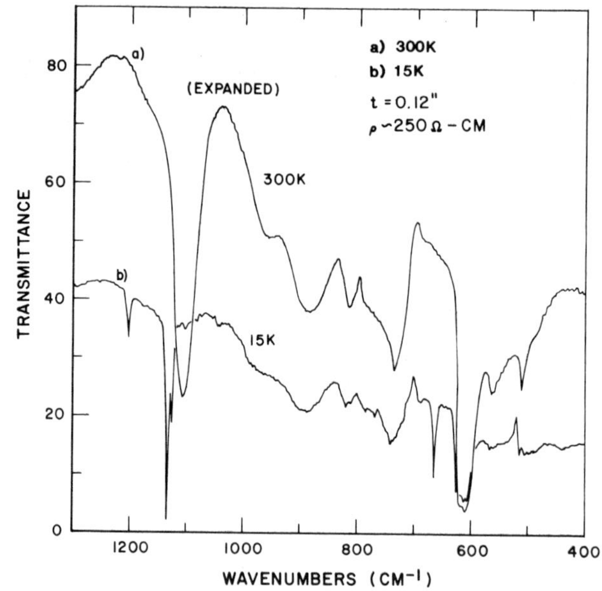

FIG. 6—*Effects of cooling a 3-mm (0.12 in.) thick, high-resistivity,* p-*type doped silicon crystal to 15 K.*

in Fig. 6*b* is a sharp peak situated at 669 cm^{-1}. Figure 7*a* and *b* show in the 650 to 200 cm^{-1} region the net transmission values at room temperature and 20 K of a silicon wafer 0.4 mm (0.015 in.) thick containing boron. Comparison with the corresponding spectra for the 400 to 600 cm^{-1} region of Fig. 7*a* and *b* indicate the same qualitative features as found in Fig. 6*a* and *b*, respectively. The absorption features are now weaker for the thinner silicon sample (Fig. 6*a* and *b*). The band at 519 cm^{-1} (the oxygen bending mode) in Fig. 6*a* is not present, but a band at about 572 cm^{-1} is observed along with the antiresonance. The band at 570 cm^{-1} is believed to be the ^{14}C local mode band. The most significant change in the low-temperature spectrum is the addition of intense bands in the 250 to 400 cm^{-1} region. These bands arise from the excited electron or hole transitions from the neutral impurity centers, as outlined in the theoretical section. As a further example, Figs. 8*a* and *b* are expanded plots of two samples 3 and 4 mm (0.12 and 0.16 in.) thick, respectively, in the 450 to 150 cm^{-1} region. The spectra of Fig. 8 confirm the presence of both boron and phosphorus in the specimens and confirm the suspicion that these samples were deliberately compensated. The frequency positions of the boron and phosphorus bands are indicated by the annotated lines in Figs. 8*a* and *b*. Figure 9*a* illustrates the net transmittance spectrum of a 3-mm (0.12-in.) sample. The presence of phosphorus is not observed (with no light present on sample). The boron peak at approximately 319 cm^{-1} dominates the spectrum. However, when the sample is irradiated with white light the presence of the phosphorus peak as the compensating impurity peak at about 316 cm^{-1} is evident. The free electrons so created recombine with a fraction of the phosphorus impurities. Figure 9*b* illustrates the total impurity spectrum in which the white light is turned on. Figure 10 shows the net transmission versus wavenumber plot in the 650 to 200 cm^{-1} region for a sample suspected of being doped with arsenic. The spectrum reveals a band at 382 cm^{-1} and confirms the presence of arsenic.

In all cases studied as yet, the spectra were from the center of the silicon

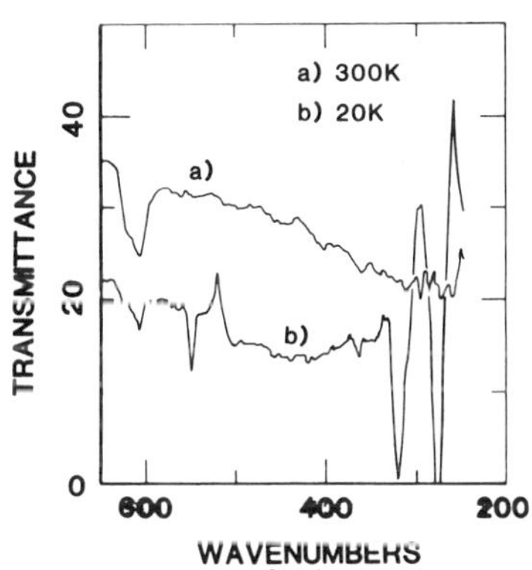

FIG. 7 *Effect of cooling a silicon wafer from 300 K to 20 K.*

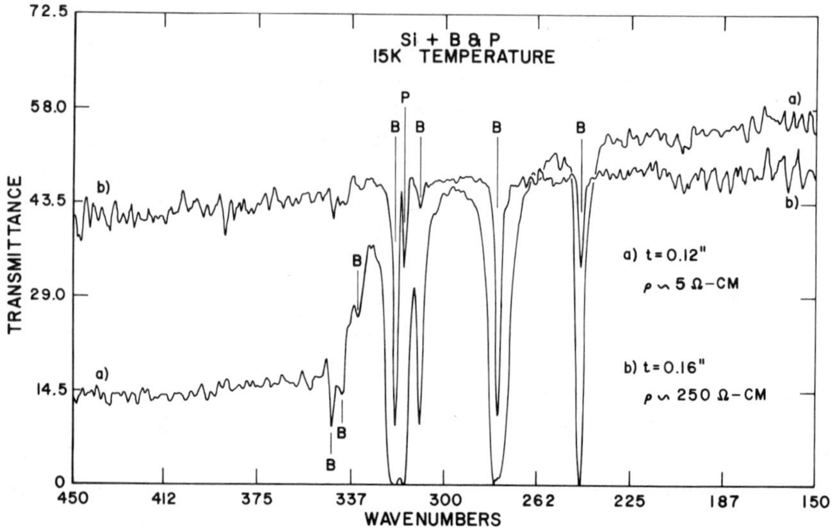

FIG. 8—*The far infrared excitation spectra for two silicon crystals of differing resistivities and thicknesses.*

specimens studied, and the impurity distribution was assumed homogeneous.

As a final example, Fig. 11 illustrates the integrated absorbance values for the carbon and oxygen bands versus distance across a single-sided polished 76-mm (3-in) diameter Czochralski-grown silicon wafer (0.3 mm [0.015 in.] thick). The wafer was automatically raised into the infrared beam and was at liquid nitrogen temperature (77 K). The results are the average of three inde-

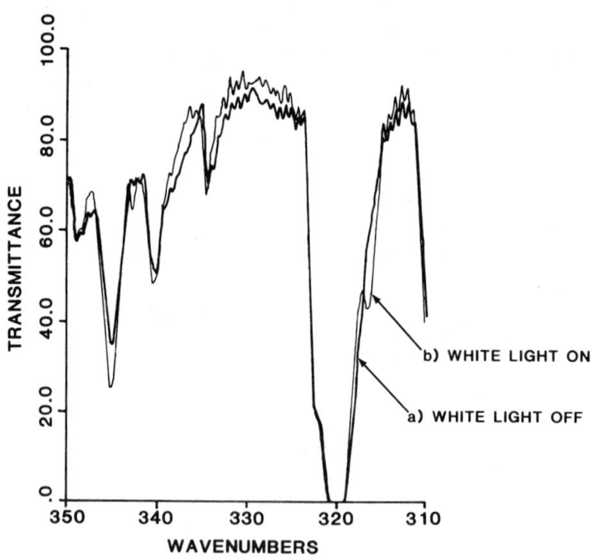

FIG. 9—*Effect of white light excitation on the net spectrum at 15 K.*

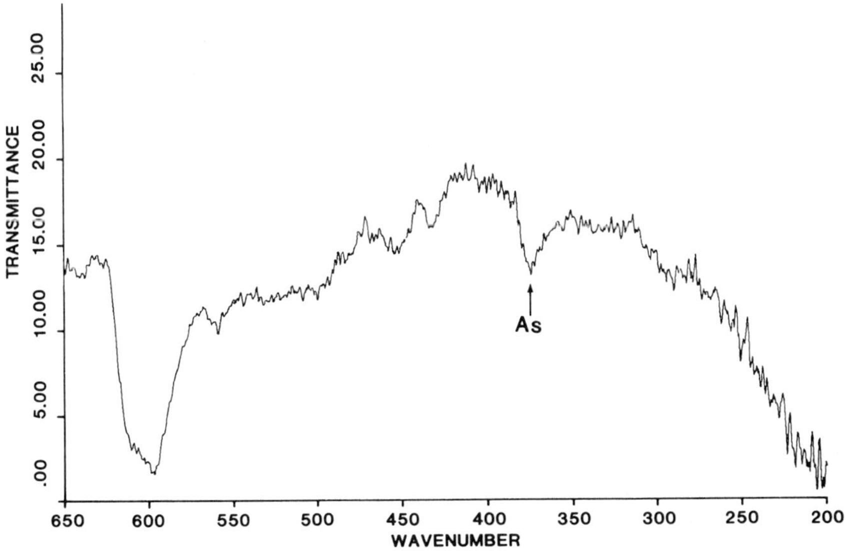

FIG. 10—*Infrared detection of arsenic in a silicon slice 0.16 mm thick at 15 K.*

pendent sets of measurements. Integrated absorbance values were measured for the oxygen split bands centered at 1136 and 1127 cm^{-1} and for the ^{12}C peak at 607.5 cm^{-1}.

Discussion

For the cases studied in this paper, we deal mostly with lightly compensated samples (~10^{16} impurity centers · cm^{-3}), so the low-temperature spectra should be characteristic of optical excitations from the ground state of

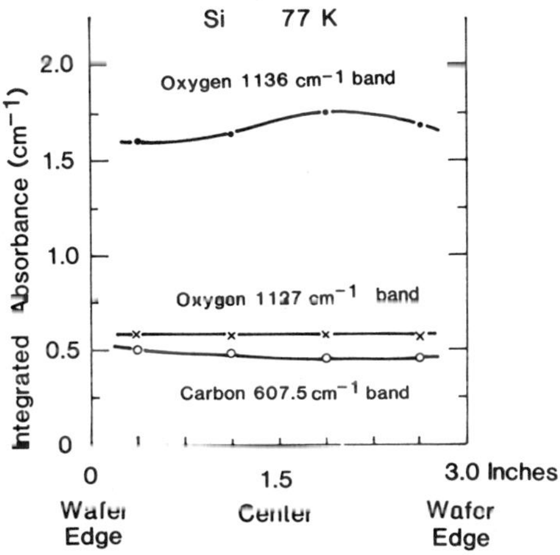

FIG. 11—*Profile of carbon and oxygen lines at liquid nitrogen temperature across a 76-mm (3-in.) diameter silicon wafer.*

both neutral donor and acceptor centers. With increasing impurity densities ($\sim 10^{18}$ cm^{-3}), the associated free carrier absorption makes infrared investigations impracticable. This problem has been overcome by counter-doping silicon [*19*] so that various localized vibrational modes have been observed. For example, after compensation with electrons, a sample containing both B and P (impurity densities of 5×10^{18} cm^{-3}) showed local mode peaks due to the isotopes of B^{10} and B^{11} at 617.2 cm^{-1} and 641.2 cm^{-1}, respectively. A weak feature present at 641.0 cm^{-1} in Fig. 6*b*, and 641.2 cm^{-1} in Fig. 7*b*, was seen in both the liquid nitrogen and helium spectra. The sharp peak at 669 cm^{-1} in Fig. 6*b* has been attributed to the excitation of a hole from the B ground state to the first resonant excited state associated with the $P_{1/2}$ valence band [*20*]. Unfortunately, the peak position of this line occurs where a feature due to carbon dioxide associated with the purge system of the spectrometer exists, so that it would not be useful for quality control measurements. Careful purging with dry nitrogen or spectral subtraction of the carbon dioxide present would eliminate such effects for the 669 cm^{-1} line.

With increasing concentrations (as in highly compensated material), there are naturally interactions or pairing effects between neighboring donors and acceptors. The presence of boron-donor pairs in a sample should lead to four infrared active bands since there are two boron isotopes. Newman and Smith [*3*] have found lines at 622, 631, 653, and 655 cm^{-1} due to B^{10}-P and B^{11}-P interactions, and a line at 615 cm^{-1} due to the B^{11}-B^{11} vibration. Although no such pairing modes were seen for the samples studied here, one should be aware of such effects if the silicon sample under study has a relatively high doping concentration ($\gtrsim 10^{19}$ atoms $\cdot$ cm^{-3}).

Returning to the observed donor level transitions, Table 2 lists the previously measured results by workers using grating spectrometers. (Also included are results obtained from aluminum-, antimony-, arsenic-, boron-, gallium-, and indium-doped silicon specimens.) Experimental errors are included. The line observed at 316.1 in Figs. 8*a* and *b* is that of phosphorus. Picus, Burstein, and Henvis [*23*] observed phosphorus lines at 318.5, 359.7, and 381.7 cm^{-1} for a sample doped only with phosphorus. The latter two lines were not observed in the compensated samples studied herein. The line strengths of the phosphorus line relative to the boron line at 319.4 cm^{-1} indicates only the approximate degree of compensation (N_A/N_D) achieved. The 3-mm (0.12-in.) thick sample in the cassette holder was believed to have been only lightly doped with boron, but the results also indicate the presence of phosphorus.

In all samples studied, both carbon and oxygen impurities were present in varying amounts. Carbon and oxygen impurities in silicon have recently been studied by using Fourier transform techniques [*24–27*]. The room temperature band at 1104 cm^{-1}, the most prominent line for interstitial oxygen, is assumed to arise from the asymmetric stretching mode of Si_2O [*4–6*]. The 300 K band at 1204 cm^{-1} is due to the oxygen symmetric stretch.

TABLE 2—*Comparison of some of the $P_{1/2}$ and $P_{3/2}$ experimental positions for boron and phosphorus with silicon as the host lattice in the 4.2 to 20 K temperature range.*[a]

Transition	Shallow Impurity	Type	Frequencies	
			Dispersive	Interferometry
$P_{3/2}$	boron	A	(±0.2)[b]	(±0.1)
			245.0	244.9_4
			278.5	278.2_2
			309.3	309.7_0
			319.7	319.4_4
			321.9	not observed
			334.9	340.0_7
			342.8	342.3_2
			345.1	344.5_0
$P_{1/2}$	boron	A	668.6	669.0
			692.7	692.6
$P_{3/2}$	aluminum	A	471.8	
$P_{3/2}$	gallium	A	548.1[c]	548.1
$P_{3/2}$	indium	A	1175.9[d]	1174.9
$P_{3/2}$	arsenic	D	384.0	382.0
$P_{3/2}$	antimony	D	295.1[e]	293.5
$P_{3/2}$	phosphorus	D	318.5	316.09
			359.7	not observed
			381.7	not observed

[a] All values are in cm^{-1}. Also included are the strongest absorption lines for aluminum, antimony, arsenic, gallium, and indium impurities.
[b] Ref *10*.
[c] Ref *21*.
[d] Ref *22*.
[e] Ref *23*.

Band splitting occurs at lower temperatures. Figure 6*b* shows the oxygen band has split into two bands due to the rotational motion of the small bivalent oxygen atom. At helium temperatures, these bands show finer structure under higher resolution. At 1 cm^{-1} resolution five bands were observed at 1123.0, 1127.0, 1130.5, 1132.2, and 1137.4 cm^{-1}. Pajot [*5*] has also studied the infrared absorption of the SiO_2 group at low temperature and under higher resolution (0.5 cm^{-1}). A phase transition was observed, so that at 23 K seven lines appeared, while at 6 K only four lines were present.

The infrared intensities of each fine structure line were assigned [*6*] in terms of the relative abundance for the silicon isotopes (^{28}Si, ^{29}Si, and ^{30}Si). The band splitting (with peaks at 1127 and 1136 cm^{-1}) seen from the results at 77 K and from the 15 K measurements (see for example Fig. 6*b*) are due to the averaging of the isotopes of silicon involved in the Si-O^{16}-Si vibration. For example, the variation in the 1136 cm^{-1} band observed may be due to a grouping of ^{28}Si-O-^{28}Si with ^{28}Si-O-^{29}Si and ^{28}Si-O-^{30}Si, to give the relative strength of the line shown for a particular resolution.

It is noteworthy that for oxygen the infrared results of Fig. 11 give only the variation of the Si-O-Si band and not a variation of the total oxygen present in the crystal. Similarly for the carbon band, the results are only indicative of the integrated absorbance of the substitutional band.

In cooled crystals containing oxygen, extra structure has been found on the high-frequency side of the oxygen band. Such features were observed (in the 1150–1270 cm^{-1} range) even in the samples studied here, but were not reproducible and are not given here. Such weak features may be indicative of weak platelet precipitation [*28*].

Acknowledgments

We are grateful to J. Covey for his support, L. Johnson and A. Zerbel for help in the construction of the cryostat, and S. Weibel for technical assistance.

References

[*1*] Baker, J. A., *Solid State Electronics,* Vol. 13, 1970, p. 1431.
[*2*] Newman, R. C. and Willis, J. B., *Journal of Physics and Chemistry of Solids,* Vol. 26, 1965, p. 373.
[*3*] Newman, R. C. and Smith, R. S., *Journal of Physics and Chemistry of Solids,* Vol. 30, 1969, p. 1493.
[*4*] Hrostowski, H. J. and Alder, B. J., *Journal of Chemical Physics,* Vol. 33, 1960, p. 980.
[*5*] Pajot, B., *Journal of Physics and Chemistry of Solids,* Vol. 28, 1967, p. 73.
[*6*] Pajot, B. and Deltour, J. P., *Infrared Physics,* Vol. 7, 1967, p. 195.
[*7*] Kaiser, W., Keck, P. H., and Lange, C. F., *Physics Review,* Vol. 101, 1959, p. 1264.
[*8*] Burstein, E., Oberly, J. J., Davisson, J. W., and Henvis, B. W., *Physics Review,* Vol. 82, 1951, p. 764.
[*9*] Colbow, K., *Canadian Journal of Physics,* Vol. 41, 1963, p. 1801.
[*10*] Onton, A., Fischer, P., and Ramdas, A. K., *Physics Review,* Vol. 163, 1967, p. 686, and references therein.
[*11*] White, J. J., *Canadian Journal of Physics,* Vol. 45, 1967, p. 2797.
[*12*] Kolbesen, B. O., *Applied Physics Letter,* Vol. 27, 1975, p. 353.
[*13*] Lifshitz, T. M. and Ya Nad, F., *Soviet Physics Doklady,* Vol. 10, 1965, p. 532.
[*14*] Sidorov, V. I. and Lifshitz, T. M., *Soviet Physics Solid State,* Vol. 8, 1967, p. 2000.
[*15*] Baber, S. C., *Thin Solid Films,* Vol. 72, 1980, p. 201.
[*16*] Pajot, B. and Debarre, D. in *Proceedings,* Third International Conference on Neutron Transmutation Doped Silicon, Copenhagen, Aug. 1980, in press.
[*17*] Johnson, F. A., *Proceedings of the Physics Society* (London), Vol. 73, 1959, p. 265.
[*18*] Watkins, G. D. and Fowler, W. B., *Physics Review B,* Vol. 16, 1977, p. 4524.
[*19*] Angress, J. F., Goodwin, A. R., and Smith, S. D., *Proceedings of the Royal Society* (London), Vol. 287A, 1965, p. 64.
[*20*] *Proceedings,* First International Conference on Localized Excitations in Solids, University of California, 1967, Wallis, R. F. Ed., Plenum Press, New York, 1968.
[*21*] Burstein, E., Picus, G., and Henvis, B., *Journal of Physics and Chemistry of Solids,* Vol. 1, 1956, p. 65.
[*22*] Hrostowski, H. J. and Kaiser W., *Journal of Physics and Chemistry of Solids,* Vol. 7, 1958, p. 236.
[*23*] Picus, G., Burstein, E., and Henvis, B., *Journal of Physics and Chemistry of Solids,* Vol. 1, 1956, p. 75.
[*24*] Mead, D. G. and Lowry, S. R., *Applied Spectroscopy,* Vol. 34, 1980, p. 167.
[*25*] Mead, D. G., *Applied Spectroscopy,* Vol. 34, 1980, p. 171.
[*26*] Vidrine, D. W., *Analytical Chemistry,* Vol. 52, 1980, p. 92.
[*27*] Baghdadi, A. and Forman, R. A., *Applied Spectroscopy,* Vol. 35, 1981, p. 473.
[*28*] Baber, S. C. et al, *Proceedings of Electrochemical Society Meetings,* Extended Abstract 498, Vol. 79-2, Oct. 1978, p. 1262.

Robert K. Graupner[1]

Analysis of Infrared Spectra for Oxygen Measurements in Silicon

REFERENCE: Graupner, R. K., "**Analysis of Infrared Spectra for Oxygen Measurements in Silicon,**" *Silicon Processing, ASTM STP 804,* D. C. Gupta, Ed., American Society for Testing and Materials, 1983, pp. 459–468.

ABSTRACT: The effect of multiple reflections when analyzing the infrared spectra for oxygen measurements in silicon wafers is important. It may cause errors up to 20% when ignored. The procedure and calculations to derive oxygen concentration from the spectra by taking into account this effect is described in detail. The effect of back-surface finish on the measurements is also discussed.

KEY WORDS: Oxygen concentration in silicon, silicon wafers, multiple reflection, infrared spectra, absorption coefficient, back-surface finish

The simplest formula for converting a transmittance measurement to an absorption coefficient for a sample polished on both sides is

$$\begin{aligned} T &= (1 - R)^2 e^{-\alpha d} \\ \ln T &= \ln (1 - R)^2 - \alpha d \\ \alpha &= -\frac{1}{d} \ln (T/(1 - R)^2) \end{aligned} \tag{1}$$

where

T = transmittance,
d = slice thickness,
R = reflectance (~0.3 for silicon), and
α = absorption coefficient.

The absorption coefficient so obtained for the 9-μm absorption band is the sum of two contributions: (*1*) the absorption at 1106 cm^{-1} due to any interstitial oxygen content of the silicon, and (*2*) overlap from the intrinsic absorption band centered at approximately 1124 cm^{-1}. The combined absorption band and the intrinsic band is shown in Fig. 1. To obtain the oxygen

[1] Wacker Siltronic Corporation, Portland, Ore. 97203.

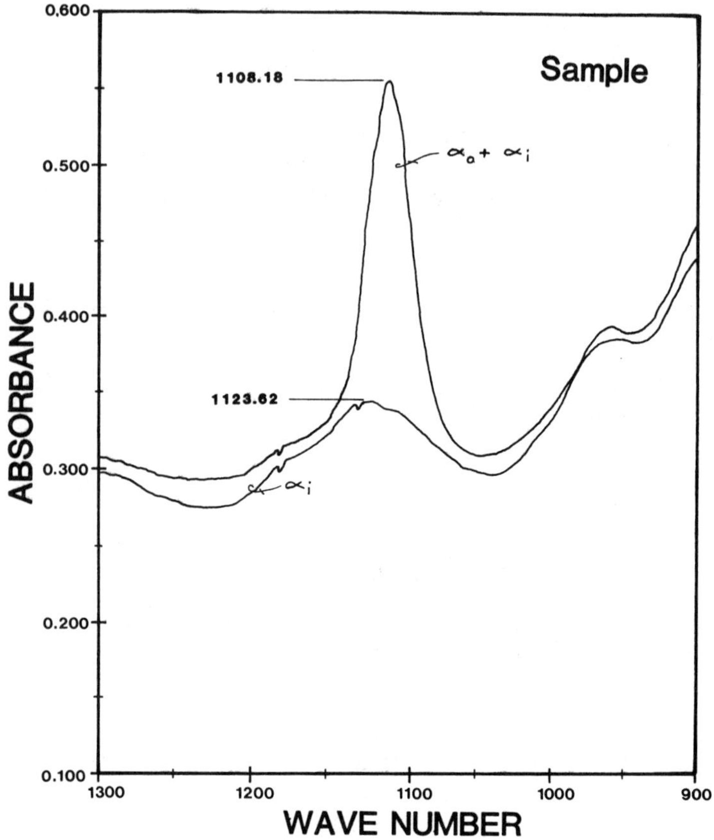

FIG. 1—*Spectra of FZ and CZ process silicon.*

absorption coefficient α_o, the intrinsic absorption coefficient α_i at 1106 cm^{-1} with a value of 0.82 ± 0.05 cm^{-1} must be subtracted.[2] This step may be avoided if a float-zone (FZ) oxygen sample is available as a reference. In this case, the absorbance of the reference is subtracted from the sample absorbance after multiplying by a constant to account for thickness differences. This leaves only α_o, as shown in the derivation

$$T_s = (1 - R)^2 \, e^{-(\alpha_o + \alpha_i)d_s}$$

$$T_r = (1 - R)^2 \, e^{-\alpha_i d_r}$$

$$\ln T_s - (d_s/d_r) \ln T_r = \ln (1 - R)^2 - (\alpha_o + \alpha_i)d_s - \ln (1 - R)^{(2d_s/d_r)}$$

$$- (d_s/d_r)(-\alpha_i d_r) = \ln (1 - R)^2 - (d_s/d_r) \ln (1 - R)^2 - \alpha_o d_s$$

where d_s and d_r are sample and reference slice thickness respectively, and T_s and T_r are sample and reference slice transmittance respectively.

[2] Pajot, B., *Analusis*, Vol. 5, 1977, p. 293.

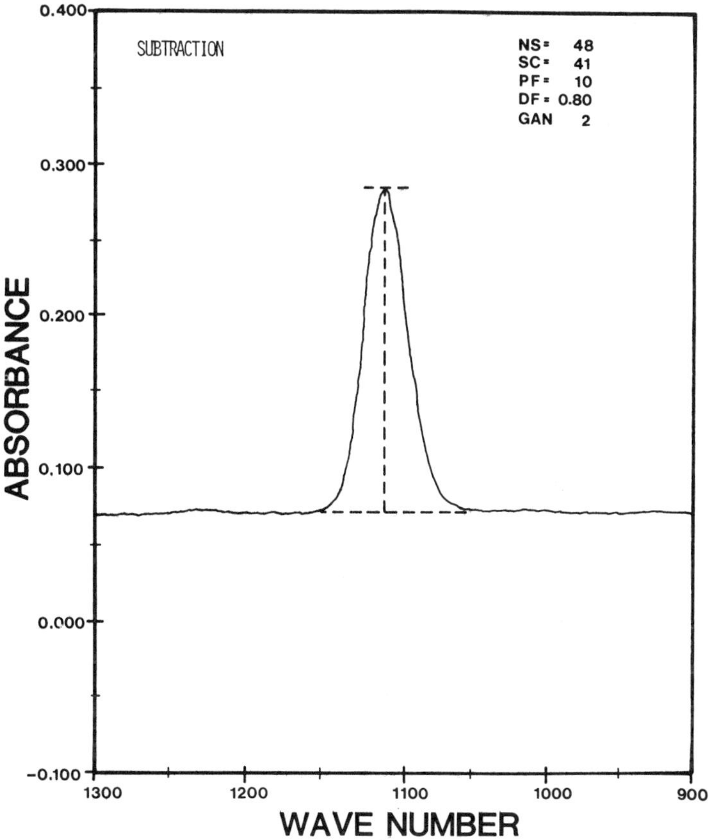

FIG. 2—*Spectra after subtraction of FZ reference.*

A plot after such a subtraction is shown in Fig. 2. If a measurement is now made from the obvious baseline, the ln terms drop out since they are weak functions of frequency and only displace the plot from zero absorbance, yielding

$$\begin{aligned}\alpha_0 &= \frac{-1}{d_s}(\ln T_s - (d_s/d_r)\ln T_r) \\ &= \frac{2.303}{d_s}\text{(Peak height from baseline in absorbance units)}\end{aligned} \qquad (2)$$

The 2.3 factor appears since absorbance is defined as $-\log_{10} T$.

The previous formulas assume that the light reflected at each air-silicon interface is lost. Some of this light may be reflected back towards the detector and contribute to the spectrum. The derivation for a formula in this case is shown in Fig. 3. This formula [from ASTM Practice for Infrared Absorp-

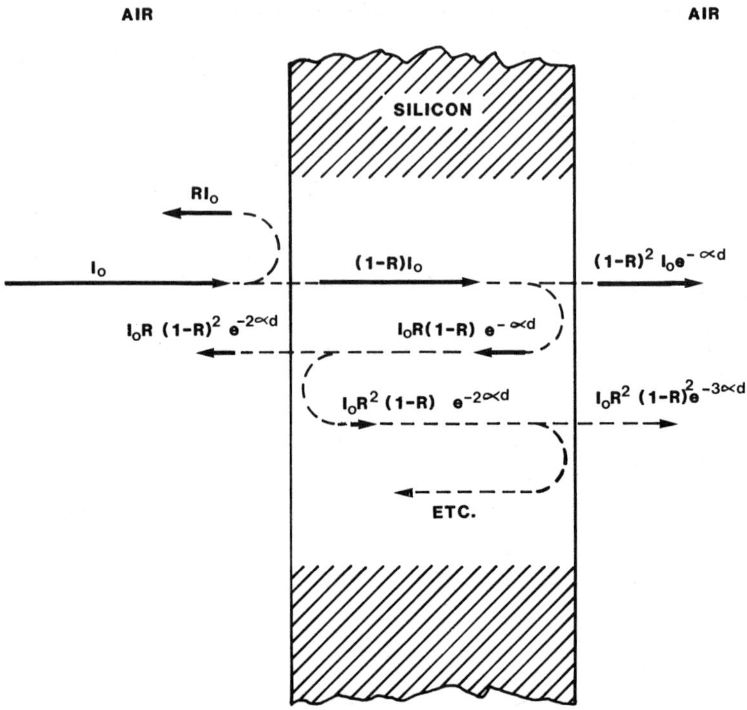

$$T = (1-R)^2 e^{-\alpha d} + (1-R)^2 R\, e^{-3\alpha d} + (1-R)^2 R^4 e^{-5\alpha d} + \ldots$$

$$= (1-R)^2 e^{-\alpha d} (1 + R^2 e^{-2\ d} + R^4 e^{-4\alpha d} + \ldots)$$

$$T = (1-R)^2 e^{-\alpha d} / (1-R^2 e^{-2\alpha d}) \qquad (3)$$

FIG. 3—*Derivation of multiple-reflection formula.*

tion Analysis of Impurities in Single Crystal Semiconductor Materials (F 120)] assumes that both sides are optically flat so that no light is scattered or refracted out of the beam.

The multiple reflections increase the measured transmittance without changing peak shape (Fig. 4). This effect, which results in an increase in the measured absorbance peak height, is shown in Fig. 5 where

$$\alpha d \text{ (measured)}/\alpha d \text{ (actual)} = \text{enhancement factor}$$

is plotted versus absorbance. Note that for small values of absorbance substantial errors can occur. To obtain the actual absorption coefficient in cases involving multiple reflections, Eq 3 in Fig. 3 must be inverted to yield

$$\alpha = \frac{-1}{d} \ln \left(\frac{-(1-R)^2 + \sqrt{(1-R)^4 + 4R^2T^2}}{2R^2T} \right) \tag{4}$$

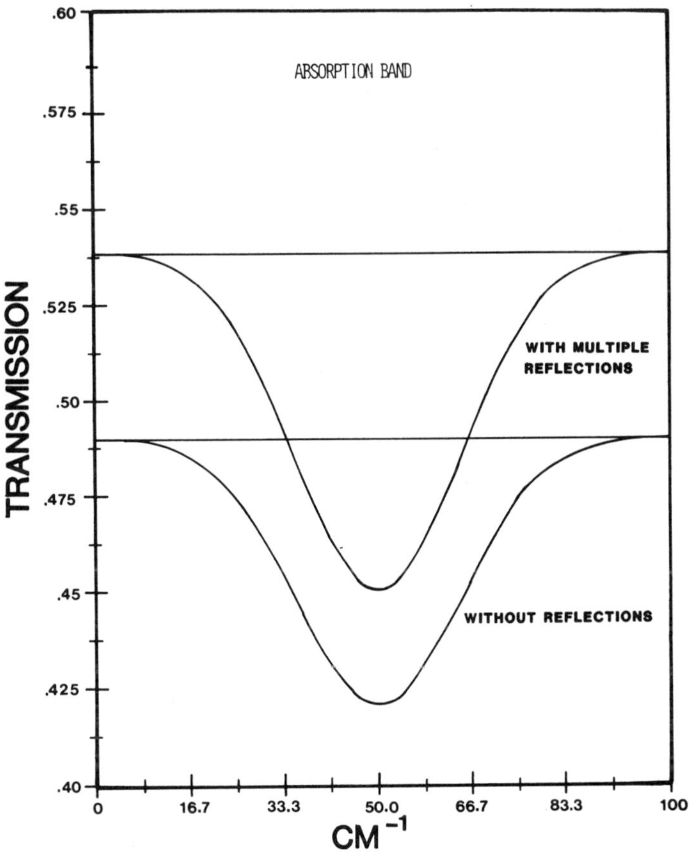

FIG. 4—*Comparison of spectra with and without multiple reflections.*

Here again, α for the 9-μm absorption band calculated by using Eq 4 includes a contribution from the intrinsic absorption band which must be subtracted before converting the absorbance to an interstitial oxygen concentration.

As before, a reference slice is usually used to provide good baselines. The formula for this case is derived below, where it is assumed that multiple reflections occur in both sample and reference.

$$-\ln \frac{(1-R)^2 c^{-(\alpha_o - \alpha_i)d_s}}{(1-R^2 e^{-2(\alpha_o - \alpha_i)d_s})} + \frac{d_s}{d_r} \ln \frac{(1-R)^2 e^{-\alpha_i d_r}}{(1-R^2 e^{-2\alpha_i d_r})}$$

$$= -\ln (e^{-(\alpha_o + \alpha_i)d_s}) - \ln (1-R)^2 + \ln (1 - R^2 e^{-2(\alpha_o + \alpha_i)d_s})$$

$$+ \frac{d_s}{d_r} \ln (1-R)^2 + \frac{d_s}{d_i} \ln (e^{-\alpha_i d_r}) - \frac{d_s}{d_r} \ln (1 - R^2 e^{-2\alpha_i d_r})$$

$$= \alpha_o d_s + \ln (1 - R^2 e^{-2(\alpha_o + \alpha_i)d_s}) - \frac{d_s}{d_r} \ln (1 - R^2 e^{-2\alpha_i d_r})$$

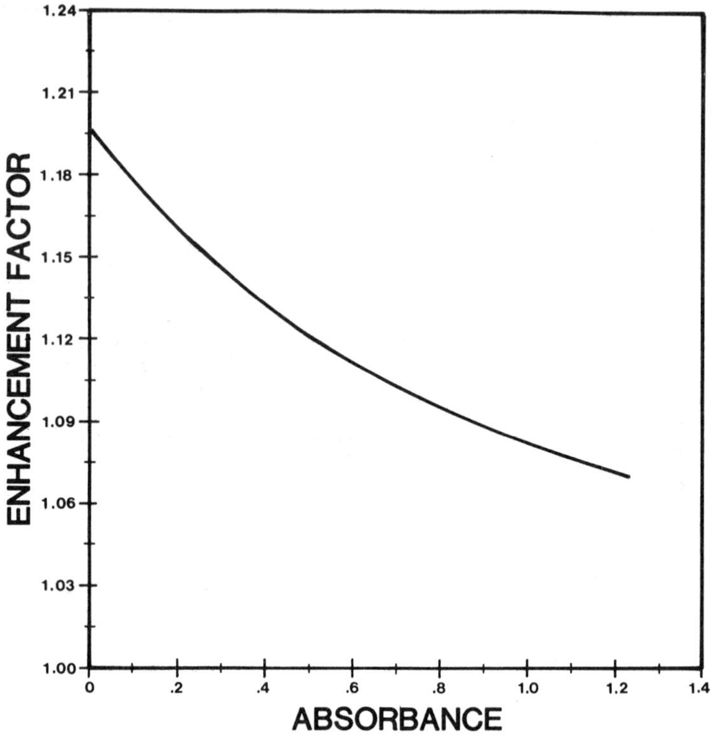

FIG. 5—*Error caused by multiple reflections when using Eq 1.*

Eliminate the ln $(1 - R)^2$ terms by making measurements from a baseline. Then

$$\alpha_o = \frac{2.30}{d_s}\text{(peak height from baseline in absorbance units)}$$
$$- \frac{1}{d_s} \ln \left(1 - R^2 e^{-2(\alpha_o + \alpha_i)d_s}\right) + \frac{\ln \left(1 - R^2 e^{-2\alpha_i d_r}\right)}{d_r} \quad (5)$$

The two ln-terms may be evaluated by using Eq 4 on the sample and reference scans separately and entering the values of $(\alpha_o + \alpha_i)$ into Eq 5. The effect of this correction for various oxygen contents for a 525-μm slice is shown in Fig. 6. A similar treatment is given by Thurber.[3]

If the derivation in Fig. 3 is carried out by using electric fields rather than

[3] Thurber, W. R., "Determination of Oxygen Concentration in Silicon and Germanium by Infrared Absorption," NBS Technical Note 529, National Bureau of Standards, Washington, D.C., 1970.

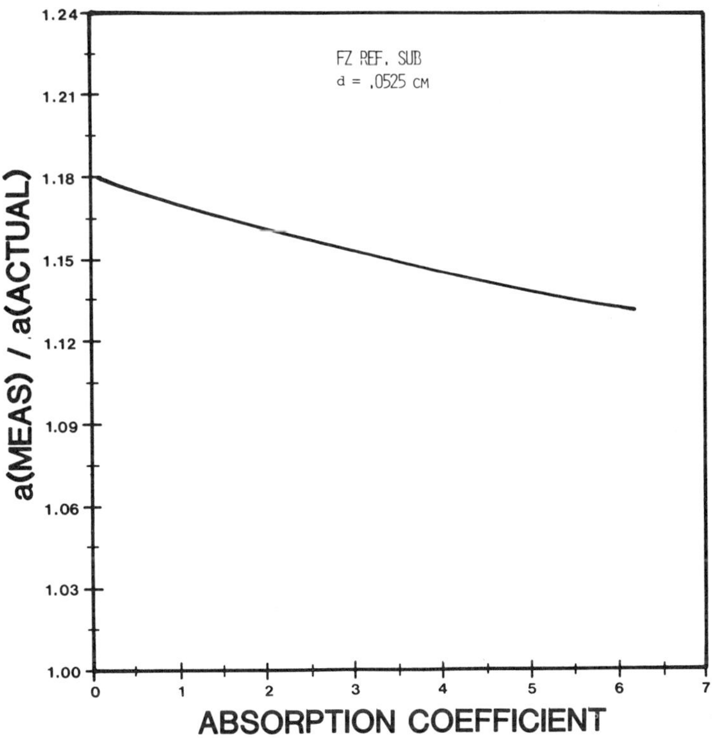

FIG. 6—*Error caused by multiple reflections in both sample and reference for 525-μm sample and reference slices.*

intensity, which is proportional to the square of the electric field, a formula describing the "interference fringes" is obtained:[4]

$$T = \frac{e^{-\alpha d}(1 - R)^2(1 + (k/n)^2)}{(1 - Re^{-\alpha d})^2 + 4e^{-\alpha d}R\sin^2\delta} \qquad \delta = 2\pi nd/\lambda \qquad (6)$$

where $(n - iK)$ is the complex index of refraction.

While this formula is of theoretical interest, the previous results are unchanged since it can be shown by explicit integration of Eq 6 over a fringe cycle that the average of this equation over one fringe cycle is identical to Eq 3, if we assume that α remains constant over the fringe cycle. Since the FWHM of the 9-μm band is 32 cm^{-1} while the fringe spacing for a 525 μm slice is 2.9 cm^{-1}, this last assumption is reasonable. It should be noted that FTIR software programs which "remove" the fringes do not eliminate the enhancement of the absorption coefficient due to the multiple reflections. After such

[4] Heavens, O. S., *Optical Properties of Thin Solid Films*, Butterworths, London, 1955.

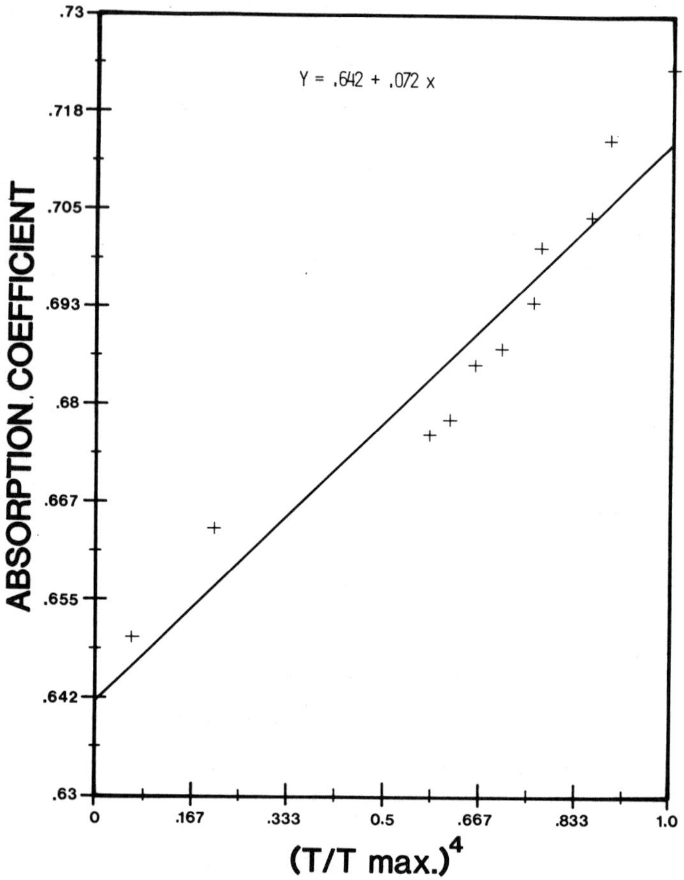

FIG. 7—*Baseline to peak absorption coefficient for 1124 cm*$^{-1}$ *band with different backside finishes by using Eq 1.*

a fringe removal operation the transmission is described by Eq 3, assuming no losses due to scattering.

The previous results were derived by assuming either no internal reflections or optically polished front and back surfaces. Most silicon slices fall somewhere between those extreme cases. If a fraction $(1 - f)$ of both the transmitted and internally reflected light is scattered out at each backside incidence, the multiple-reflection formula (Eq 3) may be modified to yield

$$T = \frac{(1 - R)^2 e^{-\alpha d} f}{1 - R^2 e^{-(2\alpha d)} f}$$

Since $R^2 e^{-(2\alpha d)} f \leq 0.09$, we may use a series approximation and a value for

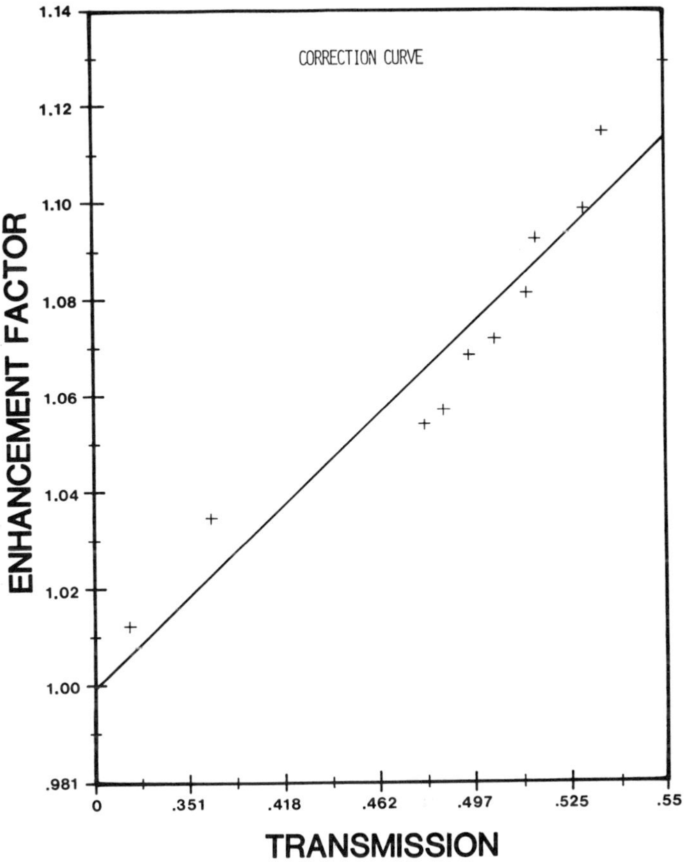

FIG. 8—*Figure 7 relabeled to show relative error at various transmissions.*

αd corresponding to a 525 μm slice with an oxygen content of 8×10^{17} cm^{-3}:

$$T = (1 - R)^2 e^{-\alpha d} f(1 + R^2 e^{-2\alpha d} f + \ldots)$$
$$\simeq (1 - R)^2 e^{-\alpha d}(f + 0.072 f^2)$$

Measurements were made of the intrinsic absorption coefficient on a high-purity FZ slice after lapping and sequential etching. The best curve fit suggests an f^4 dependence. The importance of the multiple-reflection correction diminishes rapidly with decreasing transmittance (Figs. 7 and 8). Since the transmittance of different back-surface finishes varies considerably, each vendor's surface finish should be evaluated to determine the importance of multiple reflections.

Conclusion

Ignoring the effect of multiple reflections can result in a measured oxygen level differing from the actual concentration by up to 20%. The back-surface finish of the slice can affect the oxygen measurements; however, for baseline transmissions of less than 0.2, the error that can be expected by ignoring multiple reflections is 3% maximum.

Acknowledgments

This study was, in part, based on previous work performed by Dr. P. Stallhofer at Wacker Chemitronic, Burghausen, West Germany.

Takao Abe,[1] *Shozaburo Gotoh,*[1] *Norihide Ozawa,*[1] *and Tsumoru Masui*[1]

Determination of Oxygen Concentration in Silicon by Infrared Absorption

REFERENCE: Abe, T., Gotoh, S., Ozawa, N., and Masui, T., **"Determination of Oxygen Concentration in Silicon by Infrared Absorption,"** *Silicon Processing, ASTM STP 804,* D. C. Gupta, Ed., American Society for Testing and Materials, 1983, pp. 469–476.

ABSTRACT: Infrared absorption measurement and activation analyses were carried out for various CZ crystals with different oxygen concentrations. Data for infrared (IR) absorption were found to vary depending upon various factors such as wafer thickness and surface finishing. Under well-defined experimental conditions, however, an accurate linear correlation between the IR absorption and oxygen concentration measured by activation analysis was obtained: 6.0 × absorption coefficient = oxygen concentration in parts per million atomic (ppma).

KEY WORDS: oxygen concentration, infrared absorption, charged particle activation analysis, absorption coefficient, interstitial oxygen, difference method

Intrinsic gettering (IG) [*1*][2] has been used in device production. By annealing Czochralski (CZ) wafers at a suitable temperature cycle, oxygen precipitates occur inside the wafers, and simultaneously a precipitation-free "denuded zone" is formed near the surface by outdiffusion of oxygen. Oxygen precipitates inside the wafers act to absorb metal impurities from the active surfaces, similar to backface damage gettering. Recently, IG has been considered to be an effective means of absorbing excess carriers formed by α particles. For such purposes, the depths of denuded zones must be controlled precisely. The zone depth depends mainly on the initial oxygen concentrations in crystals for a certain annealing condition, and therefore precise measurements of oxygen concentration are required.

Oxygen concentrations in silicon have been measured by the gas fusion [*2*] and vacuum fusion [*3,4*] methods. These methods might be usable for high concentrations, but not in a range of one or two parts per million atomic (ppma), which is often important in materials characterization for electronic

[1] R&D Center, Shin-Etsu, Handotai Company, Annaka-Shi, Gunma-Ken, Japan.
[2] The italic numbers in brackets refer to the list of references appended to this paper.

devices. Charged particle activation analysis [*5*] is capable of detecting light elements such as carbon, nitrogen, and oxygen in the range of 0.01 ppma. However, it is an expensive and destructive measurement method. Therefore infrared (IR) absorption measurement [*3*] is widely employed, using a correlation between the total oxygen concentration and the absorption coefficient at 9 μm due to interstitial oxygen atoms in solid solution. Unfortunately, the correlation changes depending upon equipment, calculation, specimen thickness, surface finishing, etc., and a procedure for accurate measurement has not yet been established.

The purpose of this paper is to show that an accurate correlation can be obtained under well-defined experimental conditions.

Experimental Procedure

Two popular methods of IR absorption measurement were examined. The base difference line method given in the ASTM Test for Interstitial Atomic Oxygen Content of Silicon by Infrared Absorption (F 121) was employed for measuring the oxygen absorbance at 1106 cm^{-1} at room temperature, using a Fourier transform infrared spectroscope (Digilab FTS-15C). The other is a method developed by Nicolet Co. [*6*], where $[O_i]$ is obtained from the peak-to-peak ratio of the absorbance at 1106 cm^{-1} due to interstitial oxygen to the absorbance at 625 cm^{-1} due to the silicon lattice. This method is referred to as the Nicolet method for convenience and was compared with the conventional ASTM method by using the same sample.

Specimens were sliced from conventional CZ crystals with various oxygen concentrations. To obtain data for low oxygen concentration, specimens were also prepared from a oxygen-doped float-zone (FZ) crystal.

To examine the effects of specimen thickness and surface finishing on the absorption data, slices with thicknesses of 2 and 0.625 mm were prepared from the same ingot.

In order to measure the correlation between [O] and IR absorption, the absorption coefficients at 9 μm in infrared were measured, and then activation analysis of oxygen was carried out by Nozaki's method [*7*], using reaction of $^{16}O(^{3}He,P)^{18}F$. Specimens were 2-mm thick slices polished on both sides. The eight specimens were divided into two pieces each; one was used for activation analysis, and the other was subjected to annealing at 1000°C for 16 h in an oxygen atmosphere, followed by the infrared absorption measurement.

Results and Discussion

Accuracy of the IR Absorption Measurement

Interstitial oxygen concentrations $[O_i]$ obtained by IR measurement depend upon methods used, crystal thickness, and surface finishing of specimens. First, the difference (ASTM F 121) and Nicolet methods were compared.

Profiles of absorbance for crystal thicknesses of 2 and 0.625 mm were measured by the difference method (Fig. 1), and the proportional values to absorption coefficient, which are divided by their thicknesses for the height of the peaks from their base lines, were obtained as 0.170 and 0.184 (Fig. 1*a* and *b*), respectively. The Nicolet method shows a different thickness dependence. Since the Nicolet instrument does not provide absorbance profiles, the profiles were measured with a Digilab FTS-15C (Fig. 2). Two peaks, left and right, are seen due to interstitial oxygen and silicon lattice, respectively. The ratio of the two peak heights from the base lines was calculated to be 0.6855 and 0.6645 for the crystal thicknesses of 2 and 0.625 mm, respectively. This thickness dependence is opposite to the case of the difference method.

These results are shown in Fig. 3 and may be summarized as follows:

1. Both the methods give different values of absorption coefficient, even for a 2-mm thick crystal.
2. The differences are larger for thinner crystals: The difference and Nicolet methods for a thickness of 0.625 mm give ~8% higher and ~4% lower values for double side polish, compared with the case of 2-mm thick crystals.

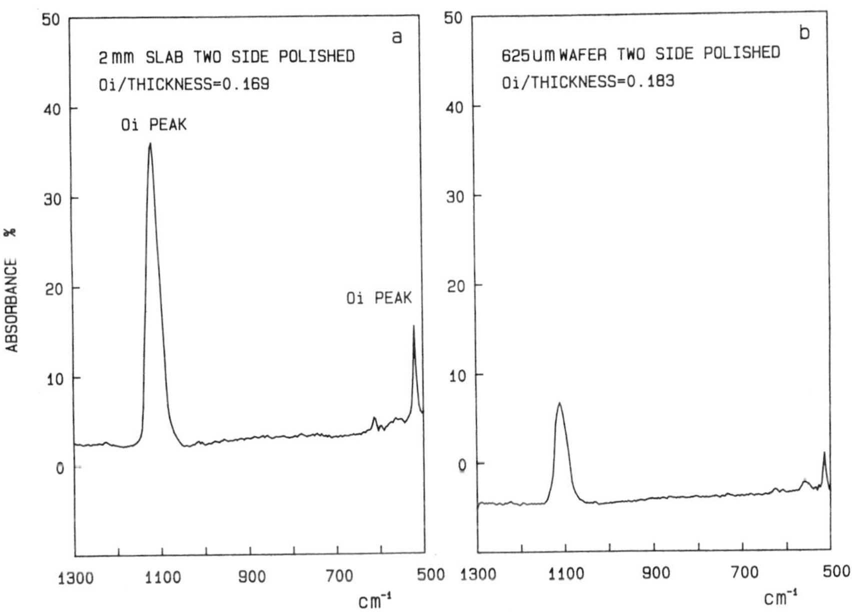

FIG. 1—*Absorption spectra of wafers with different thicknesses sliced from the same CZ crystal (a) 2 mm thick, (b) 0.625 mm thick. The difference method was employed.*

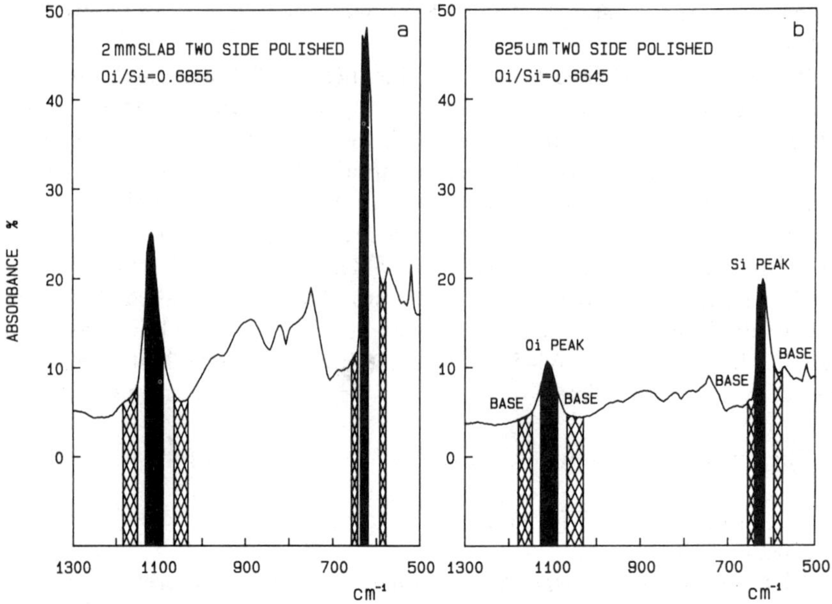

FIG. 2—*Absorption spectra of the same wafers that were used in Fig. 1.* (a) *2 mm thick.* (b) *0.625 mm thick. The Nicolet method was used.*

3. Both methods give smaller values of absorption coefficient for one side–etched specimens, and also show a similar thickness dependence.
4. The absorption depends upon the degree of backface damage.

In view of thee experimental results, the IR measurement must be made with the same instrumentation and specimen preparation. In the measurement of correlation between [O] and IR absorption, as described in the next section, therefore, IR measurement was carried out for 2-mm thick double side–polished wafers by the difference method.

Conversion Factor Between [O] and IR Absorption

Figure 4 shows correlation lines between [O] and absorption coefficients at 9 μm at 300 K as reported by various investigators [*2–5*]. The data by Baker [*2*] and Graff et al [*4*] were obtained by the gas fusion and vacuum fusion methods, respectively. They scatter around the respective correlation lines. Yatsurugi et al [*5*] measured [O] by an activation analysis. Their data for conventional CZ and FZ crystals are indicated by the squares and open circles in Fig. 4, respectively. Their correlation line was drawn from the data indicated by the open circles, which were obtained for the lower oxygen concentration range. Since oxygen precipitation occurs in the cooling process of CZ crystals from the crystallization temperature to room temperature, the

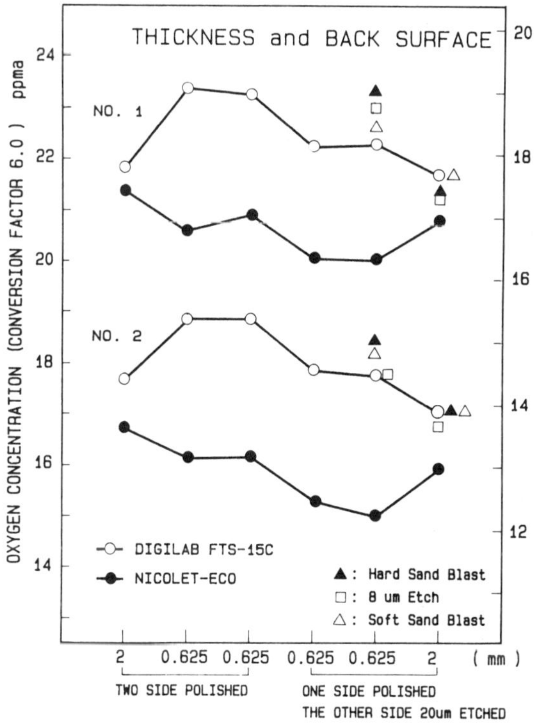

FIG. 3—*Effects of specimen thickness and surface finishing on oxygen concentration measurement. Two CZ wafers, No. 1 and No. 2, were measured by the difference and Nicolet methods, using one side-polished specimens cut from the same wafers. Two grade surface of back-damaged and a flash (8 μm) etched surface were examined by the difference method.*

data for the conventional CZ crystals (squares) are located above the correlation line.

In the present experiment, CZ crystals with various oxygen concentrations were examined by activation analysis [7]. Figure 5 shows the results with the correlation line having a conversion factor of 6.0. By taking into account the accuracy of the activation analysis we can say that all the data are on the line. This means that our as-grown CZ crystals are free from any appreciable oxygen precipitates. In fact, the conversion factor agrees with Yatsurugi's that was obtained for the lower oxygen concentration range in Fig. 4. It should be noted that the data for the samples annealed at 650°C for 8 h in Ar are also located on the line; annealing at 650°C for such short period does not cause any appreciable oxygen precipitation. However, annealing at 1000°C for 16 h in dry oxygen caused a large reduction of interstitial oxygen concentration [O_i], as indicated by the arrows in Fig. 5. Carbon concentrations are shown in parentheses; the lines without parentheses indicate concentrations less than 0.1 ppm (the detection limit of IR measurements). The

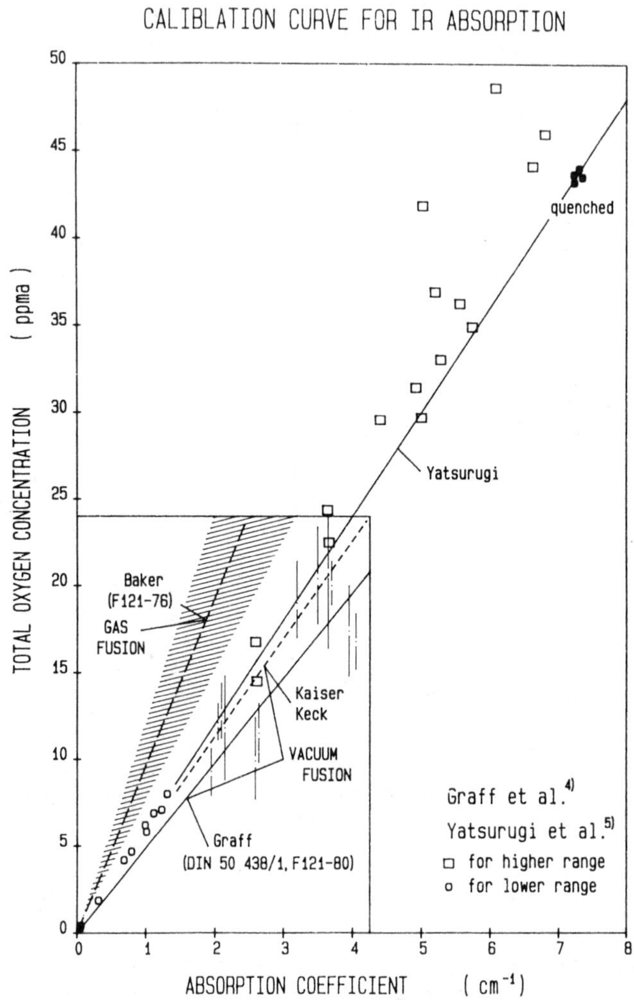

FIG. 4—*Oxygen concentration versus absorption coefficient at 9 μm at 300 K obtained by various methods. The data with the error bars are reported by Graff et al* [4]. *The data by Yatsurugi et al* [5] *are indicated by squares.* (*The closed circles are for the quenched samples.*)

reduction in [O_i] does not depend on carbon concentrations in such high-temperature annealing. The oxygen precipitation may result from various origins, such as thermal history [*8*] and temperature fluctuations near growth interfaces [*9*].

The data reported by Yatsurugi et al [*5*] in 1973 (see Fig. 4) show that as-grown CZ crystals contain oxygen precipitates, especially in the high concentration range. The CZ crystals of those days were 51 to 63 mm (2 to 2½ in.) in diameter. Recently, CZ crystals have a large diameter (102 to 127 mm [4 to 5 in.]) and their growth conditions have been changed. The oxygen concentration introduced from the melt may be lower due to suppressed thermal con-

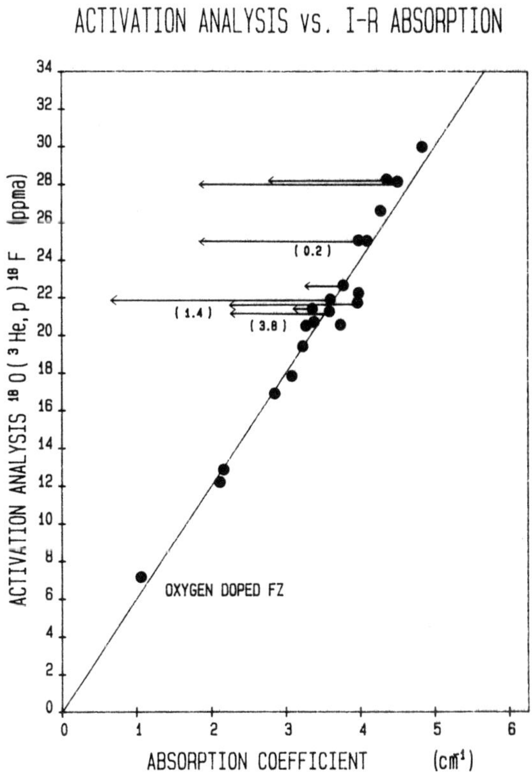

FIG. 5—*Oxygen concentration versus absorption coefficient at 9 μm at 300 K obtained in the present experiment. Activation analysis* $^{16}O(^{3}He,P)^{18}F$ *was employed. Open circles are the data for as-grown CZ crystals. The eight arrows from open circles indicate the reduction in absorption coefficients by annealing at 1000°C for 16 h in a dry oxygen atmosphere. In these specimens carbon concentrations are less than 0.1 ppma, unless otherwise specified in the parentheses.*

vection, and also rapid cooling of the crystal may be responsible for the precipitation free crystal growth.

Conclusions

The IR absorption measurement must be carried out with the same instrument and specimen preparation to evaluate oxygen concentrations [O] in Si. By activation analysis, we obtained the correlation line with a conversion factor of 6.0 between [O] and the IR absorption coefficient, which is applicable for as-grown crystals in a range of 2×10^{17} to 2×10^{18} oxygen atoms/cm^3.

Acknowledgments

The authors express their thanks to Dr. T. Nozaki and Y. Itoh of the Institute of Physical and Chemical Research for their activation analysis.

References

[1] Tan, T. Y., Gardner, E. E., and Tice, W. K., *Applied Physics Letters,* Vol. 30, No. 4, 1977, p. 175.
[2] Baker, J. A., *Solid State Electronics,* Vol. 13, No. 11, 1970, p. 1431.
[3] Kaiser, W. and Keck, P. H., *Journal of Applied Physics,* Vol. 28, No. 8, 1957, p. 882.
[4] Graff, K., Grallath, E., Ades, S., Goldbach, G., and Tolg, G., *Solid State Electronics,* Vol. 16, No. 8, 1973, p. 887.
[5] Yatsurugi, Y., Akiyama, N., Endo, Y., and Nazaki, T., *Journal of the Electrochemical Society,* Vol. 120, No. 7, 1973, p. 975.
[6] Vidrine, D. W., *Analytical Chemistry,* Vol. 52, No. 1, 1980, p. 92.
[7] Nozaki, T., Yatsurugi, Y., and Akiyama, N., *Journal of Radioanalytical Chemistry,* Vol. 4, No. 1, 1970, p. 87.
[8] Tsuya, H., Shimura, F., Ogawa, K., and Kawamura, T., *Journal of the Electrochemical Society,* Vol. 129, No. 2, 1982, p. 374.
[9] Shirai, S., *Applied Physics Letters,* Vol. 36, No. 2, Jan. 1980, p. 156.

S. P. Weeks[1]

Thickness Measurement of Thin (1.0-μm) Epitaxial Silicon Layers by Infrared Reflectance

REFERENCE: Weeks, S. P., "**Thickness Measurement of Thin (1.0-µm) Epitaxial Silicon Layers by Infrared Reflectance,**" *Silicon Processing, ASTM STP 804,* D. C. Gupta, Ed., American Society for Testing and Materials, 1983, pp. 477–491.

ABSTRACT: Current infrared reflectance techniques for measurement of the thickness of silicon epitaxial layers are generally unreliable for thicknesses below 2.5 μm. In order to extend these methods to thinner layers, an increased understanding of the physics of infrared reflectance from thin silicon layers is required. A model has been developed to calculate the complex refractive index as a function of impurity concentration and wavelength. It has been coupled with a multiple-layer treatment of reflectivity, allowing the simulation of the optical properties of an epitaxial layer above a substrate with arbitrary impurity distribution. The model is used to evaluate the shortcomings of both dispersive and interferometric measurement techniques.

KEY WORDS: silicon epitaxy, thin epitaxial layer, epi thickness measurement, infrared reflectance, dispersive measurement, interferometric measurement, refractive index

With the decreasing thickness of epitaxial silicon layers used in bipolar integrated circuits, and the increasing control of epitaxial properties required for this and other technologies, the ability to accurately and precisely measure epitaxial layer thickness is gaining increasing attention. In this paper, the extension of infrared reflectance techniques to measurement of layers on the order of 1 μm thick will be discussed. It will be shown that a multiple-layer reflectance model including calculation of optical constants as a function of resistivity can be used to understand and improve current thickness measurement capabilities.

Epitaxial layer thicknesses were originally determined from the wavelengths of the interference peaks, as suggested by Spitzer and Tannenbaum [*1*].[2] In their technique, which was subsequently adopted as ASTM Test for Thickness of Epitaxial Layers of Silicon on Substrates of the Same Type by Infrared Reflectance (F 95), the thickness of an epitaxial silicon layer depos-

[1] Member of Technical Staff, Bell Laboratories, Allentown, Pa. 18103.

[2] The italic numbers in brackets refer to the list of references appended to this paper.

ited on a substrate of lower resistivity is related to the wavelength separation of peaks arising from interference between rays reflected at the air-epi and epi-substrate interfaces, according to the formula

$$T = \frac{[(P - \frac{1}{2}) + (\phi_2/2\pi)]}{2(n_1^2 - \sin^2\theta)^{1/2}}\lambda$$

where

T = epitaxial thickness,
P = orders of the interference fringe,
ϕ_2 = phase shift at the epi-substrate interface,
n_1 = index of refraction of the epitaxial layer,
θ = angle of incidence of the infrared beam, and
λ = wavelength of interference fringe.

The geometry and resulting reflectance spectrum used for this measurement are shown in Fig. 1. In order for this technique to succeed, accurate values of

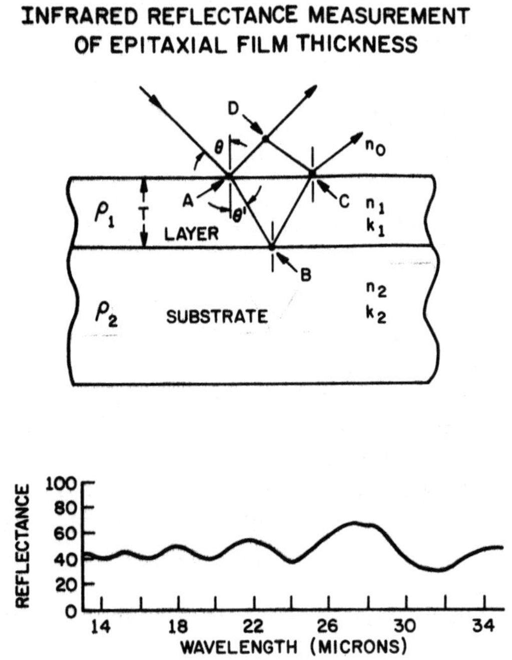

$$T = \frac{((P - \frac{1}{2}) + (\phi_2/2\pi))}{2\,(n_1^2 - \sin^2\theta)^{1/2}}\lambda$$

FIG. 1—*Geometry of the reflectance measurement and typical wavelength dispersive spectrum determined by using ASTM F 95.*

the phase shift ϕ_2, which depends on both the wavelength of the radiation and the dielectric constants of the substrate, are required. For this method, the infrared reflectance spectrum of the epitaxial wafer is measured using a suitable spectrophotometer and the thickness is calculated manually.

Commercial instruments are available for automatic epitaxial layer thickness measurements. In the majority of these, an interferometer is used to directly measure the effects of interference in the epilayer subjected to polychromatic (broadband) illumination. The technique is illustrated in Fig. 2. When the path length between the moving mirror of the interferometer and the beam splitter is equal to the path length to the epitaxial surface (zero path difference or zeroth order phase coincidence), radiation at all wavelengths is in phase, and a large signal due to constructive interference is measured at the detector. When the moving mirror is displaced by a distance equal to the additional path length encountered by the beam in passing through the epitaxial layer to the second boundary, the radiation of nearly all wavelengths is again in phase (first order phase coincidence), and a large signal is again measured at the detector. This displacement is, to first order, equal to

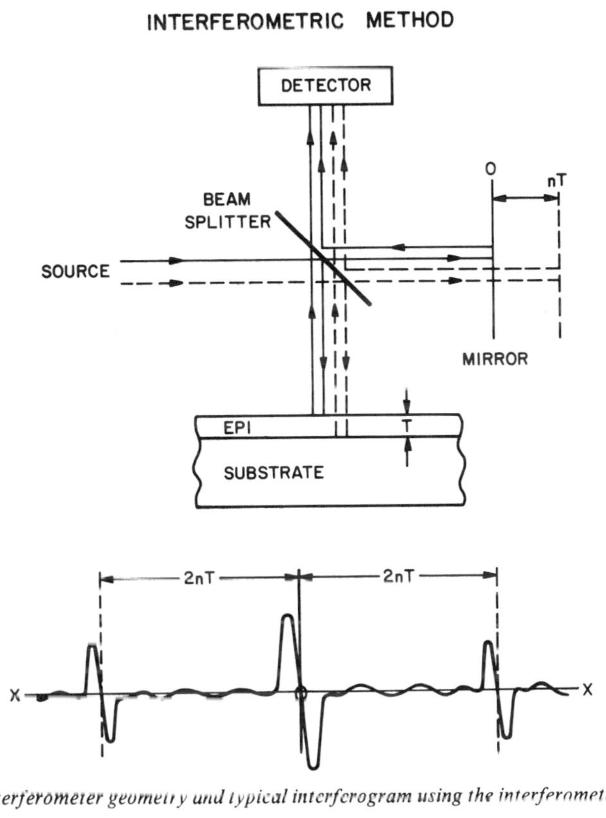

FIG. 2—*Interferometer geometry and typical interferogram using the interferometric technique.*

$2n_1T \cdot \cos\theta$. The thickness of the layer is measured according to

$$T = \frac{x}{2n_1 \cos\theta}$$

where x is the position of the movable mirror for which the first order phase coincidence is strongest, and the other variables are as before. The effect of the phase shift at the epi-substrate interface is to retard the onset of phase coincidence. Thus it is important for this technique as well that the magnitude of ϕ_2 be known accurately. Since the interferometric technique measures the signal due to all wavelengths at once, it is more difficult to include the effects of ϕ_2 in a straightforward way, due to its variation as a function of wavelength. It is for this reason that in most interferometric thickness instruments, an empirical relationship is established between thickness measured by the interferometer and that measured by ASTM F 95, and this relationship is used to calculate thicknesses from the interferometric pattern. In order to avoid interaction between the zero-path-difference and first-order phase coincidence peaks, it is advantageous to remove the zero-path-difference peak from the interferogram by subtracting the signal measured on a reference with a much thicker epitaxial layer on a similar substrate to that of the sample. Since the first instruments of this type did not employ a Fourier transform of interferometric data to produce a spectrum in frequency space, we refer to this as the *interferometric method.*

The effects of ϕ_2, the phase shift at the epi-substrate interface, are graphically illustrated in Figs. 3 and 4. Figure 3 shows frequency-dispersive reflectance measurements for three identical epitaxial layers grown on substrates with different resistivities. Not only do the interference extrema occur at different frequencies, but the separations between adjacent extrema are also different. In addition, it is noted that the magnitude of the oscillations varies as a function of substrate resistivity, with stronger oscillations associated with lower resistivity (higher carrier concentration). A similar situation is evident in interferometric measurements of the same samples (Fig. 4). In this case, the position and magnitude changes as a function of resistivity, with stronger signals again associated with lower resistivity.

These differences, as well as the more subtle effects due to depth variation of the optical properties of the substrate such as with an implanted subepitaxial structure, become quite significant when measuring reflectance from epitaxial layers of the order of 1 μm thick. Figure 5 shows data for both types of measurements made on thin layers. The upper part of the figure shows reflectance data for three 1.2-μm-thick simultaneously grown epitaxial layers on substrates containing different carrier concentrations. Even variations small enough to be within the specifications of a given device code can lead to differences in the data large enough to cause significant error in thickness measurement. The lower part of the figure shows thickness mea-

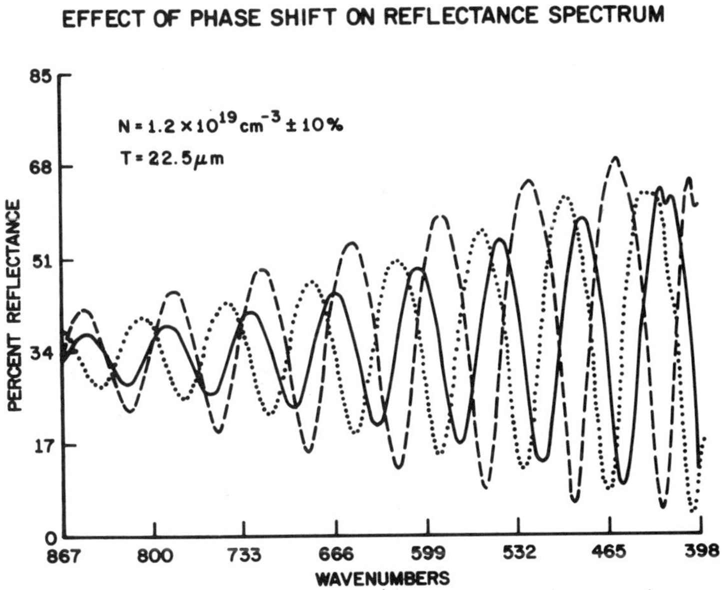

FIG. 3—*Frequency dispersive reflectance spectrum of three samples of low (solid), medium (dot), and high (dash) carrier concentration. Implant doses were 1.9, 2.1 and 2.3 × 10^{15} atom/cm^2 respectively, followed by a heat treatment of 150 min at 1250°C.*

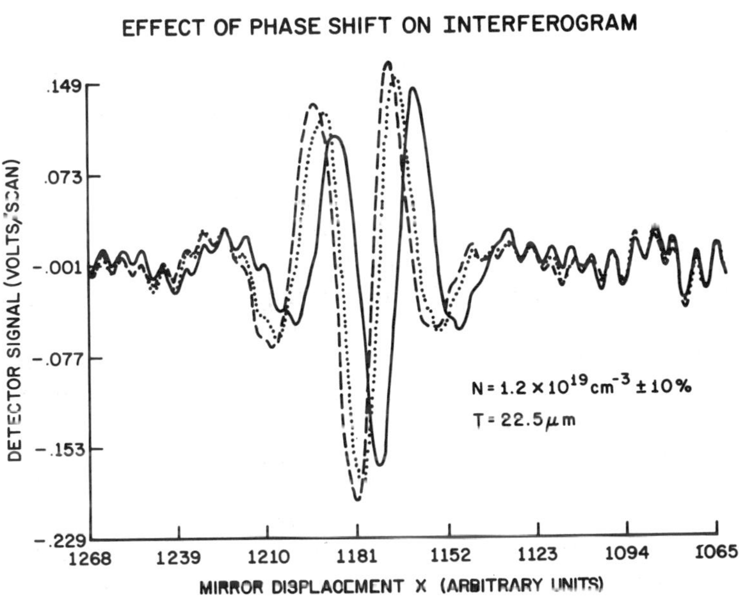

FIG. 4—*Detailed view of first phase coincidence peak in interferometric measurement of the three samples described in Fig. 3.*

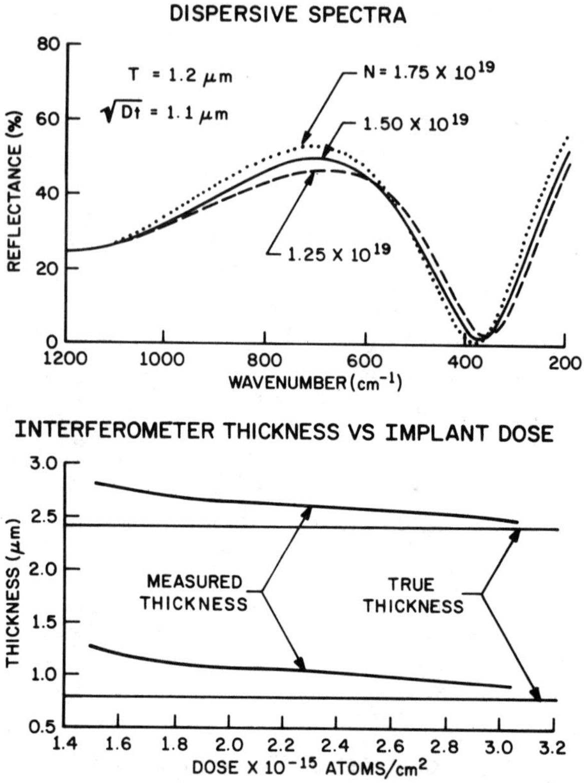

FIG. 5—*Dependence of thickness measurements on substrate doping.* (a) *reflectance spectra for three 1.2-μm thick epitaxial films on substrates of varying peak buried layer concentration, and* (b) *interferometric epitaxial thickness as a function of implant dose for a 2.4-μm and a 0.8-μm-thick epitaxial layer.*

sured by a commercial interferometric instrument as a function of substrate implant dose at two different epitaxial thicknesses. All samples received equivalent implant drive-in cycles. A change by a factor of two in dose can lead to a measurement error of nearly 0.5 μm, which for the thinner layer amounts to more than half of the actual thickness represented by the solid line. It is these effects that must be accurately accounted for in order to assure accurate measurement in this thickness range, especially when wafers with different buried layer parameters are measured under the same conditions.

In an attempt to resolve some of these uncertainties, we have investigated the reflectance spectrum of thin (1 μm) epitaxial layers deposited on high-resistivity silicon substrates containing highly doped buried layers of finite thickness. In these studies [*2*], the impurity profile at the substrate interface was sufficiently steep, as measured by secondary ion mass spectroscopy (SIMS), so that it could be represented by an abrupt step. From measurements of reflectance spectra a determination of the epitaxial thickness, peak concentration, and diffusion depth could be made. In the next section, we will describe our measurements to determine an appropriate model for the optical constants. These constants will then be used to compare the theoretical and experimental spectra for thin epitaxial layers. The theoretical understanding of the reflectance properties will then be used to determine the accuracy and precision with which the thickness of these layers can be measured using the traditional techniques discussed.

Choice of Optical Constants

In the wavelength region of 5 to 50 μm (2000 to 200 cm^{-1}) the relative permittivity κ of silicon consists of a wavelength-independent contribution of the intrinsic crystal κ_∞, and a varying contribution of free carriers arising from impurities [*3*].[3] For an applied electric field varying harmonically as exp ($j\omega t$), we can treat the electron as a classical point charge subject to collisonal damping (Drude model) and write

$$\kappa = \kappa_\infty - (\omega_p^2/\omega^2)\,[(1 - j(\nu/\omega))^{-1}] \tag{1}$$

The plasma frequency ω_p is equal to $(Ne^2)/(m^*\epsilon_0)$, where N, in this case, is the donor impurity concentration, e is the electronic charge magnitude, m^* is the electron effective mass, and ϵ_0 is the vacuum permittivity. The collision frequency ν is the inverse of the mean free time between collisions and the brackets [ ] designate an average over the electron energy distribution. As the applied frequency ω approaches zero, the d-c resistivity ρ can be expressed as $(j\omega\kappa\epsilon_0)^{-1}$ so that Eq 1 yields the relation

$$\rho^{-1} = \omega_p^2\epsilon_0[\nu^{-1}] \tag{2}$$

To determine the permittivity κ as a function of impurity concentration N, we will use the known values of $\kappa_\infty = 11.7$ and $m^* = 0.26\,m$ [*3,4*], and evaluate the term in brackets in Eq 1 for the two special cases described below.

In Case I, we will assume, as did Schumann and Phillips [*5*], classical Boltzmann statistics and ionized impurity scattering with a frequency ν that varies with the conduction electron energy E as $\nu_0 E^{-3/2}$. The integration in Eq 2 can then be carried out in terms of tabulated function [*6*] and the constant ν_0 can be expressed as a function of resistivity ρ, which can be determined for a given impurity concentration from National Bureau of Standards data [*7*]. The expression for ν is then used to evaluate the term in brackets in Eq 1 numerically.

In Case II we will assume a high doping level and degenerate statistics so that only electrons in a narrow band of energies around the Fermi level contribute to the permittivity. We assume further that the collision frequency

[3] A discussion of previous work by other researchers can be found in Ref *2*.

remains approximately constant over this energy interval so that we can omit [8] the averaging brackets in Eqs 1 and 2.

The constant collision frequency ν can then be found as a function of resistivity from Eq 2. In this case, which corresponds to the Drude model with a constant relaxation time, we can express the relative permittivity as a function of materials parameters as

$$\begin{aligned}
\kappa &= \kappa' - j\kappa'' \\
\kappa' &= 11.7 - (\omega_p{}^2/\omega^2)\,(1 + \nu^2/\omega^2)^{-1} \\
\kappa'' &= (\omega_p{}^2/\omega^2)\,(\nu/\omega)\,(1 + \nu^2/\omega^2)^{-1} \\
(\omega_p/\omega) &= 5.87 \times 10^{-11}\,\lambda\sqrt{N} \\
(\nu/\omega) &= 5.74 \times 10^{-19}\,\lambda N \rho
\end{aligned}$$

where λ is the wavelength in micrometres, N is the impurity concentration in cm^{-3}, and ρ is the resistivity in Ω-cm.

In order to study the range of validity of the above models, the degree to which they fit the reflectance spectra of uniformly doped silicon was investigated. The relative permittivity can be related to the reflectance by noting that when a plane electromagnetic wave in medium i is normally incident on a semi-infinite medium $i + 1$, the reflection coefficient Γ_i can be written as [9]

$$\Gamma_i = \frac{1 - (\kappa_{i+1}/\kappa_i)^{1/2}}{1 + (\kappa_{i+1}/\kappa_i)^{1/2}}$$

The reflectance from a bulk silicon sample in air is then equal to $|\Gamma_0|^2$ where $\kappa_0 = 1$ and $\kappa_i = \kappa$, as given in Eq 3.

A comparison of the above models with experimental data [2] is shown in Fig. 6 for 0.0094 Ω-cm n-type material ($N = 4.93\text{E}18\ cm^{-3}$). It is evident that for this resistivity Schumann's model (Case I) is a better fit to the experiment than Drude's model (Case II). This result was also noted by Saifi and Stolen [10] for the same resistivity sample and presented in their Fig. 3. Although our experimental values agree with those of Ref *10*, the theoretical curves differ slightly due to the different N versus ρ data. Case I corresponds to the use of an isotropic, energy-dependent collision time, which Saifi and Stolen found differs only slightly from their more exact calculation using an anisotropic, energy-dependent collision time [11]. Measurements performed on samples of 0.014 Ω-cm resistivity ($N = 2.6\text{E}18\ cm^{-3}$) gave similar results (not shown).

When the above comparison was made with 0.006 Ω-cm ($N = 9.6\text{E}18\ cm^{-3}$) and 0.003 Ω-cm ($N = 2.2\text{E}19\ cm^{-3}$) material, different results were obtained, as seen in Fig. 7. Here Drude's model yields the better fit. The Drude model was also in excellent agreement with measurements of 0.0017 Ω-cm material. Optical constants of samples with still higher impurity concentrations were investigated by E. Barta [12] at a wavelength of 40 μm and were also found to be in agreement with the predictions of the Drude model.

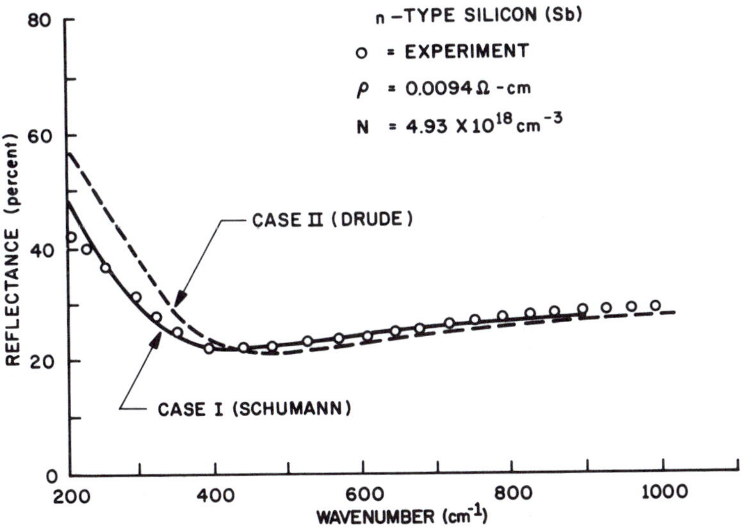

FIG. 6—*Reflectance versus wavenumber for uniformly doped 0.0094 Ω-cm material.*

On the basis of these measurements, the following qualitative statements hold for the wavelength range of 5 to 50 μm and n-type concentration range of 2E18 to 4E19 cm^{-3}:

1. From 5 to 10 μm both models will be in approximate agreement.
2. In the 10 to 50 μm range, the Drude model should be used for concentrations above 7.5E18 cm^{-3} and Schumann's model for lower values. Results using the two models are roughly equivalent at 7.5E18 cm^{-3}.

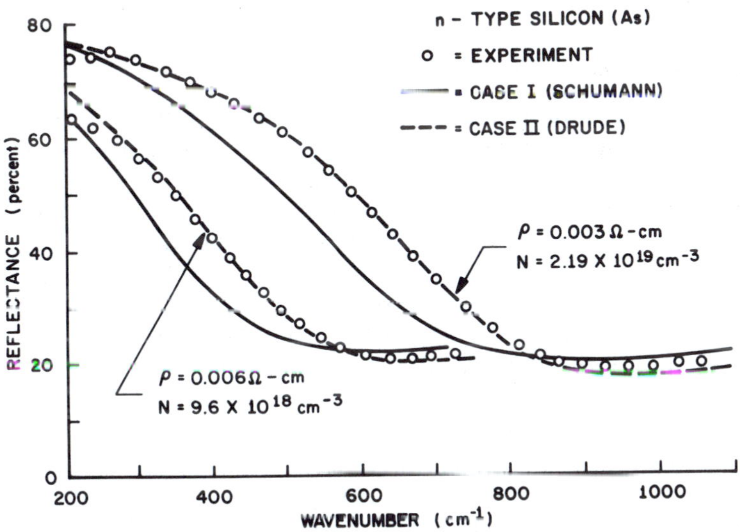

FIG. 7—*Reflectance versus wavenumber for uniformly doped 0.006 and 0.003 Ω-cm material.*

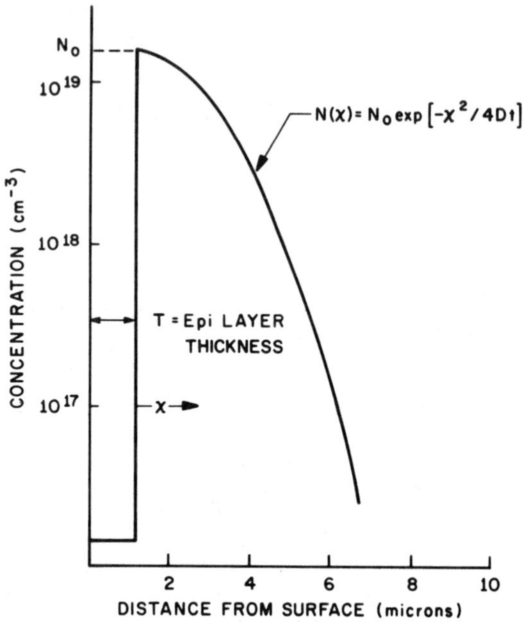

FIG. 8—*Impurity profile used in computations of infrared spectra of epitaxial structures.*

In the next section, it will be shown that the reflectance spectrum from the impurity distribution modeled in Fig. 8 is primarily dependent on the epitaxial layer thickness T and maximum buried layer concentration N_0. From practical considerations, we will be studying epitaxial wafers with values of N_0 greater than 1E19 cm^{-3}, so the Drude model will be used, and optical constants will be evaluated from Eq 3.

Reflectance Spectra of Epitaxial Wafers

The overall goal of the type of measurements considered here is to deduce the complete impurity profile of an inhomogeneous medium from the reflectance spectrum. This will be achieved by performing a calculation of reflectance from an epitaxial film on a substrate with an impurity profile described by a limited number of parameters. By varying the parameters, the calculated spectrum will be determined that best matches the experimental data.

To compute the reflectance spectra of these samples, we will use the three-parameter (T, N_0, $\sqrt{Dt}$) profile model shown in Fig. 8 and compute the optical constants using the Drude mode. This approach is valid for the following reasons. Firstly, the transition region between the epitaxial and buried layers for these particular samples is much smaller than the wavelengths used in the analysis. Therefore, unlike some previous investigations [*13*], the abrupt interface shown in Fig. 8 should be a valid approximation, alleviating the need to model this interface. Next, the following factors, based on the measurements above, tend to make the Drude model a good approximation: (*1*) the main reflection, which occurs at the abrupt interface, is well described by this model; (*2*) most of the reflection within the buried layer comes from the high concentration regions where the Drude model is specifically valid; and (*3*) the Drude model will yield optical constants for both high *and* low concentrations that are in reasonable agreement with the measured reflectance if the 200 to 250 cm^{-1} region is disregarded. Lastly, by comparison with calculations using more sophisticated modeling of the impurity profile [*13,14*], it was found that a Gaussian approximation for the profile shape yields equivalent results for this application.

Reflectance spectra are computed for a given epitaxial thickness T, peak concentration N_0, and diffusion depth $\sqrt{Dt}$ by approximating the dopant distribution as a series of thin, homogeneous slabs and performing a numerical evaluation of a chain matrix, as outlined in Ref *2*. The complex magnitude of the incident and reflected radiation at the boundary of a slab are written as a two-dimensional vector. A 2 by 2 matrix describing the propagation for the wave across the slab to the next boundary operates on the vector, producing a new vector for that boundary. A separate matrix is written for each slab, describing its optical properties. When these operations are chained together, the resultant vector describes the reflectance properties of the entire system taken as a whole. Typically, 60 slabs of 0.1 μm thickness are used. The thickness of 0.1 μm does not apply to the epitaxial layer, whose thickness T can be adjusted in whatever increment required to yield a good fit to the data. The three parameters can then be adjusted to achieve the best fit to the experimental data, as shown in Fig. 9 for two wafers, one that underwent epitaxial deposition for 3.2 min (Wafer 1), and another that underwent deposition for 7.2 min (Wafer 6) at the same temperature and pressure. The close agreement between experiment and theory over a major portion of

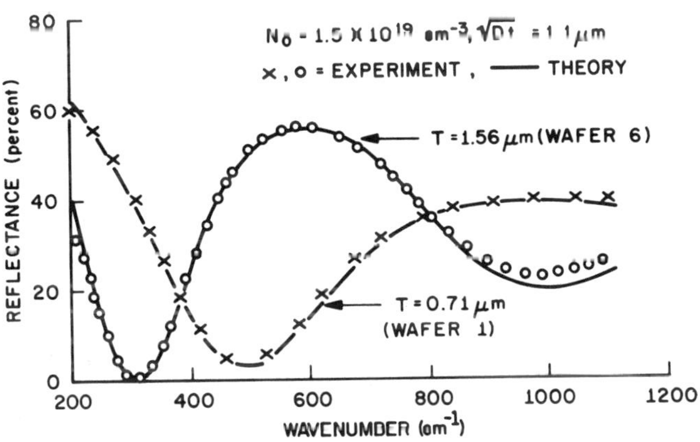

FIG. 9 *Theoretical and experimental reflectance spectra of typical epitaxial film/buried layer structures.*

the reflectance spectrum studied indicates that the model is valid over that spectral range. Wafer 1 is characterized by the parameters $T = 0.71$ μm, $N_0 = 1.5E19$ cm^{-3}, and $\sqrt{Dt} = 1.1$ μm, and Wafer 6 by $T = 1.56$ μm, $N_0 = 1.5E19$ cm^{-3}, and $\sqrt{Dt} = 1.1$ μm. These values are in excellent agreement with values determined by SIMS.[4]

Based on the calculations, the following observations can be made. As the buried layer diffusion depth is increased, the rate of change of the optical constants in the buried layer decreases and the resultant reflection within the buried layer also decreases. This effect is illustrated by the computed spectra in Fig. 10 of a 1.2-μm thick epitaxial wafer, where it is seen that as $\sqrt{Dt}$ increases, the spectrum approaches that of a thin film on a uniform substrate ($\sqrt{Dt} = \infty$). Thus for values of $\sqrt{Dt}$ greater than 2 μm, the effect of this parameter is weak, and the reflectance spectrum is primarily determined by the epitaxial thickness and the maximum buried layer concentration N_0, resulting in a two-parameter measurement. We can then match our data with a much simpler analytic expression [*15*] to obtain T and N_0.

Discussion

The knowledge gained from successful application of the preceding calculation can be used to enhance the capability of infrared reflectance measurements to determine the thickness of thin epitaxial layers. We have found that when using conventional techniques, an error of a factor of two in the measured substrate carrier concentration can lead to a 50% error in mea-

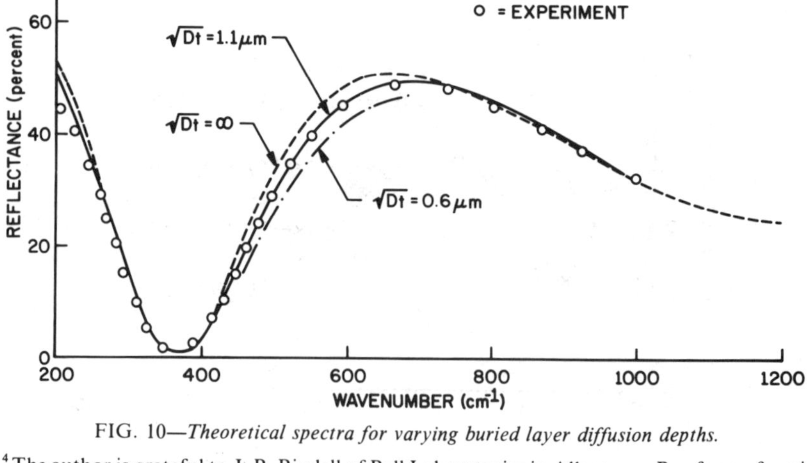

FIG. 10—*Theoretical spectra for varying buried layer diffusion depths.*

[4] The author is grateful to J. B. Bindell of Bell Laboratories in Allentown, Pa., for performing the SIMS measurements.

sured epitaxial thickness. This is further complicated by the fact that phase shifts calculated according to this optical model fit the experimental data, while the values published in ASTM F 95 do not. The calculated phase shifts differ by as much as 12% from those published by ASTM. In addition, failure to account for the finite thickness of a buried layer structure can cause further deviation from true thickness values. In general, conventional methods require prior knowledge of the substrate impurity profile in order to assure accuracy. Lastly, the precision depends upon the degree to which the position of a single broad minimum can be precisely determined. Fitting the entire spectrum instead of the extrema alone greatly increases accuracy, and can improve precision to the order of 0.01 μm for 1-μm, scale layers.

Detailed analysis of the frequency dependence of the phase shift ϕ_2 provides insight into the dependence of interferometric thickness measurements on substrate and epilayer properties. The interference pattern is essentially the Fourier transform of the reflectance spectrum, and is roughly proportional to the derivative with respect to frequency ω of the total phase of the wave reflected from the substrate, $d(\delta_1 - \phi_2)/d\omega$, where δ_1 is the phase lag associated with traversing the epitaxial layer. Since $d\phi_2/d\omega$ is negative, the thicknesses determined interferometrically will be greater than true values for all thicknesses and resistivities. Measurements have shown, for example, that for a 1-μm layer, an error in carrier concentration of a factor of two can lead to a 50% or greater error in epilayer thickness. The precision of this technique depends on the precision of determination of the interference peak. In presently available commercial instruments, precision of approximately 0.01 μm in epitaxial thickness can be achieved, but accuracy remains dependent upon substrate properties.

In addition to dependences on substrate carrier concentration, interferometric measurements are also subject to errors due to the limitations of the technique itself. As layers become thinner, the separation between the zero-path-difference peak and the first interference peak move closer together. As layers approach the 1-μm range, these two peaks interact with each other. This can lead to nonlinearity in the relationship between measured and true (metallurgical) layer thickness, causing an even larger error. This problem, which in general requires transforming the interference pattern into a frequency spectrum using a high-speed computer, is difficult to treat, but is currently being addressed by instrument manufacturers.

Additional factors that are not yet critical but will eventually affect the measurement of thin epitaxial layers include the lateral geometry of patterned implanted or diffused layers and the presence of multiple diffusions. Assuming the substrate is lightly doped except for a diffused layer that is patterned (buried layer structure), the infrared reflectance spectra depend upon the fractional area occupied by the diffusion. For the ASTM F 95 and interferometric methods, the minimum requirement is that the substrate surface contains sufficient diffused regions to yield a detectable interference signal. In the event of lateral diffusion from these regions, however, a fraction of the surface area is occupied by regions of varying carrier concentration. This leads to a substrate whose optical properties are difficult to characterize.

An equally difficult situation arises when multiple diffusions are present in the substrate. These may be both overlapping and nonoverlapping, depending upon device requirements. Optical constants in such substrates would vary both in depth and laterally, and would again be quite difficult to characterize.

A reasonable approach to accurate and precise measurement of epitaxial thickness of 1-μm layers with present techniques would be to calibrate the instrument by performing a reflectance measurement on one sample from a given epitaxial run using a technique such as curve-fitting. For technologies involving lateral variation in carrier concentration, this may require performing the measurement on a special area set aside for this purpose and containing only one diffusion. Cases where sample-to-sample variations in resistivity profiles are large may require each measurement to be made using the curve-fitting method, or other techniques that are insensitive to such variations.

Summary

We have shown that infrared reflectance spectra of epitaxial layers over *n*-type buried layers can be accurately characterized over the wavelength range of 10 to 50 μm by using simple models for the optical properties of the conduction electrons. The spectra can then be used to determine the epitaxial thickness, peak buried layer concentration, and, to a lesser accuracy, the buried layer diffusion depth. The values obtained from the optically measured epitaxial thickness and peak buried layer concentration are in good agreement with SIMS measurements. The information gained through this technique can be used to understand and improve present methods of epitaxial thickness measurement.

Acknowledgments

The author would like to acknowledge the collaboration of Prof. Ben Senitzky of New York Polytechnical Institute, without whose important contributions this work would not have been completed.

References

[*1*] Spitzer, W. G. and Tannenbaum, M., *Journal of Applied Physics,* Vol. 32, 1961, p. 744.
[*2*] Senitzky, B. and Weeks, S. P., *Journal of Applied Physics,* Vol. 52, 1981, p. 5308.
[*3*] Spitzer, W. G. and Fan, H. Y., *Physics Review,* Vol. 105, 1957, p. 882.
[*4*] Dexter, R. N., Zeiger, H. J., and Lax, B., *Physics Review,* Vol. 104, 1956, p. 637.
[*5*] Schumann, P. A., Jr. and Phillips, R. P., *Solid State Electronics,* Vol. 10, 1967, p. 943.
[*6*] Schumann, P. A., Jr. *Journal of the Electrochemical Society,* Vol. 116, 1969, p. 409.
[*7*] Thurber, W. R., Mattis, R. L., Liu, Y. M., and Filliben, J. J., *Journal of the Electrochemical Society,* in press.

[*8*] Lyden, H. A., *Physics Review*, Vol. 134, 1964, p. A 1106.
[*9*] Ramo, S., Whinnery, J. R., and Duzer, T. V., *Fields and Waves in Communication Engineering*, Wiley, New York, 1967.
[*10*] Saifi, M. A. and Stolen, R. H., *Journal of Applied Physics*, Vol. 43, 1972, p. 1171.
[*11*] Holm, J. D. and Champlin, K. S., *Journal of Applied Physics*, Vol. 39, 1968, p. 275.
[*12*] Barta, E., *Infrared Physics*, Vol. 17, 1977, p. 319.
[*13*] Antoniadis, D. A., Hansen, S. E., and Dutton, R. W., "SUPREM II-A Program for IC Process Modeling and Simulation," Stanford University, Stanford, Calif., June 1978.
[*14*] Fair, R. B., *Proceedings*, Third International Symposium on Silicon Materials Science and Technology (77-2), The Electrochemical Society, May 1977.
[*15*] Vasicek, A., *Optics of Thin Films*, North Holland, Amsterdam, The Netherlands, 1960, p. 115.

Lee K. Galbraith[1]

Automated Detection of Wafer Surface Defects by Laser Scanning

REFERENCE: Galbraith, L. K., "**Automated Detection of Wafer Surface Defects by Laser Scanning,**" *Silicon Processing, ASTM STP 804,* D. C. Gupta, Ed., American Society for Testing and Materials, 1983, pp. 492–500.

ABSTRACT: Several systems are presently being marketed that automatically detect and map surface defects on unpatterned wafers by means of light scattering. Applications of these wafer inspection systems in process control and incoming inspection are described. Systems based on scanned laser beams, discussed here in detail, have considerably higher dynamic range and give more reproducible results with non-isotropically-scattering defects than systems using television cameras.

KEY WORDS: surface defects, light scattering, scanned-laser systems, automated inspection

In spite of the great inroads made by automation in almost all areas of semiconductor microcircuit processing, wafer inspection for surface flaws has until recently defied practically all attempts to put the operation under automatic control. The semiconductor industry is thus still largely dependent on human operators for a task that requires considerable training and experience and is generally conceded to be nonreproducible.

The term *wafer inspection* as used here means the critical examination with visible light of blank, unpatterned wafers (which may be coated with photoresist, insulating or conducting films, polysilicon, etc.) such as would be produced at the wafer polishing stage or after being sent through various processes as unpatterned monitor wafers. Instruments designed for detailed comparison of patterns on wafers or masks are not discussed here, since the high cost and low throughput of present systems makes them unsuitable for the production-environment inspection tasks described below.

Over the past ten years a number of automatic wafer inspection systems have been developed by various semiconductor manufacturers and laboratories for in-house use,[2–4] but none of these have, to the author's knowledge,

[1] Senior Scientist, Tencor Instruments, Mountain View, Calif. 94043.
[2] Oswald, D. R. and Munro, D. F., *Journal of Electronic Materials,* Vol. 3, No. 1, 1974, pp. 225–241.
[3] Martin, M. and Williams, H., *Electro-Optic Systems Design,* Vol. 12, No. 9, Sept. 1980, pp. 45–49.
[4] Gara, A. D., *Electronics Test,* Vol. 4, No. 5, May 1981, pp. 60–70.

appeared in the general marketplace. In the last couple of years, however, development activity in the wafer defect scanner field has greatly increased and instrumentation designers have become involved. One recent design will be described in some detail later in this report.

Operating Principles

All systems presently on the market detect defects by the light they scatter from an intense illuminating beam directed at the substrate surface. This is illustrated in Fig. 1. The illuminating beam is shown directed at the substrate nearly perpendicular to its surface. A beam of reflected light, which we shall call the *specular beam,* leaves the surface at a small angle to the surface perpendicular; the intensity of the specular beam is often a large fraction of the incident beam intensity. If a surface defect is present, a much smaller amount of light is scattered in a wide range of directions and some of it encounters a detector situated to miss the specular beam. This is the basic configuration for visual inspection of polished wafers in accordance with ASTM Practice for Unaided Visual Inspection of Polished Silicon Slices (F 523). In this case, the scattered light detector is the human eye, and the efficiency of the inspection procedure depends on the inspector's ability to observe light scattered

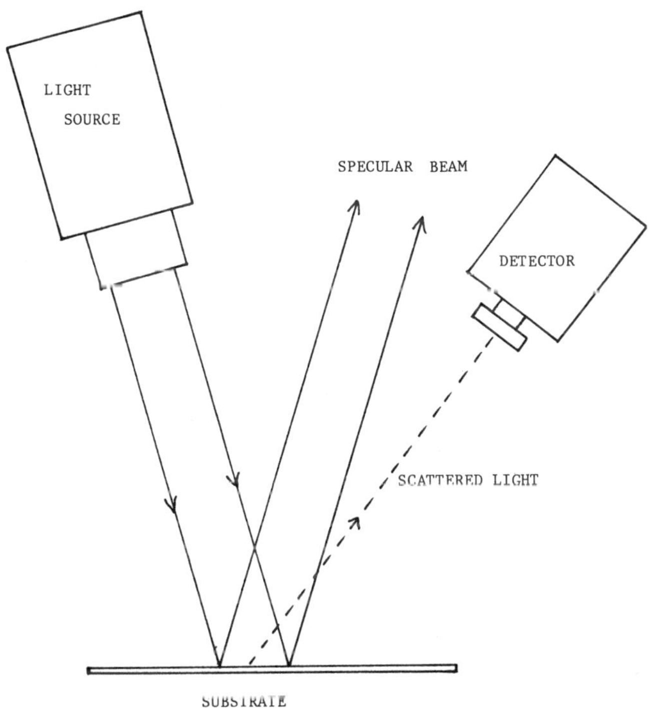

FIG. 1—*Basic setup used for the detection of surface flaws by light scattering.*

from defects without encountering the orders-of-magnitude more intense specular beam.

Automatic defect scanners presently available fall into two basic categories: the camera type, which uses a video camera to image the substrate defects, and the scanned-laser type, in which a small spot of laser light is optically scanned across the substrate.

The basic configuration of a camera type system is as shown in Fig. 1 with a television camera as the detector. The defects are imaged on an ordinary TV monitor as bright areas on a dark background. Special circuitry digitizes the intensity of the video signal containing the defect information and stores the data in a large memory as a function of horizontal and vertical coordinates on the wafer surface. Software routines can then perform calculations on the video data to extract information about the defects such as size distribution and carry out elementary pattern-recognition algorithms such as line and area defect discrimination. This information can then be used to sort wafers according to operator-entered pass/fail criteria.

The scanned-laser beam system utilizes a sharply focused spot of laser light incident normally (or nearly so) on the wafer and scanned rapidly in one direction, while the wafer moves slowly in the perpendicular direction (Fig. 2). Thus the illuminating light spot traverses the substrate in a raster pattern much like that in an ordinary TV system, except that the time scale is considerably longer (several seconds per complete frame). Light scattered by defects is collected and integrated by an optical system covering a range of different scattering angles (but intercepting as little as possible of the much more intense specularly reflected beam) and is directed to a sensitive high-speed detector such as a photomultiplier tube. The signal from the photomultiplier is then digitized directly and the data are stored in a memory organized (as in the camera systems) by horizontal and vertical coordinates on the substrate. A television display of the defects is obtained from a video graphics board loaded with processed data from the detector signal memory, whose data can also be manipulated by software to obtain defect information and classification and to sort wafers.

Limitations of Defect Analysis in Available Systems

Of the two basic configurations mentioned previously, the camera type is easier to construct, mechanically simpler, and needs fewer critical internal adjustments. However, the range of defect scattering angles to which it is sensitive is limited by available camera lenses to a few degrees around the camera optical axis. The solid angle subtended by the camera lens is a small fraction of the total available solid angle (2π steradians) into which a defect can scatter light. In a scanned-laser system, on the other hand, the scattered-light collector can subtend a large fraction of the total solid angle (as great as 70% of 2π steradians). For isotropic scatterers such as particles the lower

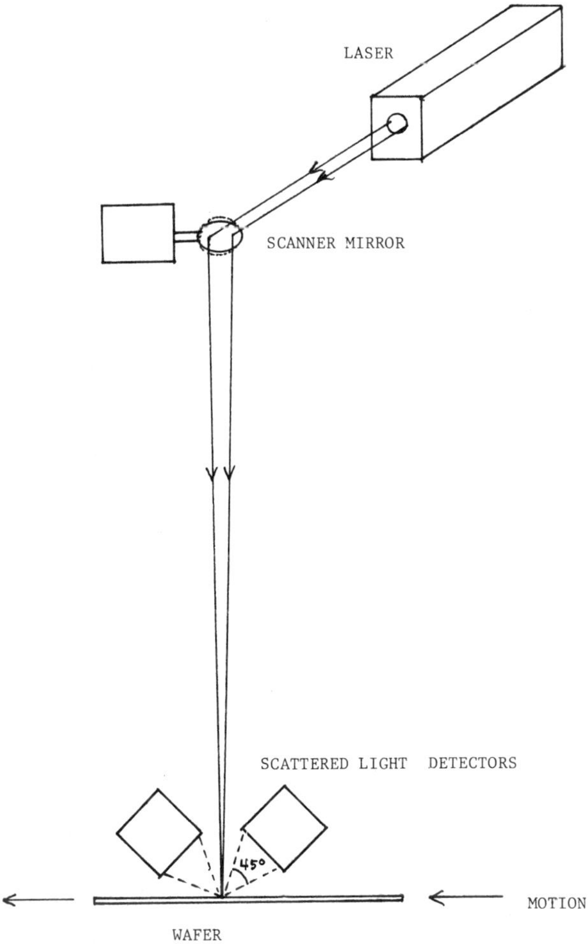

FIG. 2 *Simplified schematic representation of scanned-laser defect detection systems.*

light-gathering capability of a video camera relative to that of the light collector in a scanned-laser system can be compensated for by adjusting the detector sensitivity or integration time. However, there is a wide range of possible wafer defects that do not scatter isotropically. Experiments have shown that scratches, pits in epitaxial layers, and stacking faults, for example, scatter light preferentially in certain directions relative to the crystallographic planes of the wafer, or in the case of scratches, perpendicular to the scratch direction.[5] The limited range of scattering angles detectable in a camera-type system makes its defect detection sensitivity critically dependent on wafer and defect orientation. Since the detector in a scanned-laser system can be

[5] Adley, J. M. and Gorey, E. F., *Journal of the Electrochemical Society,* Vol. 117, No. 7, July 1970, pp. 971–975.

made to cover a large range of scattering angles, such a system will be able to measure nonisotropic defects with greater reliability and reproducibility.

Another limitation of camera-type systems is the relatively low dynamic range of the detectors used. In the absence of light, vidicons produce a constant nonzero signal that varies over the image plane in a manner that differs among individual vidicons. In order to measure reliably a low level of scattered light such as would arise from haze on a substrate, the dark response of the camera tube would have to be accurately known as a function of position on the wafer surface, and small errors or drift in the dark response would make haze measurements meaningless. In a scanned-laser system employing a photomultiplier tube as the scattered-light detector, the dark response of the detector is several orders of magnitude lower than that of a vidicon when the photomultiplier gain is set to a level that produces the same sensitivity to defects as the vidicon in a camera-type system. Therefore, photomultiplier-equipped scanned-laser systems can measure lower levels of haze more accurately than camera-type systems and can detect a wider range of defect sizes.

Operation of a Wafer Scanner System

The scanned-laser wafer inspection system discussed below detects surface defects on wafers down to the submicrometre range in the case of particles and counts the number of particles at or above a given optical scattering cross-section. It also produces independent measurements of defect area for large-area defects such as fingerprints, stains, and gross haze. Overall surface scatter is monitored separately and read out as average haze, independently of discrete defect detection. The user sets limits on the allowable levels of four defect parameters: point defect number, line defect total length, area defect total extent, and haze averaged over the measurement area. The system then automatically sorts wafers into accept or reject categories in separate cassettes. A defect map of each wafer is displayed on a video monitor to allow an operator to visualize defects without having to examine the wafers under high-intensity light. A printer makes available hard copy of the four measured defect parameters for each wafer and shows which wafers exceeded one or more of the set parameter limits. The complete system is shown in Fig. 3.

A block diagram of the system is shown in Fig. 4. Light from an HeNe laser at 632.8 mm is focussed onto the wafer surface and scanned by a galvanometer mirror across the surface 800 times per second while the wafer travels slowly through the system on belts. The collector of scattered light is based on the integrating sphere principle and allows uniform response over a large range of scattering angles using a single photomultiplier tube as a detector. The scanning beam enters the collector and the specularly reflected light exits through a slot (not shown) so that only light scattered by defects and by the wafer edge is detected by the photomultiplier. The scanning beam

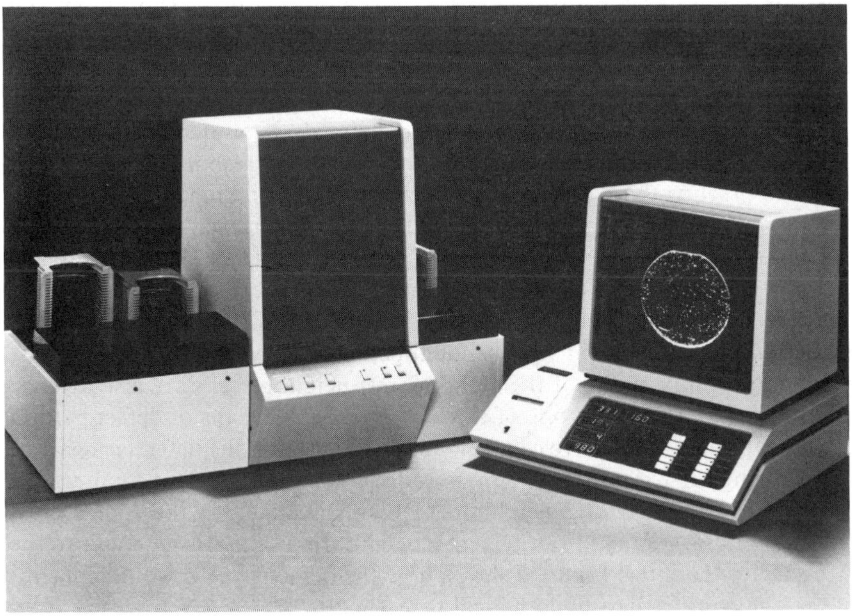

FIG. 3 *Typical scanned-laser defect detection system.*

also enters two other detectors, designated *auto-zero* and *auto-calibrate.* The auto-zero detector is designed to scatter as little light as possible into the photomultiplier and thereby provides a pulse signal designating a time period when light scattering is at a minimum. The system uses the photomultiplier output during this time as the zero signal reference. The system sensitivity is thus virtually independent of ambient light entering the light collector. The auto-calibrate detector is designed to scatter a constant, known amount of light into the collector. The photomultiplier signal during the auto-calibrate time interval, corrected by the signal during auto-zero, gives a measurement of photomultiplier gain that is kept constant by a feedback loop,

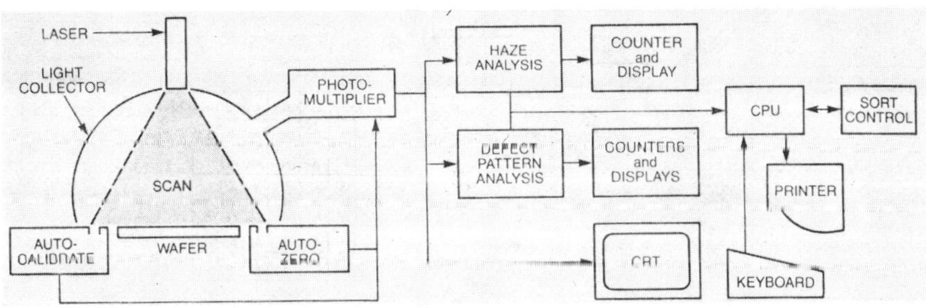

FIG. 4—*Block diagram depicting components of the system in Fig. 3.*

regardless of tube temperature and gain variation over time. Thus the system is kept at a constant level of sensitivity despite changes in ambient lighting and component variation with aging.

The photomultiplier signal is sent into two parallel signal processing channels. One channel processes transient signals generated by point, line, and area defects. If the signal pulse height exceeds a threshold (designated by *defect size*) set by the user, a memory location is written high in a video memory mapping the x and y coordinates of the wafer surface. The video memory maps the wafer into square "cells" 0.5 mm on a side, with 256 by 256 cells available to map wafers up to 125 mm in diameter. If a pulse exceeding the defect size threshold occurs anywhere in a cell, the entire cell is written high and a corresponding bright spot appears on the video display. The system analyzes how many of the eight nearest neighbor cells are also at logic high and adds a count to one of three counters as a means of defect pattern classification. If a defective cell (cell at logic high) has one defective nearest-neighboring cell (or none), a count is added to the Point Defect counter. If it has two to four defective neighbors, a count is added to the Line Defect counter, and more than four defective neighboring cells adds a count to the Area Defect counter. Figure 5 shows how the defective cells would contribute to the counter in a hypothetical defect map.

The other signal processing channel takes the average photomultiplier current, representing overall scatter from the wafer, sends it to a voltage-to-frequency converter and accumulates the pulses in the haze counter after division by a factor representing the wafer area. The haze counter reading represents the fraction of the incident laser beam scattered into the light collector in parts per million, averaged over the wafer surface.

The edges of even a perfect wafer scatter considerable amounts of light, and in order to avoid classifying the wafer edge as a line defect, an electronic

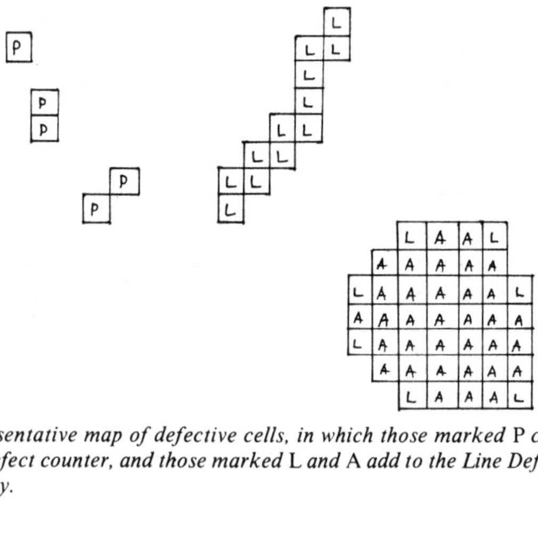

FIG. 5—*A representative map of defective cells, in which those marked* P *contribute one count each to the Point Defect counter, and those marked* L *and* A *add to the Line Defect and Area Defect counters respectively.*

blanking circuit gates off the photomultiplier signal when the beam gets within a certain distance of the wafer edge. This distance, called the edge exclusion, is variable from zero to 7 mm.

The system has been calibrated so that the user-adjustable defect size parameter corresponds to the minimum light scattering cross-section in square micrometres that a point defect must have in order to indicate a defective cell. Instrument calibration is done with a diffuse white substrate that scatters 90% of the incident laser light. The laser beam is attenuated with a neutral-density optical filter to bring the photomultiplier response within the system measurement range, and the resulting detector signal level is observed on an oscilloscope. Then the photomultiplier gain is adjusted to bring the signal level to a standard value. This procedure standardizes the pulse height produced by a hypothetical particle with 1 μm^2 of optical scattering cross-section to agree with the defect size threshold level corresponding to Defect Size 1.

The range of defect sizes programmable by the operator is 1 to 255, corresponding roughly to a particle size range of 1- to 16-μm diameter. In practice it has been found that a significant percentage of 0.5-μm diameter polystyrene spheres deposited on a silicon wafer will be detected at Defect Size 1. Experiments to determine the detection sensitivity limits more quantitatively are now in progress.

System Applications

Incoming wafer inspection is an obvious area of application for defect scanning systems. Wafers can be classified according to the severity of point, line, and area defects and haze in a manner that is more reliable and less subjective than inspection by eye under high-intensity light. Although extremely fine scratches occasionally escape detection by the automatic system, and particles and pits cannot be distinguished from each other, submicrometre defects are readily detected. Although the system cannot as yet replace the skilled wafer inspector, the most tedious aspects of the task can be done more reliably and reproducibly, freeing inspectors for less fatiguing work and increasing the overall throughput per person of the incoming inspection process.

Another area of application for wafer inspection systems is in process control. The instrument can be used to monitor the appearance of spikes and stacking faults in epitaxial deposition, the generation of particulates by process machinery, particles and bubbles in photoresist layers (the scanned-laser systems using HeNe lasers will not expose the photoresist), particle contamination during metal film deposition or chemical vapor deposition (CVD), residue left by a chemical cleaning or etching process, improper polishing, etc. Since one usually has some prior knowledge of what kind of defects are likely to show up in a given process, the inability of the inspection system to distinguish certain kinds of defects (for example, particles, pits, and epitaxial spikes) is not particularly serious. An in-line wafer inspection system in a particular critical process can sound an alarm when defective wafers appear and allow process to be stopped in time to prevent further yield losses.

Wafer inspection systems are primarily designed for polished, unpatterned silicon wafers, but they are not limited to these substrates. Any flat plate with a polished surface that is reasonably opaque (attenuating 90% or more of the incident light through its thickness) can be tested with useful results, including semiconductors other than silicon, polished metal, chrome-coated mask substrates, and certain colored filter glasses. Transparent substrates are problematic in that defects on both sides will be seen with no indication by the system of which surface the defects are on. If the underside of the transparent substrate is rough in texture (as in the case of sapphire and silicon-on-sapphire), light scattering from the roughened side will overload the system. Nevertheless, a wide range of objects and substrates can be inspected automatically with instruments presently available.

Conclusion

Recent developments in automatic wafer inspection systems have brought them to the point where they can play a valuable role in incoming inspection and process control. Although the systems cannot as yet recognize everything that a skilled visual inspector can, their accuracy and stability make them superior for many inspection tasks that are especially fatiguing to human inspectors. As semiconductor manufacturers become more aware of the potential capabilities of these systems, further areas of application will open up. Progress in inspection system development will depend critically on potential users communicating their requirements clearly to the instrument designers.

Tom F. Pomposo[1] *and Vincent J. Coates*[1]

Computerized Electron-Beam Linewidth Measuring and Inspection: A New Tool

REFERENCE: Pomposo, T. F. and Coates, V. J., **"Computerized Electron-Beam Linewidth Measuring and Inspection: A New Tool,"** *Silicon Processing, ASTM STP 804,* D. C. Gupta, Ed., American Society for Testing and Materials, 1983, pp. 501–508.

ABSTRACT: This paper describes a field emission linewidth measuring and inspection instrument. The low-voltage operation, which allows nondestructive measurement of critical dimensions on photoresist with good precision, is described. The effect on photoresist of nonpenetrating electrons in the low-voltage mode versus penetrating electrons in the high-voltage mode is discussed.

KEY WORDS: scanning electron microscope, linewidth measurement, photoresist critical-dimension measurement, field emission source, field emission gun

Scanning electron microscopy (SEM) has been widely used for the inspection and study of microelectronic circuits [*1–11*],[2] for example, in the metal step coverage, the analysis of defects, bonding studies, etc. Conventional SEM has a disadvantage, however, because it uses highly energetic electrons (15 to 30 kV) in order to achieve the necessary small scanning spot size. High-energy electrons of deep penetration can create damage in the circuit. To minimize this damage, the maximum allowable electron accelerating voltage must be about 1.5 kV; and conventional SEM operating at such a low voltage has poor resolution. Other disadvantages of conventional SEM are its use of an oil diffusion pump, whereby the backstreamed oil from the pump cracked by the electron beam causes a contaminating carbon layer on the scanned area, and the defocusing of the incoming beam ("charging") in certain applications, which can reduce the clarity of the image.

In 1971, Welter and Coates [*10*] indicated that a new type of SEM using a high-brightness field emission gun and a clean vacuum system could avoid all these problems. It was capable of focusing with electron voltages down to 600 V and lower, and produced high resolution, nondestructive SEM images

[1] Product Marketing Manager and President, respectively, Nanometrics, Inc., Sunnyvale, Calif. 94086.

[2] The italic numbers in brackets refer to the list of references appended to this paper.

at commercial television scan rates. Furthermore, at these voltages, charging effects are minimized because equilibrium is achieved at the nonconductive surfaces and they can be studied in great detail. The instrument described in this paper has evolved from that work and has quantitative critical dimension measurement capability. The instrument (Fig. 1) utilizes a high–current density field emission electron source having its own ion pump and a titanium sublimator. The specimen chamber consists of an ion pump, a cryopump, a titanium sublimator, and a molecular sieve trapped direct-drive roughing pump. Results are viewed either on a cathode ray tube (CRT) or a photographic monitor. A single crystal tungsten field emission tip having a radius of approximately 1000 Å is used for the electron beam source (Fig. 2). The tip is spot-welded to a tungsten filament (Fig. 3). The column (Fig. 4) consists of a field emission gun and the final magnetic lens. The gun utilizes an electrostatic lens.

Table 1 compares the field emission source to the tungsten filament and lanthanum hexaboride sources (the latter two sources being thermionic). The lower energy spread of the field emission source decreases the chromatic aberration of the magnetic lens and provides a smaller beam size than a thermionic source. A value of 150 Å has been achieved at 1 kV by the former.

FIG. 1—*CWIKSCAN system.*

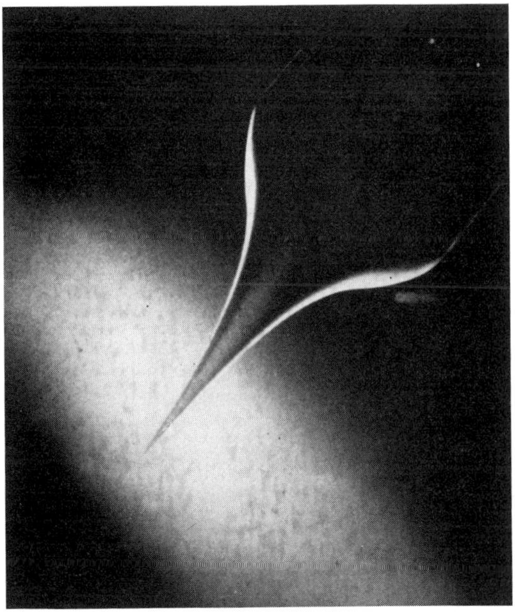

FIG. 2—*Field emission tip.*

FIG. 3—*Field emission tip in holder.*

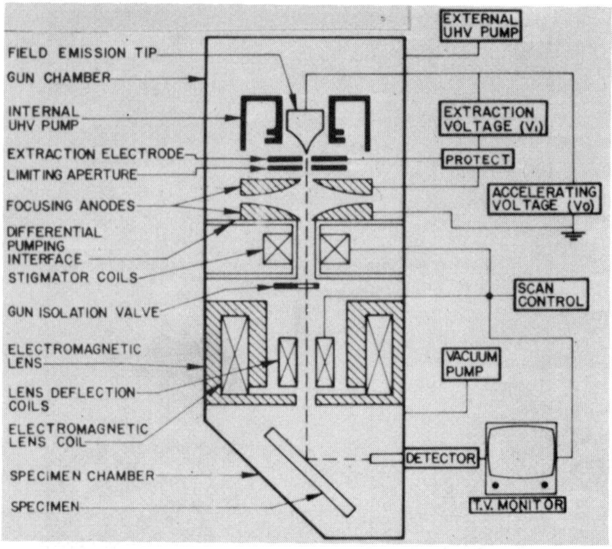

FIG. 4—*Block diagram showing the electron optical column of a field emission source SEM with a field emission electron gun and single electromagnetic lens designed to produce high-resolution images for both high and low accelerating voltages.*

The high-speed internal vacuum pump in the field emission gun provides the required tip emission stability. The low-voltage secondary electron image formed by electrons released from the specimen provides a true image of the surface topography. Figure 5 is a low-voltage photomicrograph of resist structure showing the surface detail. The high-voltage micrograph of the same resist structure shows very little surface detail due to high-voltage beam penetration.

The effect of high-voltage beam penetration can also be demonstrated on a graphite sample. Figures 6 and 7 show the effect of beam penetration. The bright washed-out area (Fig. 6) is caused by an excess of secondary electrons emitted from both sides of the thin edge of the graphite. The low-voltage

TABLE 1—*Comparative electron source characteristics.*

Characteristic	Tungsten Filament	Lanthanum Hexaboride	Field Emission
Type of emission	thermal	thermal	field
Source temperature, K	2650 to 2900	1750 to 2000	300 to 1000
Brightness, $A/cm^2/sr$	10^4 to 10^5	10^5 to 10^6	10^7 to 10^9
Effective source size, Å	1 000 000	200 000	100
Energy spread, eV	1 to 5	0.5 to 3	0.2 to 0.3
Operating vacuum, torr[a]	10^{-4} to 10^{-3}	10^{-6} to 10^{-5}	10^{-10} to 10^{-8}
Operating life, h	30 to 10	100 to 500	300 to 1000

[a] 1 torr = 133 Pa.

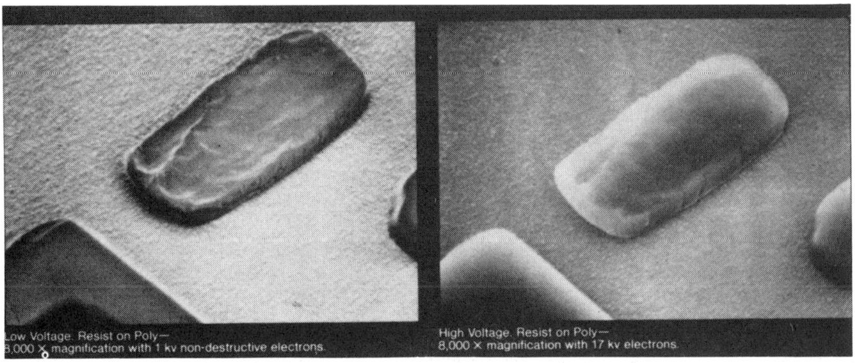

FIG. 5—*Comparison showing superior surface detail in the low-voltage mode. Left micrograph shows true resist structure; right micrograph shows distorted image of resist structure in high-voltage mode due to excessive beam penetration effects.*

beam does not penetrate through the sample and electrons are emitted from the top surface only. Figure 7 thus shows topographic detail.

A unique stage, which is motor driven and computer controlled, allows tilts from 0 to 60 deg. This stage, which has 125 mm (5 in.) of x-travel, 125 mm (5 in.) of y-travel, and 360-deg rotation, when combined with the acute angle conical final lens, allows the entire wafer to be examined with a minimum

FIG. 6—*Graphite at high voltage (18 kV).*

FIG. 7—*Graphite at low voltage (1 kV).*

working distance, and consequently optimum resolution over the entire wafer. The bottom of the wafer is viewed with the same optimum resolution as the top of the wafer.

The computer allows preselection of any location or a series of locations on wafers up to 125 mm (5 in.) in diameter. These locations can be preprogrammed by the operator. A joystick control is also provided which allows manual operation and can be used to find the proper location when creating a program.

A test specimen, namely small gold particles ranging in size from 30 to 400 Å on a carbon substrate, is used to check low-voltage resolution (Fig. 8).

The maximum secondary electron yield for most materials varies from 400 to 4000 V. Maximum resolution in the field emission SEM occurs at 1 kV, which makes it ideally suited for observation of metals and metal oxides at low voltages.

The effects of beam penetration in the high-voltage mode versus low-voltage mode are shown in Figs. 9 and 10. Figure 9 shows the line of demarcation between the base of the resist and the oxide substrate, while Fig. 10 does not have a sharp demarcation between the base of the resist and the oxide substrate, making a linewidth determination at high-voltage implausible.

A typical standard deviation 1σ of 0.01 μm is obtained on lines with measured mean values of 1-μm line width.

FIG. 8—*Resolution test.*

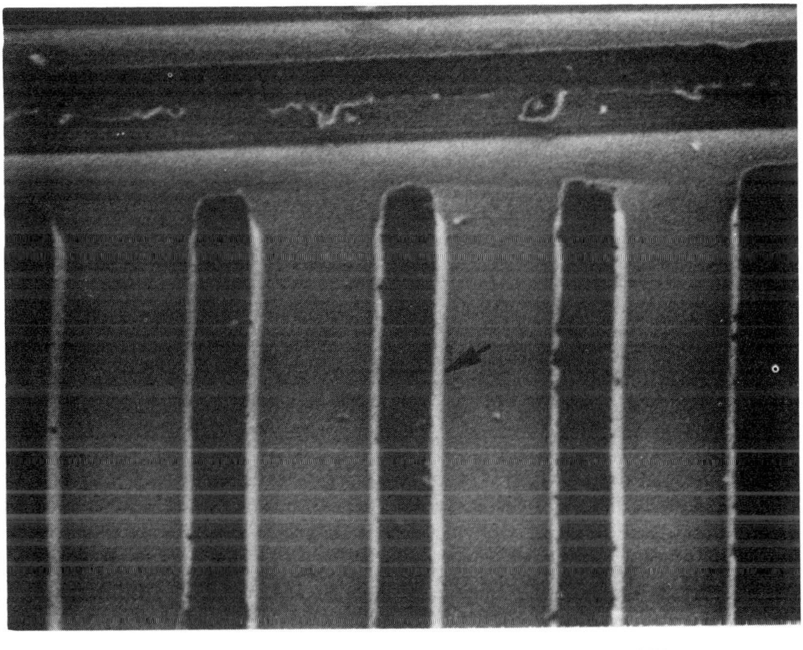

FIG. 9—*Resist lines on oxide, ×7000, low voltage (1.6 kV).*

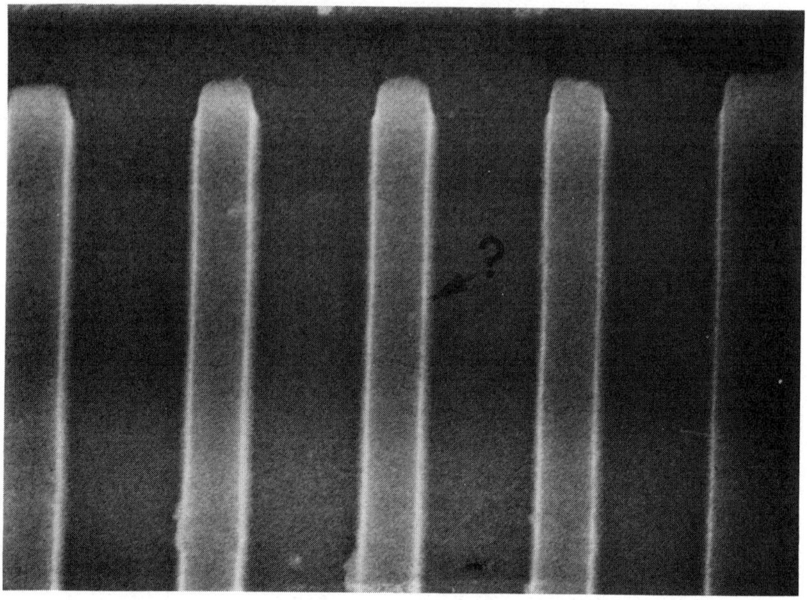

FIG. 10—*Resist lines on oxide, ×7000, high voltage (19 kV).*

References

[*1*] Broers, A. N., "High Brightness Electron Gun Using a Field Emission Cathode," in *Proceedings,* 26th Annual Meeting of the Electron Microscopy Society of America, Claitor's Publishing Division, Baton Rouge, La., 1968, p. 294.

[*2*] Crewe, A. V., Eggenberger, D. N., Wall, J., and Welter, L. M., *Review of Scientific Instruments,* Vol. 39, No. 4, April 1968, pp. 576–583.

[*3*] Crewe, A. V., Isaacson, M., and Johnson, D., *Review of Scientific Instruments,* Vol. 40, No. 2, Feb. 1969, pp. 241–246.

[*4*] Crewe, A. V. and Wall, J., "High Resolution Scanning Microscopy," in *Proceedings,* 27th Annual Meeting of the Electron Microscopy Society of America, Claitor's Publishing Division, Baton Rouge, La., 1969, p. 172.

[*5*] Crewe, A. V., Wall, J., and Welter, L. M., "A High Resolution Scanning Microscope," in *Proceedings,* 26th Annual Meeting of the Electron Microscopy Society of America, Claitor's Publishing Division, Baton Rouge, La., p. 356.

[*6*] Pfeiffer, H. C., "Basic Limitations of Probe-Forming Systems Due to Electron-Electron Interaction," in *Proceedings,* Illinois Institute of Technology Research Institute, Chicago, 1972, p. 113.

[*7*] Ploc, R. A., "Resolution Test for the SEM," in *Proceedings,* Electron Microscopy Society of America, Miami Beach, 1976, p. 510.

[*8*] Postek, M. T., Howard, K. S., Johnson, A. H., and McMichael, K. L., *Scanning Electron Microscopy,* Ladd Research Inc., Burlington, Vt., 1980.

[*9*] Roth, T. A., Munch, J., and Zeitler, E., "The Effect of Tip-Anode Distance on Field Emission current," in *Proceedings,* Electron Microscopy Society of America, St. Louis, 1974, p. 420.

[*10*] Welter, L. M. and Coates, V. J., "A Compact Field Emission Scanning Electron Microscope," in *Proceedings,* Electron Microscopy Society of America, Boston, EMSA, 1971, p. 32.

[*11*] Welter, L. M. and McKee, A. N., "Observations on Uncoated Non-conducting or Thermally Sensitive Specimens Using a Fast Scanning Field Emission Source SEM," in *Proceedings,* Illinois Institute of Technology Research Institute, Chicago, 1972, p. 161.

Arthur K. Hochberg[1]

Measurement of Ion Implantation Doses in Silicon by Ellipsometry and Spectral Reflectance

REFERENCE: Hochberg, A. K., "**Measurement of Ion Implantation Doses in Silicon by Ellipsometry and Spectral Reflectance,**" *Silicon Processing, ASTM STP 804,* D. C. Gupta, Ed., American Society for Testing and Materials, 1983, pp. 509–533.

ABSTRACT: Results of ellipsometric measurements are presented for arsenic implants in silicon at doses from 2×10^{13} to 8×10^{15} cm^{-2} and energies from 30 to 180 keV. For implanters which mechanically scan wafers through the ion beam, it is possible to construct curves relating ellipsometric parameters to implant parameters. Reproducible ellipsometric results are strongly dependent on wafer preparation and precise setting and control of implant energies. Spectral reflectance measurements of implanted samples show that this method may be also used for implant monitoring.

KEY WORDS: silicon wafer, ion implantation, dose measurements, ellipsometric measurements, spectral reflectance, arsenic implants in silicon

Ellipsometry has been used to study the changes in optical characteristics of silicon wafer surfaces owing to crystalline damage produced by ion implantation. This rapid, nondestructive evaluation method exhibits sufficient sensitivity to variations in implant parameters and has adequate measurement reproducibility to be considered for process control monitoring of dosimetry within certain limits. The residual damage measured, after an implant, is dependent on ion species, ion fluence, total dose, accelerating voltage, and instantaneous wafer temperatures during an implant. For reproducible ellipsometric results, identical implant conditions must be provided. This criterion is met over a wide range of doses and energies in an ion implanter that mechanically scans wafers through an ion beam of fixed fluence during the implant.

Previous investigations of implant damage in silicon by ellipsometry [*1–10*] and spectral reflectance [*11–17*] concentrated on determining damage profiles and effective optical constants of the damaged layer.[2] Some studies

[1] TRW LSI Products, La Jolla, Calif. 92038; formerly Senior Engineer, Burroughs Corporation, San Diego, Calif. 92127.

[2] The italic numbers in brackets refer to the list of references appended to this paper.

assumed an optically homogeneous damaged layer with or without a nascent oxide overlayer [*5–10*]. The effective thickness for such a layer has been estimated as Rp + 1.08 ΔRp for P_{31} implant damage [*10*]. Rp and ΔRp are the projected range and straggle, respectively, of the ion according to theory [*18*].

Except for "high" doses, with the term "high" dependent on ion species and wafer temperature, the damage distributions after implantation resemble skewed gaussians [*1,4,13,14,19–22*], with perhaps, a similar distribution of optical constants [*1,4*]. The damage peaks are dependent on species and energy with high dose P_{31} damage peaks at 0.7 Rp, As_{75} peaks at 0.8 Rp, and damage extending to 2 Rp [*19,22*]. To date, however, ellipsometric measurements have not provided unambiguous determinations of implant results. McCrackin and Colson [*23*] have reported that ellipsometer measurements are insufficient to distinguish homogeneous from nonhomogeneous layers, even using multiple angles of incidence.

Initial investigations of implant damage were performed on high-dose implantations into thermally grown silicon dioxide (SiO_2). The effective refractive index changes could be correlated with implant parameters when damage did not extend into the underlying silicon. Results are presented which show that damage depths in SiO_2 are significantly greater than for similar implants in silicon.

More reliable process control may be obtained from the results of ellipsometric measurements on "bare" silicon wafers, presented for As_{75} implants from 2×10^{13} to 8×10^{15} cm^{-2} and for effective energies from 30 to 180 keV. As a check on ellipsometer results, pseudo-substrate refractive indexes were obtained from ellipsometric parameters in order to calculate reflectivities for comparison with spectral reflectance measurements. The good agreement between calculated and measured reflectivities at 632.8 nm encouraged further studies of spectral reflectance. The resulting spectra from 480 to 780 nm are presented for implants of various doses and energies and show that reflectance may be another technique for process control. The major deterrent to obtaining reproducible ellipsometric parameters and reflectance spectra for damaged silicon from supposedly identical implant conditions has been the poor precision in setting and maintaining an exact accelerating voltage.

Experimental Procedure

Ion Implantation

Samples used for these experiments were boron-doped 76.2 mm (3 in.) and 100 mm diameter wafers with (100) oriented surfaces and 10 to 20 Ω·cm resistivity. The majority of the As_{75} implants were performed in a Lintott Series III high-current implanter that mechanically scans wafers through the ion beam. Additional implants were obtained by using a Lintott Series IIIA and an Extrion 200-1000, both mechanical scan implanters. As_{75} was obtained from vaporization of solid sources in the Lintotts and from arsine (AsH_3) in the Extrion.

In order to estimate current density distributions at the wafers, beam current profiles at various total currents and accelerating voltages were obtained by fixing wafers in the beam line behind a mechanical shutter and opening the shutter for a time sufficient to produce a visual damaged area. The wafers were then annealed, and subsequent sheet and spreading resistance measurements were compared with results from scanned implants in order to obtain beam distributions.

All implants were performed with wafer surfaces oriented 7 deg off the beam direction in order to minimize ion channeling. All implanters have aluminum pallets with recesses and clips to hold the wafers. Each implant was performed with wafers loaded on pallets that had cooled to room temperature (23°C). Wafer temperatures during implant were estimated from temperature tapes placed on backsides of wafers in positions that should not have interfered with normal thermal contacts of wafers to the pallets.

The background pressures in the processor chambers during implants were usually less than 2×10^{-6} torr. Two sets of implants were made when processor pressures were oscillating near 1×10^{-5} torr due to leaking bellows. Ellipsometric measurements of these samples indicated higher than expected doses and severe nonuniformities, later confirmed after anneals and sheet-resistance measurements. Generally, the implant uniformities determined by sheet-resistance measurements had standard deviations of $\pm 1\%$ across a wafer and from wafer to wafer. Each implant condition had 2 to 10 wafers measured with the ellipsometer in 5 to 10 places each.

Wafer preparation prior to implant ranged from no additional clean (as obtained from the manufacturers), through acid and solvent cleans, to hydrofluoric acid stripping of the nascent oxides followed by deionized water rinses. The most reproducible results occurred with wafers receiving no prior clean except for a high-pressure deionized water scrub. These wafers had an average oxide thickness of 1.7 nm, assuming a refractive index of 1.464 at 632.8 nm. Chemically cleaned wafers produced a great amount of scatter in ellipsometric measurements of unimplanted samples, perhaps because of trapped chemicals. Wafers stripped of nascent oxide and rinsed to achieve 20 MΩ·cm effluent water resistivity had approximately 0.7 to 0.9 nm of oxide (ignoring a possible interfacial film of higher refractive index [*24*] or other nonstoichiometries [*25*]). The vagaries of implanter operation and scheduling did not always allow "stripped" wafers to be implanted within a set time from final rinse. Also, the large numbers of wafers implanted in a session caused varying time lags from implant to measurement. Within 24 h from final rinse, unimplanted stripped wafers again had from 1.3 to 1.7 nm of oxide. In order to alleviate measurement and implant scheduling problems and to allow for remeasurements of wafers over a period of time to assure ellipsometer calibration, the 1.7-nm oxide-covered samples were standardized for the experiments. Additionally, this may allow for monitoring of device wafers that have oxide-free, unimplanted areas available for each implant. Generally, measurements of implanted "stripped" wafers or wafers stripped after implant differed little in ellipsometric parameter Ψ but had a wide scatter in Δ, indicative of different oxide thicknesses only.

Implanted Oxides

For an estimation of implant damage distributions, and to determine if ellipsometric results are sensitive to oxide damage, wafers with thermally grown oxides were implanted and then etched in a sequence of steps. For deep damage profiles, oxides were etched in a solution of 6 parts ammonium fluoride (NH_4F) to 1 part (49%) HF for 30 s at a time. This was followed by deionized water rinses, spin drying, and film thickness determinations by ellipsometry and spectral reflectance. For greater detail near the oxide surfaces, oxides were etched in a solution of 40 H_2O to 1 part (49%) hydrofluoric acid (HF) for 10 s at a time in a temperature-controlled bath at 23°C. Unimplanted thermally oxidized wafers were etched with each group of implanted oxides in order to obtain relative etch rates independent of the accuracy of the etch times.

Measurements

Ellipsometric measurements were made with a Rudolph Research Auto EL III; this has a configuration light source (632.8 nm), polarizer, compensator, sample, analyzer, and detector. The angle of incidence (ϕ) for all measurements was 69.98 ± 0.02 deg checked by alignment with a 70-deg prism and by comparison of measurements of thin SiO_2 samples with expected ellipsometric solutions from a computer program [*26*]. An autocollimator alignment of the samples prior to each measurement assured the repeatability of ϕ. The waveplate constants were obtained from measurements on a series of SiO_2 films from 10 to 400 nm thick in accordance with the method of McCrackin [*26*]. The ratio of fast-to-slow axis transmittance ($T_c = 0.9993$) and the phase difference ($\Delta_c = 89.733$ deg) indicated no corrections were required in subsequent calculations. All measurements were made in two zones to reduce the inaccuracies associated with azimuthal calibration errors and the small imperfections in the waveplate. Both the polarizer and analyzer are rotated to obtain nulls with angular resolution and repeatability of 0.10 deg for parameter Δ and 0.05 deg for Ψ.

Specular reflectance measurements were made with a Nanometrics Nanospec AFT system in the range of 480 to 800 nm. For oxide thickness measurements, the AFT and ellipsometric results agreed within ±1% for films greater than 45 nm. For thinner films, the AFT errors increased as the film thicknesses decreased. For reflectance measurements of implanted surfaces, the AFT was operated in a single wavelength mode with a resolution of 2 nm.

Ellipsometry

The theory and background for ellipsometric measurements has been covered in recent reviews [*27,28*]. The ellipsometric parameter Δ, obtained from polarizer azimuths, represents the difference in phase shifts after reflection from the sample for components of the electric vector of the light waves polarized parallel and perpendicular to the plane of incidence. The parameter Ψ, obtained from analyzer readings at nulls, is the arc tangent of the ratio of the reflection coefficients of parallel and perpendicular polarized light components. The complex reflection coefficient is $\rho = \tan\Psi \exp(i\Delta)$.

Solutions of the ellipsometric equations for obtaining components of the complex dielectric constant $\epsilon = (\epsilon_1 + i\epsilon_2)$ of a substrate, assuming no covering oxide, are given by

$$\epsilon_1 = \sin^2\phi\left\{1 + \tan^2\phi\left[\frac{\cos^2 2\Psi - \sin^2 2\Psi \sin^2\Delta}{(1 + \sin 2\Psi \cos\Delta)^2}\right]\right\} \quad (1)$$

$$\epsilon_2 = \sin^2\phi \tan^2\phi\left[\frac{\sin 4\Psi \sin\Delta}{(1 + \sin 2\Psi \cos\Delta)^2}\right] \quad (2)$$

The corresponding complex refractive index, $\bar{n} = (n - ik)$, components are obtained from

$$\epsilon_1 = n^2 - k^2$$

$$\epsilon_2 = 2nk$$

To correct this result for an oxide covering of fixed thickness and refractive index, new pseudo-substrate constants may be obtained from [*27*]

$$\epsilon_s' = \epsilon_s + \left[\frac{4\pi d i}{\lambda}\right]\cdot\left[\frac{\epsilon_s(\epsilon_s - \epsilon_o)(\epsilon_o - \epsilon_a)}{\epsilon_o(\epsilon_s - \epsilon_o)}\right]\cdot\left[\frac{\epsilon_s}{\epsilon_a} - \sin^2\phi\right]^{1/2} \quad (3)$$

using numerical iteration to obtain n_s, k_s given n_s', k_s'. The measured dielectric constant is ϵ_s', the corrected constant is ϵ_s, the overlayer has constant ϵ_o and thickness d, ambient is ϵ_a, and λ is the measurement wavelength. The resulting k_s is in good agreement with the substrate constants required to obtain a solution for assumed top film parameters. The resulting n_s is in error for $d > 0.5$ nm, and errors become larger as d increases. For As_{75} implants, the damaged layers were not sufficiently thick to use n_s, k_s as effective substrate constants which would result in calculated Δ, Ψ matching values. An iteration method was employed to obtain effective refractive indexes, assuming a mix of phases similar to the "effective medium theory" of Wood and Ashcroft [*29*].

The reflectivity is obtained from

$$R = [(n_s - 1)^2 + k_s^2]/[(n_s + 1)^2 + k_s^2] \tag{4}$$

R-values using n_s, k_s and n_s', k_s' are within 0.2% of each other for the assumed 1.7-nm covering oxides.

Results and Discussion

Beam Current Distributions

Beam profile measurements showed that beam distributions became more nonuniform with increasing current. For 1-mA and greater beams, the current densities at the wafers varied by a factor of three within the beam area, which produced "hot spots" covering about 5% of the total area. For mechanical wafer scanning through the beam, the "hot spots" should cause no problems if they are spatially stable during an implant, unless the higher than expected localized charging of isolated structures causes oxide damage. A point on a wafer passes through a 6-cm-high beam about 85 times during a scan.

The beam current densities ranged from an average of 5×10^{-6} A/cm^2 for 5×10^{13} cm^{-2} doses to 1.2×10^{-4} A/cm^2 for doses of 1×10^{15} cm^{-2} and greater at 100 keV. The current densities decreased for implant energies less than 40 keV where post-acceleration was not used, and the additional focusing of the post-acceleration gap was suppressed. At a fixed voltage, the beam areas shrink with decreasing current.

Temperature tape indications on implanted wafers showed that no significant heating occurred for implant energies less than 100 keV. For 1×10^{15} cm^{-2} implants at 100 keV, the wafer backside temperatures reached 80°C. For similar implants at 180 keV, the temperatures exceeded 130°C. Surface temperatures, however, may have been high enough to cause significant annealing. Doses greater than 1×10^{15} cm^{-2} were obtained by multiscans of 1×10^{15} cm^{-2} (or less) per scan.

Implanted Oxides

The ability to implant through SiO_2 is not without consequences. The oxide is damaged and some oxygen may be implanted into the silicon (oxygen knock-ahead) [*30–33*]. These damaged oxide areas have greater etch rates than unimplanted regions, and this effect has been used to advantage in tapering oxide steps [*34*].

Figure 1 shows the etch ratio of implanted to unimplanted SiO_2 for doses

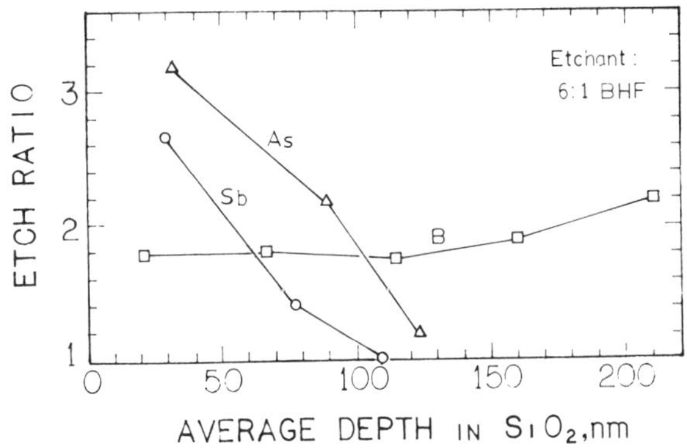

FIG. 1—*Ratio of implanted to unimplanted SiO_2 etch rates versus depth for 1×10^{15} cm^{-2} As_{75} and Sb_{121} at 35 keV and 6.5×10^{14} B_{11} at 180 keV.*

of 1×10^{15} cm^{-2} of As_{75}^{+} and Sb_{121}^{+} at 35 keV and for 6.5×10^{14} B_{11}^{+} cm^{-2} at 180 keV. The etchant was 6 to 1 buffered HF at 23°C. The unimplanted SiO_2 etch rate was approximately 100 nm/min and resulted in poor depth resolution of the faster etching damaged oxides. The damaged regions extended to approximately 6.5 Rp for antimony and 8 Rp for arsenic at 35 keV. The boron damage extended into the underlying silicon as determined by ellipsometer measurements after complete oxide removal. These damage depth determinations are essential for monitoring implant results in oxides, since the oxides must be thick enough to prevent damage from reaching the underlying silicon. If the silicon is damaged, the substrate optical constants are no longer valid for use in calculating the oxide parameters.

These damage distributions are markedly different from those estimated for silicon, where the damage penetrates to approximately 2 Rp [*22*]. The extended damage in SiO_2 is owing, possibly, to recoiling oxygen atoms. Hirao et al [*33*] have estimated that recoils may penetrate the SiO_2 about 2 Rp past the projected range for their 315 keV As implants. Their work [*35*] with nitrogen recoils caused by phosphorous implantations through Si_3N_4 showed increasing recoil penetration with decreasing Rp in silicon as the incident ion energy decreased. Moline et al [*36*] indicated that recoil oxygen penetration could be significantly deeper than the incident ion, and that the ratio of recoils to incident ions increases with decreasing incident energy. This is consistent with results for 1×10^{15} As_{75} cm^{-2} at 50 keV, where the etch-defined damage extended only to approximately 5 Rp.

To obtain better depth resolution, the etchant was changed to 40 to 1 unbuffered HF; this resulted in 1.2 to 2 nm removals of unimplanted SiO_2 during each etch cycle. Figure 2 shows the etch ratios of 370-nm-thick thermal oxides implanted with 4×10^{15} As cm^{-2} at 40, 80, and 174 keV. The data points represent average etch ratios in a cycle at the midpoint of the removed layer. The damage peaks appear to be deeper than corresponding

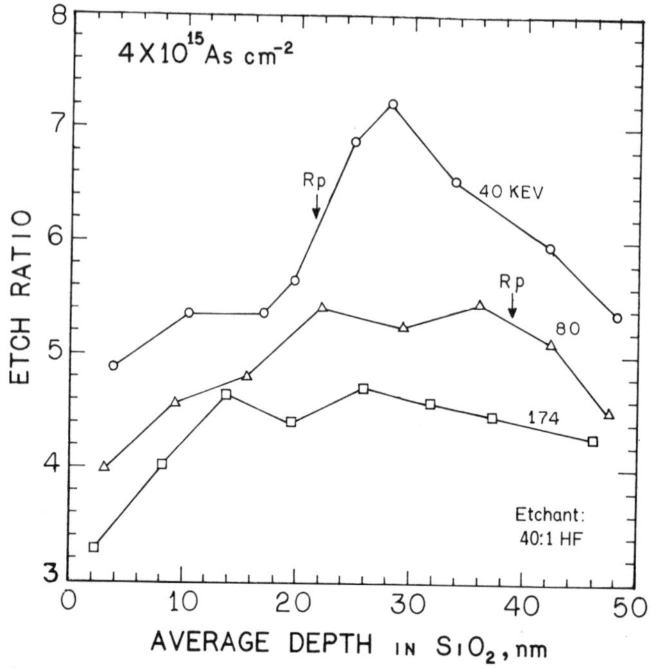

FIG. 2—*Ratio of implanted to unimplanted SiO_2 etch rates versus depth for 4×10^{15} As cm^{-2} at 40, 80, and 174 keV.*

damage in silicon. This may be the result of more beam annealing of oxide surfaces than of silicon, because of the poorer thermal conductivity of the former.

Ellipsometric measurements of the damaged oxides showed some sensitivity of Δ and Ψ to implant parameters, when the damage was confined to the oxide. The effective film refractive indexes were changed slightly by the implants. The index changed from 1.464 unimplanted to 1.47 for 1×10^{15} As cm^{-2} at 35 keV. For 4×10^{15} cm^{-2} implants, the index was 1.49 at 40 keV and decreased at higher energies to 1.469 at 174 keV. The latter implant is believed to have produced oxide damage to the silicon interface.

Accelerating Voltage Variations

For ellipsometer measurements of implanted wafers with approximately 1.7 nm of SiO_2, the parameters Δ and Ψ were very consistent from wafer to wafer in a run. The standard deviation of Δ was usually less than 0.1% of the average and for Ψ was less than 0.5%. However, ellipsometric parameters obtained from supposedly identical implants were not consistent with sheet resistance results from run to run. Subsequent anneals and sheet resistance measurements showed run to run standard deviations in dose of only $\pm 1\%$.

The Lintott implanters are post-acceleration machines with a variable 10 to 40 keV extraction supply referenced to the high-voltage terminal. For low-voltage implants, where extraction only is required, it is necessary either to apply a small amount of post-acceleration voltage to prevent the terminal from drifting negative with respect to ground or to ground the terminal. The procedures give equivalent results.

Figure 3 shows Δ versus Ψ for 5×10^{14} and 1×10^{15} As cm^{-2} in the energy range from 30 to 40 keV. (Not all data are shown due to considerable overlap.) The nonreproducibility of Δ,Ψ results for "identical" implants appears to have been caused by variable post-acceleration voltage settings. When the terminal was grounded to assure 35-keV implants, the scatter in Δ,Ψ measurements decreased significantly, as shown in the figure. The remaining scatter may be due to voltage settability to only ±50 V, variations in nascent oxide thicknesses, implant dose nonuniformities, and voltage variations during implants. (Monitoring of voltages during implants has shown variations of up to 700 V, independent of the starting voltage.)

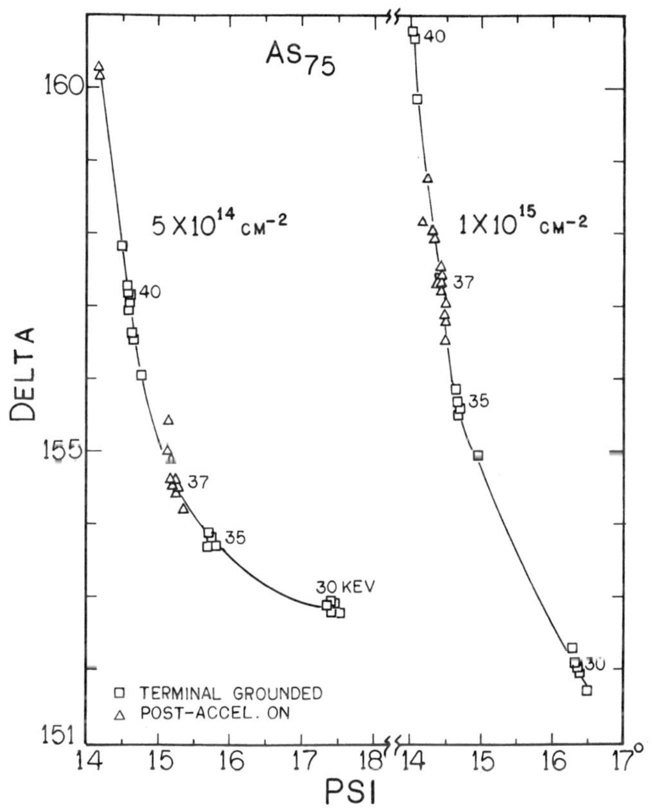

FIG. 3—*Ellipsometric parameter measurements of 5×10^{14} and 1×10^{15} As cm^{-2} implants at energies from 30 to 40 keV, showing scatter due to imprecise control of the extraction energies.*

Table 1 shows the results for consecutive implants at 40 keV with the terminal grounded. The 5×10^{14} cm^{-2} results, plotted in Fig. 3, show that slight parameter variations are caused by small average voltage differences during each run. The data for consecutive implants of 7.5×10^{13} As cm^{-2} show that the variations in Runs 5 to 7 are consistent with voltage drift from plots similar to Fig. 3. The result of Run 8 does not fall directly on such a curve (not shown in detail). Runs 1 to 7 were implants on 100-mm wafers from the same wafer lot with the same oxide thicknesses. Run 8 was on 76.2-mm (3-in.) wafers of similar orientation and diameter. These slight parameter variations are the result of oxide thickness differences; around 1.7 nm the parameter Δ decreases while Ψ increases for slight increases in film thickness. Also shown in Table 1 are reflectivities calculated according to Eq 4 and measured by optical reflectance.

Generally, the apparent differences in parameters measured on 76.2 mm (3 in.) and 100 nm wafers in the same implant run can be explained by oxide thickness variations for implants using 2.5 mA and less up to 100 keV. At higher beam currents and higher energies there were significant variations caused by wafer size, these variations will be reported elsewhere. The measurements for doses greater than 5×10^{14} cm^{-2} at energies greater than 100 keV in this paper were taken on 100-mm wafers.

At implant energies requiring post-acceleration voltage (above 40 keV), a similar Δ,Ψ inconsistency problem occurs, because the total voltage readout for the post-acceleration supply is to only ± 0.5 keV. In addition, the stability of the supply voltage may be inadequate to maintain ± 0.5 keV under loading, and implant operators occasionally set the voltage slightly higher than required to compensate for some voltage sag which occurs. This is shown in Fig. 4 for 5×10^{14} and 1×10^{15} As_{75}^+ implants at approximately 70 keV performed over several months. The true 70-keV points were established by

TABLE 1—*Ellipsometric parameters and reflectances for consecutive implants of* 5×10^{14} *and* 7.5×10^{13} *As cm*$^{-2}$ *at 40 keV with the high-voltage terminal grounded. Delta and psi values are average plus-or-minus one standard deviation for four wafers per run and ten measurements per wafer.*

			Relative Reflectance at 632.8 nm	
Lot	Delta	psi	Calculated	Measured
5×10^{14} #1	157.096 ± 0.185	14.614 ± 0.019	1.142	1.125
#2	157.204 ± 0.128	14.596 ± 0.023	1.142	1.124
#3	157.168 ± 0.192	14.600 ± 0.023	1.142	1.123
#4	157.356 ± 0.150	14.584 ± 0.008	1.142	1.126
7.5×10^{13} #5	156.172 ± 0.088	18.028 ± 0.034	1.283	1.276
#6	155.907 ± 0.220	17.924 ± 0.067	1.278	1.269
#7	155.880 ± 0.169	17.882 ± 0.055	1.276	1.268
#8	155.920 ± 0.098	17.898 ± 0.029	1.277	1.269

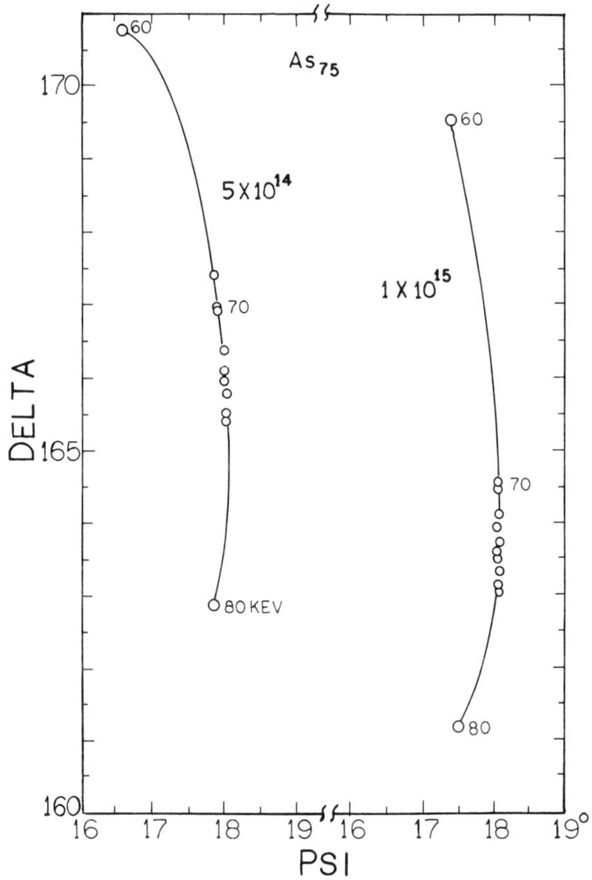

FIG. 4—*Ellipsometric parameter measurements of 5×10^{14} and 1×10^{15} As cm^{-2} implants at energies from 60 to 80 keV, showing scatter due to imprecise control of the post-acceleration energies.*

monitoring the implants and maintaining 70 keV. Some of the scatter is owing to slight differences in nascent oxide thicknesses.

Ellipsometric Parameters versus Dose and Energy

Ellipsometric measurements were made on wafers implanted with doses of 2×10^{13} to 8×10^{15} As cm^{-2} at energies from 30 to 180 keV in order to establish the sensitivity of parameters Δ and Ψ to variations in implant parameters. For nonabsorbing films on silicon, the Δ versus Ψ plots form continuous curves for each film refractive index. For absorbing films, such as damaged silicon layers, the plots form nonrepeating spirals for each complex film index. For a fixed complex index, there are unique spiral plots for different nascent oxide overlayers. The measurements of implanted wafers show that a Δ,Ψ data point is repeatable for identical implants but is not unique to that implant. The following data are for Lintott III implants.

Figure 5 shows the results of Δ versus Ψ for 5×10^{14} As cm^{-2} at implant energies from 30 to 180 kV. The data points in the figures have been connected by "smooth" curves in order to show trends. The dashed lines point toward lower doses at the indicated energies. This curve is indicative of an absorbing film with each increasing energy related to increasing film thickness. A match of this curve to a theoretical Δ versus Ψ plot has not been successful, since the effective refractive indexes of the damaged layers differ at each energy.

Figure 6 illustrates similar results for 1×10^{15} As cm^{-2} for energies from 30 to 180 keV. The beam currents were 2.5 mA for each implant shown, but the beam densities varied with voltage to a greater extent than for lower doses. For 100-mm-diameter wafers, there were no significant differences in Δ,Ψ results for implants using 0.83, 1.25, or 2.5 mA (3 scans, 2 scans, or 1

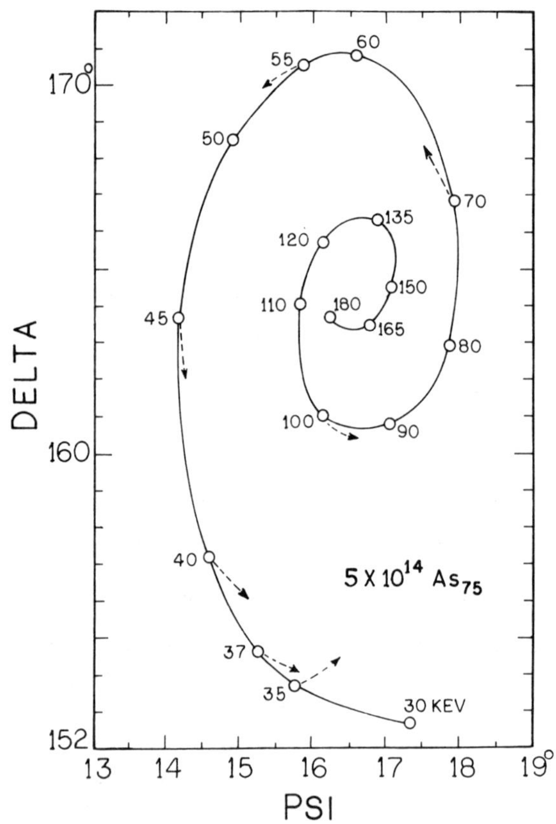

FIG. 5—*Ellipsometric parameter measurements of 5×10^{14} As cm^{-2} implants at energies from 30 to 180 keV. Dashed lines point towards lower doses at indicated voltages.*

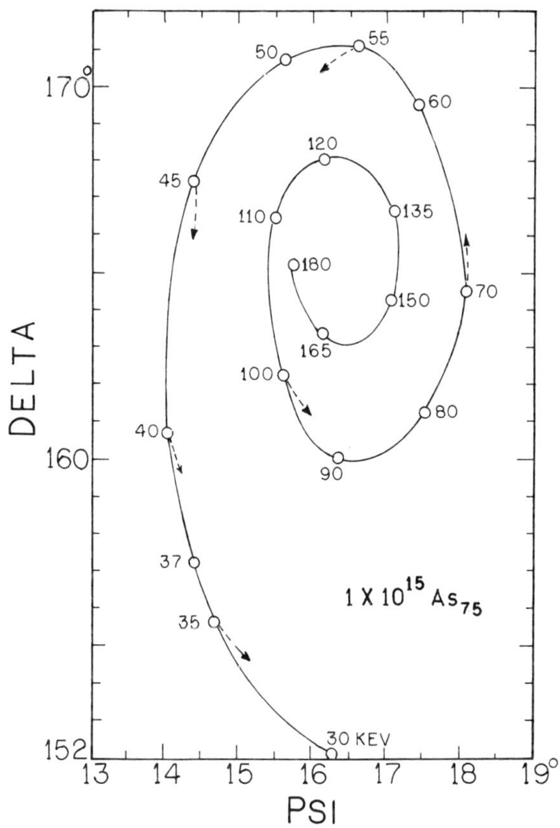

FIG. 6—*Ellipsometric parameter measurements of 1 × 10^{15} As cm^{-2} implants at energies from 30 to 180 keV. Dashed lines point towards lower doses at indicated voltages.*

scan, respectively). Again, the dashed lines point in the directions of lower doses.

The results for 7.5 × 10^{13} As cm^{-2} are of the expected spiral form from 30 to 100 keV (Fig. 7). Reflectance measurements made with a Nanospec AFT show that interference spectra for 80 to 130 keV implants cross-over near 632.8 nm, thereby causing the tight grouping of corresponding Δ,Ψ values shown in the figure.

Figure 8 shows Δ versus Ψ results for 37-keV implants of doses from 2 × 10^{13} to 8 × 10^{15} As cm^{-2}. For these implants, the damage is confined to a surface layer of less than 50 nm depth [*22*]. Previous plots indicated increasing layer thicknesses around the spirals. This plot may be explained by assuming each damaged layer is of approximately the same thickness with varying effective optical constants.

Figures 9, 10, and 11 show Δ versus Ψ results for 70, 100, and 180 keV implants, respectively. The dashed lines point in the direction of parameter

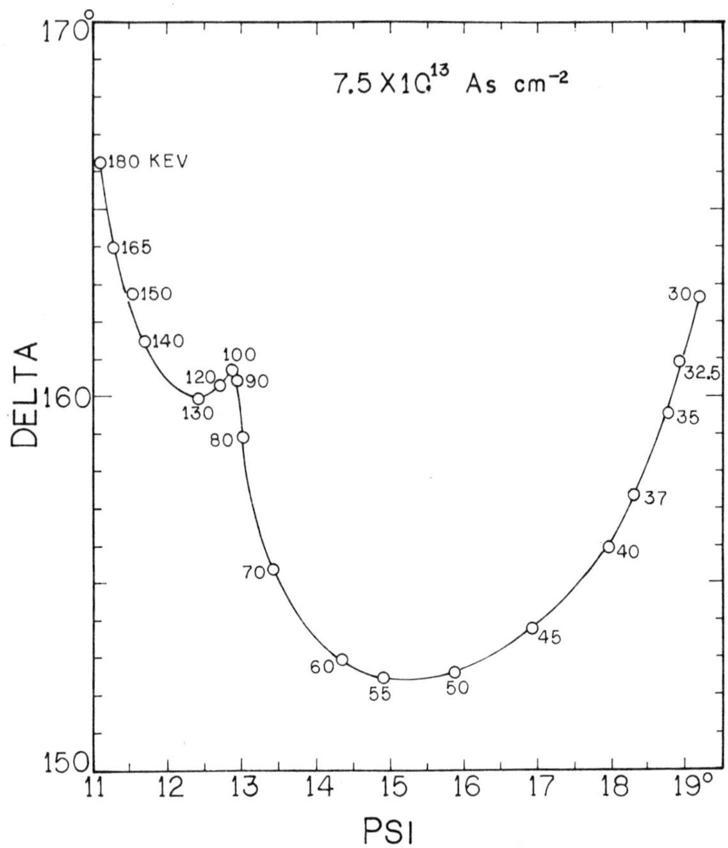

FIG. 7—*Ellipsometric parameter measurements of 7.5 × 10^{13} As cm^{-2} implants at energies from 30 to 180 keV.*

changes for identical doses at lower voltages. Spectral reflectance curves, described later, may explain the change in curvature of the 100-keV results between 1.75 and 2 × 10^{14} cm^{-2}. For 180-keV implants the standard deviations in Ψ-parameter measurements exceeded 1% at doses greater than 9 × 10^{14} cm^{-2}. Similar implants at 165 keV had deviations less than 0.4%. Variability in wafer heat sinking on different pallets may be responsible for these results at 180 keV, since wafer standard deviations were less than 0.5%. Although the results presented are machine dependent, similar results should be obtained with other implanters.

Ellipsometer measurements of wafers implanted in the Extrion 200-1000 can be correlated with those from the Lintott III. The effective doses, for the same settings, differed by about 5% between the two machines. Measurements of wafers implanted by the Lintott IIIA fall on the preceding curves if accelerating voltage differences are assumed. This latter machine has silver-

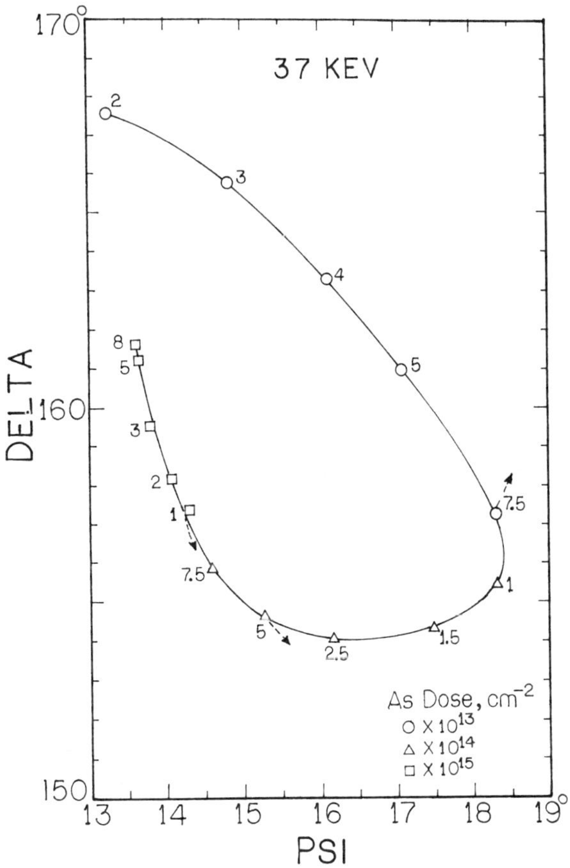

FIG. 8—*Ellipsometric parameter measurements of 37-keV implants for doses from 2 × 10^{13} to 8 × 10^{15} As cm^{-2}. Dashed lines point in direction of lower energies at indicated doses.*

impregnated silicone disks between wafers and pallets to improve heat sinking. Once the III and IIIA voltage settings are cross-calibrated, we may be able to determine the effect of improved heat-sinking on ellipsometric parameters (and residual damage).

Optical Reflectance

For an alternative method of estimating damaged layer optical constants, spectral reflectance measurements of the implanted wafers were performed from 480 to 780 nm. The method also was investigated in order to determine if the reflectance curves were unique to the implant parameters, and if these results were consistent with ellipsometer measurements.

Figure 12 shows experimental results for 7.5 × 10^{13} As cm^{-2} implants normalized to reflectances from unimplanted wafers. For clarity, only seven

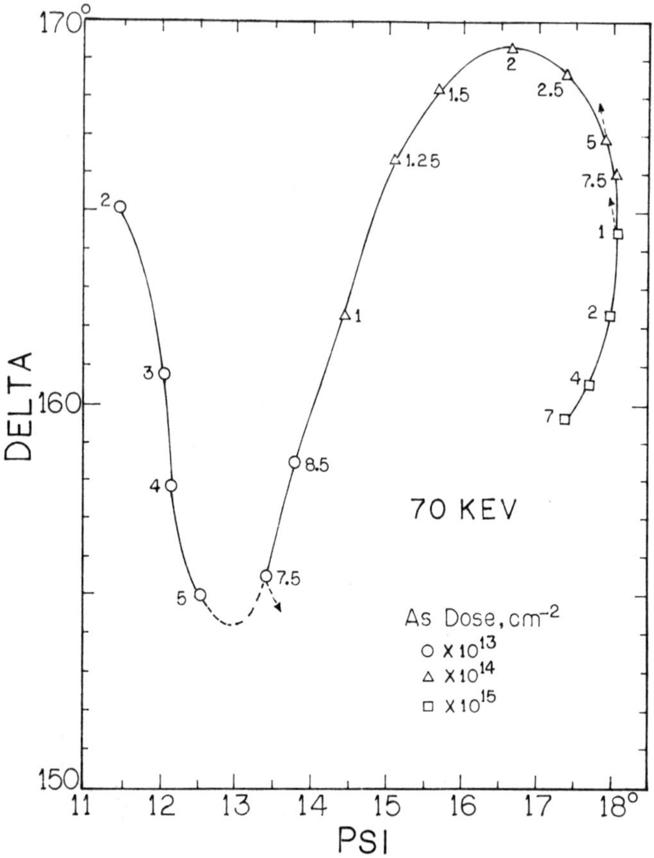

FIG. 9—*Ellipsometric parameter measurements of 70-keV implants for doses from* 2×10^{13} *to* 7×10^{15} *As* cm^{-2}*. Dashed lines point in direction of lower energies at indicated doses.*

of the curves are presented, but each implant from 30 to 180 keV had a unique spectrum. The data points were obtained at discrete wavelengths for subsequent calculations. The measured reflectances at 633 nm were within ±5% of the corresponding values calculated from Eq 4 using n_s', k_s' values obtained from Eqs 1 and 2. The curves show wavelengths of reflectance maxima increasing with increasing implant energy; this is indicative of thicker damaged layers. The small difference in reflectances at 632.8 nm for 80 to 130 keV implants is similar to the tight grouping of ellipsometric parameters for these samples.

Figures 13 and 14 illustrate reflectance results for doses of 5×10^{14} and 1×10^{15} As cm^{-2}, respectively. These data are also consistent with reflectances calculated according to Eq 4 and again show increasing film thicknesses with implant energy.

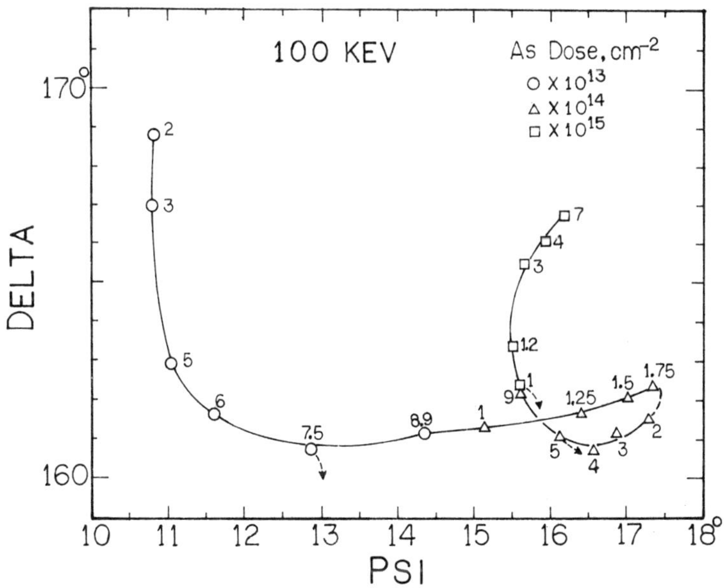

FIG. 10—*Ellipsometric parameter measurements of 100-keV implants for doses from 2 × 10^{13} to 7 × 10^{15} As cm^{-2}. Dashed lines point in direction of lower energies at indicated doses.*

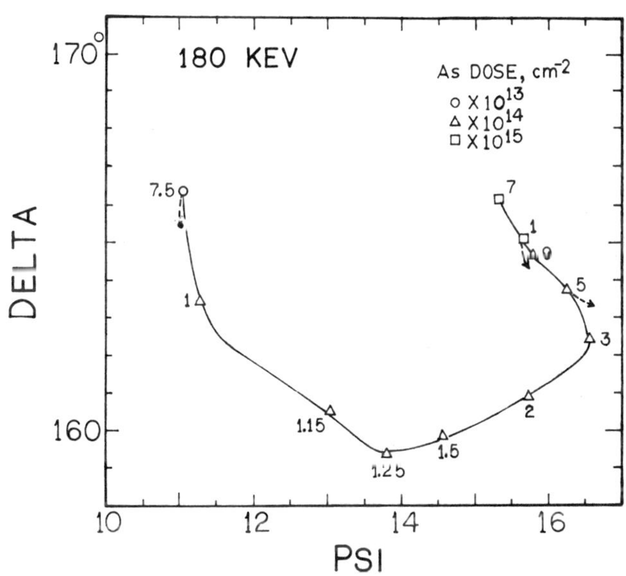

FIG. 11—*Ellipsometric parameter measurements of 180-keV implants for doses from 7.5 × 10^{13} to 7 × 10^{15}. Dashed lines point in direction of lower energies at indicated doses.*

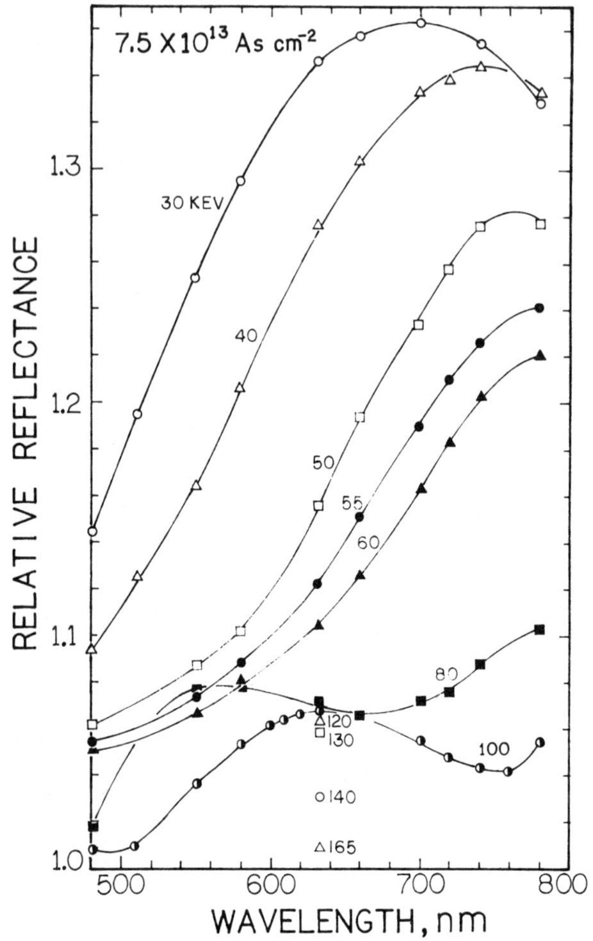

FIG. 12—*Reflectance measurements of 7.5 × 10^{13} As cm^{-2} implants relative to unimplanted silicon.*

Figure 15 shows the relative reflectances of 100-keV implants of doses from 2 × 10^{13} to 4 × 10^{14} As cm^{-2}. The 632.8-nm reflectances increase with dose to 1.75 × 10^{14} and then decrease. This reversal correlates with the abrupt change in curvature (between 1.75 and 2 × 10^{14} cm^{-2}) in the corresponding Δ versus Ψ plot of Fig. 11. These reflectance plots indicate that a wavelength of around 740 nm would be preferable for ellipsometer measurements of these implants in order to expand the Δ versus Ψ plots so as to minimize the tight groupings such as occurred in Figs. 7 and 10.

The small reflectances for low doses limit this method of monitoring implant results, just as Δ,Ψ values approaching those of unimplanted silicon limit the ellipsometric method. Although low dose limits of detectability by

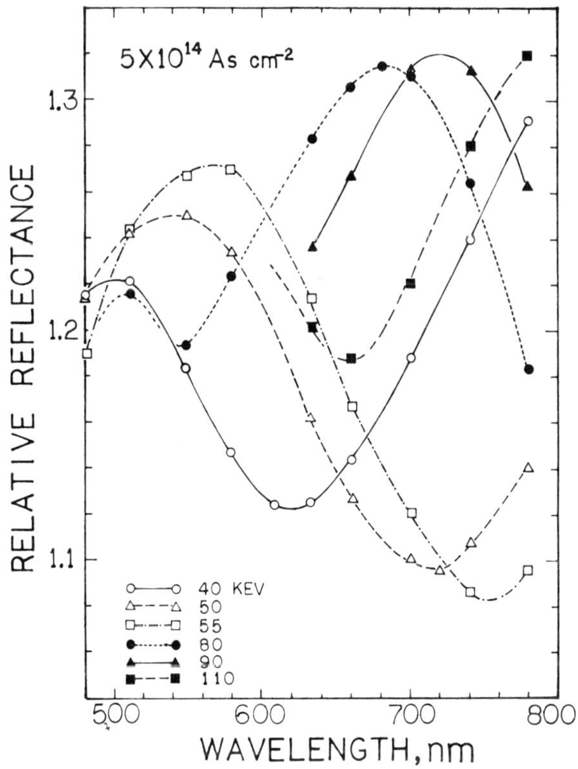

FIG. 13—*Reflectance measurements of 5×10^{14} As cm^{-2} implants relative to unimplanted silicon.*

ellipsometry and reflectance have not been determined to date, the preceding data show that the lower limits decrease with decreasing implant energies.

Modeling

Simple models were explored in order to explain the plots. Figure 16 shows calculated thicknesses and effective dielectric constants for the damaged layers at 37 keV. The solutions were obtained by assuming homogeneous damaged layers with dielectric constants obtained by a mix of those of crystalline silicon (c-Si) and amorphous silicon (a-Si) and iterating the mix until calculated Δ,Ψ values matched experimental results (method I) [*29*]. At 632.8 nm, the refractive index of c-Si is approximately 3.858-*i*0.018, while various values for a-Si have been published [*37–42*].

Since we have only two known parameters (Δ and Ψ) and three unknowns (layer thickness and complex dielectric constant components), there are no unique solutions. The calculation procedure eliminated solutions outside of the range of thicknesses from Rp + ΔRp to 2 Rp. The results, denoted I in Fig. 16, were obtained by using the approximate a-Si constant of Ref *41* (4.36-

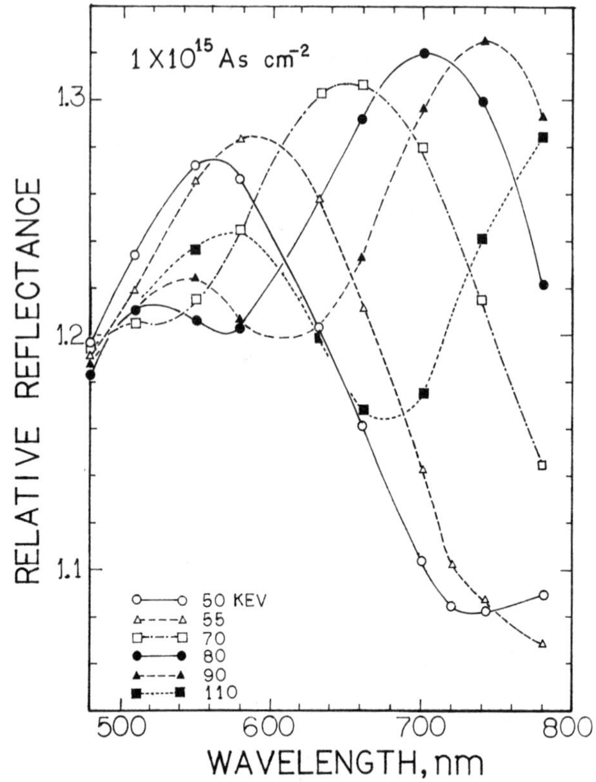

FIG. 14—*Reflectance measurements of* 1×10^{15} *As* cm^{-2} *implants relative to unimplanted silicon.*

*i*0.86). The next group of solutions outside the foregoing thickness range was obtained for layers several tens of nanometres thicker, where there was no correlation with dose (and no solutions for several of the implants). The corresponding effective dielectric constants are denoted by the percentage of the amorphous constant used in the mix. Using the a-Si constant of Philip [*40*], which has a smaller imaginary component, the layer thicknesses ranged from 40 to 50 nm, but some solutions required effective constants greater than that of a-Si.

The damaged layers were also investigated by a second procedure, which approximated the actual nonhomogeneous distribution by ten or more piecewise continuous homogeneous sublayers. The optical constants of the sublayers were varied according to the arsenic damage distribution of Brice [*22*]. The sublayer constants were then entered in a multilayer program [*26*] in order to calculate Δ and Ψ. For this distribution of optical constants, with one sublayer completely amorphous and damage depth of 2 Rp, the Δ and Ψ results for 37 keV were approximately equal to experimental values obtained

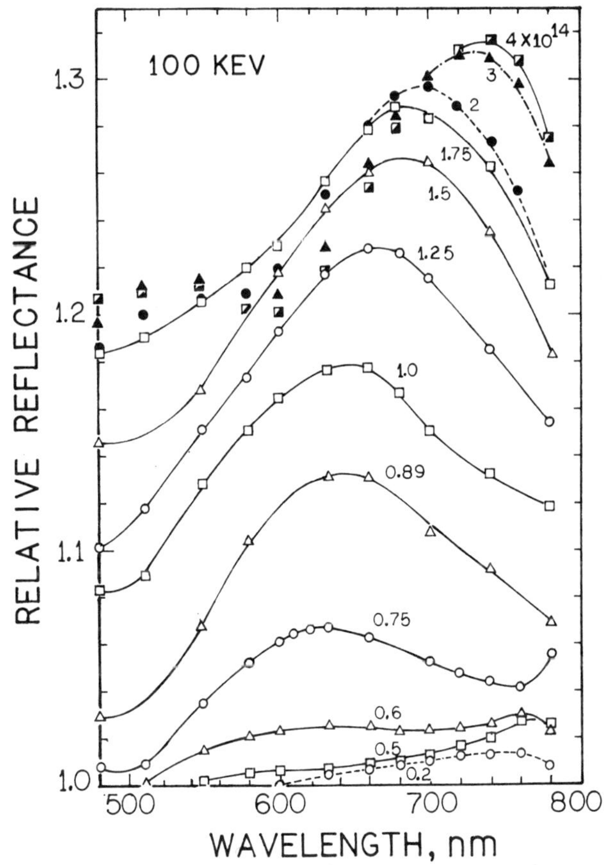

FIG. 15—*Reflectance measurements of 100-keV implants for doses from* 2×10^{13} *to* 4×10^{14} *As* cm^{-2} *relative to unimplanted silicon.*

from a dose of 2×10^{14} cm^{-2}. This Δ, Ψ result was employed in Method I and showed that this nonhomogeneous distribution corresponded to a homogeneous layer 38.8 nm thick with the optical constants of an 80% amorphous mix (Brice-homogeneous equivalent).

To obtain Δ, Ψ values from this second approximation, which matched experimental values for other doses, the sublayer constants were normalized to the maximum constant, and the dielectric constant corresponding to the maximum damage was varied. Additionally, the distribution depth was varied by varying the sublayer thicknesses. This procedure produced the curves denoted II in Fig. 16. These solutions show that the damage depth increases with dose to approximately 5×10^{14} cm^{-2}, and that above 1×10^{14} cm^{-2} there appears to be some beam annealing of the damage. To maintain peak damage not exceeding a-Si, the distribution was modified by expanding the

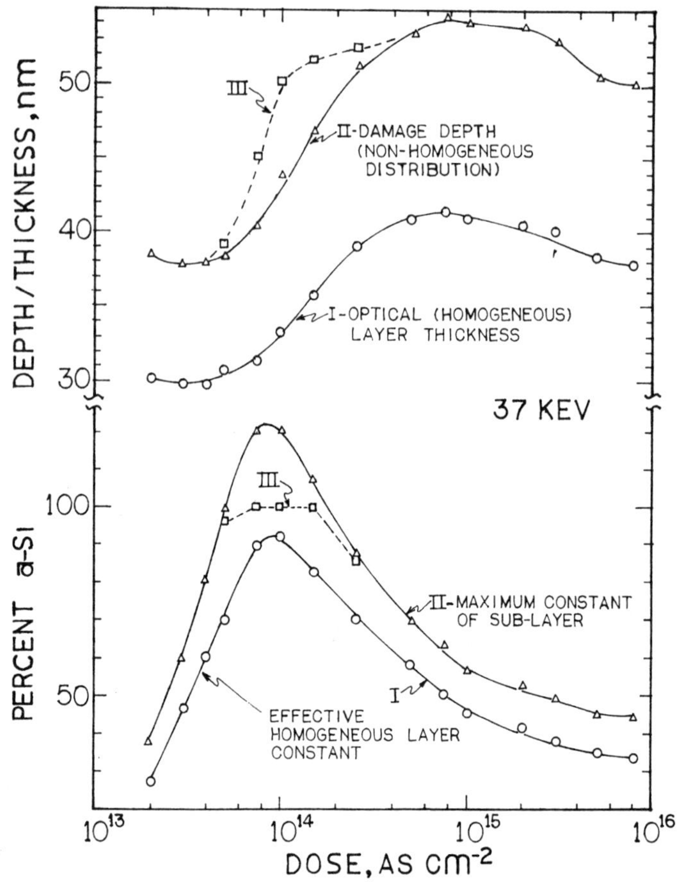

FIG. 16—*Calculated layer thicknesses and effective optical constants for 37-keV As implants.* (I) *Match of experimental data to calculated homogeneous damaged layer.* (II) *Piecewise continuous multilayer solution matching experimental results.* (III) *Same as II, but restricting sublayers with greatest residual damage in order to have optical constants of amorphous silicon (a-Si).*

sublayer thickness; the 100% a-Si constants produced the curves denoted III in Fig. 16.

This model (III) was applied to the Δ versus Ψ curves at other implant energies and showed that experimental Δ,Ψ-values along relatively linear portions of the curves could be matched by similar distributions with varying sublayer peak damage. At doses increasing from 2×10^{13} cm^{-2}, the damage increases with dose as expected. Along the highest dose portions of the curves, however, the effective peak damage (defined by the highest percent a-Si constant used for a sublayer) decreases with increasing dose, owing, perhaps, to beam annealing. Changes in curvature required distribution changes (that is, expansion or contraction of maximum damage sublayers).

For the 70, 100 and 180 keV implant results shown in Figs. 9, 10, and 11, Model III showed that the effective thicknesses of corresponding homogeneous damaged layers (Brice-homogeneous equivalent) are 70, 84, and 140 nm, respectively. These latter values are approximately equal to 1.5 Rp.

Conclusions

For implanters which mechanically scan wafers through the ion beam, it is possible to develop curves relating ellipsometric parameters to implant parameters. This must be done for each machine, since voltages, heat-sinking, and dose may differ sufficiently between implanters to result in different Δ versus Ψ plots before annealing, although sheet resistance measurements may be similar. For reproducible ellipsometric results, all monitor wafers should have similar nascent oxide thickness, and implanter energies must be set and controlled precisely. The sensitivity of implant parameters to slight variations in ion energy is to be expected. Since the effective optical thickness of a damaged layer is approximately equal to 1.5 Rp, a 1 keV change in energy corresponds to a 0.7 nm change in layer thickness. The measuring instruments are used, routinely, to detect such changes in other types of films. A Δ,Ψ value not coincident with such curves does not necessarily indicate that implant parameters were in error; Δ is very sensitive to slight variations in oxide thickness and surface cleanliness. A Ψ-variation, however, would indicate that further checks are required.

The unique spectra, associated with different implant parameters, obtained from optical reflectance suggest that this measurement method may be also used for implant process monitoring. However, further work is required to produce a "number" indicative of the implant from such scans in a manner similar to that obtained for transparent films on silicon. The simple models explored to explain Δ versus Ψ plots may be useful in developing algorithms for determining multi-sublayer reflectance spectra. For the instruments used in this study, ellipsometric results are more sensitive than those of reflectance to changes in dose at the 10^{13} cm^{-2} level. The ellipsometric and reflectance methods are applicable to other species implanted in silicon, such as phosphorus, antimony, and argon, which produce residual damaged layers.

Work is continuing on developing other models for the implanter layers incorporating the reflectance data. The distribution of optical constants similar to silicon damage distributions gives reasonable explanations for changes in curvature of Δ versus Ψ plots. The results of any such model, however, are dependent on the amorphous silicon dielectric constant used. The literature values are strongly dependent on sample preparation and history [*43*], and implanted layers may have defect structures producing different optical constants than those of thin film samples.

Acknowledgments

The author expresses his sincere appreciation to M. Knight for numerous implants and measurements and to J. Rostvet for implanter calibration and maintenance.

References

[*1*] Adams, J. R. and Bashara, N. M., *Surface Science,* Vol. 49, 1975, p. 441.
[*2*] Kurcirkova, A., *Radiation Effects,* Vol. 28, 1976, p. 129.
[*3*] Marton, J. P. in *Optical Polarimetry,* Society of Photo-Optical Instrumentation Engineers, Vol. 112, 1977, p. 82.
[*4*] Mardezhov, A. S. and Seryapin, V., *Society Physics–Semiconductors,* Vol. 13, 1979, p. 1374.
[*5*] Watanabe, K., Miyao, M., Takemoto, I., and Hashimoto, *Applied Physics Letters,* Vol. 34, 1979, p. 518.
[*6*] Nakamura, N., Gotoh, T., and Kamoshida, M., *Journal of Applied Physics,* Vol. 50, 1979, p. 3985.
[*7*] Nakamura, N. and Kamoshida, M., *Radiation Effects,* Vol. 42, 1979, p. 29.
[*8*] Watanabe, K., Motooka, T., Hashimoto, N., and Tokuyama, T., *Applied Physics Letters,* Vol. 36, 1980, p. 451.
[*9*] Motooka, T. and Watanabe, K., *Journal of Applied Physics,* Vol. 51, 1980, p. 4125.
[*10*] Delfino, M. and Razouk, R. R., *Journal of Applied Physics,* Vol. 52, 1981, p. 386.
[*11*] Kurtin, S., Shifrin, G. A., and McGill, T. C., *Applied Physics Letters,* Vol. 14, 1969, p. 223.
[*12*] McGill, T. C., Kurtin, S., and Shifrin, G. A., *Journal of Applied Physics,* Vol. 41, 1970, p. 246.
[*13*] Miyao, M., Miyazaki, T., and Tokuyama, T., *Japanese Journal of Applied Physics,* Vol. 17, 1978, p. 955.
[*14*] Miyao, M., Yoshihiro, N., Tokuyama, T., and Mitsuishi, T., *Journal of Applied Physics,* Vol. 49, 1978, p. 2573.
[*15*] Wensch, W. and Gotz, G., *Radiation Effects,* Vol. 49, 1980, p. 137.
[*16*] Hubler, G. K., Waddell, C. N., Spitzer, W. G., Fredrickson, J. E., Prussin, S., and Wilson, R. G., *Journal of Applied Physics,* Vol. 50, 1979, p. 3294.
[*17*] Hubler, G. K., Malmberg, P. R., and Smith, T. P., III, *Journal of Applied Physics,* Vol. 50, 1979, p. 7147.
[*18*] Johnson, W. S. and Gibbons, J. F., "Projected Range Statistics in Semiconductors," Stanford University, Stanford, Calif., 1970.
[*19*] Crowder, B. L. and Title, R. S., *Radiation Effects,* Vol. 6, 1970, p. 63.
[*20*] Brice, D. K., *Radiation Effects,* Vol. 6, 1970, p. 77.
[*21*] Mayer, J. W., *Radiation Effects,* Vol. 8, 1971, p. 269.
[*22*] Brice, D. K., *Journal of Applied Physics,* Vol. 46, 1975, p. 3385.
[*23*] McCrackin, F. L. and Colson, J. P. in *Ellipsometry in Measurement of Surfaces and Thin Films,* E. Passaglia, R. Stromberg, and J. Krueger, Eds., National Bureau of Standards Miscellaneous Publication 256, 1961, p. 61.
[*24*] Taft, E. and Cordes, L., *Journal of the Electrochemical Society,* Vol. 126, 1979, p. 131.
[*25*] Raider, S. I. and Flitsch, R., *Journal of Vacuum Science and Technology,* Vol. 13, 1976, p. 58.
[*26*] McCrackin, F. L., National Bureau of Standards Technical Note 479, 1979.
[*27*] Azzam, R. and Bashara, N. M., *Ellipsometry and Polarized Light,* North-Holland, New York, 1977.
[*28*] Rzhanov, A. V. and Svitashev, K. K. in *Advances in Electronics and Electron Physics,* Vol. 49, Academic Press, New York, 1979, p. 1.
[*29*] Wood, D. W. and Ashcroft, N. W., *Philosophical Magazine,* Vol. 35, 1977, p. 269.
[*30*] Chu, W. K., Muller, H., Mayer, J. W., and Sigmon, T. W., *Applied Physics Letters,* Vol. 25, 1974, p. 297.

[31] Moline, R. A. and Cullis, A. G., *Applied Physics Letters,* Vol. 26, 1975, p. 551.
[32] Goetzberger, A., Bartelink, D. J., McVittie, J. P., and Gibbons, J. F., *Applied Physics Letters,* Vol. 29, 1976, p. 259.
[33] Hirao, T., Inouc, K., Yaegashi, Y., and Takuyangi, S., *Japanese Journal of Applied Physics,* Vol. 18, 1979, p. 647.
[34] Moline, R. A., Buckley, R. R., Hasko, S. E., and MacRae, A. U., *Transactions on Electron Devices,* Institute of Electrical and Electronics Engineers, Vol. 20, 1973, p. 840.
[35] Hirao, T., Inoue, K., Takayangi, S., and Yaegashi, Y., *Applied Physics Letters,* Vol. 31, 1977, p. 505.
[36] Moline, R. A., Reutlinger, G. W., and North, J. C. in *Atomic Collisions in Solids,* Vol. 1, S. Dotz, B. R. Appleton, and C. D. Moak, Eds., Plenum, New York, 1975, p. 159.
[37] Bennett, J. M. and Booty, M. J., *Applied Optics,* Vol. 5, 1966, p. 41.
[38] Grigorovici, R. and Vancu, A., *Thin Solid Films,* Vol. 2, 1968, p. 105.
[39] Brodsky, M. H., Title, R. S., Weisner, K., and Petit, G. D., *Physical Review B,* Vol. 1, 1970, p. 2632.
[40] Philip, H. R., *Journal of Physics and Chemistry of Solids,* Vol. 32, 1971, p. 1935.
[41] Beaglehole, D. and Zavetova, M., *Journal of Non-Crystalline Solids,* Vol. 4, 1970, p. 272.
[42] Thutapalli, G. K. M. and Tomlin, S. G., *Journal of Physics C, Solid State Physics,* Vol. 10, 1977, p. 467.
[43] Theye, M.-L. in *Optical Properties of Solids—New Developments,* B. O. Seraphin, Ed., North-Holland, New York, 1976, p. 353.

Appendixes

APPENDIX I

The material presented in this appendix did not go through the review process and is presented for information only.

The Role of Standards in the Electronics Industry

Standards are important because they are an essential part of the socio-economic development of all nations. They open and improve channels of communications. This improved communication aids commerce because it promotes a real understanding of the products of our technology. Standards form the basis for achieving health, safety, and a higher quality of life.

Standards in the United States are generated through a voluntary system. Several hundred organizations are involved in this system, which includes branches of government, professional and technical societies, manufacturing and nonmanufacturing trade associations, public service and consumer groups, testing and inspection bodies, and organizations such as ASTM whose main purpose is the development of standards. Although participation in the system and use of standards is on a voluntary basis, many standards prepared for voluntary use have been made mandatory by governmental bodies. The coordinating body in the United States is the American National Standards Institute (ANSI).

Electronics standards in ASTM are generated through Committee F-1 on Electronics. The purpose of the committee is defined in its scope:

> The development of methods of test, specifications, recommended practices, and nomenclature, the promotion of knowledge, and the stimulation of research relating to electronic devices and the materials, process controls, and specialized test equipment used in their fabrication.

The work of Committee F-1 covers a wide range of activities. Table 1 lists present technical activities; however, these change from time to time to meet the current needs of the industry.

To fulfill the commitment defined in scope, two important criteria—knowledge of industry needs and availability of technical people in the field—must be met. Because ASTM is a voluntary consensus organization, these two criteria require the support of technical volunteers knowledgeable in the field. With the knowledge of industry needs and the dedication of technical volunteers, ASTM can help make the electronics industry in the United States, a strong and dynamic part of world commerce.

Kenneth E. Benson

Bell Laboratories, Allentown, Pennsylvania; and Vice Chairman, ASTM Committee F-1 on Electronics

TABLE 1—*Technical activities of ASTM Committee F-1 on Electronics.*

Subcommittee	Descriptive Title
1.02	Lasers, Light-Emitting Diodes, and Other Optically Related Materials and Devices
1.03	Metallic Materials: Lead Frames, Glass-to-Metal Seals, Wire for Electron Devices, Strip
1.04	Semiconductor Materials and Physical Properties, including Crystalline Perfection
1.05	Encapsulants
1.06	Electrical and Optical Measurements Used in the Preparation and Processing of Semiconductor Materials
1.07	Interconnection/Bonding (Wire and Tape Carrier)
1.08	Microelectronic Imaging: Materials, Processes, and Metrology Related to Photolithography
1.09	Enclosures, Substrates, and Films: Materials and Processes Related to Hybrid Technology
1.10	Processing Environments
1.11	Quality and Hardness Assurance
1.12	Microwave Passive Components
1.14	Gadolinium/Gallium/Garnet (GGG) Materials for Bubble Memories
1.15	Gallium Arsenide

NBS Work for the Semiconductor Industry

ASTM Committee F-1 on Electronics recently celebrated its 25th anniversary; for nearly all that time it has been developing standard measurement methods needed by the semiconductor industry. The National Bureau of Standards (NBS) has been working on measurement science and technology in the same field for almost as long a period. It is no coincidence that both groups have been active for the same length of time. The need for better measurements was apparent to both groups from the beginning, and that same need still exists. Today most of the variables to be measured are different, the instrumentation is much more sophisticated, and the required sensitivity and precision of the measurements are greater. The challenges are immense.

The Semiconductor Technology Program at NBS includes the work of two divisions, one emphasizing metrology for materials and processes and the other concerned with metrology for semiconductor devices. In all, there are 36 professionals in the program. Our work always has measurement as a goal, since that is the principal mission of NBS. Indeed, we focus on measurements needed for commercial transactions, although many such measurements are useful for process control within a company's operations as well. NBS does not develop semiconductor technology, processes, devices, or equipment except as occasional by-products of our metrological research. Those are the province of the industry, and form the products which it sells.

We do not compete with industry but support it.

There are presently six main areas of work in the Semiconductor Technology Program. In *Defect Metrology* we are developing ways of measuring deep levels, such as by deep-level transient spectroscopy and isothermal transient capacitance; carbon and oxygen in silicon; and resistivity, with emphasis on artifacts for calibration of resistivity and spreading resistance instruments. We are also doing research to gain a better understanding of the physics of deep levels and their origins and to apply some of our metrological experience on silicon to compound semiconductors, particularly gallium arsenide.

In *Micrometrology* we are developing the optical physics for measuring linewidths on silicon wafers, with the aim of developing wafer artifacts for transferring this measurement capability to the industry. This effort follows the same pattern as that for the optical measurement of linewidths on masks. The research on the latter is complete.

The *Process Metrology* group operates our processing facility and is developing an advanced-design ellipsometer for research on the measurement of thin films. This area of work also includes hermeticity measurements.

A group on *Device and Process Modeling* is working to understand the physics for such models and to improve the capabilities of widely-used codes such as SPICE and SUPREM.

Test structures for measuring sheet resistance, linewidth, interlayer alignment, minority carrier lifetime, charge-carrier distribution, and a host of other variables are being developed by the *Integrated Circuit Test Structures* group.

Finally, the *Device Metrology* group, which has worked mainly on second breakdown and thermal impedance measurement problems in the past, is extending its work to deal with thermal measurements on integrated circuit packages.

Many ASTM standards developed by Committee F-1 are based on NBS work. That work has often been guided by the dialogue and exchange in standards meetings, where we can learn directly from the industry about their measurement needs. The interactions between NBS and ASTM F-1 have resulted in the issue of a number of Standard Reference Materials for calibration of resistivity and linewidth measuring equipment, and others are being developed. Most of the ASTM test methods for wire bond, hermeticity, resistivity, and linewidth evaluation are either the result of NBS work or were strongly influenced by it.

There are opportunities for more extensive interactions, however. Guest workers from industry or universities can work on nonproprietary measurement problems at NBS, and under certain circumstances NBS staff can consult with commercial firms on their problems of measurement. Guest workers and research associates can come to NBS for any period of time and under a wide variety of arrangements. To explore such possibilities, please contact the undersigned.

Robert I. Scace
Chief, Semiconductor Materials and Processes Division, National Bureau of Standards, Washington, D.C.; and Chairman, ASTM Committee F-1 on Electronics

NSF Role in Microelectronics

The National Science Foundation (NSF) was established in 1950 to advance scientific progress in the United States by providing support for basic and applied research in the mathematical, physical, medical, biological, and social sciences, and in engineering. In 1981 the role of engineering was enhanced by establishing the Engineering Directorate as the fourth research directorate at NSF. Engineering support in FY 81 was only approximately 10% of the overall research budget, however, indicating that basic science remains the cornerstone of NSF research support.

There is a significant amount of funding support in the Microelectronics/VLSI area in both the Mathematical and Physical Sciences and the Engineering Directorates. There is approximately 19 million dollars of annual grant support relevant to Microelectronics/VLSI.

Most of this support resulted from a natural evolution of research in the related disciplines, but a significant part resulted from an NSF emphasis on microscience and microstructures between 1975 and 1980. Part of this effort consisted of the formation of the National Research and Resource Facility for Submicron Structures (NRRFSS) at Cornell University in 1976. This facility has a charter to perform research in submicron fabrication technology, to provide a resource for the general research community, and to stimulate innovative interdisciplinary research involving submicron technology.

Since NRRFSS was established there has been a rapidly growing university involvement in the Microelectronics/VLSI program. University-based centers in microelectronics and related fields have been established in several parts of the United States with regional industry and state government support to provide a local base of suitably educated personnel and to attract and maintain an industry base.

The growth and development of these centers will have a major impact on university education and research in semiconductor electronics. This research impact will be significantly enhanced if recent plans by the Semiconductor Industry Association and the American Electronics Association for direct support of university research programs are implemented and sustained over a long time period. Industry–University–Government interaction is evolving rapidly, giving the promise of alleviating the shortage of engineers and sustaining our economic and defense position in this critical technology.

Industrial involvement with NSF takes many forms. Established methods of interaction involve individual professional service and information exchange. Newer methods of interaction involve funding of research in two programs: the Industry–University Cooperative (IUC) and the Small Business Innovative Research (SBIR) grant programs. The IUC program encourages research done in a collaborative fashion between a university and industrial organization. Unsolicited proposals are technically evaluated by the appropriate NSF program in the normal peer review system. The SBIR program has as its purpose the evaluation of innovative research concepts with market potential.

Ronald J. Gutmann

Solid State and Microstructures Engineering, National Science Foundation, Washington, D.C.

Future VLSI Technology: A Review

The future of VLSI technology will be determined by the broad needs it fills, by the technical possibilities it offers, by the demands implicit in the actualization of the possibilities, by the requirements imposed on engineering, manufacturing, and investment, and by organizational responses to these needs. Let us focus first on the rationale for VLSI, then on its technical potential, and next on the problems and pressures these will create for the people and organizations committed to VLSI. The dominating theme and conclusion is that from raw materials to finished products and at every step of design and process, improved control is the central technical issue in bringing into being the possibilities of VLSI.

VLSI exists and will exist in increasingly complex forms to provide added technical capabilities at reduced cost. Basically, it is to save money overall, which is another way of saying "achieve more for less human effort". VLSI contributes primarily by providing computing capability, whether of memory or logic. It will play a steadily growing part not only by increasing computing power but also by simplifying the task of programming computers and other related forms of electronic logic and memory.

VLSI Potential

Table 1 outlines the manufacturable complexity of VLSI a decade hence, and Table 2 translates these into product capabilities. Minimum geometries should be less than 0.5 μm and defect levels below 1/cm^2 total, making possible large and dense arrays of logic memory The capability for providing

TABLE 1—*VLSI manufacturable complexity ca. 1990.*

$\leq$ 0.5-μm geometry
$\geq$ 1-cm^2 area
$\geq$ 3-layer metal
$\geq$ 10^6 logic gates
$\geq$ 10^7 memory bits
$\geq$ 25-W dissipation
$\geq$ 300 pin connections

several layers of metal interconnect and much-improved heat removal characteristics should allow operation at VLSI gate delays of less than 100 ps. The practical consequences are random access memories of 2 megabytes and microcomputer chips individually capable of upward of 10^9 instructions per second.

Design Capabilities

Exploitation of the product potential inherent in VLSI technology is already forcing the creation of very complex design systems. In their maturity these systems will provide integrated, inter-level-coupled chip simulation all the way from product concept to silicon structure (Fig. 1). These systems will permit modelling and semiautomatic design from algorithms to mask layout. In other words, it will be possible to implement desired mathematical signal transforms without major human design effort specific to the particular product. Hardware to solve an enormous range of problems, most of them not yet defined, will be practical to design and economical to manufacture and test. The need for direct human effort will, however, limit VLSI in some areas.

Market Limits

VLSI memory has, so far as we understand it today, no easily predictable demand limit, although it obviously will not become as large a fraction of gross world product as food, energy, or shelter. If it can be maintained and powered, additional memory at steadily falling cost would be useful.

Logic, as suggested earlier, falls into two broad classes. Firstly, there are the desired complex functions for which programming costs are less than chip development cost. For such functions limits are imposed only by perceived needs and the potential of the VLSI technology. In contrast, some VLSI structures, particularly new computers, require extensive program-

TABLE 2—*VLSI product capabilities ca. 1990.*

$\cong 16 \times 10^6$ bits RAM
$\cong 10^3$ MIPS
$\geq$ 64 bit hardware FP

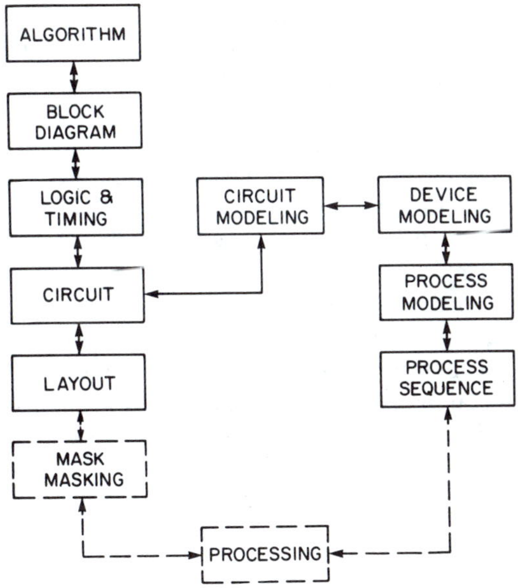

FIG. 1—*VLSI CAD flow.*

ming, and programming costs substantially exceed the cost of chip development. The programming referred to is that for operating systems, compilers, and the like rather than for applications programs. Obviously, as the ratio of programming cost to chip development cost rises, such programming cost imposes a limit on VLSI use.

Technical Issues

Major technical challenges must be met to bring these projections to shippable, economical realties. The central issues in the evolution of VLSI are reduction of dimensions, cutting of contamination, and raising of interconnectability. In each area, the basic advances to be made are clear, and work on appropriate technology is under way.

Dimensions are being slashed by successive generations of increasingly refined optical exposure equipment with X-ray and electron-beam equipment aimed at still higher performance under development in many organizations. Structures already built in laboratories show that 0.5 μm on a consistent basis with adequate overlay precision will become an industrial reality.

The fact that airborne contaminants increase in concentration as the reciprocal of the cube of particle diameter is an intimidating practical statistic. One thousand times more particles at one tenth the diameter is not an attractive control problem. Improvement of filters for gases and liquids coupled with minimization, at critical steps, of wafer exposure to even highly controlled atmospheres are part of the technology which is overcoming the inherent difficulties. At the same time, methods for cleaning wafers as well as for overriding contamination are also advancing. The results already in sight at the 2-μm geometry level argue strongly that the improvements needed for high-yield manufacture of 0.5-μm geometries will be realized.

Sophisticated bipolar devices have for many years used two layers of metal. MOS devices have used one layer of polysilicon and one layer of metal, and more recently, two layers of polysilicon and one layer of metal. The complexity of future VLSI logic will require additional interconnections. Silicides are already being introduced into VLSI memories to circumvent speed limitations imposed by the use of polysilicon. Even MOS memories may be forced to use two layers of metal as well as one or more layers of polysilicon. VLSI logic devices, particularly bipolar, will require three or more layers of metal in order to provide properly for interconnection of signals and for distribution of power and ground with satisfactory uniformity over large chips.

Common to all these technical advances are major steps forward in control and complexity.

Control: The Key Issue

These predicted accomplishments depend on complicated and highly precise equipment, which does not yet exist, utilized in processes rigorously controlled, with maintenance and human discipline in the operations at a level not yet commonplace in even the highest of technologies. The equipments are so designed, built, and self-calibrated that they are intrinsically capable of the required control. The processes necessarily rely on external controls, monitors, and feed-back mechanisms which allow these built-in capabilities to be used effectively. The workers are provided with motivation and means both to avoid error and to reproduce faithfully those process steps where human experience and judgment remain essential.

Table 3 displays typical modern process equipment, including types still under development, and enumerates key technical and productivity aspects. Those familiar with semiconductor wafer fabrication equipment of a decade ago will recognize the great advances in precision, productivity, and cost of the equipment, and, for some operations, the large increase in cost per wafer resulting from the increased unit cost.

TABLE 3—*VLSI key processing equipment.*

Type	Per Unit Cost, \$M	Wafers/h	Key Features
Ion Implanter	~ 0.5	~300	High vacuum, high voltage
Plasma Etcher	~ 0.3	~ 50	High vacuum, elec. fields
Optical Exposure	~ 0.5 to 0.8	~ 50	Optics, mech. precision
X-ray Exposure	~ 0.5 to 0.8	~ 50	High vacuum, mech. prec.
Electron-Beam Exposure	~ 1.5 to 2.0	~ 20	High vacuum, mech. prec.

TABLE 4—*VLSI key processes.*

Process	Key Control Issues	Values
Ion Implant	Uniformity; Absolute Value	±2%; ±4%
Wafer Exposure	Uniformity of Lateral Dimensions	±10%
	Layer-to-Layer Registration	±20%
Plasma Etching	Selectivity Between Layers	≥10:1
	Uniformity of Lateral Dimensions	±10%
	Slope of Contours	≤60 deg

Table 4 lists modern processes and emphasizes the control which is both required and possible. Historically electron devices have shown manufacturing variations of critical characteristics which are very large compared with those of simple mechanical objects such as nuts and bolts. This has been true in transistors and integrated circuits, with, for example, spreads of 3:1 in the current gain (amplification factor) of transistors being commonplace. It can be seen from Table 4 that the process variations, particularly of dimensions, have been and are being reduced commensurate with the reduction of dimensions. It is anticipated that adequate control can be obtained down to the limits on device size imposed by electrical design considerations which are inherent in the physical properties of silicon and related insulators.

Achievement of high-yield VLSI requires that large, organized groups of human beings process large quantities of material through sophisticated equipment in essentially the same way each day. To accomplish this, much of the control must be automatic, and the individuals involved must act with responsibility and self-discipline. In some sense they must also act bureaucratically, working according to the rules of the system rather than substituting their own judgments. Table 5 shows some of the relationships involved in satisfactory execution of high-yield VLSI production. The design engineer rightly sees his choice constrained almost totally by needs of the customer as expressed in the specifications, by the product technology chosen, and by the laws of nature, to which structure, design, and process must conform in order for the final device to meet the specifications. As the design moves from paper through to actual manufacturing, these basic constraints remain, but often other engineers, managers, and workers involved regard them as somewhat arbitrary and attempt to bring about "improvements" without an understanding of the total consequences of changes made. At the same time, almost inevitably, the designers of structures and process will have planned structures or procedures which are not fully compatible with high yield manufacturing. This inherent conflict must be resolved consultatively and experimentally; however, a time should arrive when the factory is truly a factory. At this point it can function as a self-correcting feed-back system without engineering intervention because the operating group has the tools and self-discipline to carry out the same tasks in the same way with the same results consistently.

TABLE 5—*VLSI discipline.*

Performance	Specifiers	Constraints
Manufacturing Operations	Manufacturing Engineer	"The Factory"
Manufacturing Engineering	Design Engineer	Available Equipment
Design Engineering	Customer	Laws of Nature
		State of Technology

Summary

Through advances in control of dimensions, structures, materials, design technology, and working organizations, VLSI technology will produce in a decade from now individual memory chips of 2 megabytes and individual microprocessors an order of magnitude or more more powerful than any central processor yet built. This control will be embodied in production and test equipment, with costs per unit running in some cases well beyond a million dollars. Particularly critical areas of control will be dimensions, vertical as well as lateral, contamination of both materials and processes, and chip intraconnection technology. The individual organizations producing with such technology must be characterized by systematic thoroughness, predictable regularity, and comprehending self-discipline.

James M. Early
Fairchild Advanced R&D Laboratory,
Palo Alto, California

APPENDIX II

Panel and Workshop Discussions: A Synopsis

The material presented in this appendix is based on the best recollections of editor and chairmen. The material did not go through the review process and is presented for information only.

Panel Discussions: D. C. Gupta, chairman

A panel discussion was held on Semiconductor Business and Education Requirements in the Eighties. The panel comprised spokesmen from industry, education, and government: W. C. Benzing, Applied Materials Incorporated; C. B. Drexel, Tylan Corporation; R. J. Gutmann, National Science Foundation; A. B. Holbrook, Advanced Micro Devices; R. E. Lorenzini, Siltec Corporation; D. O'Keefe, California State Senator; C. A. Mead, California Institute of Technology; J. D. Meindl, Stanford University; H. G. Molyneux, Siliconix Incorporated; J. E. Springgate, Monsanto Company; A. Varadi, National Semiconductor Corporation; and F. G. Vieweg-Gutberlet, Wacker-Chemitronic.

The questions put forth to the panel from the audience related to the material, equipment, devices, education, and legislation required for the semiconductor industry. A synopsis of the comments made by the panel on these topics is presented below.

Silicon Material

Originally the silicon material business was quite straightforward, being essentially limited to the three fundamental process steps, namely the growing, slicing, and polishing of the silicon wafers. At present, the technology has advanced to the point where not only have the basic process steps become extremely complex, but also the manufacturing cycle now includes many sophisticated subprocesses. In order for the material customers to obtain maximum performance from silicon wafers, certain wafer properties must be customized. This has led manufacturers to provide product differentiation in a competitive marketplace.

Present capacities seem to be adequate to meet near-future routine requirements. As wafer specifications change, most silicon manufacturers modify and update their equipment and processes to meet the new demand. As diameters become larger, and requirements tighter, it is most likely that much of the standing capacity will be replaced or substantially updated within the next few years. There is no reason why growth to larger diameter wafers will not continue as it has in the past.

Equipment

Automation is the wave of the future. The semiconductor device industry demands it. The realization of the precision and reduced defect density of complex VLSI structures is not possible without a high degree of automation. Besides automatic wafer handling, automation in future equipment must also include precise in-process sensing and control, end point sensing, full computer data logging, and feedback control. All this will add to equipment cost, but the cost will be offset by device yield improvement. Modeling will permit effective control and, through the use of interactive standard interfaces, sequential process steps will adjust to each other to maximize yields.

Increasing Capital Requirements

If an acceptable return on capital can be achieved, there may not be any problem with the increasing capital requirements. If, however, a group of companies with sufficient resourses chooses to significantly reduce or eliminate profit margins, then the smaller semiconductor companies will have difficulties remaining competitive in technology. Acceptable return on capital is the challenge faced today, not the increasing capital requirements. The priorities for new enterprises and smaller companies may be the new products in the high-growth areas or the quality and cost improvements needed to generate comfortable profits.

Vertical Integration and Silicon Foundry

The evolving process technology presents substantial problems for the nonmerchant vertically integrated supplier:

1. Poor amortization of production costs because of low volume.
2. New technologies that require ever-increasing investment costs.
3. Focusing as a team and remaining at the forefront of technology.
4. Providing management and recruiting capabilities in such a competitive industry.

A vertically integrated manufacturer is as vulnerable, if not more so, to design and technology obsolescense problems and production costs as the merchant supplier. What is actually required is a closer engineering relationship with key suppliers to ensure availability of the latest technologies and competitive costs. A compromise for the vertically integrated companies may be the focusing of their resources on custom and quasi-custom products while depending on the merchant supplier for standard products.

This leads to the idea of the silicon foundry. The concept, whose time is right, requires established manufacturing technology and rapid accurate processing with intrinsic parametric and logistics control, probably with built-in feedbacks. Best results are obtained by dedicated facilities; with the addition of variants, the chances of malfunctions increase and yields and efficiency decrease. The jeopardy of sole sources with a vendor may be minimized if there are designs with two or more foundries or an agreement exists to transfer technology to the customer under specified default conditions. In a foundry relationship, keeping information proprietary or revealing the design rules is not considered to be a major problem. There is a major benefit, however, accruing from consistent automated processing, namely an improved quality level.

Education and Legislative Requirements

The shortage of engineers and technical staff in the disciplines needed in the semiconductor industry appears to be caused by two primary factors: the growth of our industry itself, which has outpaced the issuance of graduates, and the exodus of teachers and professors from schools to industry, for both technical challenge and higher salary. Educational problems are compounded by the inability of the universities to maintain adequate equipment and facilities for research and teaching. It does not appear that this situation will change soon. Probably the situation will cause a general decline in the quality of engineers and technical people. Whenever there has been a severe engineering shortage, there has been a tendency for people without degrees or with only trade school educations to get promoted both in position and title in most major firms. The result was an overall lowering of standards in almost all technical areas.

Although there has been a significant increase in cooperation between industry, academia, and state government in the last few years, resulting in the launching or enhancement of large-scale university Microelectronics/VLSI programs such as at Arizona State, University of California at Berkeley, California State Polytechnic University, Cornell, Minnesota, MIT, North Carolina, Rensselaer Polytechnic Institute, and Stanford, there remains a great need for further research and educational efforts.

There is also a need to create an effective voice for the electronics industry that will propose and support effective legislation which will respond to the needs of new technologies and help achieve new scientific and intellectual horizons by removing the lack of commitment to scientific and engineering education.

Workshop Discussions

Chairmen: C. A. Germano, W. H. Kroeck, J. W. Lampe, and P. H. Langer.

Moderators: J. T. Clemens, J. R. Ehrstein, G. A. Gruber, T. I. Kamins, E. Mendel, C. W. Pearce, R. F. W. Pease, M. Powell, S. Prussin, K. C. Saraswat, J. Stach, and F. G. Vieweg-Gutberlet.

Six workshops were held, one after each technical session. The interest at these workshops was overwhelming and the topics discussed were of a wide variety. It being impossible to present each discussion in its full form, a short summary of the topics discussed at each workshop is presented.

Crystal Growth and Material Shaping

The main discussion at this workshop concerned oxygen and its role in silicon crystal growth dynamics. Three of the papers (Bonora, Lin and Hill, Murgai) touched on this subject. It was evident that there was considerable interest in crystal growth methodology to achieve uniform oxygen distributions in narrow concentration domains custom-designed to user needs. Also discussed was the influence of laser marking on silicon slices. How to make the laser marking more useable and practical for the user was the main concern of the group.

Microelectronics Imaging and Patterning

Various techniques employing projection and proximity printing were discussed. The emphasis was on reducing feature size, improving device yields, and lowering device fabrication costs for VLSI applications. Discussion was extended from optical lithography to more advanced electron-beam and X-ray lithographic techniques.

Key parameters and process tradeoffs in improving resist profile quality and linewidth control for optical projection printing were of interest. Advantages of positive resist over negative resist were pointed out. Discussions were held on the considerations pertinent to process optimization of resist relating to sensitivity, resolution, and defect levels.

Silicon CVD Techniques

After a brief description on the evolution of silicon epitaxy, discussions were held on the current problems faced from a fast-growing epitaxial technology and quality requirements for VLSI applications. Recent developments were mentioned on producing low-temperature, defect-free, thin silicon epitaxial layers related to bipolar applications. Improvements in MOS VLSI device characteristics built on P/P+ epitaxial structures were outlined. The primary advantages of using the epitaxial technology for NMOS VLSIs were said to include the damping of substrate voltage transients and efficient collection of minority carriers injected into the substrate. The possibility of achievement of noise attenuation up to two orders of magnitude was mentioned.

Discussions were extended to reduced-pressure CVD techniques for polysilicon and plasma-enhanced chemical vapor depositions of silicon and silicon-containing films such as silicon dioxide and silicon nitride used as intermetallic and passivation layers in integrated circuit applications.

Dielectrics and Junction Formation

This workshop covered a wide area of interest. Starting from the fabrication technology of high-density, high-performance NMOS circuits, discussions were held on various oxidation techniques, processing of polyimide films, and junction formation techniques utilizing implantations.

The advantages of high-pressure steam oxidation were mentioned. The existence of a strong interaction between chlorine and hydrogen, with the chlorine being an effective gettering agent for hydrogen, was revealed. Interface, breakdown, and trapping properties of very thin gate oxides were discussed.

There was considerable interest in the application of polyimides as insulators in multilevel interconnect systems. In an ever-growing implantation technology, there is always a concern for defect-formation and other anomalies due to the high currents and high energies associated with the process. Making use of cross-contamination of a different species during the implantation of a certain species was a topic of substantial interest.

Defects, Oxygen Precipitation, and Carrier Lifetime

The large turnout for this workshop ended in considerable controversy about the influence of oxygen in Czochralski-grown silicon. It was thought that the defect formation during processing, effectiveness of internal gettering, and mechanical strength of silicon was mainly controlled by the characteristics of oxygen in silicon. It was widely discussed and believed that the oxygen precipitation kinetics depended not only on the crystal growth conditions but also on the interactions during device processing.

Discussions were extended to the effects of oxygen precipitation on minority carrier lifetime and the use of DLTS technique for detection of deep-level impurities in silicon.

Measurements for Process Control

After a brief review of the recent progress in the field of process modeling and measurements, discussions were focussed on the novel statistical and graphical display methods in characterizing process nonuniformities on silicon wafers using two- and three-dimensional mapping techniques and inside the wafers using advanced spreading resistance measurement techniques. A procedure for obtaining spatial resolutions of 30 to 50 Å in spreading resistance measurements was described.

Discussions included the use of infrared measurements to characterize doping impurities and analysis of infrared spectra for determining oxygen concentrations in silicon. Difficulties envisaged in deriving the ion implant doses using the ellipsometric and spectral reflectance techniques were also discussed.

Dinesh C. Gupta

Siliconix Incorporated, Santa Clara, California; symposium chairman and editor

Index

S